# POROUS SILICON

## From Formation to Application

*Formation and Properties, Volume One*

Porous Silicon: From Formation to Application

*Porous Silicon: Formation and Properties, Volume One*

*Porous Silicon: Biomedical and Sensor Applications, Volume Two*

*Porous Silicon: Optoelectronics, Microelectronics, and Energy Technology Applications, Volume Three*

Edited by
Ghenadii Korotcenkov

# POROUS SILICON

## From Formation to Application

*Formation and Properties, Volume One*

**CRC Press**
Taylor & Francis Group
Boca Raton  London  New York

CRC Press is an imprint of the
Taylor & Francis Group, an **informa** business

CRC Press
Taylor & Francis Group
6000 Broken Sound Parkway NW, Suite 300
Boca Raton, FL 33487-2742

First issued in paperback 2020

© 2016 by Taylor & Francis Group, LLC
CRC Press is an imprint of Taylor & Francis Group, an Informa business

No claim to original U.S. Government works

Version Date: 20151012

ISBN 13: 978-0-367-57532-8 (pbk)
ISBN 13: 978-1-4822-6454-8 (hbk)

**Library of Congress Cataloging-in-Publication Data**

Names: Korotchenkov, G. S. (Gennadiæi Sergeevich), editor.
Title: Porous silicon : formation and properties, volume one / editor, Ghenadii Korotcenkov.
Description: Boca Raton : Taylor & Francis, 2015-<2016> | Includes bibliographical references and index.
Identifiers: LCCN 2015017925 | ISBN 9781482264548 (alk. paper)
Subjects: LCSH: Soluble glass. | Porous silicon. | Glass--Microstructure. | Glass manufacture. | Gas detectors. | Medical technology. | Optoelectronic devices.
Classification: LCC TP862 .P67 25015 | DDC 666/.1--dc23
LC record available at http://lccn.loc.gov/2015017925

Visit the Taylor & Francis Web site at
http://www.taylorandfrancis.com

and the CRC Press Web site at
http://www.crcpress.com

# Contents

## SECTION I—Introduction

## SECTION II—Silicon Porosification

## SECTION III—Properties and Processing

# Contents for *Porous Silicon: Biomedical and Sensor Applications, Volume Two*

## SECTION III—Biomedical Applications

# Contents for *Porous Silicon: Optoelectronics, Microelectronics, and Energy Technology Applications, Volume Three*

# Preface

In recent decades, porous silicon has been regarded as a means to further increase the functionality of silicon technology. It was found that silicon porosification is a simple and cheap way of nanostructuring and bestowing to silicon properties, which are markedly different from the properties of the bulk material. Because of this, increased interest in porous silicon appeared in various fields, including optoelectronics, microelectronics, photonics, medicine, chemical sensing, and bioengineering. It was established that this nanostructured and biodegradable material has a range of properties, making it ideal for indicated applications. As a result, during the last decade we have been observing an extremely fast evolution of porous Si-based optoelectronics, photonics, microelectronics, sensorics, energy technologies, and biomedical devices and applications. It is predicted that this growth will continue in the near future. However, despite the progress achieved in the field of design of porous silicon–based devices and their applications, it is necessary to note that there is a very limited number of books published in the field of silicon porosification and especially porous silicon (PSi) applications. No doubt such situations should be recognized as unsatisfactory. Thus, it was decided to prepare a set of books devoted to the analysis of the current state of the technology of silicon porosification and the use of these technologies in the development of devices for different applications. While developing the concept of this series, the objective was to collect in one edition information concerning all aspects of the formation and the use of porous silicon. This is of great importance nowadays, due to the speed of technological development and the rate of an appearance of new fields of PSi-based technology applications

This set, Porous Silicon: From Formation to Application, prepared by an international team of expert contributors, well known in the field of porous silicon study and having high qualifications, represents the most recent progress in the field of porous silicon and gives a fascinating report on the state-of-the-art in silicon porosification and the valuable perspective one can expect in the near future.

The set is divided into three books by their content. Chapters in *Porous Silicon: Formation and Properties, Volume One* focus on the fundamentals and practical aspects of silicon porosification by anodization and the properties of porous silicon, including electrical, luminescence, optical, thermal properties, and contact phenomena. Processing of porous silicon, including drying, storage, oxidation, etching, filling, and functionalizing, are also discussed in this book. Alternative methods of silicon porosification using chemical stain and vapor etching, reactive ion etching, spark processing, and so on are analyzed as well. *Porous Silicon: Biomedical and Sensor Applications, Volume Two* describes applications of porous silicon in bioengineering and various sensors such as gas sensors, biosensors, pressure sensors, optical sensors, microwave detectors, mechanical sensors, etc. The chapters in this book present a comprehensive review of the fabrication, parameters, and applications of these devices. PSi-based auxiliary devices such as hotplates, membranes, matrices for various spectroscopies, and catalysis are discussed as well. Analysis of various biomedical applications of porous silicon including drug delivery, tissue engineering, and *in vivo* imaging can also be found in this book. No doubt, porous silicon is rapidly attracting increasing interest in this field due to its unique properties. For example, the pores of the material and surface chemistry can be manipulated to change the rate of drug release from hours to months.

Finally, *Porous Silicon: Optoelectronics, Microelectronics, and Energy Technology Applications, Volume Three* highlights porous silicon applications in opto- and microelectronics, photonics, and micromachining. Features of fabrication and performances of photonic crystals, fuel cells, elements of integral optoelectronics, solar cells, LED, batteries, cold cathodes, hydrogen generation and storage, PSi-based composites, etc. are analyzed in this volume.

I believe that we have prepared useful books that could be considered a real handbook encyclopedia of porous silicon, where each reader might find the answers to most questions related to the formation, properties, and applications of porous silicon in practically all possible fields. Previously published books do not provide such an opportunity. Recently, several interesting books became available to readers, such as *Porous Silicon in Practice: Preparation, Characterization and Applications* by M.J. Sailor (Wiley-VCH 2011), *Porous Silicon for Biomedical Applications* by Santos H.A. (ed.) (Woodhead Publishing Limited 2014), and *Handbook of Porous Silicon*, by Canham L. (ed.) (Springer 2015). However, in *Porous Silicon in Practice: Preparation, Characterization and Applications*, M.J. Sailor describes mainly features of silicon porosification by electrochemical etching without any analysis of the correlation between parameters of porosification and properties of porous silicon. In the *Handbook of Porous Silicon* most attention was paid to the properties of porous silicon and, as well as in the book of M.J. Sailor, the consideration of PSi applications in devices was brief. At the same time, *Porous Silicon for Biomedical Applications* focuses on only the analysis of biomedical applications of porous silicon. I hope that our books, which cover all of the above-mentioned fields and provide a more detailed analysis of PSi advantages and disadvantages for practically all possible applications, will also be of interest to the reader. Our books contain a great number of various figures and tables with necessary information. These books will be a technical resource and indispensable guide for all those involved in the research, development, and application of porous silicon in various areas of science and technology.

From my point of view, our set will be of interest to scientists and researchers, either working or planning to start activity in the field of materials science focused on multifunctional porous silicon and porous silicon–based semiconductor devices. It also could be useful for those who want to find out more about the unusual properties of porous materials and about possible areas of their application. I am confident that these books will be interesting for practicing engineers or project managers working in industries and national laboratories who intend to design various porous silicon–based devices, but don't know how to do it. They might help select an optimal technology of silicon porosification and device fabrication. With many references to the vast resources of recently published literature on the subject, these books can serve as a significant and insightful source of valuable information and provide scientists and engineers with new insights for better understanding of the

process of silicon porosification, for designing new porous silicon–based technology, and for improving performances of various devices fabricated using porous silicon.

I believe that these books can be of interest to university students, post docs, and professors, providing a comprehensive introduction to the field of porous silicon application. The structure of these books may serve as a basis for courses in the field of material science, semiconductor devices, chemical engineering, electronics, bioengineering, and environmental control. Graduate students may also find the books useful in their research and for understanding that porous silicon is a promising multifunctional material.

Finally, I thank all contributing authors who have been involved in the creation of these books. I am also thankful that they agreed to participate in this project and for their efforts in the preparation of these chapters. Without their participation, this project would have not been possible.

I also express my gratitude to Gwangju Institute of Science and Technology, Gwangju, Korea, which invited me and gave me the ability to prepare these books for publication, and especially to Professor Beongki Cho for his fruitful cooperation. Many thanks to the Ministry of Science, ICT, and Future Planning (MSIP) of the Republic of Korea for supporting my research. I am also grateful to my family and my wife, who always support me in all undertakings.

# Editor

**Ghenadii Korotcenkov** earned his PhD in physics and the technology of semiconductor materials and devices from Technical University of Moldova in 1976 and his DrSci degree in the physics of semiconductors and dielectrics from the Academy of Science of Moldova in 1990 (Highest Qualification Committee of the USSR, Moscow). He has more than 40 years of experience as a teacher and scientific researcher. He was a leader of a gas sensor group and manager of various national and international scientific and engineering projects carried out in the Laboratory of Micro- and Optoelectronics, Technical University of Moldova. In particular, during 2000–2007 his scientific team was involved in eight international projects financed by EC (INCO-Copernicus and INTAS Programs), United States (CRDF, CRDF-MRDA Programs), and NATO (LG Program). In 2007–2008, he was an invited scientist at the Korea Institute of Energy Research (Daejeon) in the Brain Pool Program. Since 2008, Dr. Korotcenkov has been a research professor in the Department of Materials Science and Engineering at Gwangju Institute of Science and Technology (GIST) in Korea.

Specialists from the former Soviet Union know Dr. Korotcenkov's research results in the field of study of Schottky barriers, MOS structures, native oxides, and photoreceivers on the base of III-Vs compounds very well. His present scientific interests include material sciences, focusing on metal oxide film deposition and characterization, surface science, porous materials, and gas sensor design. Dr. Korotcenkov is the author or editor of 29 books and special issues, including the 11-volume Chemical Sensors series published by Momentum Press, the 10-volume Chemical Sensors series published by Harbin Institute of Technology Press, China, and the 2-volume *Handbook of Gas Sensor Materials* published by Springer. He has published 17 review papers, 19 book chapters, and more than 200 peer-reviewed articles (h-factor = 33 [Scopus] and h = 38 [Google scholar citation]). A citation average for his papers, included in Scopus, is higher than 25. He is a holder of 18 patents. In most papers, Dr. Korotcenkov is the first author. He has presented more than 200 reports on national and international conferences, and was the co-organizer of several conferences. His research activities were honored by an award of the Supreme Council of Science and Advanced Technology of the Republic of Moldova (2004), a prize of the Presidents of Ukrainian, Belarus and Moldovan Academies of Sciences (2003), a Senior Research Excellence Award of the Technical University of Moldova (2001, 2003, 2005), a fellowship from International Research Exchange Board (1998), and the National Youth Prize of the Republic of Moldova (1980), among others.

# Contributors

**Giampiero Amato**
Quantum Research Laboratory
INRIM
Torino, Italy

**Dmitriy Andrusenko**
Faculty of Physics
Taras Shevchenko National University of Kyiv
Kyiv, Ukraine

**Sukumar Basu**
IC Design and Fabrication Center
Department of Electronics and Telecommunication
    Engineering
Jadavpur University
Kolkata, India

**Rabah Boukherroub**
Institute of Electronics, Microelectronics
    and Nanotechnology (IEMN)
Villeneuve d'Ascq, France

**Vladimir Brinzari**
Laboratory of Physics of Nanostructures
State University of Moldova
Chisinau, Republic of Moldova

**Roman Burbelo**
Faculty of Physics
Taras Shevchenko National University of Kyiv
Kyiv, Ukraine

**Beongki Cho**
School of Material Science and Engineering
Gwangju Institute of Science and Technology
Gwangju, Republic of Korea

**Yannick Coffinier**
Institute of Electronics, Microelectronics
    and Nanotechnology (IEMN)
Villeneuve d'Ascq, France

**Helmut Föll**
Institute for Materials Science
Christian-Albrechts-University of Kiel
Kiel, Germany

**Bernard Jacques Gelloz**
Department of Applied Physics
Graduate School of Engineering
Nagoya University
Chikusa-ku, Nagoya, Japan

**Mykola Isaiev**
Faculty of Physics
Taras Shevchenko National University of Kyiv
Kyiv, Ukraine

**Jayita Kanungo**
IC Design and Fabrication Center
Department of Electronics and Telecommunication
    Engineering
Jadavpur University
Kolkata, India

**Kurt W. Kolasinski**
Department of Chemistry
West Chester University
West Chester, Pennsylvania

**Ghenadii Korotcenkov**
School of Material Science and Engineering
Gwangju Institute of Science and Technology
Gwangju, Republic of Korea

**Gilles Lérondel**
Laboratory for Nanotechnology and Optical
    Instrumentation
Charles Delaunay Institute
Université de Technologie de Troyes
Troyes, France

**Pascal J. Newby**
Centre de Collaboration MiQro Innovation (C2MI)
Université de Sherbrooke
Bromont, Quebec, Canada

**Enrique Quiroga-González**
Institute of Physics
Benemérita Universidad Autónoma de Puebla
San Manuel, Puebla, Mexico

**Kateryna Voitenko**
Faculty of Physics
Taras Shevchenko National University of Kyiv
Kyiv, Ukraine

# I

# Introduction

# Porous Silicon Characterization and Application

*General View*

**Ghenadii Korotcenkov**

## CONTENTS

## 1.1 INTRODUCTION

For the first time, porous silicon (PSi) was discovered in 1950s by Uhlir (1956) and Turner (1958), while performing electropolishing experiments on silicon wafers using a HF-containing electrolyte. They found that under the appropriate conditions of applied current and solution composition, the silicon did not dissolve uniformly but instead fine holes were produced. PSi formation was then obtained by electrochemical dissolution of Si wafers in aqueous or ethanoic HF solutions (Smith and Collins 1992; Lehmann 1993; Halimaoui 1995, 1997). However, until the 1990s, PSi had always been considered a somewhat esoteric material, of interest to primarily advanced research and development centers and companies dealing with silicon electronics (Watanabe and Sakai 1971; Watanabe et al. 1975; Imai 1981). In particular, in 1971, Watanabe and Sakai (1971) demonstrated the first application of PSi in electronics, the so-called full isolation by the porous oxidized Si process (FIPOS), where the PSi layers were used for device isolation in integrated circuits. In the 1980s, the silicon-on-insulator (SOI) in integrated circuits technology (Takai and Itoh 1986), the silicon on sapphire technology (SOS), and silicidation of PSi (Ito et al. 1989) were introduced as well. The high surface area of porous Si was found to be useful as a model of the crystalline Si surface in spectroscopic studies (Dillon et al. 1990), and as a sensing layer in chemical sensors (Anderson et al. 1990). Only in the last decades after the papers of Canham (1991) and Lehmann and Gosele (1991), who demonstrated room temperature light-emitting properties of PSi, electrochemically produced microporous silicon ($\mu$PSi) became one of the most studied materials in the field of material science. Although it should be noted that Pickering et al. (1984) observed low temperature photoluminescence in 1984. The discovery of room-temperature photo- and electroluminescence boosted research because of the huge potential in silicon-based integrated optoelectronics applications. As it is known, devices based on a normal single crystalline Si substrate cannot emit light efficiently. The appearance of PSi-based light emitting devices (LEDs) gave the hope that porous Si will make possible the integration of passive optical devices like gratings, waveguides, and so on, on the same substrate with Si LEDs. Such integration would make true an old dream—to produce integrated optoelectronic circuits on cost-effective and well-investigated Si substrates. In the same years, Lehmann and Foll (1990) showed the possibility to etch deep and straight macropores with diameters in the micrometer region with a predefined lateral arrangement into *n*-type Si by electrochemical processing (see Table 1.1). Later, this approach became the basis for the development of photonic crystals. In the same time period, the unique features of the material—its large surface area within a small volume, its controllable pore sizes, its convenient surface chemistry, and its compatibility with conventional silicon microfabrication technologies—inspired research into applications far outside optoelectronics. Subsequently, pores have been found in most if not all single crystalline semiconductors (Foll et al. 2003a,b; Rittenhouse et al. 2003; Fang et al. 2006; Santinacci and Djenizian 2008; Shen et al. 2008).

The chart presented in Figure 1.1 illustrates the evolution of the growth of the overall number of publications dedicated to PSi during the last 30 years (as it follows from the Scopus Citation data). It is seen that if before 1990 there was a slight interest of researchers in PSi, which was reflected in the small number of publications, 10–20 publications per year (Parkhutik 2000), then in 1995 the number of publications increased 10 times, and in 2010, 100 times in relation to 1985.

It should be noted that during the last decades it was established that besides conventional electrochemical etching, many other methods such as chemical stain etching, chemical vapor etching, spark processing, laser-induced etching, and reactive ion etching can be used for Si porosification (see Figure 1.2). For example, stain etching, in which the silicon sample is simply immersed in an HF-based solution, is the easiest way of PSi producing. It has been reported that the physical structure of those layers was similar to the one fabricated by the anodization method. The resulting pores are in the range of 1 nm up to the micrometer. Stain etching is an electroless process and, therefore, it has several advantages in comparison with electrochemical etching. However, layers formed by this method have low photoluminescence efficiency, deficient homogeneity, and poor reproducibility (Cullis et al. 1997). HF spray and vapor etching methods are also electroless processes. Therefore, vapor etching was investigated to address the difficulty of isolating metal contacts of devices from the electrolyte solution during anodic etching (Kalem

**TABLE 1.1    Stages in the Development of Technologies of PSi Formation and Theories Describing Phenomena of Silicon Porosification in a Time Frame**

| Time | Phenomena Discovered | Theories and Models Suggested |
|---|---|---|
| 1950–1960 | Discovery of PSi | PSi is a subfluoride ($SiF_2$) layer grown during anodic dissolution |
| | | PSi is product of dissolution/precipitation |
| 1960–1970 | | PSi is a result of disproportionation reaction |
| 1970–1980 | Macro pores formation on $n$-Si | Pores on $n$-Si appear due to local breakdown of depletion layer |
| | Crystallinity of PSi on $n$- and $p$-Si identified | |
| | Co-relation of PSi growth to anodization current, HF and doping concentrations | |
| | Formation of two layer PSi structure on $n$-Si | |
| 1980–1990 | Anisotropy of pore growth (preferred growth in the <100> direction on different types of Si substrate) | Pores formation due to appearance of a silicic layer on the wall |
| | Formation conditions of PSi layer generalized | PSi formation due to enhanced dissolution by intensified field at pore tip |
| | | PSi formation due to diffusion of holes |
| | | The presence of passive layer between PSi and Si |
| | | Pore formation in $p$-Si is mainly determined by the charge transfer |
| 1990–2000 | Deep straight array of macro pores on $n$-Si with back illumination | PSi formation due to sensitivity of reactions to surface curvature |
| | Macro pores formation on $p$-Si | Nano pores on $p$-Si due to quantum confinement |
| | Formation of micro PSi on macro pore walls on $n$- and $p$-Si | Macro pores on $p$-Si controlled by space charge effects |
| | | Macro pores on $p$-Si due to hole diffusion in space charge layer |
| 2000–2014 | Dependence of material reactivity on the confinement size | Current burst theory |
| | Three-dimensional macropore arrays in $p$-Si | Geometry relativity theory |
| | | Marcus electron transfer theory |

*Source:*  With kind permission from Springer Science+Business Media: *Modern Aspects of Electrochemistry*, Vayenas, C.C. and White, R.E. (eds.), Porous silicon: Morphology and formation mechanisms, vol. 39. 2006, pp. 65–133, Zhang, G.X.

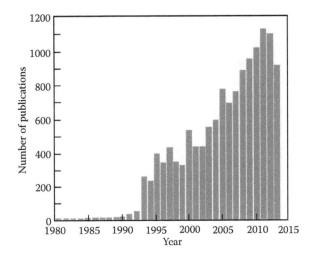

**FIGURE 1.1**    Number of publications in refereed journals dedicated to PSi as a function of time calculated using "Scopus" data.

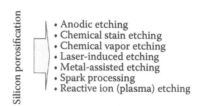

**FIGURE 1.2**   Methods used for silicon porosification. (Data extracted from Korotcenkov, G. and Cho, B.K., *Crit. Rev. Sol. St. Mater. Sci.* 35(3), 153, 2010.)

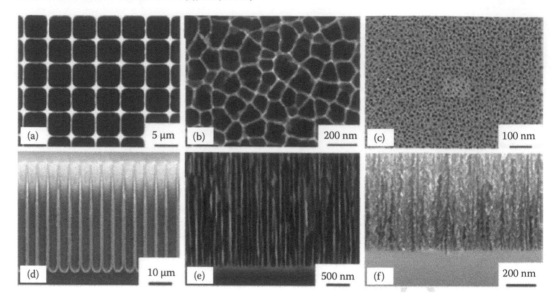

**FIGURE 1.3**   Top views (a, b, c) and cross-sectional views (d, e, f) of PSi with different pore diameters: (a, d) ordered macropores, (b, e) medium-sized pores, and (c, f) mesopores. Actual average diameters of the pores, which are determined from the SEM images, are 20 nm (mesopore), 120 nm (medium-sized pore), and 5 µm (macropore), respectively. (Reprinted from *Electrochem. Commun.* 10, Fukami K. et al., 56, Copyright 2008, with permission from Elsevier.)

and Yavuzcetin 2000). Spark erosion was also tested for preparing porous Si (Hummel and Chang 1992). Noble metal assisted etching of Si is another method suitable to form PSi. This method is also a simple process that does not require the attachment of any electrodes and can be performed on objects of arbitrary shape and size (Cullis et al. 1997; Kolasinski 2005).

However, despite the wide variety of methods that can be used for silicon porosification, the electrochemical (anodic) etching is the most successful and used technique for producing pores in semiconductors. All above-mentioned methods allow forming PSi with its unique parameters, but none of them has reached a stage of maturity, similar to anodization. Besides, many methods have essential limitations on reproducibility, on attainment of required porosity and thickness of formed layer. Therefore, the attention of developers is attracted mainly to anodic oxidation, which is the most controllable process, and provides reproducible parameters. We must note that the standard method of electrochemical etching has appreciably more resources for fabrication of high quality PSi layers in comparison with the above-mentioned methods, such as stain etching, metal-assisted stain etching, laser-induced etching, and spark processing. Electrochemical etching is simple, inexpensive, and gives designers a sufficiently free hand for fabrication PSi layer with required structure (see Figure 1.3).

## 1.2  PORE CLASSIFICATION

According to the International Union of Pure and Applied Chemistry (IUPAC), conventional porous materials need to be geometrically classified as micropores (diameter, <2 nm), mesopores (diameter, 2–50 nm), or macropores (diameter, >50 nm). In the frame of the above-mentioned classification porous Si (PSi) with micro-, meso-, and macropores need to have corresponding abbreviations such as µPSi, meso-PSi, and macro-PSi. However, it is necessary to note that the describing of pores in semiconductors by just one qualifier ("micro," "meso," or "macro") does not contribute to correct characterization of such a complex subject. Too many words would be needed for an exhaustive verbal description of some of the pores formed in semiconductors. As a rule, there are straight cylindrical pores with very large aspect ratios (>1000) on the one extreme, and three-dimensional sponge-like structures on the other extreme, with anything in between, for example, two-dimensional fractal arrangements or self-organized structures exhibiting some periodicity in one, two, or three dimensions (Zhang 2005). Use of various terms by numerous authors has caused some additional confusion with the IUPAC terminology. The formally undefined but rather clear term "nanopores" is often used for all "small" pores, and the term "mesopore" is often used for pores that are technically macropores but with the typical wavy and heavily branched morphology associated with mesopores.

Besides the above-mentioned classification of pores, some authors propose additional subclasses, which allow specifying peculiarities of their structure. In particular, besides the term "nanopores," which is already sufficiently clear, Foll and co-workers (Foll et al. 2002) used terms such as "bsi-macropores" addressing the pronounced pores with diameters approximating 1 µm obtained in backside illuminated (bsi) $n$-type Si at medium potentials, "breakthrough pores" denote the narrow (diameter <100 nm) pores often obtained in $n$-type Si in the dark at high potentials (Hejjo Al Rifai et al. 2000), "fsi-macropores," that is, macropores obtained in $n$-Si under frontside illumination (fsi) conditions (Osaka et al. 1997), and "litho-pores," where "litho" indicates nucleation sites for pores with a lateral arrangement defined by using lithography methods (Foll et al. 2002). Last, macropores are usually long and cylindric. Rather large "p-macropores" (diameters in the micrometer range), first described by Propst and Kohl (1994), also can be considered a subclass of macropores.

Other classifications have been proposed by Langa et al. (2000) in order to define two different pore morphologies. When pores grow along the current lines, that is, termed perpendicularly to the surface, they are called current line oriented (CLO), but when they propagate along specific crystallographic orientations, they are crystal oriented (CO). These two varieties are also called "curro" and "crysto" pores, respectively (Langa et al. 2003). Note also that porous layers are sometimes referred to as "nanocrystals" because the crystal structure of the pore walls is not modified during porous etching (Asano et al. 1992). The presence of such various approaches to pores' classification in PSi attests to a large variety of pore shapes and to the absence of a common approach to analysis of a porous material structure. Such a situation essentially hampers the analysis and summary of results obtained by different groups in this field of science and technology.

## 1.3  POROUS SILICON CHARACTERIZATION

Real views of meso-porous and macro-porous Si are shown in Figures 1.3 and 1.4.

The figures show that PSi is composed of a silicon skeleton permeated by individual pores or a network of pores. Describing the geometry and morphology includes the average pore size and the distance between pores (= geometry), the branching parameters, pore and branch orientation, layer thickness, and general porosity (= percentage of dissolved Si). PSi as a representative of the class of porous materials can be also characterized by such parameters as roughness, pore distribution, pore volume $V_P$, pore density $N_P$, permeability, homogeneity, and surface area.

There are many various methods such as optical and electronic microscopy, x-ray scattering, liquid extrusion, liquid intrusion, capillary condensation method, permeametry, counter-diffusion, ultrasonic method, thermoporometry, immersion calorimetry, gas adsorption, and so on, which can be used for PSi characterization (see Figure 1.5). These techniques and their derivatives can measure a variety of properties. one can find the description of some of them in reviews (Lowell and Shields 1984; Rouquerol et al. 1994; Leofanti et al. 1997; Jena and Gupta 2002; Denoyel et al. 2004; Lowell et al. 2006).

However, one should note that it is general practice to determine the morphological characteristics of a particular PSi layer only by high-resolution light and scanning electron microscopy in

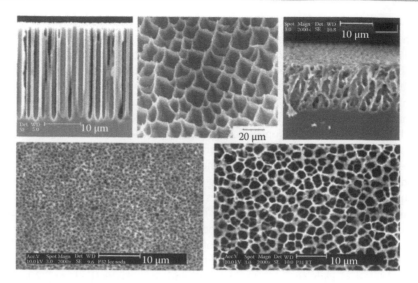

**FIGURE 1.4**    Possible morphologies of macroporous Si layer. (Reprinted from *Mater. Sci. Eng. R* 39, Foll, H. et al., 93, Copyright 2002, with permission from Elsevier; and *Electrochim. Acta* 48, Blackwood D.J. and Zhang Y., 623, Copyright 2003, with permission from Elsevier.)

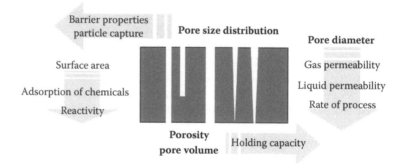

**FIGURE 1.5**    Parameters, which can be used for PSi characterization.

cross-section and in top-views. The wide application of this technique is based on the advantages such as a high magnification, large depth of focus, great resolution, and ease of sample preparation and observation. Moreover, in many cases, again in particular for micropores and mesopores silicon, researchers content themselves by simply referring to a few parameters such as porosity and the pore size, pore width, or pore diameter.

### 1.3.1 PORE TYPE AND PORE SHAPE

In the most general sense, a "pore" is an etch pit whose depth, $d$, exceeds its width, $w$. Most porous Si layers are a few micrometers deep, and individual pores are generally closed at one end (Figure 1.6a) and interconnected to some degree. "Closed" porosity is created via capping or by thermally induced reconstruction of the pore network (Labunov et al. 1986; Searson 1991). Pores open at both ends (see Figure 1.6b) can be realized in porous Si membrane structures by either extended anodization of wafers (Anderson et al. 2003), anodization of prethinned areas, or the Turner liftoff technique (Anderson et al. 2003). This generates so-called free-standing porous Si films typically of 5 to 50 μm thickness that have received much study (Feng et al. 1994; Maly et al. 1994). However, experiment has shown that the most common shape of pores in PSi by far is that of cylindrical pores with varying degrees of branching (see Figure 1.6a and c) and "necking." As we indicated earlier, pore diameter in porous Si can vary widely from <2 nm to >50 nm.

(a)                                    (b)                                    (c)

**FIGURE 1.6**  Types of pores formed in PSi: (a) closed pores at one end, (b) pores opened at both ends, and (c) pores with branching. (Idea from Allongue, P., In: *Properties of Porous Silicon*, Canham, L. (ed.). INSPEC, London, p. 3, 1997.)

### 1.3.2  PORE SIZE AND PORE DENSITY

The pore size is a property of major importance in practical applications of porous materials. However, it should be borne in mind that only if the pores were cylindrical tubes of uniform diameter, or spherical bodies, would the pore size would be unique. As a rule, this case is realized only in the macroporous silicon. In other cases, implemented in micro- and mesoporous silicon, it is necessary to talk about some effective pore size because whereas the volume parameter is usually measured directly, the characteristic pore size is always calculated from some measured physical parameter in terms of the arbitrary model of pore structure. Owing to the complexities of pore geometry, the characteristic pore size is often not at all characteristic of the pore volume to which it has been assigned. Usually, effective pore size can be calculated using the formula for hydraulic diameter ($D_h$):

(1.1)
$$D_h = \frac{2V_p}{A}$$

where $V_p$ is the pore volume determined at saturation and $A$ is the surface area, for example, determined by the BET method (Lowell et al. 2006).

The pore density in contrast to porosity is also quantitied directly related to the actual size of pores and pore walls. The pore density $N_p$ is defined as the number of pores per unit area and it usually refers to a plane normal to the pore axis. For (100) oriented substrates, this plane is parallel to the electrode surface, but for other orientations or strongly branched pores, there is no preferred plane orientation and $N_p$ refers to an average of the pore density of different planes. For arrays of straight pores, the pore density can be directly calculated from the array geometry. For cylindrical pores of diameter $d$ orthogonal to the electrode surface, for example, the average pore density $N_p$ is given by

(1.2)
$$N_p = \frac{P}{\pi(d/2)^2}$$

where $P$ is porosity.

### 1.3.3  SURFACE AREA

The specific surface area is defined as the accessible area of solid surface per unit mass of material. For an array of macropores, the specific surface area can be calculated directly from its geometrical dimensions (Lehmann 2002). For example, an orthogonal array of circular macropores with a pitch of 2.5 μm and a diameter of 2 μm has a specific surface area of 0.859 $m^2/g$ (or 1 $m^2/cm^3$).

The porosity of such an array is 50%. While for macroporous structures, the inner surface can be calculated from the geometry, meso and micro PSi layers require other methods of measurement. First evidence, surface area of through and blind pores can be measured accurately by gas adsorption technique and calculated using Equation 1.3 obtained in the frame of Brunauer–Emmett–Teller (BET) theory (Brunauer et al. 1938; Lowell and Shields 1984).

(1.3)
$$S = \frac{W_m N \alpha}{m}$$

where $S$ is the surface area per unit mass, $W_m$ is the amount of adsorbed gas that can form a monolayer ($W_m$ is in moles), $N$ is the Avogadro's number, $\alpha$ is the cross-sectional area of an adsorbed molecule, and $m$ is the mass of the sample. Regarding pore size distribution, which is the density function given the distribution of pore volume by a characteristic pore size, we have to note that this parameter is a poorly defined quantity, partly because it depends, sometimes very markedly, on the particular method used in its determination. The general procedure used for the determination of a pore size distribution consists of measuring some physical quantity in dependence on another physical parameter under the control of the operator and varied in the experiment. For example, in vapor sorption, the volume of gas absorbed is measured as a function of the gas pressure (Storck et al. 1998). The isotherm is usually constructed point-by-point by the admission and withdrawal of known amounts of gas, with adequate time allowed for equilibration at each point. For removal of all physisorbed gases, the outgassing to a residual pressure of ~$10^{-4}$ Torr is considered acceptable. Automated techniques in some cases involve the slow continuous admission of the adsorptive and thus provide a measure of the adsorption under quasi-equilibrium conditions. Alternatively, a carrier gas technique, along with conventional gas chromatographic equipment, may be employed. However, in this case, the adsorption of the carrier gas should be negligible. In most applications, nitrogen (at 77 K) is the recommended adsorptive for determining the surface area and mesopore size distribution (Rouquerol et al. 1994). However, experiment has shown that it is necessary to employ a range of probe molecules to obtain a reliable assessment of the micropore size distribution. For example, Storck et al. (1998) have found that the pore sizes in micropore-size distributions derived from the adsorption isotherm are very sensitive to adsorbate gas and adsorption temperature. They established that only with argon as the adsorbate at the temperature of liquid argon, reliable pore sizes in the micropore range are being obtained. For operational reasons, krypton adsorption (at 77 K) is usually adopted for the determination of relatively low specific surface areas (say, < ~2 m²/g), but this technique cannot be employed for the study of porosity. An alternative technique to gas adsorption (e.g., mercury porosimetry) must be used for macropore size analysis.

### 1.3.4 POROSITY

The main parameter of micro- and mesoporous material is the porosity $P$(%):

(1.4)
$$P = \frac{V_p}{V}$$

where $V$ is the volume of sample and $V_p$ is the volume of the pores. This parameter determines what percentage of the material is "filled" by pores. However, analyzing the porosity of the PSi formed during anodization we have to take into account that depending on the PSi structure, the porosity can be "open," "closed," or "total." "Open porosity" is the volume of pores accessible to a given probe molecule; "closed porosity" is the volume of pores that are inaccessible to the probe molecule; and "total porosity" is the volume of pores accessible to a given probe molecule plus the volume of pores that are inaccessible to the probe. Since PSi is prepared by electrochemical etching of solid crystalline silicon, one can assume that all the pores in an as-formed PSi sample had to be accessible to the electrolyte at the time of Si porosification. Therefore, as-prepared PSi is an open porous material with no inaccessible voids. However, if the PSi has been annealed, we can receive situations when some of the pore mouths are closed, that is, PSi becomes material with closed porosity. This effect could be because of the conversion of Si to $SiO_2$, which is accompanied by the increase of the volume. A similar situation can be also observed during surface functionalizing of the inner pore walls of a PSi sample. Finally, it is important to keep in mind that the term "open porosity" is defined relative to a given probe molecule; a small pore that accommodates a molecule like ethanol may not accept a larger protein molecule (Sailor 2012). This means that the meaning of the term "total pore volume" depends not only on the treatment used in the manufacturing process, but also on the method used for volume $V_p$ and porosity control (Rouquerol et al. 1994). For example, some methods, indeed, only have access to "open pores," whereas others may also have access to "closed pores." Moreover, as mentioned previously, for a given method, the value of $V_p$ depends on the size of the molecular probe (fluid displacement, adsorption). Thus, a recorded value of porosity can be expected to reflect not only a physical property of the material, but also the experimental method used for its determination. Thus, the pore volume $V_p$ used in the above relationship may be that of the open pores (leading to the "open porosity") or that of the closed pores (leading to the "closed porosity") or that of both types of pores together (leading to the "total porosity").

In real experiments with PSi, the total porosity can be easily determined by weight measurements before and after etching. This is the so-called gravimetric method (Halimaoui 1997; Du Plessis 2007). The porosity ($P$) is given simply by the following equation:

(1.5)
$$P(\%) = \frac{(m_1 - m_2)}{(m_1 - m_3)}$$

where $m_1$ is wafer weight before anodization, $m_2$ is just after anodization, and $m_3$ is finally after dissolution of the whole porous layer in a molar NaOH aqueous solution. Uniform and rapid stripping in the NaOH solution is being obtained when the PSi layer is covered with a small amount of ethanol, which improves the infiltration of the aqueous NaOH in the pores. From these measured masses, it is also possible to determine the thickness ($L$) of the layer according to the following formula:

(1.6)
$$L = \frac{(m_1 - m_3)}{S \cdot d}$$

where $d$ is the density of bulk silicon ($d = 2.33$ g/cm³) and $S$ is the wafer area exposed to HF during anodization. Du Plessis (2007) has shown that surface area can be determined also using the gravimetric method. It is important to note that the above gravimetric determinations are independent of the morphology or structure of the porous layer. The thickness can be directly determined as well by scanning electron microscopy (SEM) or by step measurement after a part of the layer has been completely dissolved to generate a step, corresponding to the layer thickness (Halimaoui 1997).

It is important to note that the error of gravimetric method for small thicknesses (up to 10 µm) of the porous layer and high porosity (over 70%) can reach more than 15–20%. In addition, the use of such method of porosity monitoring leads to destruction of the sample because the porous layer is removed during measurement. The X-ray method, based on measuring the intensities of X-rays passing through a single-crystal silicon wafer, $I_1$, through the sample with a porous layer, $I_2$, and through a silicon sample without a porous layer, $I_3$, has these same disadvantages. The filling or infiltration of pores by liquids or condensed vapors also can be used for porosity determination. For example, Janshoff et al. (1998) proposed the procedure of porosity calculation based on the change of PSi optical properties after solvent impregnation or exposure in saturated vapor of organic solvents. However, gas and liquid porosimetry methods mentioned above do not have high accuracy as well.

For macroporous silicon, the porosity can be calculated using geometrical parameters of formed structure. For a square and hexagonal pore arrangement, the porosity is given as

(1.7)
$$\text{square}: P = \frac{\pi r^2}{a^2}; \quad \text{hexagonal}: P = \frac{2\pi r^2}{\sqrt{3} \cdot a^2}$$

where $a$ is interpore distance and $r$ is a pore radious.

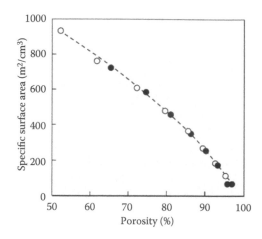

**FIGURE 1.7**    Specific surface area as a function of PSi porosity. Layer thickness ~1 μm. (Data extracted from Halimaoui A., In: *Properties of Porous Silicon,* Canham, L. (ed.). INSPEC, London, p. 12, 1997.)

One should note that for PSi usually used in experiments, the porosity spans in unusually wide ranges from 5% to 95% (0–30% is low porosity; 30–70% is medium porosity, and 70–95% is high porosity). Depending on the value of the porosity and pore geometry, the surface area of PSi ranges from 10 to 100 $m^2/cm^3$ for macroporous silicon, from 100 to 300 $m^2/cm^3$ for mesoporous silicon, and from 300 to 1000 $m^2/cm^3$ for microporous silicon. Later it will be mentioned that the surface area is not strictly proportional to the porosity as one might expect. As a rule, the dependence of surface area on the porosity has the maximum, and at a certain value of the porosity the surface area begins to decrease with increasing porosity (see Figure 1.7).

### 1.3.5  OTHER METHODS OF STRUCTURE AND SURFACE CHARACTERIZATION OF POROUS SILICON

As for the other parameters of the PSi, in particular the surface properties, one needs to note that the control and analysis of these properties is a difficult task. The difficulty with surfaces in most practical situations is that the processes controlling surface properties are extremely complex, and furthermore, surface properties depend strongly on the surrounding conditions, which can vary widely. At the same time, control of these properties is necessary because the surface plays a crucial role in many thermal, chemical, physical, and mechanical processes, taking place in PSi. The properties of PSi surface in practical applications can be determined by material and surface analytical techniques. The main techniques are listed in Table 1.2. Such techniques include a variety of physical, chemical, and material characterizations. As it is seen, surface characterization (diagnostic) techniques are now available for measuring the crystallographic structure, chemical and elemental composition, chemical states of any solid surface, adsorbed species and its bonding state, phase transitions, and so on. Detailed description of these methods can be found in reviews (Leofanti et al. 1997) and books (Brundle et al. 1992; Milling 1999; Brune et al. 2007; Che and Vedrine 2012).

## 1.4  FIELDS OF POROUS SILICON APPLICATIONS

It is known that microporous Si became famous because of its unexpected optical ability to show strong luminescence from yellow to blue, depending on its precise structure. Therefore, porous material first became very promising for various optoelectronic applications (Canham et al. 1996; Canham 1997; Bisi et al. 2000; Foll et al. 2006; Kochergin and Foll 2006). It was expected that PSi would form the basis of both effective silicon light-emitting diodes (LED) and devices capable of guiding, modulating, and detecting light. The rationale for realizing PSi-based optoelectronics has been described in a review article by Canham et al. (1996). However, exploiting luminescence

**TABLE 1.2    Main Techniques, Which Can Be Used for Surface Characterization of PSi**

| Technique | Information |
|---|---|
| **Optical Methods** | |
| IR (infrared) and FTIR spectroscopy | Surface functional groups; bonding state; chemical structure |
| Raman spectroscopy | Identification of unknown compound; chemical state; bonding state; structural order; phase transitions; imaging and mapping |
| PAS (photoacoustic spectroscopy) | Local environment; functional groups; structure |
| UV-Vis (ultraviolet-visible) spectroscopy | Local environment of surface groups perturbed by adsorption |
| **X-Ray, Electron, and Neutron Radiation** | |
| XRD (X-ray diffraction) | Oxidation degree; crystalline species; crystalline degree; crystallite size; film thickness |
| EXAFS (extended X-ray absorption fine structure) | Local structure (coordination number, inter- atomic distances); adsorbate bonding |
| XANES (X-ray absorption near edge structure) | Local structure |
| NMR (nuclear magnetic resonance | Chemical state; phase identification; disordered state |
| EPR (ESR; electron paramagnetic spin resonance) | Oxidation degree; symmetry; nature of ligands |
| Mossbauer | Local environment; oxidation degree (few elements) |
| XRF (X-ray fluorescence) | Atomic composition |
| NS (neutron scattering) | Crystal structure; molecular diffusion |
| LEED (low-energy electron diffraction) | Surface crystallography and microstructure; surface cleanliness; surface disorder; imaging |
| **Adsorption/Desorption Methods** | |
| Volumetric adsorption; gravimetric adsorption | Amount of adsorbate as a function of pressure |
| Dynamic adsorption | Amount of irreversible adsorbate |
| Calorimetry | Heat of adsorption as a function of coverage |
| TPD (temperature programmed desorption) | Amount of desorbed species as a function of temperature |
| TPD-MS (TPD coupled with mass spectrometry | Amount and composition of desorbed species as a function of temperature |
| TPD-IR (TPD coupled with IR spectroscopy) | Composition of desorbed species as a function of temperature |
| TPSR(MS) (temperature programmed surface reactions with mass spectrometry) | Competitive interaction of different probe molecules- Reactive of surfaces |
| **Vacuum Methods** | |
| XPS (X-ray photoelectron spectroscopy) | Elemental composition; chemical state; depth profiling; imaging and mapping |
| EPMA (electron probe microanalysis) | |
| AES (Auger electron spectroscopy) | |
| SIMS (secondary ions mass spectrometry) | |
| ISS (ion scattering spectroscopy) | Elemental composition (outermost monatomic layer); atomic structure |

property of PSi for devices proved to be difficult, if not impossible, until now (Lehmann 1996; Bisi et al. 2000; Kochergin and Foll 2006).

It was established that µPSi in unmodified form is fragile, easily oxidized, and open to chemical attack. As a result, as-prepared µPSi exhibits unstable photoluminescence (PL) (Tischler et al. 1992; Schuppler et al. 1995; Harper and Sailor 1996; Torchinskaya et al. 1997, 2001; Kim et al. 2001). It is also known that the internal surface of the luminescent microporous silicon reaches values of (600 to 1000) $m^2/cm^3$. This means that almost every second atom in the porous skeleton pertains to the surface (Lehmann 1996). From this point of view, it is not surprising that the properties of µPS are sensitive to the state of its surface and changes with time due to, for example, hydrogen desorption, absorption of organic vapors, or oxidation producing the so-called native oxide.

Numerous post formation treatments of µPSi, aiming at stabilizing the PL, have been reported (Schuppler et al. 1994; Bjorkqvist et al. 2003; Vrkoslav et al. 2005). Thus far, however, this issue has not been completely resolved and the early enthusiasm for a key role of µPSi in advanced photonic and electrooptical applications has disappeared in view of the practical difficulties. At present, the interest extends to other applications of PSi (see Table 1.3), including micromachining (Lang et al. 1994; Steiner and Lang 1995; Lang 1996; Hedrich et al. 2000) and sensor applications (Vial et al. 1992; Koch et al. 1993; Ben-Chorin and Kux 1994; Huanca et al. 2002, 2008; Marsh 2002; De Stefano et al. 2004; Ozdemir and Gole 2007; Korotcenkov and Cho 2010a,b).

PSi layers are very attractive from a sensor point of view because of the unique combination of a rather perfect single-crystalline structure, a huge internal surface area of up to 200 to 1000 $m^2/cm^3$ with a concomitant enhancement of the adsorbate effects, and often a very high activity in surface chemical reactions (Korotcenkov and Cho 2010a). For comparison, the specific surface of monocrystalline silicon is only 0.1–0.3 $m^2/cm^3$. For that reason, PSi films constitute an ideal matrix for the inclusion of catalytic materials. Numerous researchers have shown that many parameters of a PSi vary as a function of a change in the gas and liquid environment in which the PSi is being kept (Anderson et al. 1990; Vial et al. 1992; Vial and Derrien 1995; Canham 1997; De Stefano et al. 2004; Blackwood and Akber 2006). For example, reproducible changes of the photoluminescence intensity of microporous silicon may be observed on both changing the surrounding gas atmosphere and cycling the sample from an electrolyte to air and back.

Soaking microporous silicon with chemical reagents strongly alters its photoluminescence intensity and conductance features as well. Several investigations showed that the electrical and optical characteristics of mostly micro- and mesoporous semiconductors, such as the capacity of the porous layer, the conductance, the reflection coefficient, the IR absorption, and the resonance frequency of a Fabry-Perot resonator made of the PSi also may change considerably on adsorption of molecules on its surfaces and/or by filling the pores (Vial et al. 1992; Mares et al. 1995; Canham 1997; Cullis et al. 1997). Consequently, the surface adsorption and capillary condensation effects of PSi layers can be used for the development of effective sensor systems (Anderson et al. 1990; Parkhutik 1999; Korotcenkov and Cho 2010a). Numerous studies carried out in this direction confirmed the availability of the above-mentioned approach. The literature contains a great number of works focused on research in this field (Korotcenkov and Cho 2010a,b). The most interesting approaches used while designing sensors on the base of porous semiconductors (mostly on Si) are given in the following list.

**TABLE 1.3    Porous Silicon as a Multidisciplinary Material**

| Scientific Area | Key Property | Application Example |
|---|---|---|
| Optoelectronics | Light emission | Displays |
| Analytical chemistry | Porosity | Sensors |
| Optics | Tunable refractive index | Optical filter |
| Ultrasonics | Low thermal conductivity | Transducer |
| Surface science | High surface area | Catalyst |
| Microbiology | Biocompatibility | Bioreactor |
| Energy conversion | Low reflectivity | Solar cell |
| Astrophysics | Cosmic abundance | Dust chromophore |
| Microengineering | Lithographic patterning | Microsystems |
| Nuclear science | Conversion to oxide | Radiation-hard circuits |
| Electronics | Dielectric properties | Microwave circuits |
| Signal processing | Dynamic processes | Process control |
| Medicine | Resorbability | Drug delivery |

*Source:*  Parkhutik, V.P. and Canham, L.T., *Phys. Stat. Sol. (a)* 182, 591, 2000. With permission.

Approaches used for sensor design on the base of porous semiconductors (Korotcenkov and Cho 2010a).

1. Capacitance type
2. Photoluminescence quenching
3. Conductometric types
4. Optical measurements
5. Schottky barriers and heterostructures
6. Contact Potential Difference (CPD) measurement
7. Field ionization
8. Combined approach

Sensors based on these approaches were intended as biosensors, electrochemical sensors, gas sensors, mechanical sensors, humidity sensors (see Figure 1.8), and so on. The approaches indicated in the list above for designing PSi-based sensors have both advantages and shortcomings discussed in detail in reviews (Korotcenkov and Cho 2010a,b). Read also Chapters 1–9 in *Porous Silicon: Biomedical and Sensor Applications*. Sberveglieri and co-workers (Di Francia et al. 1998) noted the following advantages of PSi over other porous materials such as ceramics or nano- and polycrystalline films of metal oxides that are used for sensor design: (1) PSi is basically a crystalline material and thus, in principle, is perfectly compatible with common microelectronic processes devices; (2) PSi can be electrochemically fabricated with very simple and inexpensive equipment; and (3) PSi can be produced in a large variety of morphologies, all of which exhibit large values of surface-to-volume ratio. In addition, one can also take advantage of other properties related to PSi, such as the possibility to generate three-dimensional structures or the possible design of multisensors, based on the use of various registration techniques for gas detection (optical, electrical, luminescent, etc.) in parallel. A low power consumption compared with metal oxide gas sensors, due to operation at room temperature, is another important advantage of PSi gas sensors.

In summary, because PSi sensors are based on silicon wafers and can be manufactured by using established integrated circuit production techniques and because they can be operated at room temperature using relatively low voltages, they are particularly well suited for compact and low-cost sensor systems on a chip, where both the sensing element and the read-out electronics can be effectively integrated on the same wafer. Some authors contend that the sensors based on PSi are so simple that they could ultimately be mass-produced for a few cents apiece, and this perspective is a powerful inducement for further development (Gole et al. 2000). These new sensors could be integrated into electronic equipment and used, for example, in building sensing arrays necessary for environmental monitoring, laboratory testing, process control, chemical warfare, biochemistry, and medicine.

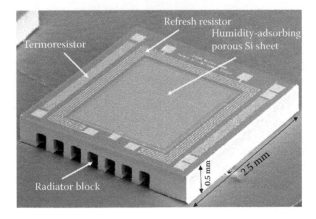

**FIGURE 1.8**    SEM micrograph of the PSi humidity sensor of capacitive type. (Reprinted from *Sens. Actuators B* 68, Rittersma, Z.M. et al., 210, Copyright 2000, with permission from Elsevier.)

Among other applications that are drawing increased attention, one should also mark the advent of PSi as an active biomaterial (Canham 1995; Buckberry and Bayliss 1999; Angelescu et al. 2003). It was established that PSi is a material that is biocompatible, bioactive, and biodegradable. Biocompatibility is the ability of a material to interface with natural biological substances or tissues without provoking a biochemical response. For example, living neurons have been cultured on microporous Si where they stay alive for quite some time (in contrast to solid Si) and the tissue compatibility of PSi has thus been demonstrated (Mayne et al. 2000). PSi as a biomaterial offers a new and dynamic field of research with important potential applications (Kingman 2001; Li et al. 2003). The ability to culture mammalian cells directly onto PSi, coupled with the material's apparent lack of toxicity, offers exciting possibilities for the future of biologically interfaced sensing. Buckberry and Bayliss (1999) believe that PSi could prove to be the bridge that allows signals and information to be transmitted directly between a semiconductor device and a biological system. In addition, due to the large surface area of the material, large amounts of biomolecular interactions can take place over a small working area, thereby facilitating the miniaturization of the sensor.

Drug delivery is another very promising field for PSi applications. The low toxicity of porous Si and $SiO_2$, the high porosity, and the relatively convenient surface chemistry has spurred interest in the use of this system as a host, or "mother ship" for therapeutics, diagnostics, or other types of payloads (Figure 1.9). As it is known, providing a controlled and localized release of therapeutics within the body are key objectives for increasing efficacy and reducing the risks of potential side effects (Brigger et al. 2002). It should be noted that already there aremany technologies allowing loading a molecular payload into a porous Si host (Anglin et al. 2008). It is important that PSi's ability to degrade in the body presents fewer challenges during chronic use of PSi-based drug delivery materials than, for example, carbon nanotubes, which are not metabolized and so must be excreted after administration.

*In vivo* monitoring using the optical properties of PSi is another approach based on the advantages of PSi. Many material hosts have been developed for drug delivery, but few can "self-report" on the amount of drug loaded or released. It is important to know these quantities when determining the efficacy of a treatment to identify when it is time to administer a new dose (Anglin et al. 2008). The unique optical properties that can be engineered into PSi provide a mechanism to perform such assays *in vivo*. Incorporation of molecules into a PSi layer alters its index of refraction, and the spectrum obtained from a thin film or multilayer structure provides a measure of loading in the nanostructure.

One might say that the suitability and efficacy of various forms of PSi are being assessed for medical applications, and some are currently in clinical trials (Anglin et al. 2008). The incorporation of anti-cancer therapeutics (Vaccari et al. 2006), anti-inflammatory agents (Salonen et al. 2005), analgesics (Salonen et al. 2005), and medicinally relevant proteins and peptides has been

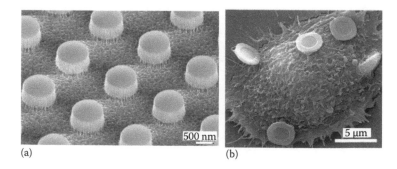

(a)                                        (b)

**FIGURE 1.9**   Schematic overview presenting PSi microparticle fabrication and applications in drug delivery and immunotherapy. (a) SEM image showing a patterned PSi wafer prior to sonication-based release of the patterned particles. (Reprinted from *Curr. Opin. Pharmacol.* 13, Savage, D.J. et al., 834, Copyright 2013, with permission from Elsevier.) (b) The scanning electron micrograph shows an endothelial cell with 3-µm hemispherical PSi particles associated with the cell surface. (Reprinted with permission from Serda, R.E. et al., *Nanoscale* 1, 250, 2009. Reproduced by permission of the Royal Society of Chemistry.)

demonstrated (Foraker et al. 2003). The oral administration of PSi to provide a dietary supplement of silicon has also been assessed (Macdonald et al. 2005; Canham 2007). Some of the most advanced clinical studies have been performed by pSiMedica inc., using PSi as a brachytherapy device for the treatment of cancer (Goh et al. 2007). In this work, percutaneous implants of PSi particles (on the order of 20 μm in size) containing radioactive $^{32}P$ provide local radiation to a tumor. The radioactive isotope is being synthesized in the PSi material by elemental transmutation of Si, induced by exposure to high energy neutrons emanating from a nuclear reactor. More detailed description of PSi biomedical applications can be found in Chapters 17–20 in *Porous Silicon: Biomedical and Sensor Applications.*

Another benefit is that the optical and optoelectronic properties of μPSi could enable linkage to a data logger by optical fibers. This would remove the risk of electromagnetic forces influencing the responses of cells. Such devices could, for example, receive optical information and convert this to a biological signal that would be passed into neural tissue as a substitute for "sight" sensation. Alternatively, PSi could be used for building either environmental or pharmaceutical sensing systems. Optical signals to and from the material could be used to sense wavelength shifts, which would correspond to changes in attached cells caused by the presence of chemicals or drugs (Bayliss et al. 1997; Buckberry and Bayliss 1999).

An antireflection coating of solar cells with micro- or mesoporous Si is another interesting direction for PSi application (Strehlke et al. 1997; Striemer and Fauchet 2003). A controlled change of the porosity along the depth introduces a concomitant change for the index of refraction to suppress reflection to a large degree. Alternatively, μPSi layers may also act as a conventional antireflection layer.

In the same vein of thought, micro- and mesoporous silicon is also an interesting material for making integrated optical devices because of its easily modified refractive index. In fact, by changing the anodization current density during its production process, meso-PSi layers with different porosities and therefore with different refractive indexes (Guerrero-Lemus et al. 1999; Pirasteh et al. 2006) can be obtained. This possibility opens new and interesting fields in the development of optically filtered mirrors based on micro- or meso-PSi, with selected bands in the visible and infrared spectra (Berger et al. 1994). Thus, a periodic variation in depth and porosity is possible, resulting in PSi multilayers or Bragg mirrors (Frohnhoff et al. 1995). The modulation of the PSi refractive index is still present after an oxidation process and can be applied for increasing the optical efficiency of PSi multilayers working in the visible range of the spectrum. In this way, for example, optical waveguide structures could be constructed (Pickering et al. 1984). Recent developments like a silicon Raman laser (Rong et al. 2005) or high-speed optical silicon modulators (Liao et al. 2005) also further the integration of electronic and photonic active devices on the same silicon substrate, furthering the cause of PSi. Passive photonic devices designed in silicon such as waveguides, couplers, and splitters have been reported by Agarwal et al. (1996). Detailed discussion of optical, optoelectronic, and photonic applications of PSi is also presented in Chapters 1–4 and 10 of *Porous Silicon: Opto- and Microelectronic Applications.*

It has been shown that micro- and meso-PSi layers are excellent materials for local thermal isolation on bulk silicon (Nassiopoulou and Kaltsas 2000; Lysenko et al. 2002) and for fabrication of micro-hotplates for low-power sensors (Maccagnani et al. 1998, 1999; Tsamis et al. 2003). Experiment has shown that the technology of silicon porosification is alternative for silicon on insulator (SOI) and silicon on sapphire (SOS) technologies. Full isolation by porous oxidized silicon (FIPOS) method is based on the oxidation of PSi to isolate predefined islands of crystalline silicon from the bulk silicon substrate. Methods of implementing FIPOS technique are shown in Figure 1.10. Compared to the conventional methods, the FIPOS offered the simplicity of processing and low leakage current (Thomas et al. 1989). Micro- and meso-PSi could also be used as isolation material for radio frequency (rf) applications because of its high resistivity and compatibility with modern very-large scale-integration IC technology (Kim et al. 2003). These applications of PSi will be analyzed in Chapters 8 and 11, in *Porous Silicon: Biomedical and Sensor Applications,* and Chapters 5 and 6 in *Porous Silicon: Opto- and Microelectronic Applications.*

Studies have shown that PSi is a good support not only for growth of monocrystalline silicon films. Low porosity layers (P <30%) proved to be effective as a buffer layer for growing single-crystal (epitaxial) films of other semiconductors as well. Using the intermediate layers of PSi allowed solving the problem of the growth on silicon substrate of high-quality films of GaAs, PbS,

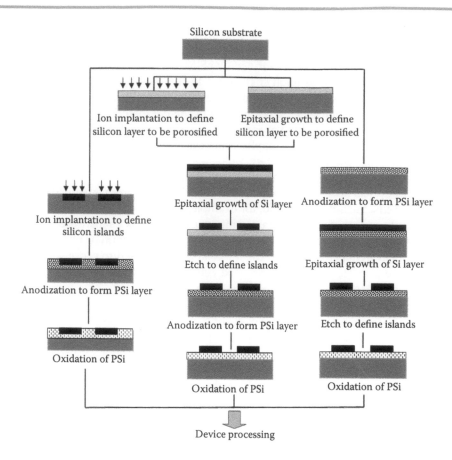

**FIGURE 1.10** Methods of implementing FIPOS technique.

PbTe, and other semiconductor compounds even with large mismatch in the lattice parameters. This topic is discussed in Chapter 7 of *Porous Silicon: Opto- and Microelectronic Applications*.

The list of possible applications is not yet exhausted; there are many more possible applications of porous semiconductors covered in the reviews (Halimaoui 1995; Steiner and Lang 1995; Parkhutik 1999; Bisi et al. 2000; Muller et al. 2000; Kleps et al. 2001, 2007; Foll et al. 2002, 2003a, 2006; Marsh 2002; Angelescu et al. 2003; Zhu et al. 2005; Kilian et al. 2007; Mizsei 2007; Anglin et al. 2008), which we will be discussed in *Porous Silicon: Biomedical and Sensor Applications* and *Porous Silicon: Opto- and Microelectronic Applications*. This includes the many possible applications of macroporous Si, which has a few outstanding properties on its own. In particular, the most thoroughly investigated macropores have diameters in the 0.5–10 μm range, perfectly matched to the size regime of today's photolithographic patterning techniques (Lehmann 1996). This allows producing porous structures with a given geometry, for example, for photonic crystals (Muller et al. 2000; Birner et al. 2001; Kilian et al. 2007; Huanca et al. 2008; Wehrspohn et al. 2008). Moreover, the mechanism of macropore formation is understood or at least experimentally established in some detail, allowing for easy production. PSi-based photonic crystals, for example, can be fabricated by using conventional microelectronic manufacturing technologies (plus pore etching) inherently compatible with CMOS technology (Sze 2001). The well-controlled manner in which macroporous silicon can be structured allows envisioning a new class of photonic materials—from "photonic band gap crystals" ("photonic band gap" refers to a frequency range, where photons are not allowed to propagate in the material) working for light propagation perpendicular to the pores, to a large range of filters and polarizers working for light propagation parallel to the pores with optical properties quite generally tied to the geometry of pore arrays, spanning the range from the far infrared (IR) to the ultraviolet (UV) (Gruning et al. 1996; Kochergin and Foll 2006). Capabilities of this technology for the formation of such structures are illustrated in Figure 1.11. Experiment has shown that the use of PSi in the design of Li batteries, supercapacitors, thermoelectrical generators, and fuel cells is also a very promising direction

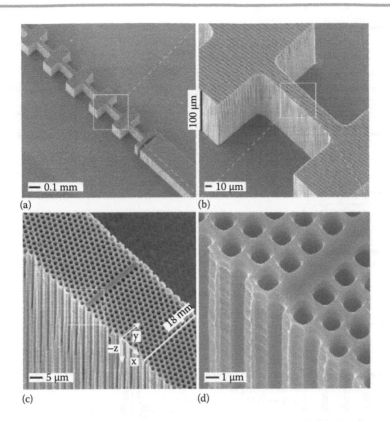

**FIGURE 1.11** Scanning electron microscope image of a bar of macroporous silicon the lateral thickness of which is periodically altered. The thinner parts consist of distinct numbers of crystal rows whereas the thicker ones provide mechanical stability (a). The wall that is marked by the square consists of 13 crystal rows (bi-layers) along the Γ-K direction of the hexagonal lattice. The height of the porous structure, which is fixed on a substrate of silicon, is 100 mm (b). By omitting some of the pores, defect structures of various geometries can be introduced (c). Cutting along the Γ-M direction provides sidewalls with a grating-like surface (d). (Reprinted from *Mater. Sci. Semicond. Process* 3(5–6), Birner, A. et al., 487, Copyright 2000, with permission from Elsevier.)

in the development of new energy technologies. Chapters 10–18 of *Porous Silicon: Opto- and Microelectronic Applications* are devoted to the consideration of these approaches.

It should be noted that there are unexpected applications of technology of silicon porosification as well. For example, Rodriguez et al. (2005) suggested using technology of macroporous silicon forming for fabrication of silicon dioxide ($SiO_2$) microneedles (see Figure 1.12). Persson et al. (2007) and Balucani et al. (2011) believe that PSi has benefits for use as electrodes in biomedical applications. Cytrynowicz et al. (2002) and Kan et al. (2004) have shown that PSi membrane structure could be incorporated into the closed-loop cooling system, which in turn could be used in laptop computer, space electronic application, or in high-energy physics experiments. The principle of cooling system operation is based on the use of a heat pipe and a two-phase thermal device, which utilizes the significant heat of evaporation when the operating fluid changes to the vapor phase. Canham (2007) stated that PSi has potential for useto improve the shelf life and bioavailability of specific nutrients in functional foods. Since silicon is a very hard material, therefore Langner et al. (2011) believe that it can be used as a micrometer-sized structuring tool, for example, as a stamp for imprint lithography.

As it was shown in numerous articles (van den Meerakker et al. 2003; Qu et al. 2011), the technology of electrochemical etching, designed for silicon porosification, can also be successfully applied for Si nanowires fabrication. For example, van den Meerakker et al. (2003) have shown that the length of these wires can reach values up to 100 μm, while their width can be as small as 30 nm (see Figure 1.13). Such nanowires can be used in solar cells, fuel cells, Li ion batteries, photocatalysis, gas sensors, drug delivery, and so on (Qu et al. 2011; Ge et al. 2012; Najar et al. 2012).

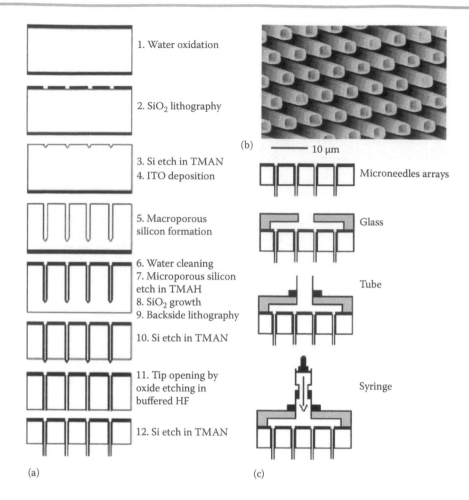

**FIGURE 1.12**    (a) Process flow for microneedles fabrication. (b) SEM micrograph (bottom view) of an array of microneedles as released by TMAH etch with 70-nm wall thickness and square arrangement. (c) Schematics of the injection system. (Reprinted from *Sens. Actuators B* 109, Rodriguez, A. et al., 135, Copyright 2005, with permission from Elsevier.)

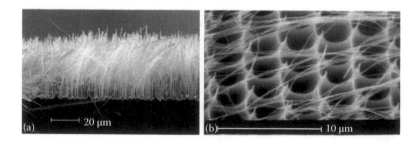

**FIGURE 1.13**    SEM pictures of nanowires obtained after etching *p*-type Si with a current density close to the peak current density. (a) Cross-section of wafer with nanowires with lengths of up to 80 μm. (b) The surface of a wafer with the smallest nanowires obtained (30 nm). (van den Meerakker, J.E.A.M. et al.: *Phys. Stat. Sol. (a)*. 2003. 197(1). 573. Copyright Wiley-VCH Verlag GmbH & Co. KGaA. Reproduced with permission.)

## ACKNOWLEDGMENTS

This work was supported by the Ministry of Science, ICT and Future Planning (MSIP) of the Republic of Korea.

## REFERENCES

Agarwal, A.M., Liao, L., Foresi, J.S., Black, M.R., Duan, X., and Kimerling, L.C. (1996). Low-loss polycrystalline silicon waveguides for silicon photonics. *J. Appl. Phys.*, **80**, 6120–6123.

Allongue, P. (1997). Porous silicon formation mechanisms. In: Canham, L. (ed.) *Properties of Porous Silicon.* INSPEC, London, pp. 3–11.

Anderson, R.C., Muller, R.S., and Tobias, C.W. (1990). Investigation of porous silicon for vapour sensing. *Sens. Actuators, A* **21–23**, 835–839.

Anderson, M.A., Tinsley-Bown, A., Allcock, P. et al. (2003). Sensitivity of the optical properties of porous silicon layers to the refractive index of liquid in the pores. *Phys. Stat. Sol. (a)*, **197**(2), 528–533.

Angelescu, A., Kleps, I., Mihaela, M. et al. (2003). Porous silicon matrix for application in biology. *Rev. Adv. Mater. Sci.*, **5**, 440–449.

Anglin, E.J., Cheng, L., Freeman, W.R., and Sailor, M.J. (2008). Porous silicon in drug delivery devices and materials. *Adv. Drug. Deliv. Rev.*, **60**, 1266–1277.

Asano, T., Higa, K., Aoki, S., Tonouchi, M., and Miyasato, T. (1992). Effects of light exposure during anodization on photoluminescence of porous Si. *Jpn. J. Appl. Phys.*, **31**(Part 2), L373–L375.

Balucani, M., Nenzi, P., Crescenzi, C. et al. (2011). Technology and design of innovative flexible electrode for biomedical applications. In: *Proceedings of IEEE 61st Conference on Electronic Components and Technology,* May 31–June 3, Lake Buena Vista, FL, pp. 1319–1324.

Bayliss, S., Buckberry, L., Harris, P., and Rousseau, C. (1997). Nanostructured semiconductors: Compatibility with biomaterials. *Thin Solid Films*, **297**, 308–309.

Ben-Chorin, M. and Kux, A. (1994). Adsorbate effects on photoluminescence and electrical conductivity of porous silicon. *Appl. Phys. Lett.*, **67**, 481–483.

Berger, M.G., Dieker, C., Thonissen, M. et al. (1994). Porosity superlattices: A new class of Si heterostructures. *J. Phys. D: Appl. Phys.*, **27**, 1333–1336.

Birner, A., Li, A.P., Müller, F. et al. (2000). Transmission of a microcavity structure in a two-dimensional photonic crystal based on macroporous silicon. *Mater. Sci. Semicond. Process*, **3**(5–6), 487–491.

Birner, A., Wehrspohn, R.B., Gösele, U., and Busch, K. (2001). Silicon-based photonic crystals. *Adv. Mater.*, **13**(6), 377–388.

Bisi, O., Ossicini, S., and Pavesi, L. (2000). Porous silicon: A quantum sponge structure for silicon based optoelectronics. *Surf. Sci. Rep.*, **38**, 1–126.

Bjorkqvist, M., Salonen, J., Paski, J., and Laine, E. (2003). Comparision of stabilizing treatments on porous silicon for sensor applications. *Phys. Stat. Sol. (a)*, **197**, 374–377.

Blackwood, D.J. and Zhang, Y. (2003). The effect of etching temperature on the photoluminescence emitted from, and the morphology of, p-type porous silicon. *Electrochim. Acta*, **48**, 623–630.

Blackwood, D.J. and Akber, M.F.B.M. (2006). *In-situ* electrochemical functionalization of porous silicon. *J. Electrochem. Soc.*, **153**(11), G976–G980.

Brigger, I., Dubernet, C., and Couvreur, P. (2002). Nanoparticles in cancer therapy and diagnosis. *Adv. Drug Deliv. Rev.*, **54**, 631–651.

Brunauer, S., Emmett, P.H., and Teller, E. (1938). Adsorption of gases in multimolecular layers. *J. Am. Chem. Soc.*, **60**(2), 309–331.

Brune, D., Hellborg, R., Whitlow, H.J., and Hunderi, O. (eds.). (2007). *Surface Characterization: A User's Sourcebook.* Wiley-VCH, Weinheim, Germany.

Brundle, C.R., Evans, C.A., and Wilson, S. (eds.). (1992). *Encyclopedia of Materials Characterization: Surfaces, Interfaces, Thin Films.* Reed Publishing, Stoneham, MA.

Buckberry, L. and Bayliss, S. (1999). Material for melding humans and machines. *Mater. World*, **7**(4), 213–215.

Canham, L.T. (1991). Silicon quantum wire array fabrication by electrochemical and chemical dissolution of wafers. *Appl. Phys. Lett.*, **57**, 1046–1048.

Canham, L.T. (1995). Bioactive silicon structure fabrication through nanoetching techniques. *Adv. Mater.*, **7**, 1033–1037.

Canham, L. (ed.). (1997). *Properties of Porous Silicon.* INSPEC, London.

Canham, L.T. (2007). Nanoscale semiconducting silicon as a nutritional food additive. *Nanotechnology*, **18**, 185704 (1–6).

Canham, L.T., Cox, T.I., Loni, A., and Simons, A.J. (1996). Progress towards silicon optoelectronics using porous silicon technology. *Appl. Surf. Sci.*, **102**, 436–441.

Che, M. and Vedrine, J.C. (eds.). (2012). *Characterization of Solid Materials and Heterogeneous Catalysts from Structure to Surface Reactivity*, Vols. 1 and 2. Wiley-VCH, Weinheim, Germany.

Cullis, A.G., Canham, L.T., and Calcott, P.D.G. (1997). The structural and luminescence properties of porous silicon. *J. Appl. Phys.*, **82**, 909–965.

Cytrynowicz, D., Hamdan, M., Medis, P., Shuja, A., Henderson, H.T., Gerner, F.M., and Golliner, E. (2002). MEMS loop heat pipe based on coherent porous silicon technology. In: *AIP Conf. Proc.*, **608**(1), pp. 220–232.

Denoyel, R., Llewellyn, P., Beurroies, I., Rouquerol, J., Rouquerol, F., and Luciani, L. (2004). Comparing the basic phenomena involved in three methods of pore-size characterization: Gas adsorption, liquid intrusion and thermoporometry. *Part. Part. Syst. Charact.*, **21**, 128–137.

De Stefano, L., Moretti, L., Rendina, I., and Rossi, A.M. (2004). Time-resolved sensing of chemical species in porous silicon optical microcavity. *Sens. Actuators B*, **100**, 168–172.

Di Francia, G., De Filippo, F., La Ferrara, V. et al. (1998). Porous silicon layers for the detection at RT of low concentrations of vapoura from organic compounds. In: *Proceeding of European Conference Eurosensors XII*, September 13–16, 1998, Southampton, UK. IOP, Bristol, vol. 1, pp. 544–547.

Dillon, A.C., Gupta, P., Robinson, M.B., Bracker, A.S., and George, S.M. (1990). FTIR studies of water and ammonia decomposition on silicon surfaces. *J. Electron. Spectrosc. Relat. Phenom.*, **54–55**, 1085–1095.

Du Plessis, M. (2007). A gravimetric technique to determine the crystallite size distribution in high porosity nanoporous silicon. *ECS Trans.*, **9**(1), 133–142.

Fang, C., Foll, H., and Carstensen, J. (2006). Electrochemical pore etching in germanium. *J. Electroanal. Chem.*, **589**, 259–288.

Feng, Z.C., Wee, A.T.S., and Tan, K.L. (1994). Surface and optical analyses of porous silicon membranes. *J. Phys. D*, **27**, 1968–1975.

Foll, H., Christophersen, M., Carstensen, J., and Hasse, G. (2002). Formation and application of porous silicon. *Mater. Sci. Eng. R*, **39**, 93–141.

Foll, H., Langa, S., Carstensen, J., Lolkes, S., Christophersen, M., and Tiginyanu, I.M. (2003a). Engineering porous III-Vs, III-Vs REVIEW. *Adv. Semicond. Mag.*, **16**(7), 42–43.

Foll, H., Langa, S., Carstensen, J., Christophersen, M., and Tiginyanu, I.M. (2003b). Pores in III-V semiconductors. *Adv. Mater.*, **15**(3), 183–198.

Foll, H., Carstensen, J., and Frey, S. (2006). Porous and nanoporous semiconductors and emerging applications. *J. Nanomater.*, **2006**, 91635 (1–10).

Foraker, A.B., Walczak, R.J., Cohen, M.H., Boiarski, T.A., Grove, C.F., and Swaan, P.W. (2003). Microfabricated porous silicon particles enhance paracellular delivery of insulin across intestinal Caco-2 cell monolayers. *Pharm. Res.*, **20**, 110–116.

Frohnhoff, St., Berger, M.G., Thonissen, M. et al. (1995). Formation techniques for porous silicon superlattices. *Thin Solid Films*, **255**, 59–62.

Fukami, K., Harraz, F.A., Yamauchi, T., Sakka, T., and Ogata, Y.H. (2008). Fine-tuning in size and surface morphology of rodshaped polypyrrole using porous silicon as template. *Electrochem. Commun.*, **10**, 56–60.

Ge, M., Rong, J., Fang, X., and Zhou, C. (2012). Porous doped silicon nanowires for Lithium ion battery anode with long cycle life. *Nano Lett.*, **12**, 2318–2323.

Goh, A.S.-W., Chung, A.Y.-F., Lo, R.H.-G. et al. (2007). A novel approach to brachytherapy in hepatocellular carcinoma using a phosphorus$^{32}$ ($^{32}$P) brachytherpay delivery device—A first-in-man study. *Intl. J. Radiation Oncology Biol. Phys.*, **67**, 786–792.

Gole, J.L., Seals, L.T., and Lillehei, P.T. (2000). Patterned metallization of porous silicon from electroless solution for direct electrical contact. *J. Electrochem. Soc.*, **147**(10), 3785–3789.

Gruning, U., Lehmann, V., Ottow, S., and Busch, K. (1996). Macroporous silicon with a complete two-dimensional photonic band gap centered at 5 μm. *Appl. Phys. Lett.*, **68**, 747–749.

Guerrero-Lemus, R., Ben-Hander, F.A., Moreno, J.D. et al. (1999). Anodic oxidation of porous silicon bilayers. *J. Lumin.*, **80**, 173–178.

Halimaoui, A. (1995). Porous silicon: Material processing, properties and applications. In: Vial, J.C. and Derrien, J. (eds.) *Porous Silicon Science and Technology*. Springer-Verlag, New York.

Halimaoui, A. (1997). Porous silicon formation by anodisation. In: Canham, L.T. (ed.) *Properties of Porous Silicon*. IEE INSPEC, The Institution of Electrical Engineers, London, pp. 12–23.

Harper, J. and Sailor, M.J. (1996). Detection of nitric oxide and nitrogen dioxide with photoluminescent poprous silicon. *Annal. Chem.*, **68**, 3713–3717.

Hedrich, F., Billat, S., and Lang, W. (2000). Structuring of membrane sensors using sacrificial porous silicon. *Sens. Actuators B*, **84**, 315–323.

Hejjo Al Rifai, M., Christophersen, M., Ottow, S., Carstensen, J., and Föll H. (2000). Dependence of macropore formation in n-Si on potential, temperature, and doping. *J. Electrochem. Soc.*, **147**(2), 627–635.

Huanca, C., Faglia, G., Sberveglieri, G. et al. (2002). Multiparametric porous silicon sensors. *Sensors*, **2**, 121–128.

Huanca, D.R., Ramirez-Fernandez, F.J., and Salcedo, W.J. (2008). Porous silicon optical cavity structure applied to high sensitivity organic solvent sensor. *Microelectron. J.*, **39**, 499–506.

Hummel, R.E. and Chang, S.-S. (1992). Novel technique for preparing porous silicon. *Appl. Phys. Lett.*, **61**, 1965–1967.

Imai, K. (1981). A new dielectric isolation method using porous silicon. *Solid-State Electron.*, **24**, 159–164.

Ito, T., Yamama, A., Hiraki, A., and Satou, M. (1989). Silicidation of porous silicon and its application for the fabrication of a buried metal layer. *Appl. Surf. Sci.*, **41–42**, 301–305.

Janshoff, A., Dancil, K.-P.S., Steinem, C. et al. (1998). Macroporous p-type silicon Fabry-Perot layers. Fabrication, characterisation, and applications in biosensing. *J. Am. Chem. Soc.*, **120**, 12108–12116.

Jena, A.K. and Gupta, K.M. (2002). Characterization of pore structure of filtration media. *Fluid/Particle Separation J.*, **14**(3), 227–241.

Kalem, S. and Yavuzcetin, O. (2000). Possibility of fabricating light emitting porous silicon from gas phase etchants. *Opt. Express*, **6**, 7–11.

Kan, P.Y.Y., Finstad, T.G., Kristiansen, H., and Foss, S.E. (2004). Porous silicon for chip cooling applications. *Phys. Scripta. T*, **114**, 77–79.

Kilian, K.A., Böcking, T., Gaus, K., Gal, M., and Gooding, J.J. (2007). Si–C linked oligo(ethylene glycol) layers in silicon-based photonic crystals: Optimization for implantable optical materials. *Biomater.*, **28**(20), 3055–3170.

Kim, S.-J., Lee, S.-H., and Lee, C.-J. (2001). Organic vapour sensing by current response of porous silicon layer. *J. Phys. D: Appl. Phys.*, **34**, 3505–3509.

Kim, H., Chong, K., and Xie, Y.-H. (2003). The promising role of porous Si in mixed-signal integrated circuit technology. *Phys. Stat. Sol. (a)*, **197**(1–2), 269–274.

Kingman, S. (2001). Holey chips for drug delivery. *Drug Discovery Today*, **6**(23), 1186–1187.

Kleps, I., Angelescu, A., Samfirescu, N., Gil, A., and Correia, A. (2001). Study of porous silicon, silicon carbide and DLC coated field emitters for pressure sensor application. *Solid-State Electron.*, **45**(6), 997–1001.

Kleps, I., Miu, M., Ignat, T., and Simion, M. (2007). Nanocomposite porous silicon layers. In: *Proceedings of the International Offshore and Polar Engineering Conference*, July 1–6, 2007, Lisbon, Portugal, pp. 2916–2923.

Koch, F., Petrova-Koch, V., and Muschik, T.J. (1993). The luminescence of porous Si: The case for the surface state mechanism. *J. Lumin.*, **57**, 271–281.

Kochergin, V.R. and Foll, H. (2006). Novel optical elements made from porous Si. *Mater. Sci. Eng. R*, **52**, 93–140.

Kolasinski, K.W. (2005). Silicon nanostructures from electroless electrochemical etching. *Curr. Opin. Solid State Mater. Sci.*, **9**(1–2), 73–83.

Korotcenkov, G. and Cho, B.K. (2010a). Porous semiconductors: Advanced material for gas sensor applications. *Crit. Rev. Sol. St. Mater. Sci.*, **35**, 1–23.

Korotcenkov, G. and Cho, B.K. (2010b). Silicon porosification: State of the art. *Crit. Rev. Sol. St. Mater. Sci.*, **35**(3), 153–260.

Labunov, V., Bondarenko, V., Glinenko, L., Dorofeev, A., and Tabulina, L. (1986). Heat-treatment effect on porous silicon. *Thin Solid Films*, **137**, 123–134.

Lang, W. (1996). Silicon microstructuring technology. *Mater. Sci. Eng. R*, **17**, 1–54.

Lang, W., Steiner, P., Richter, A., Marusczyk, K., Weimann, G., and Sandmaier, H. (1994). Applications of porous silicon as a sacrificial layer. *Sens. Actuators A*, **43**, 239–242.

Langa, S., Tiginyanu, I.M., Carstensen, J., Christophersen, M., and Foll, H. (2000). Formation of porous layers with different morphologies during anodic etching of *n*-InP. *Electrochem. Solid-State Lett.*, **3**, 514–516.

Langa, S., Tiginyanu, I.M., Carstensen, J., Christophersen, M., and Foll, H. (2003). Self-organized growth of single crystals of nanopores. *Appl. Phys. Lett.*, **82**, 278–280.

Langner, A., Frak Müller, F., and Gösele, Y. (2011). Macroporous silicon. In: Hayden, O. and Nielsch, K. (eds.), *Molecular- and Nano-Tubes*. Springer Science+Business Media, NewYork, pp. 431–460.

Lehmann, V. (1993). The physics of macropore formation in low doped *n*-type silicon. *J. Electrochem. Soc.*, **140**(10), 2836–2843.

Lehmann, V. (1996). Developments in porous silicon research. *Mater. Lett.*, **28**, 245–249.

Lehmann, V. (2002). *Electrochemistry of Silicon: Instrumentation, Science, Materials and Applications*. Wiley-VCH Verlag GmbH, Weinheim.

Lehmann, V. and Foll, H. (1990). Formation mechanism and properties of electrochemically etched trenches in n-type silicon. *J. Electrochem. Soc.*, **137**, 653–659.

Lehmann, V. and Gosele, U. (1991). Porous silicon formation: A quantum wire effect. *Appl. Phys. Lett.*, **58**(8), 856–858.

Leofanti, G., Tozzola, G., Padovan, M., Petrini, G., Bordiga, S., and Zecchina, A. (1997). Catalyst characterization: Characterization techniques. *Catal. Today*, **34**, 307–327.

Li, Y.Y., Cunin, F., Link, J.R. et al. (2003). Painting a rainbow on silicon—A simple method to generate a porous silicon for sensing and drug delivery applications. *Science*, **299**, 2045–2047.

Liao, L., Samara-Rubio, D., Morse, M. et al. (2005). High-speed silicon Mach-Zehnder modulator. *Opt. Express*, **13**, 3129–3135.

Lowell, S. and Shields, E. (1984). *Powder Surface Area and Porosity*. Chapman & Hall, London.

Lowell, S., Shields, J.E., Thomas, M.A., and Thommes, M. (2006). *Characterization of Porous Solids and Powders: Surface Area, Pore Size and Density*. Springer, Dordrecht, the Netherlands.

Lysenko, V., Perichon, S., Remaki, B., and Barbier, D. (2002). Thermal isolation in microsystems with porous silicon. *Sens. Actuators A*, **99**, 13–24.

Maccagnani, P., Angelucci, R., Pozzi, P. et al. (1998). Thick oxidized porous silicon layer as a thermo-insulating membrane for high-temperature operating thin- and thick-film gas sensors. *Sens. Actuators B*, **49**, 22–29.

Maccagnani, P., Dori, L., and Negrini, P. (1999). Thick porous silicon thermo-insulating membranes for gas sensor applications. *Sens. Mater.*, **11**, 131–147.

Macdonald, H.M., Hardcastle, A.E., Jugdaohsingh, R., Reid, D.M., and Powell, J.J. (2005). Dietary silicon intake is associated with bone mineral density in premenopausal women and postmenopausal women taking HRT. *J. Bone Miner. Res.*, **20**, S393–S393.

Maly, P., Trojanek, F., Hospodkova, A., Kohlova, V., and Pelant, I. (1994). Transmission study of picosecond photocarrier dynamics in free-standing porous silicon. *Solid State Commun.*, **89**(8), 709–712.

Mares, J.J., Kristofik, J., and Hulicius, E. (1995). Influence of humidity on transport in porous silicon. *Thin Solid Films*, **255**, 272–275.

Marsh, G. (2002). Porous silicon. A useful imperfection. *Mater. Today*, January, 36–41.

Mayne, A.H., Bayliss, S.C., Barr, P., Tobin, M., and Backberry, L.D. (2000). Biologically interfaced porous silicon devices. *Phys. Stat. Sol. (a)*, **182**, 505–513.

Milling, A.J. (ed.). (1999). *Techniques, and Applications Surface Characterization Methods: Principles*. Marcel Dekker Inc., Basel, Switzerland.

Mizsei, J. (2007). Gas sensor applications of porous Si layers. *Thin Solid Films*, **515**, 8310–8315.

Muller, F., Birner, A., Schilling, J., Wehrspohn, R.B., and Gosele, U. (2000). Photonic crystals from macroporous silicon. *Adv. Solid State Phys.*, **40**, 545–559.

Najar, A., Charrier, J., Pirasteh, P., and Sougrat, R. (2012). Ultra-low reflection porous silicon nanowires for solar cell applications. *Opt. Express*, **20**(15), 16861–16870.

Nassiopoulou, A.G. and Kaltsas, G. (2000). Porous silicon as an effective material for thermal isolation on bulk crystalline silicon. *Phys. Stat. Sol. (a)*, **182**, 307–312.

Osaka, T., Ogasawara, K., and Nakahara, S. (1997). Classification of the pore structure of n-type silicon and its microstructure. *J. Electrochem. Soc.*, **144**, 3226–3237.

Ozdemir, S. and Gole, J.L. (2007). The potential of porous silicon gas sensors. *Curr. Opin. Solid State Mater. Sci.*, **11**, 92–100.

Parkhutik, V. (1999). Porous silicon-mechanisms of growth and applications. *Sol. State Electron.*, **43**, 1121–1141.

Parkhutik, V. (2000). Analysis of publications on porous silicon: From photoluminescence to biology. *J. Porous Mater.*, **7**, 363–366.

Parkhutik, V.P. and Canham, L.T. (2000). Porous silicon as an educational vehicle for introducing nanotechnology and interdisciplinary materials science. *Phys. Stat. Sol. (a)*, **182**, 591–598.

Persson, J., Danielsen, N., and Wallman, L. (2007). Porous silicon as a neural electrode material. *J. Biomater. Sci. Polym. Ed.*, **18**(10), 1301–1308.

Pickering, C., Beale, V., Robbins, D., Pearson, P., and Greef, R. (1984). Optical studies of the structure of porous silicon films formed in p-type degenerate and non-degenerate silicon. *J. Phys. C: Sol. St. Phys.*, **17**, 6535–6552.

Pirasteh, P., Charrier, J., Soltani, A. et al. (2006). The effect of oxidation on physical properties of porous silicon layers for optical applications. *Appl. Surf. Sci.*, **253**, 1999–2002.

Propst, E.K. and Kohl, P.A. (1994). Electrochemical oxidation of silicon and formation of porous silicon in acetonitrile. *J. Electrochem. Soc.*, **141**, 1006–1013.

Qu, Y., Zhou, H., and Duan, X. (2011). Porous silicon nanowires. *Nanoscale*, **3**, 4060–4068.

Rittenhouse, T.L., Bohn, P.W., and Adesida, I. (2003). Structural and spectroscopic characterization of porous silicon carbide formed by Pt-assisted electroless chemical etching. *Sol. State Commun.*, **126**, 245–250.

Rittersma, Z.M., Splinter, A., Bodecker, A., and Benecke, W. (2000). A novel surface-micromachined capacitive porous silicon humidity sensor. *Sens. Actuators B*, **68**, 210–217.

Rodriguez, A., Molinero, D., Valera, E. et al. (2005). Fabrication of silicon oxide microneedles from macroporous silicon. *Sens. Actuators B*, **109**, 135–140.

Rong, H., Jones, R., Liu, A. et al. (2005). A continuous-wave Raman silicon laser. *Nature*, **433**, 725–728.

Rouquerol, J., Avnir, D., Faibridge, C.W. et al. (1994). Recommendations for the characterization of porous solids. *Pure Appl. Chem.*, **66**(8), 1739–1758.

Sailor, M.J. (2012). *Porous Silicon in Practice: Preparation, Characterization and Applications*. Wiley-VCH, Weinheim, pp. 11–12.

Salonen, J., Laitinen, L., Kaukonen, A.M. et al. (2005). Mesoporous silicon microparticles for oral drug delivery: Loading and release of five model drugs. *J. Control. Release*, **108**, 362–374.

Santinacci, L. and Djenizian, T. (2008). Electrochemical pore formation onto semiconductor surfaces. *C. R. Chimie*, **11**, 964–983.

Savage, D.J., Liu, X., Curley, S.A., Ferrari, M., and Serda, R.E. (2013). Porous silicon advances in drug delivery and immunotherapy. *Curr. Opin. Pharmacol.*, **13**, 834–841.

Schuppler, S., Friedman, S.L., Marcus, M.A. et al. (1994). Dimensions of luminescent oxidized and porous silicon structures. *Phys. Rev. Lett.*, **72**, 2648–2651.

Schuppler, S., Friedman, S.L., Marcus, M.A. et al. (1995). Size, shape, and composition of luminescent species in oxidized Si nanocrystals and H-passivated porous Si. *Phys. Rev. B*, **52**, 4910–4925.

Searson, P.C. (1991). Porous silicon membranes. *Appl. Phys. Lett.*, **59**, 832–833.

Serda, R.E., Ferrati, S., Godin, B., Tasciotti, E., Liua, X.W., and Ferrari, M. (2009). Mitotic trafficking of silicon microparticles. *Nanoscale*, **1**, 250–259.

Shen, Y.C., Leu, I.C., Lai, W.H., Wu, M.T., and Hon, M.H. (2008). Varied morphology of porous GaP(111) formed by anodization. *J. Alloys Compounds*, **454**, L3–L9.

Smith, R.L. and Collins, S.D. (1992). Porous silicon formation mechanisms. *J. Appl. Phys.*, **71**(8), R1–R22.

Steiner, P. and Lang, W. (1995). Micromachining applications of porous silicon. *Thin Solid Films*, **255**, 52–58.

Storck, S., Bretinger, H., and Maier, W.F. (1998). Characterization of micro- and mesoporous solids by physisorption methods and pore-size analysis. *Appl. Catal. A: Gen.*, **174**, 137–146.

Strehlke, S., Sarti, D., Krotkus, A., Grigorias, K., and Levy-Clement, C. (1997). The porous silicon emitter concept applied to multicrystalline silicon solar cells. *Thin Solid Films*, **297**, 291–295.

Striemer, C.C. and Fauchet, F. (2003). Dynamic etching of silicon for solar cell applications. *Phys. Stat. Sol. (a)*, **197**, 502–506.

Sze, S.M. (2001). *Semiconductor Devices: Physics and Technology*, 2nd ed. John Wiley & Sons, Inc., New York.

Takai, H. and Itoh, T. (1986). Porous silicon layers and its oxide for the silicon-on-insulator structure. *J. Appl. Phys.*, **60**, 222–225.

Thomas, N.J., Davis, J.R., Keen, J.M. et al. (1989). High-performance thin-film silicon-oninsulator CMOS transistors in porous anodized silicon. *IEEE Electron Dev. Lett.*, **10**(3), 129–131.

Tischler, M.A., Collins, R.Y., Stathis, J.H., and Tsang, J.C. (1992). Luminescence degradation in porous silicon. *Appl. Phys. Lett.*, **60**, 639–641.

Torchinskaya, T.V., Korsunskaya, N.E., Khomenkova, L.Yu. et al. (1997). Complex studies of porous silicon aging phenomena. In: *Proceeding of International Semiconductor Conference*, Sinaie, Romania, October 1997. IEEE, pp. 173–176.

Torchinskaya, T.V., Korsunskaya, N.E., Khomenkova, L.Y., Dhumaev, B.R., and Prokes, S.M. (2001). The role of oxidation on porous silicon photoluminescence and its excitation. *Thin Solid Films*, **381**, 88–93.

Tsamis, C., Nassiopoulou, A.G., and Tserepi, A. (2003). Thermal properties of suspended porous silicon micro-hotplates for sensor applications. *Sens. Actuators B*, **95**, 78–82.

Turner, D.R. (1958). Electropolishing silicon in hydrofluoric. *J. Electrochem. Soc.*, **105**, 402–408.

Uhlir, A. Jr. (1956). Electrolytic shaping of germanium and silicon. *Bell Syst. Tech. J.*, **35**, 333–347.

Vaccari, L., Canton, D., Zaffaroni, N., Villa, R., Tormen, M., and di Fabrizio, E. (2006). Porous silicon as drug carrier for controlled delivery of doxorubicin anticancer agent. *Microelectron. Eng.*, **83**, 1598–1601.

van den Meerakker, J.E.A.M., Elfrink, R.J.G., Weeda, W.M., and Roozeboom, F. (2003). Anodic silicon etching; the formation of uniform arrays of macropores or nanowires. *Phys. Stat. Sol. (a)*, **197**(1), 57–60.

Vial, J.C. and Derrien, J. (eds.). (1995). *Porous Silicon: Science and Technology*. Les Editions de Physique, Les Ulis. Springer, Berlin.

Vial, J.C., Bsiesy, A., Gaspard, F. et al. (1992). Mechanisms of visible-light emission from electro-oxidized porous silicon. *Phys. Rev. B*, **45**, 14171–14176.

Vrkoslav, V., Jelinek, I., Matocha, M., Kral, V., and Dian, J. (2005). Photoluminescence from porous silicon impregnated with cobalt phthalocyanine. *Mater. Sci. Eng. C*, **25**, 645–649.

Watanabe, Y. and Sakai, T. (1971). Application of a thick avoided film to semiconductor devices. *Rev. Electron. Commun. Labs.*, **19**, 899–903.

Watanabe, Y., Arita, Y., Yokoyama, T., and Igarashi, Y. (1975). Formation and properties of porous silicon and its applications. *J. Electrochem. Soc.: Sol.-St. Sci. Technol.*, **122**(10), 1352–1355.

Wehrspohn, R.B., von Rhein, A., and Geppert, T. (2008). Photonic crystals: Principles and applications. In: Buschow, K.H.J., Cahn, R., Flemings, M.C. et al. (eds.) *Encyclopedia of Materials: Science and Technology*. Elsevier, Amsterdam, pp. 1–9.

Zhang, G.X. (2005). Porous silicon: Morphology and formation mechanisms. In: Vayenas, C. (ed.) *Modern Aspects of Electrochemistry*, Number 39. Springer, New York, pp. 65–133.

Zhu, Z., Zhang, J., and Zhu, J. (2005). An overview of Si-based biosensors. *Sensor Lett.*, **3**(2), 71–88.

# Silicon Porosification

# Fundamentals of Silicon Porosification via Electrochemical Etching

**2**

Enrique Quiroga-González and Helmut Föll

## CONTENTS

## 2.1 INTRODUCTION

Since the discovery of the interesting luminescent properties of microporous Si (Canham 1990; Lehmann and Gösele 1991), and controllable macropore etching under illumination in *n*-type silicon (Lehmann and Föll 1990), the development of porous Si systems has exponentially increased. In this regard, nowadays Si can be etched in a controlled manner, with lithiographically defined or undefined (random) patterns. Additionally, silicon can be electrochemically etched in various electrolytes, like HF-based solutions with fluorides (Hoffmann et al. 2000) and KOH-based (Mathwig et al. 2011) solutions. This chapter will deal with the basics of porosification, centering the discussion on the most common and successful etching system, the Si-HF system.

## 2.2 ELECTROCHEMISTRY OF SILICON. ELECTROCHEMICAL REACTIONS IN THE SILICON SYSTEM

### 2.2.1 ELECTROCHEMICAL ETCHING SYSTEM FOR Si

An electrochemical system for performing electrochemical etching of Si consists of the components schematized in Figure 2.1: (a) anode electrode, which is the Si piece to be etched; (b) cathode electrode, usually made of Pt, which is not corroded during the etching process; (c) electrolyte, which enables the ionic transport from the cathode to the anode, and contains etching species; and (d) reference electrode, which can be any standard electrode used as reference (Ag/AgCl, saturated calomel, etc.), or a noble metal like Pt (pseudo-reference). The reference electrode is positioned close to the samples to be etched, to keep the contribution of the electrolyte in the voltage measurements as small as possible.

In the etching system of Figure 2.1, the Si piece is at the bottom, making contact with the electrolyte through a hole in a container (commonly sealed with an O-ring). This is the simplest and usual way of arranging the system, to assure that the electrolyte will always make a homogeneous contact with Si, even when it is available in small quantities. A good material for fabricating the container is Teflon because it is chemically stable even in very concentrated HF solutions.

Optionally, but preferably, the etching system may also have a temperature control and an electrolyte pumping system (Christophersen et al. 2000a, 2001). A temperature control is important because the diffusion of the etchant and the semiconductor electrical properties greatly depend on the temperature. The pumping system avoids problems like electrolyte depletion at the interface but adds certain degrees of anisotropy due to the electrolyte flow direction. The uniform processing of large samples like complete wafers requires considerable sophistication in the set-up. See, for example, the system developed by Föll et al. (2002).

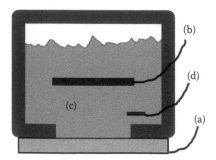

**FIGURE 2.1**    Electrochemical etching system. (a) Si sample (working electrode); (b) counter electrode, made of Pt; (c) electrolyte (etchant); and (d) reference electrode.

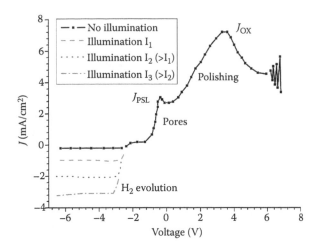

**FIGURE 2.2**    Current–voltage (I-V) characteristics of a *p*-type Si–(HF-based) electrolyte system.

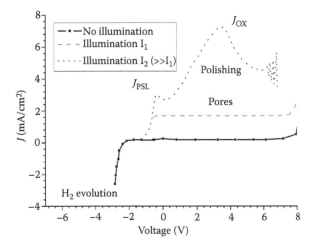

**FIGURE 2.3**    Current–Voltage (I-V) characteristics of an *n*-type Si–(HF-based) electrolyte system. Pores can be obtained just under illumination.

### 2.2.2 CURRENT-VOLTAGE CHARACTERISTICS

The electrical behavior of the electrochemical etching system behaves in a first approximation like a rectifying metal-semiconductor contact known as Schottky contact. Typical I-V curves (current-voltage characteristics) of an electrochemical Si etching system using HF-based aqueous electrolytes is shown in Figures 2.2 and 2.3 for *p*-type and *n*-type Si, respectively. The area of contact of the electrolyte with Si is defined by the etching cell, and is variable from cell to cell. Thus, it is important to use "current density" ($J$) instead of the current $I$ in all measurements and figures. Accounts that are far more detailed can be found in Lehmann (2002) and Kochergin and Föll (2009).

While the $J$-$V$ curves of the *p*- and *n*-type Si etching systems in the anodic region show some forward or reverse diode-like characteristics, respectively, there are some special features never encountered in solid-state junctions. Si may dissolve with time, invariably leading to changes in the $J$-$V$ characteristics with time. The most important features of the $J(V)$ characteristics not observed for solid state diodes are the peaks PSL (porous Si layer) at current density $J_{PSL}$, the "oxide peak" at $J_{OX}$, and current oscillations at high voltages (see Figures 2.2 and 2.3). In the case of n-Si, anodic currents need illumination, and the major deviation from a normal diode behavior is the strong nonlinearity of the response to light for current densities approaching or exceeding $J_{PSL}$.

A strong cathodic biasing of Si (forward bias for *n*-type and reverse bias for *p*-type) allows an easy transfer of electrons to the electrolyte. The electron transfer then produces $H_2$ evolution on the cathode side ($2H^+ + 2e^- \rightarrow H_2$). On the other hand, an anodic biasing will result in Si dissolution. This process requires electronic holes, thus, in $n$-type Si, for an appreciable Si dissolution, it is necessary to apply light to generate electron-hole pairs. The different sections of the $J$-$V$ curves will be described and explained in the following points, going from negative to positive voltages:

1. The cathodic part of the $J$-$V$ curves for both $p$- and $n$-type Si behaves as expected for a reverse bias in a solid-state junction. For $n$-type Si, the electrolyte-Si junction is in forward condition; the current grows exponentially with the applied voltage and essentially generates $H_2$. On the other hand, for $p$-type Si in equilibrium, the reverse current is of low intensity, and it is necessary to apply large voltages to be able to observe breakdown currents. Illuminating $p$-type Si, the current increases proportionally to the illumination intensity. An increase in the current produced by light (photocurrent) translates into an increase of $H_2$ generation, which is uninteresting when talking about microstructuring of semiconductors.

2. At around 0 A/cm² it is interesting that the voltage is not 0 V. Some "built-in" voltage depends on the nature of the reference electrode (and the surface condition of the Si sample). The needed voltages to drive some (forward) current through the etching system greatly depend on the kind of reactions taking place at the electrodes. $H_2$ evolution from HF-containing solutions needs a certain minimum (negative) "overpotential."

3. On the anodic side (i.e., positive currents), Si is always dissolved when using HF-containing electrolytes. Because of this, the surface of the Si electrode changes with time. It may at the extremes become polished or porous. This causes changes in the $J$-$V$ characteristics with time because not only is $J(V)$ small between pores and large inside pores, but also the Si area changes by orders of magnitude with time. This area change is usually neglected in calculating current densities where always the nominal area is used; nevertheless, it is of prime importance for understanding the mechanisms of pore formation.

In $p$-type Si, pores are produced in the voltage region below the *PSL* peak; typically, micropores for large current densities and macropores for lower ones. Details depend on the resistivity, the nature of the electrolyte, and the cell design, and cannot easily be generalized. When one goes beyond this limit, electropolishing is observed, and going even beyond the $J_{OX}$ peak, the polishing is always accompanied by self-induced current oscillations under potentiostatic conditions, and voltage oscillations under galvanostatic conditions. These oscillations are obvious examples of self-organized phenomena in semiconductor electrochemistry, expressing themselves as current oscillations in time. However, pore formation per definition involves a kind of current oscillation in space and thus is a self-organization feature just as well (Föll 1991).

In $n$-type Si under darkness conditions, the anodic $J$-$V$ characteristics are those for a reverse-biased junction (the current is very small), as expected. However, if illumination is applied, the current intensity increases, dissolving Si. At intermediate illumination intensities, pores are produced, independently of the voltage. Furthermore, if the illumination intensity is so high that the reaction limited maximum current density is smaller than the induced photocurrent, the $J$-$V$ characteristics look very similar to that of $p$-type Si (dotted curve of Figure 2.3); thus, the same regions of porosification and electropolishing are observed when the voltage is varied.

It is worthwhile to mention that any "working point" (i.e., the initially defined external current and voltage pair $[J, V]$) for $p$-type Si is confined in the $J$-$V$ curve of the system. In contrast, for $n$-type Si, any working point below the limiting curve for intense illumination can be chosen by simply providing suitable illumination conditions. In other words, for $p$-type Si there is only one free (prime) parameter (voltage or current), whereas for $n$-type Si there are three: voltage, current, and illumination intensity, two of which can be independently varied in principle (and within certain limits, of course).

Finally, it should be mentioned that the I-V curves shown in Figures 2.2 and 2.3 may vary depending on the level of doping of Si (e.g., $p$, $p+$, $p++$, or $n$, $n+$, $n++$) (Zhang 2001; Lehmann 2002). The same is true if one uses different solvents than water (DMA, DMF, DMSO, etc.) (Propst and Kohl 1994; Propst et al. 1994; Föll et al. 2002), and different HF concentrations.

## 2.3 Si ETCHING. CHEMICAL REACTIONS GOVERNING THE DISSOLUTION OF SILICON

Electrochemical etching of Si consists of any combinations of the following four different net chemical reactions (cf., e.g., Föll et al. 2002 for a detailed discussion):

### 2.3.1 DIRECT DISSOLUTION OF Si

(2.1)
$$Si + xh^+ - ye^- \rightarrow Si^{4+}$$

where $x$ holes ($h^+$), with $x$ being $\geq 1$, are consumed in the reactions together with y injected electrons ($e^-$) that contribute with the rest of the charge needed. This kind of dissolution makes the most efficient use of holes because it can be triggered even by only one hole.

In the case of etching in HF-containing solutions, the reaction turns to be (Anglin et al. 2008)

(2.2)
$$Si + 6F^- + 2H^+ + 2h^+ \rightarrow SiF_6^{2-} + H_2$$

which is a 2-hole process, producing an Si compound where the Si valence is 4+, as described in the general Equation 2.1. This is usually called divalent dissolution, due to the amount of necessary charges for the process (Lehmann and Gösele 1991). As described in Equation 2.1, direct dissolution could have any valence between 1 and 4; however, in most of the cases, it occurs with a valence of around 2. The final product of the etching process is the stable complex $\left[ 2H^+ - SiF_6^{2-} \right]$, or $H_2SiF_6$.

Direct dissolution consumes HF and liberates hydrogen ($H_2$), but in the intermediate reactions, there is the occurrence of $H^+$. This gives a strong hint that the reaction rate might be sensitive to the pH value of the electrolyte and that local reactions (e.g., at a pore tip) may locally change the pH value of the electrolyte.

### 2.3.2 OXIDATION OF Si

The oxidation process of Si can be described by the net reaction:

(2.3)
$$Si + 4h^+ + 2O^{2-} \rightarrow SiO_2$$

As can be observed in Equation 2.3, four holes participate in the oxidation process. In the presence of water, the oxidation reaction can be written as (Huang et al. 2011)

(2.4)
$$Si + 2H_2O + 4h^+ \rightarrow SiO_2 + 4H^+ + 4e^-$$

Much simplified, this process occurs if the current density is not compensated by the nucleophile $F^-$ coming from the dissociation of HF. As in this regime, the available HF concentration is lower than the amount of water, the water molecules take the role of nucleophile, forming Si–O bonds (Sailor 2012). Silicon oxides are easily etched by HF, at a rate depending on its concentration.

### 2.3.3 OXIDE DISSOLUTION

In the voltage range of porosification, the Si oxide dissolution process occurs mainly chemically, with no electric current involved. The net reaction can be written as Equation 2.5:

(2.5)
$$SiO_2 + 6F^- + 6H^+ \rightarrow H_2SiF_6 + 2H_2O$$

This process couples to the oxide formation and essentially limits the total current density that the system can process because the oxide generation rate cannot be larger than the dissolution rate (on average). The only exceptions are the somewhat exotic but not necessarily rare cases where the oxide does not dissolve but fractures off mechanically due to stress levels increasing with oxide thickness (Parkhutik 1999). The combined process of oxidation and oxide dissolution is known as tetravalent Si dissolution, since four holes are needed.

### 2.3.4 HYDROGEN TERMINATION OF THE Si SURFACE

The coverage of the free Si surface can be formally described by the chemical Equation 2.6:

(2.6)
$$Si^x + xH^+ \rightarrow SiH_x$$

where $x$ is the number of open covalent bonds on the Si surface (number of defects per Si atom). When Si is etched in HF-based solutions, the surface dangling bonds become terminated with H atoms (including the H atoms of alcohol or surfactant molecules). It is known through XPS and FTIR studies that hydrides like $SiH$, $SiH_2$, and $SiH_3$ are formed (Chabal et al. 1983; Yavlonovitch et al. 1986); that is why the chemical formula $SiH_x$ is used in Equation 2.6. The hydrogen termination of the surface states can be considered a passivation process, which is governed by a reaction constant. This removal of interface states is important because they introduce a certain density of states in the band-gap of the Si (Sailor 2012). The rate of passivation depends on parameters like crystal orientation, temperature, and etchant composition.

In typical pore growth situations, the dissolution of Si is a combination of divalent (direct Si dissolution) and tetravalent (dissolution of Si through the formation and dissolution of $SiO_2$). During pore formation, current flows per definition almost exclusively through the pore tip and not through the pore walls. Most of the pore surface thus must be passivated with hydrogen. The quality of this passivation determines the residual flow of current to the pore walls and consequently the enlargement of pore diameters henceforth called electrochemical "over-etching."

The four described processes are all present during Si etching. Nevertheless, they cannot by definition occur all at the same time and at the same small place increment; thus, a totally homogeneous current flow is intrinsically impossible (Föll et al. 2002). The question will be if the averages observed on a macroscopic scale are constant in time and space (producing "polishing"), constant in space but not in time (producing polishing with externally observable current/voltage oscillations), constant in time but not in space (producing pores or current oscillations in space but constant external parameters), or not at all (producing "chaos," including coupled oscillations in space and time).

## 2.4 MECHANISM OF Si POROSIFICATION DURING ANODIC ETCHING (FROM MESO- AND MICROPOROUS Si TO MACROPOROUS Si)

When talking about electrochemical pore etching in Si, a wide variety of pore diameters can be considered, from a few nanometers to tens of micrometers. Additionally, a huge range of geometries and morphologies can also be encountered. In this respect, it is hard to describe the structure of porous Si in a simple manner. However, there is a standard definition of pores considering the pore diameter only: the IUPAC convention. According to IUPAC, pores can be grouped into micropores (diameter <2 nm), mesopores (diameter 2–50 nm), and macropores (diameter >50 nm).

Even though, the name "nanopores" has been lately used to describe pores with at least one dimension smaller than 100 nm, it is not an officially accepted definition. Additionally, a plethora of nonstandardized names are used to describe the pores according to their morphology (branched pores, tree-like pores, nano-sponge, etc.), their formation mechanism (e.g., break-down pores), or their orientation (crystallographic pores and current-line pores) (cf., e.g., Kochergin and Föll 2009).

In this section, the mechanisms for the electrochemical porosification of Si in micro-, meso-, and macropores will be discussed.

## 2.4.1 PORE FORMATION MODELS

There is no generally accepted electrochemical pore formation mechanism model for Si (or other semiconductors) and what there is has no or very limited predictive power. The Si anodic porosification is a complicated process that mixes electronic, fluidic, and chemical factors. A wide range of externally accessible parameters has to be taken into account when electrochemically etching pores in Si: electrolyte composition, doping type and concentration in Si, minority carrier lifetime and lifetime uniformity, Si crystal orientation, Si sample preconditioning/patterning, applied voltage/current, temperature, illumination conditions, backside contact quality, and electrolyte flow rate. All these parameters play a role, and may affect in a higher or lower scale the porosification. In addition, the nucleation of pores is absolutely crucial and influenced critically by, for example, perfection of the lithography for "seeded" pores and perfection of the surface and starting conditions during the first few seconds/minutes of the experiment for "random" pores. Indeed, producing a uniform array of random pores usually calls for special conditions (e.g., high potential) during the pore initiation part of the experiment.

No wonder that the porous Si samples prepared by different research groups are not comparable, even for nominally identical conditions. This produces a major problem for considering mechanisms of pore formation. Nevertheless, some common features of the porosification process are known, mainly evidenced through experimental observations. Based on some points from Sailor (2012), the main features of the Si electrochemical porosification process can be summarized in the following: (1) Unless the Si wafer has been specifically prepatterned, in most of the cases the pores nucleate homogeneously with no apparent order on the Si surface; although there are some examples of self-organization by certain etching conditions (Föll et al. 2006). (2) Current by necessity flows preferentially at the pore tips and the pore walls passivate (to a certain degree, as will become clear later in this section). (3) Once formed, the pores do not rearrange.

There are different models trying to explain the pore formation in different length scales, at different growth stages. Three groups of models can be made (Parkhutik 1999)

1. Models for pore nucleation
2. Models for stationary pore growth
3. Models trying to explain the whole pore growth process

### 2.4.1.1 MODELS FOR PORE NUCLEATION

A solid representative of the first group of models is the one proposed by Kang and Jorné (1993). The nucleation of pores at the silicon surface is treated mathematically as a phenomenon of instability of a planar surface toward small perturbations. The model considers that the electron holes migrate to the surface of the semiconductor, and the $F^-$ ions from the HF-based etchant diffuse to the electrolyte-semiconductor interface. Due to the applied voltage, there exist some small perturbations at the Si surface, which can be modeled as Fourier components of the form $y = \alpha\exp(i\omega x + \beta t)$, where $\omega$ is a spatial frequency along the perturbed surface and $\beta$ is the rate of development of the perturbation. $\alpha$ is the amplitude of the perturbation. At making an analysis of stability, certain frequencies $\omega_{max}$ are found for the maximum values of $\beta$. According to this, there should be some characteristic eigenvalues for the preferential distance between pores, for a given set of experimental conditions. The prediction of this model is that the distance between pores varies with the square root of the applied voltage. This prediction is not surprising because it mirrors the behavior of a solid state junction, ruled by Poisson's equation.

The width of the junction or space charge region length (SCRL) is given by

(2.7)
$$SCRL = \sqrt{\frac{2\varepsilon V}{qN}},$$

where $\varepsilon$ is the dielectric constant of Si, $V$ is the applied voltage, $q$ is the charge of electrons, and $N$ is the doping concentration.

Another form of pore nucleation is through mechanical stress (cracks) and defects on the Si surface. These points work as easy paths for pore propagation (Parkhutik and Andrade Ibarra 1999). There is, however, no clear example for this kind of pore nucleation and it plays a minor role at best. It is also possible to introduce defects intentionally in regular arrays; pores nucleate at those defects as long as the distances between defects is not too large or small. That is, in short, what lithography aims to achieve.

Additional mechanisms of pore nucleation are saturation of vacancies at the surface (Corbett et al. 1995), surface tension (Kompan et al. 1996), and general pattern formation by current flow inhomogeneity in space (Carstensen et al. 1998, 1999).

### 2.4.1.2  MODELS FOR STATIONARY PORE GROWTH

Lehmann and Föll (1990) were the first to propose a macropore formation and growth model that centered on the SCRL. The model in its final form (Lehmann 1993) assumes that the local current density is always constant and is given by $J_{PSL}$, which is the major parameter governing pore geometry and morphology. The model does have some predictive power demonstrated by the fact that it was used to produce n-macropores for the first time. The model has been experimentally validated to some extent for macropore formation (Lehmann 1993). It is clear that the SCRL is very important for pore growth, even though it has become obvious that the SCRL is not sufficient to explain the formation of pores in general (Hejjo Al Rifai et al. 2000).

The SCRL is the central part of Lehmann's model (Lehmann 1993; Lehmann and Grüning 1997) for describing macropore formation in $n$-type Si. As for etching $n$-type material, generation of carriers with light is needed, the model considers that the current flowing through the pore tips during the etching process is mainly the photocurrent. The small dark or leakage current also flows through the pore wall, increasing the pore size with time and thus rendering the pores slightly conical in appearance. The "Lehmann formula" relates the pore diameter ($d$) with the (average) lattice constant ($a$) of the pore array and the total current density ($J$) applied during the etching:

$$(2.8) \qquad \left(\frac{d}{a}\right)^2 = \frac{J}{J_{PSL}}$$

The pores have the freedom to adjust themselves to the diameters and distances for obtaining $J_{PSL}$ in all the active areas (= pore tips) during etching. If the Si wafers are pre-structured, all pores will do the same; if not, some randomizations occur where the average distance between pores is given by 2SCRL and the pore diameters adjust accordingly. The model considers that the pore walls are passivated against the holes produced by illumination due to the space charge region. However, the SCR does not passivate against holes generated in that region (i.e., the dark current), and holes from the illumination may penetrate under certain circumstances. One reason for this is the introduction of charged surface states during the etching process.

A similar model, which is a particularization for the case of microporous Si, is the "quantum wire model" (Lehmann and Gösele 1991). It postulates that there is quantum confinement for carriers at the pore walls due to their very small thickness (in the range of nanometers). Due to this, the penetration of electrons or holes to the pore walls becomes more difficult, and can be thought of as passivated. The dissolution of the pore walls is inhibited, stabilizing the pore structure. The model defines accurately the minimum distance between micropores; however, it has nothing to say about the diameter of the pores itself.

One of the first models in providing a quantitative estimation of the pore geometry for mesopores produced by SCR failing was proposed by Beale et al. (1985). It considers that a Schottky-type barrier at the Si-electrolyte interface held at reverse conditions can be overcome by breakdown. The electron transport through the barrier is produced either by tunneling in the case of heavily doped Si or by Schottky emission in the case of lower doping. In the model, the pore tips are semispherical. The current is considered to flow preferentially at these points because the electric field is concentrated there, while the barrier height is lowered. The contribution of this model is simply the differentiation in transport mechanisms between heavily and lightly doped Si, which produce different characteristic pore sizes.

The model proposed by Zhang (1991) considers that pore growth occurs through two competing processes: anodic oxide formation and further dissolution in the HF-based electrolyte, and direct anodic dissolution of Si. The probability of occurrence of the one or the other process highly depends on the applied electric field strength and thus potential and current density. The first process mainly occurs at higher field strengths. On the other hand, it is considered that the current density and the doping concentration determine the pore morphology; these variables influence the height of the potential barrier at the pore tips, and consequently the electron tunneling probability (the carrier transport mechanism).

### 2.4.1.3 MODELS EXPLAINING THE WHOLE GROWTH PROCESS

A highly developed model for pore growth, including the stochastic nature of pore geometries and morphologies, is the "current burst model" (CBM), proposed by Föll et al. (2002). The model postulates that there are definite correlations between the four different processes occurring during the anodization of Si (see Section 2.3). It has three main postulates:

1. The current mostly flows inhomogenously through an Si sample both in time and space. There are times when no charge is transferred across some areas. The charge transfer occurs at some points of space and time (x, y, t) on the Si surface with some probability p(x, y, t) that depends on the surface state S(x, y, t), having an intrinsically stochastic probability nature. Every charge transfer event is called current burst (CB).
2. The sequence of processes occurring during a current burst is dictated by logic. On a "free" Si surface, direct dissolution of Si starts the current flow, followed by oxidation, leading to passivation and thus stop of current flow. The next CB has to wait until the oxide has been sufficiently dissolved. Hydrogen–passivation may produce a similar effect than oxidation. Figure 2.4 shows an example of two consecutive current bursts, indicating each of the processes occurring.
3. An interaction between individual CBs may occur in both space and time. This means that the nucleation probability p(x, y, t) of a CB does not only depend on S(x, y, t), but also on what has happened before (S[x, y, $t_0$]) and on what is going on in the neighborhood. Interactions in space may result in a correlation in time, and when there is a correlation in time, current or voltage oscillations result. Interactions in time produce pores.

The model is quantitative in principle and can account for all voltage and current oscillations quantitatively, applying a Monte-Carlo computer simulation (Föll et al. 2011). Simulation of pore formation is presently limited by the available computing power because it needs to consider large areas and is, by definition, three-dimensional.

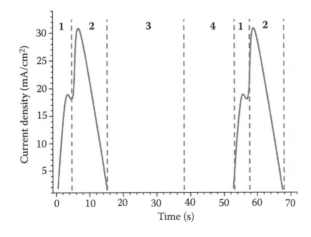

**FIGURE 2.4**    Two consecutive current bursts. They consist of a series of events: (1) Direct dissolution of Si, (2) oxidation of Si, (3) Si oxide dissolution, and (4) $H_2$ passivation.

The model works quite well in a semiquantitative way for describing the formation of different kinds of pores, depending on the kind of spatio-temporal correlations. It has predicted new kinds of pores that were then found in experiments designed with parameters demanded by the model (Leisner et al. 2010). It includes the Lehmann model as a special case.

The model works best if some oxidation occurs during the etching, the current density is not too low (a minimum spatial density of pores, for allowing interactions is required), and the oxide dissolution rate is not too fast. Thus, while it is presently the most general model, it is not a fully general model, but some of its postulates can be used independently to explain some phenomena.

The CBM allows computer simulation of pore growth. In this context, the first (and much simpler) model to allow computational simulations of pore morphology was the one proposed by Smith and Collins (Smith et al. 1988). The model deals with pore branching. It is based on the model of diffusion limited aggregation (DLA) by Witten and Sander (1983). The pore branching is explained by the random diffusion of electronic holes through depleted Si to the active etching sites.

Another important model is proposed by Parkhutik and Shershulsky (1992). It explains nucleation, pore growth, and rearrangements of the pore structure when changing the applied electric field. It has the following postulates:

- A virtual passive layer (VPL) covers the bottom of the pores. It avoids a direct contact of the electrolyte with Si.
- Pores are formed by electrochemical dissolution of the VPL. The charge transfer for the process can be tunneling through the VPL, or polarization of the VPL, which enhances the dissolution.
- The VPL growth and dissolution has an exponential dependency on the electric field strength, and it is influenced by the reactivity with the electrolyte and by other experimental variables.

The model considers a linear relationship between the pore size and the applied voltage or current. It also has a parameter containing information about the dependence of the pore diameter on the type of electrolyte. The calculation with the model results in a porosity parameter. Depending on the value of this parameter, six different pore morphologies are predicted. It is clear that this model puts parts of what should be explained into its assumptions and thus does not have much explanatory or predictive power.

## 2.4.2 FORMATION OF PORES WITH DIFFERENT DIAMETERS

In the last section, models for porosification were discussed in general. Talking about the particular cases of micropores, mesopores, and macropores, all of these pores have some particularities that are not necessarily related to their diameter scale. Specific aspects of the different models can be used for describing the specific aspects of the different pore categories. Nevertheless, there is no model fully explaining pore formation in all diameter scales; combination of models must be used for having a plausible explanation. In what follows, the CBM and the SCR models will be used for describing pore formation.

A problem for having just one model describing all kinds of pores is that there is no continuity in sizes when varying the current density or doping concentration. For example, in $p$-type Si, macropores are typically obtained at $J \ll J_{PSL}$, low HF concentrations (in the range of 5%), and relatively high Si resistivities (generally larger than 1 $\Omega \cdot$cm). Mesopores are obtained at $J \sim J_{PSL}$ and high HF concentrations (in the range of 30%), on very low resistive Si. Micropores, on the other hand, are obtained at $J \lessapprox J_{PSL}$, large HF concentrations (in the range of 30%), and low Si resistivities. It can be stated, in a very simplified way, that etching of $n+$ and $p+$ Si under almost all conditions always results in mesopores.

As an example, Figure 2.5 shows the I-V characteristics of $p$- and $n$-type Si in HF-based aqueous electrolytes, denoting the different kinds of pores that can be obtained (Föll 2015): (A) micropores, (B) mesopores, (C) macropores, and (D) breakthrough pores. The last category refers to

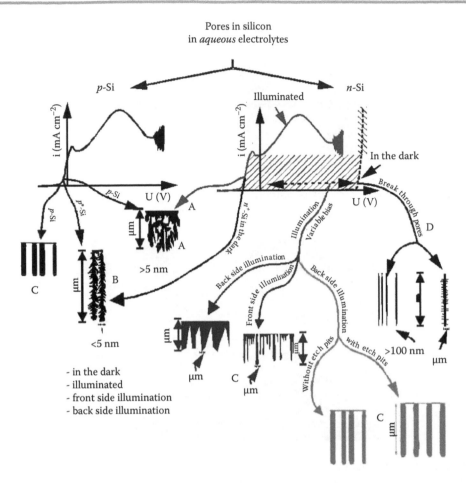

**FIGURE 2.5**    Morphologies of pores that can be obtained depending on the doping type and doping level, besides the operational point (particular I-V conditions).

macropores, in principle, according to their diameter; however, their formation mechanism is different. They are obtained at potentials of dielectric breakdown.

### 2.4.2.1 MACROPORES

According to the CBM, macropores occur when sizable domains are possible (correlations in space and time happen). A decisive factor for this is that the SCRLs of the pores are close to each other. The minimum distance between pores is given by 2SCRL; under this condition, there are theoretically no carriers flowing to the pore walls. Thus, they are passivated. However, Grüning et al. (1996) has shown that even when the distance between pores is larger than 2SCRL, the current flows preferentially at the pore tips. This would mean that there are other passivation mechanisms of the pore walls, besides the SCRL, and that is the hydrogen (or hydrogen-bearing molecules) passivation discussed before. Even though, experimentally, the distance between pores cannot be increased without limit, if the distance is too small, some pores stop growing, and if it is too large, the pore surface becomes rough or the pore branches.

In Figure 2.6, it is possible to observe how macropores are usually distributed, when they nucleate by self-organization (Föll et al. 2010). For the mechanism of nucleation, please refer to Section 2.4.1.1. The distance between pores is usually not larger than the sum of the pore diameter $d_{pore}$ plus 4SCRL. Additionally, the pores tend to close-pack (hexagonal array). Figure 2.7 shows an SEM micrograph of macropores in $n$-type Si with a self-organized hexagonal pattern. These pores are unusual and were predicted using the CBS model. They exhibit what is known as a frustrated structure—they want to assume a square lattice because of the crystal orientation constraints and a hexagonal lattice in order to be close-packed at the same time—an obvious

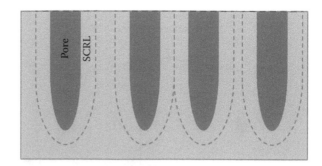

**FIGURE 2.6**   Schematic of the spatial distribution of macropores grown wild. The distance between pores is usually smaller than $d_{pore}$ + 4SCRL.

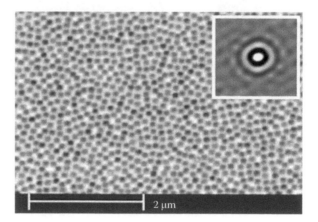

**FIGURE 2.7**   SEM micrograph of the surface of a macroporous Si sample with pores grown wild. The pores tend to grow close-packed. The inset shows the FFT of the micrograph, evidencing a predominantly hexagonal muster.

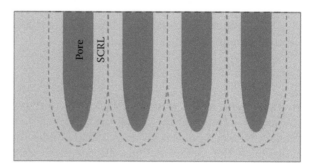

**FIGURE 2.8**   Schematic of the spatial distribution of macropores grown on pre-patterned Si. The minimum distance between pores is 2SCRL.

impossibility. Thus, no particular order is visible to the naked eye. The inset, showing the FFT of the image, however, clearly shows that while there is indeed no correlation in space between nearest neighbors, there is a clear hexagonal structure for the second-nearest one.

If the Si surface is prepatterned with an array of inverted pyramids, the pores tend to grow exactly at the positions of the pyramid tips because the probability of nucleation at those points is enhanced (the electric field strength is larger there) (Quiroga-González et al. 2011). The pores stay at those positions without branching if the distance between pyramids is not larger than $d_{pore}$ + 4SCRL. Furthermore, if the distance between pyramids is close to 2SCRL (as in Figure 2.8), the pore walls are well passivated, and their roughness is low because there is no current flow to them. $d_{pore}$ is allowed to vary if there is available space between the SCRLs of contiguous pores.

For etching macropores in *n*-type Si, photogeneration of holes is needed because the majority carriers are electrons, but holes are necessary for the dissolution reaction of Si. Backside illumination is preferred over frontside illumination because the diffusion of carriers to the pore tips has to go down into the sample and the current-focusing effect of the pore tip is lost. Pores obtained by frontside illumination thus are limited as to their depths. Pores in *n*-type Si can, within certain limits, be well predicted through the Lehmann model, described by Equation 2.8. According to the model, it is possible to vary the diameter of the pores by varying the applied current. Working at the operational point for PSL (see the *J-V* characteristics of Figure 2.3), the pore diameters adjust themselves to keep the current density at the level of $J_{PSL}$.

Working with *n*-type Si has the advantage that one can vary the photocurrent (via the intensity of the applied light) and the voltage simultaneously. This gives an additional degree of freedom. Figure 2.9 shows typical photocurrent and voltage profiles for making a cavity. Figure 2.10 (solid line) indicates the shape of pores obtained with those profiles. It is expected that if the chemical processes are fast enough to follow changes of current, diameter modulations could be obtained (as described by the dashed line of Figure 2.10). However, it has been considered a difficult task for many years because the already-etched pores were always over-etched while etching deeper, and no clear modulations could be observed (Müller et al. 2000). Nevertheless, Matthias et al. (2005), taking advantage of being able to vary voltage and current independently, reported the first macropores with sharp modulations. Figure 2.11 shows possible photocurrent and voltage profiles for accomplishing the expected pores of Figure 2.10.

In the case of *p*-type Si, the etching is relatively isotropic and it is difficult to achieve pore modulation due to the large availability of electronic holes, even though Christophersen et al. (2000b) finally found very stable growth conditions for p-type macropores, using organic electrolytes. They could grow pores as deep as 400 μm. Furthermore, adding some surfactants like polyethyleneglycol (PEG), the surface is passivated, enhancing the focusing effect at the pore tips

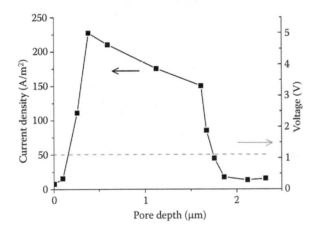

**FIGURE 2.9**   Common etching current and voltage profiles for making a pore modulation in *n*-type Si. In this case, just the photocurrent (light intensity) is varied, as one does for *p*-type Si.

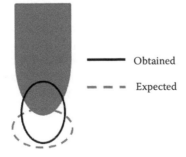

**FIGURE 2.10**   Schematic of pore obtained with the etching profiles of Figure 2.8. A modulation is obtained, but not as pronounced as desired (see dashed line).

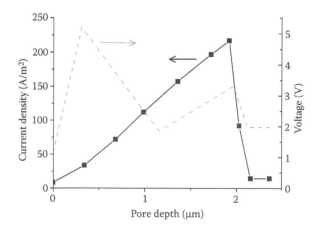

**FIGURE 2.11** Optimized etching current and voltage profiles for making a pore modulation in *n*-type Si. Varying both the current and the voltage, it is possible to enhance the modulation, and pores close to the ones of Figure 2.9 (expected) can be obtained.

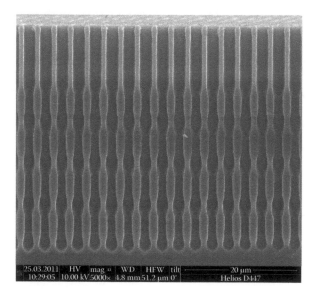

**FIGURE 2.12** SEM of modulated macropores in *p*-type Si. A good contrast between wide and narrow sections is possible by enhancing the passivation of the pore walls with surfactants.

(Quiroga-González et al. 2014). In this way, in *p*-type Si it is possible to obtain the same structures than in *n*-type Si; in fact, pore modulation in *p*-type Si adding surfactants has allowed the fabrication of arrays of microwires with an intergrown support (Quiroga-González et al. 2011). Figure 2.12 shows an example of modulated macropores in *p*-type Si, prepared using PEG simply varying the current density.

Exhaustive experimental work has shown that macropores in *n*- and *p*-type Si grow preferentially in <100> directions, but when the available <100> directions are too inclined, they grow in <113> directions (Rönnebeck et al. 2000; Christophersen et al. 2000b,c). The main pores usually grow in one of these directions, and may branch in some of the others. Only the CBS model is able to account for this peculiar behavior in a qualitative way.

### 2.4.2.2 MESOPORES

The pore growth mechanism in this kind of pore, quite different from the others, is that charge transfer for the etching process is given by barrier breakdown, for example, by tunneling (Zhang 2001). According to the CBM, these pores are produced by a positive interaction of CBs in time

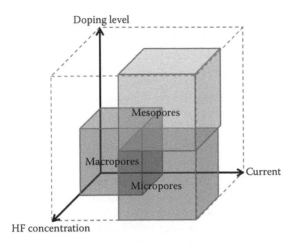

**FIGURE 2.13**   General phase diagram of pores. In this graph, the most common ways of obtaining the different dimensions of pores are depicted. The three most important parameters influencing the pore size (HF concentration, doping level, and current density) are the axes.

without the formation of synchronized domains, and a low interaction in space. The lack of domains is an indication of insufficient oxidation. The etching process is mainly produced by direct dissolution.

Mesopores can be directly prepared either in *p*- or in *n*-type Si with different doping or as by-product during the preparation of other pores:

- The most common way of preparation of mesopores is using highly doped *p*-type Si in aqueous electrolytes (Sailor 2012). In the case of the highly doped *n*-type material no illumination is required, contrary to what one expects in macropore formation, because there is enough current produced by avalanche break-down (Lehmann et al. 2000). Due to the low resistivity, this kind of process is easy to accomplish, even at small voltages. Etching in organic electrolytes also works, even at lower doping concentrations (Carstensen et al. 2000; Christophersen et al. 2000d). It seems that weak oxidizing electrolytes promote mesopore formation: As the thickness of the oxide is very small, the probability of carrier tunneling is high, and direct dissolution of Si is produced.
- Using moderately doped *n*-type Si, mesopores can be produced, too. Under darkness conditions, when high voltages are applied (as high as 100 V), dielectric breakdown occurs producing mesopores.
- Additionally, mesopores can be found on the walls of macropores grown in *n*- or *p*-type Si, using weak oxidizing electrolytes (mainly organic). This also happens when macropores stop growing (Jäger et al. 2000).

### 2.4.2.3 MICROPORES

This kind of pore occurs when an anticorrelation between nucleation points occurs. The formation of pores is also enhanced by quantum dot effects. According to the CBM, an anticorrelation happens when the nucleation probability at a certain point (x, y) gets its maximum when no previous CB has occurred there. In this case, pore nucleates as soon as possible in reduced areas, right after oxide dissolution. In this case, the current density of a CB is very large, forming pores with a sponge-like structure.

While the CBM explains the probability of micropore formation, there are no models fully explaining their growth, diameter, and morphology. For example, the model of the quantum wire effect only accounts for the distance between pores (Lehmann and Gösele 1991).

Micropores are usually obtained using large HF concentrations and high currents. This allows fast pore growth and high porosity (Lehmann 1993). If the HF concentration or the current densities are too low, macropores are produced. On the other hand, the doping concentration for obtaining micropores is usually low, as in the case of macropores. Micropores can be obtained equally in *n*- or *p*-type Si, low doped. If the doping increases, mesopores are obtained. In this way, one can situate micropores between macropores and mesopores in a phase diagram for pores. A qualitative phase diagram is shown in Figure 2.13. Here, the most common ways for preparing the different kinds of pores are presented. Transitions between pore types are possible.

## 2.5 SUMMARY

The *J-V* (current density–voltage) characteristics of the electrochemical etching system of Si behave in a first approximation like a rectifying metal-semiconductor contact, generally known as a Schottky contact. Nevertheless, the characteristics present some features not seen in a solid state. At negative potentials, $H_2$ evolution occurs, independently if one etches *p*- or *n*-type Si. At positive potentials, Si dissolution is observed, but for the case of *n*-type Si, illumination has to be applied (the dissolution process requires electronic holes, which are not available in the *n*-type material in dark). Below a current density value JPSL (current density for porous Si layer), pores are obtained. Beyond this current, electropolishing occurs. Electrochemical etching of Si with HF-based solutions consists of any combinations of four processes: direct dissolution of Si, oxidation of Si, dissolution of the Si oxides, and hydrogen passivation.

According to IUPAC, pores can be grouped into micropores (diameter <2 nm), mesopores (diameter 2–50 nm), and macropores (diameter >50 nm). There are different models trying to explain pore formation. These can be grouped in models for pore nucleation, for steady pore growth, or for the whole pore formation. Among these models, the current burst model (CBM) is one of the most complete, allowing prediction and modeling of pores, and the Lehmann model, explaining steady growth of macropores, based on the fact of pore wall passivation by a space charge region lager.

Micropores are usually obtained using large HF concentrations, high currents, and low doping, by direct dissolution or oxide formation and its dissolution. Mesopores are obtained by tunneling of carriers, using large HF concentrations, high currents, and high doping. Macropores are obtained by a similar mechanism as micropores, but using low HF concentrations, low currents, and low doping.

## DEDICATION

We nucleated as a couple,

and at the right time we branched.

With love, to my wife and children.

**E. Quiroga-González**

## REFERENCES

Anglin, E.J., Cheng, L., Freeman, W.R., and Sailor, M.J. (2008). Porous silicon in drug delivery devices and materials. *Adv. Drug Deliv. Rev., 60*, 1266–1277.

Beale, M.I.J., Chew, N.G., Uren, M.J., Cullis, A.G., and Benjamin, J.D. (1985). Microstructure and formation mechanism of porous silicon. *Appl. Phys. Lett., 46*(1), 86–88.

Canham, L.T. (1990). Silicon quantum wire fabrication by electrochemical and chemical dissolution of wafers. *Appl. Phys. Lett., 57*(10), 1046–1048.

Carstensen, J., Prange, R., Popkirov, G.S., and Föll, H. (1998). A model for current oscillations at the Si-HF-System based on a quantitative analysis of current transients. *Appl. Phys. A, 67*, 459–467.

Carstensen, J., Prange, R., and Föll, H. (1999). A model for current-voltage oscillations at the silicon electrode and comparison with experimental results. *J. Electrochem. Soc., 146*(3), 1134–1140.

Carstensen, J., Christophersen, M., Hasse, G., and Föll, H. (2000). Parameter dependence of pore formation in silicon within a model of local current bursts. *Phys. Stat. Sol. (a), 182*(1), 63–69.

Chabal, Y.J., Chaban, E.E., and Christman, S.B. (1983). High resolution infrared study of hydrogen chemisorbed on Si(100). *J. Electron. Spectrosc. Relat. Phenom., 29*, 35–40.

Christophersen, M., Carstensen, J., Feuerhake, A., and Föll, H. (2000a). Crystal orientation and electrolyte dependence for macropore nucleation and stable growth on *p*-type-silicon. *Mater. Sci. Eng. B*, **69/70**, 194–198.

Christophersen, M., Carstensen, J., and Föll, H. (2000b). Crystal orientation dependence of macropore formation in *p*-type silicon using organic electrolyte. *Phys. Stat. Sol. (a)*, **182**(1), 103–107.

Christophersen, M., Carstensen, J., and Föll, H. (2000c). Crystal orientation dependence of macropore formation in *n*-type silicon using organic electrolyte. *Phys. Stat. Sol. (a)*, **182**(2), 601–606.

Christophersen, M., Carstensen, J., and Föll, H. (2000d). Macropore formation on highly doped *n*-type silicon. *Phys. Stat. Sol. (a)*, **182**(1), 45–50.

Christophersen, M., Carstensen, J., Rönnebeck, S., Jäger, C., Jäger, W., and Föll, H. (2001). Crystal orientation dependence and anisotropic properties of macropore formation of *p*- and *n*-type silicon. *J. Electrochem. Soc.*, **148**(6), E267.

Corbett, J.W., Shereshevskii, D.I., and Verner, I.V. (1995). Changes in the creation of point defects related to the formation of porous silicon. *Phys. Stat. Sol. (a)*, **147**, 81–89.

Föll, H. (1991). Properties of silicon-electrolyte junction and their application to silicon characterization. *Appl. Phys. A*, **53**, 8–19.

Föll, H. (2015). Macroporous silicon. http://www.tf.uni-kiel.de/matwis/amat/. Accessed on January 20, 2015.

Föll, H., Christophersen, M., Carstensen, J., and Hasse, G. (2002). Formation and application of porous silicon. *Mater. Sci. Eng. R.*, **39**(4), 93–141.

Föll, H., Carstensen, J., and Frey, S. (2006). Porous and nanoporous semiconductors and emerging applications. *J. Nanomat.*, **91635**, 1–10.

Föll, H., Leisner, M., Cojocaru, A., and Carstensen, J. (2010). Macroporous semiconductors. *Materials*, **3**, 3006–3076.

Föll, H., Leisner, M., and Carstensen, J. (2011). Modeling some "meta" aspects of pore growth in semiconductors. *ECS Trans.*, **35**(8), 49–60.

Grüning, U., Ottow, S., Busch, K., and Lehmann, V. (1996). Macroporous silicon with a complete two-dimensional photonic band gap centered at 5 µm. *Appl. Phys. Lett.*, **68**(6), 747–749.

Hejjo Al Rifai, M., Christophersen, M., Ottow, S., Carstensen, J., and Föll, H. (2000). Dependence of macropore formation in *n*-Si on potential, temperature, and doping. *J. Electrochem. Soc.*, **147**(2), 627–635.

Hoffmann, P.M., Vermeir, I.E., and Searson, P.C. (2000). Electrochemical etching of *n*-type silicon in fluoride solutions. *J. Electrochem. Soc.*, **147**(8), 2999–3002.

Huang, Z., Geyer, N., Werner, P., de Boor, J., and Gösele, U. (2011). Metal-assisted chemical etching of silicon: A review. *Adv. Mater.*, **23**, 285–308.

Jäger, C., Finkenberger, B., Jäger, W., Christophersen, M., Carstensen, J., and Föll, H. (2000). Transmission electron microscopy investigations of the formation of macropores in *n*- and *p*-Si(001)/(111). *Mater. Sci. Eng. B*, **69/70**, 199–204.

Kang, Y. and Jorné, J. (1993). Morphological stability analysis of porous silicon formation. *J. Electrochem. Soc.*, **140**, 2258–2265.

Kochergin, V. and Föll, H. (2009). *Porous Semiconductors: Optical Properties and Applications*. Springer, Berlin.

Kompan, M.E., Kuzminov, E.G., and Kulik, V. (1996). Observation of a compressed state of the quantum wire material in porous silicon by the method of Raman-scattering. *JETP Lett.*, **64**(10), 748–753.

Lehmann, V. (1993). The physics of macropore formation in low doped *n*-type silicon. *J. Electrochem. Soc.*, **140**(10), 2836–2843.

Lehmann, V. (2002). *Electrochemistry of Silicon*. Wiley-VCH, Weinheim.

Lehmann, V. and Föll, H. (1990). Formation mechanism and properties of electrochemically etched trenches in *n*-type silicon. *J. Electrochem. Soc.*, **137**(2), 653–659.

Lehmann, V. and Gösele, U. (1991). Porous silicon formation: A quantum wire effect. *J. Appl. Phys. Lett.*, **58**(8), 856–858.

Lehmann, V. and Grüning, U. (1997). The limits of macropore array fabrication. *Thin Solid Films*, **297**(1–2), 13–17.

Lehmann, V., Stengl, R., and Luigart, A. (2000). On the morphology and the electrochemical formation mechanism of mesoporous silicon. *Mater. Sci. Eng. B*, **69/70**, 11–22.

Leisner, M., Carstensen, J., and Föll, H. (2010). Pores in *n*-type InP: A model system for electrochemical pore etching. *Nanoscale Res. Lett.*, **5**, 1190–1194.

Mathwig, K., Geilhufe, M., Müller, F., and Gösele, U. (2011). Bias-assisted KOH etching of macroporous silicon membranes. *J. Micromech. Microeng.*, **21**, 035015.

Matthias, S., Müller, F., Schilling, J., and Gösele, U. (2005). Pushing the limits of macroporous silicon etching. *Appl. Phys. A*, **80**(7), 1391–1396.

Müller, F., Birner, A., Schilling, J., Gösele, U., Kettner, C., and Hänggi, P. (2000). Membranes for micropumps from macroporous silicon. *Phys. Stat. Sol. (a)*, **182**(1), 585–590.

Parkhutik, V. (1999). Porous silicon—Mechanisms of growth and applications. *Solid State Electron.*, **43**, 1121–1141.

Parkhutik, V.P. and Shershulsky, V.I. (1992). Theoretical modeling of porous oxide growth on aluminium. *J. Phys. D.*, **25**(8), 1258–1263.

Parkhutik, V. and Andrade Ibarra, E. (1999). The role of hydrogen in the formation of porous structures in silicon. *Mater. Sci. Eng. B.*, **58**, 95–99.

Propst, E.K. and Kohl, P.A. (1994). The electrochemical oxidation of silicon and formation of porous silicon in acetonitrile. *J. Electrochem. Soc.*, **141**(4), 1006–1013.

Propst, E.K., Rieger, M.M., Vogt, K.W., and Kohl, P.A. (1994). Luminescent characteristics of a novel porous silicon structure formed in a nonaqueous electrolyte. *Appl. Phys. Lett.*, **64**(15), 1914–1916.

Quiroga-González, E., Ossei-Wusu, E., Carstensen, J., and Föll, H. (2011). How to make optimized arrays of Si wires suitable as superior anode for Li-ion batteries. *J. Electrochem. Soc.*, **158**(11), E119–E123.

Quiroga-González, E., Carstensen, J., Glynn, C., O'Dwyer, C., and Föll, H. (2014). Pore size modulation in electrochemically etched macroporous *p*-type silicon monitored by FFT impedance spectroscopy and Raman scattering. *Phys. Chem. Chem. Phys.*, **16**, 255–263.

Rönnebeck, S., Ottow, S., Carstensen, J., and Föll, H. (2000). Crystal orientation dependence of macropore formation in *n*-Si with backside-illumination in HF-electrolyte. *J. Por. Mat.*, **7**(1–3), 353–356.

Sailor, M.J. (2012). *Porous Silicon in Practice*. Wiley-VCH, Weinheim.

Smith, R.L., Chuang, S.-F., and Collins, S.D. (1988). A theoretical model of the formation morphologies of porous silicon. *J. Electron. Mater.*, **17**(6), 533–541.

Witten, T.A. and Sander, L.M. (1983). Diffusion-limited aggregation. *Phys. Rev. B.*, **27**(9), 5686–5697.

Yavlonovitch, E., Allara, D.L., Chang, C.C., Gmitter, T., and Bright, T.B. (1986). Unusually low surface recombination velocity on silicon and germanium surfaces. *Phys. Rev. Lett.*, **57**(2), 249–252.

Zhang, X.G. (1991). Mechanism of pore formation on *n*-type silicon. *J. Electrochem. Soc.*, **138**(12), 3750–3756.

Zhang, X.G. (2001). *Electrochemistry of Silicon and Its Oxide*. Kluwer Academic—Plenum Publishers, New York.

# Technology of Si Porous Layer Fabrication Using Anodic Etching

## General Scientific and Technical Issues

Ghenadii Korotcenkov and Beongki Cho

**3**

CONTENTS

Currently, there are a large number of reviews and specialized papers devoted to the analysis of different aspects of porous silicon formation, including discussion of the mechanisms of silicon porosification (Smith and Collins 1992; Lehmann 1993; Halimaoui 1995, 1997). However, in reality, we have a situation where most studies are focused on the decision of particular technical tasks, and most review papers are focused on the analysis of too general problems. As a result, articles do not contain the proper generalization of the results related to silicon porosification, and do not give practical advices for working with porous materials. Therefore, the present chapter is focused on the discussion of exactly practical aspects of the technology of silicon porosification.

## 3.1 SETUP FOR ELECTROCHEMICAL POROSIFICATION

The electrochemical etching cell used for porosification of semiconductors is a crucial part of an etching system.

### 3.1.1 SETUP CONFIGURATION

The simplest experimental setup for performing electrochemical measurements includes the following elements: three electrodes immersed into an electrolyte, a battery, a voltmeter, and an amperemeter. The electrode, which has to be studied or where the electrochemical reaction should take place (in our case it will be the electrode in which we intend to introduce pores), is called the working electrode (WE). The second electrode, which is closing the circuit, is called the counter electrode (CE). The third one, used to measure the voltage between the electrolyte and WE, is called the reference electrode (RE). A schematic representation of such a simple experimental set up is presented in Figure 3.1.

It should be noted that some setups do not contain the RE. However, Foell and co-workers believe that this configuration is not optimal because it does not allow controlling the voltage drop on the WE during electrochemical etching of silicon or other semiconductors. Depending on the composition of the electrolyte and the electrode material, a monolayer of adsorbates or a thin passivation layer may be formed on the electrode, and can significantly shift the electrode potential (Lehmann 2002). These effects have to be taken into account for the WE as well as for the CE. Therefore, for improvement conditions of the etching process, control is usually

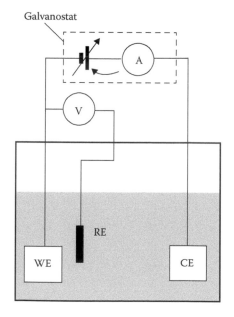

**FIGURE 3.1**   The simplest experimental set up for performing electrochemical experiments operating in galvanostatic mode.

suggested to use a device called a potentiostat (see Figure 3.2a). A potentiostat allows controlling the voltage between the working electrode and the reference electrode directly. The working principle of a potentiostat is shown in Figure 3.2b.

A potentiostat can be a three- or a four-electrode device (see Figure 3.3). Using a four-electrode potentiostat is especially important if the ohmic resistance between the WE and the sample is not sufficiently small. In this case, the use of additional sense electrode (SE) allows elimination of the influence of this contact when the potential of WE is measured. In this configuration, the contact between the sense electrode and the sample is not critical because no current is flowing through it. Now, the desired potential will be exactly applied on the sample/electrolyte junction, that is, we get an ideal tool for controlling electrochemical processes.

However, even in this case it is necessary to bear in mind that unlike the metal electrodes, the resistance of which in most cases is negligible compared to the electrolyte resistivity and space charges in the electrode do not need to be taken into account due to a high number of free charge carriers, in a semiconducting electrode the number of free carriers is orders of magnitude smaller than in a metal (Lehmann 2002). This means that possible ohmic losses in the electrode therefore have to be taken into account, especially for low-doped substrates. Under reverse bias, in addition, a significant part of the applied bias may drop across a space charge region. Another source for a potential difference is a surface passive film, for example, $SiO_2$, which may be present in the anodic regime. The correct determination of potential distribution at the interface of a silicon electrode is therefore complicated, even if a reference electrode is used. Fortunately, it is found that the electrochemical condition of the silicon electrode is, in most cases, well described by the current density across the interface (Lehmann 2002).

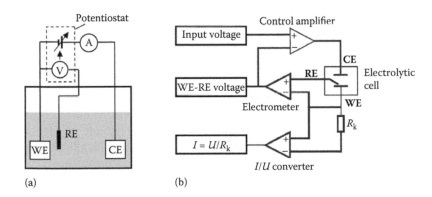

**FIGURE 3.2**  (a) Experimental setup for performing electrochemical experiments operating in potentiostatic mode; (b) a schematic representation of a three-electrode potentiostat. (Idea from http://www.porous-35.com.)

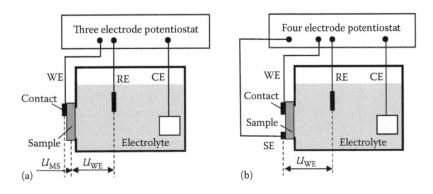

**FIGURE 3.3**  (a) A schematic representation of a setup with a three-electrode potentiostat. (b) A schematic representation of a set up with a four-electrode potentiostat. (Idea from http://www.porous-35.com.)

### 3.1.2 ELECTROLYTIC ETCH CELL USED FOR Si POROSIFICATION

It should be noted that the electrochemical cell for optimal silicon porosification is not a simple construction, as it might seem at first glance, and during its development, it is necessary to solve many technical problems. First, etching Si needs the most aggressive electrolytes. Note that the ever-present HF imposes severe limits on materials and reference electrodes and those organic electrolytes are even worse in that respect; they may require all-Teflon (polytetrafluoroethylene [PTFE]) cell designs. Lehmann (2002) believes that polyvinyl chloride (PVC), polypropylene (PP), and polyvinylidene fluoride (PVDF) can also be used for the cell body. PVC is a good choice for most designs because it is inexpensive, inert in HF, and its mechanical performance is superior to that of PTFE. In addition, PVC parts can easily be glued, which is not the case for PTFE. Note that standard plastic screws are made of polyamide, which is not resistant to concentrated HF. They should be replaced by PP or PVC screws.

Second, often this process also requires illumination. This means that cells must have a window. It is understood that windows necessary for allowing light into the cell are a challenge in either case. The transparent material used in the cell could be clear PVC, Plexiglas (polymethyl-methacrylate, PMMA), and sapphire (Lehmann 2002). PMMA shows a good transparency in the visible and the IR, it is easily machinable, and stable at low HF concentrations. In concentrated HF (>10%), however, it becomes opaque after the initial contact. Clear PVC, which is of lower transmission coefficient than PMMA, is therefore preferable for high HF concentrations. However, the use of sintered clear $Al_2O_3$ plates (sapphire) offers even better HF resistance. Therefore, one should draw a distinction between cells allowing illumination of the sample backside (or frontside in some instances), and those that do not. The latter type is much easier in manufacturing, but it has functional limitations.

Third, the porous layers, designed for use in real devices must satisfy strict requirements for uniformity and reproducibility of a layer's parameters. Therefore, we consider how to solve these problems.

To begin with, from a practical viewpoint, it makes sense to distinguish between large-area etching cells suitable for processing standard Si wafers with diameters from 150 mm to 300 mm and lab-type etching cells for samples with areas in the 1 cm$^2$ range. Large area etching cells capable of uniform etching and with intense uniform backside illumination are large and costly pieces of hardware and are not commonly used. Some information about this topic can be found in the results presented by Carstensen et al. (2005). It is necessary to note that great difficulties relating to realization of electrochemical etching with required vertical and lateral homogeneity of the PSi layer for the Si wafers with large diameter are main restraining factors for commercialization of PSi-based technology. According to Foell and co-workers (Carstensen et al. 2005), for achievement of the above-mentioned homogeneity, besides a good apparatus, which must provide uniform backside illumination, uniformity of etching conditions at the edge of the specimen, good automatic control of the etching via a feedback loop, and a high degree of safety and reliability, one must maintain high quality and homogeneity of the ohmic backside contact and uniform doping of the wafer.

Looking at particularly simple cell designs without reference electrodes but with illumination as an option, one needs to consider single- or double-cell configurations, as shown schematically in Figure 3.4. It goes without saying that for the case of backside illumination, no backside metallization can be used and that the electrolyte is usually agitated or pumped through the cell. While a double-cell design is more complicated, it is actually preferred in many cases (Canham et al. 1996; Bisi et al. 2000; Lammel and Renaud 2000) because it circumvents the problems arising from the necessity of having a good (= uniform and with low contact resistance) Si backside contact (Halimaoui 1997). In the double cell, the backside contact is electrolytic as well and not metallic. Typically, a double-cell has two platinum electrodes, one in each part of the cell, with the wafer in the middle, separating the two parts securely. The electrolytes in the two cells might be different—individually optimized for their "jobs" to allow easy current flow and for pore etching, respectively. In addition, the Si wafer might carry a third contact or it might float electrically as shown in the figure. As the current is forced to pass through the wafer, its anodic side will be etched.

The lack of need for backside metallization avoids a potential source of contamination of the porous silicon in any subsequent thermal and chemical processing; no metal on the backside might simply be necessary for thermal treatments (Halimaoui 1995). In addition, contact

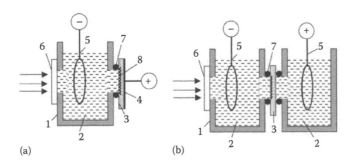

**FIGURE 3.4** Electrolytic cells for porous layer fabrication: (a) one-cell configuration, (b) two-cell configuration. 1—Teflon bath, 2—electrolyte, 3—silicon wafer, 4—metal contact, 5—Pt electrode, 6—quartz window, 7—packing ring, 8—porous layer, arrows show the direction of illumination. (From Korotcenkov, G. and Cho, B.K., *Crit. Rev. Sol. St. Mater. Sci.*, 35(3), 153, 2010. Copyright 2010: Taylor & Francis Group. With permission.)

uniformity is always rather perfect, which is often not the case with metallic backside contacts. There are drawbacks, however. In etching $p$-type Si, for example, the backside electrolyte contact is biased in the reverse direction and thus does not allow substantial current flow. The problem can be overcome by illumination of the backside or by using highly doped $p^+$-type Si that draws large reverse currents already at low potentials, if this is compatible with the pore structures desired. If the Si sample is electrically floating, the potential at the surface is not well defined. Using a three-electrode configuration with the Si sample kept, a ground potential would rectify the situation, but in this case the setup becomes considerably more involved because now two power supplies (illumination and some control circuitry) are needed.

In any case, a good uniform backside contact with low resistance is a very important requirement for obtaining a controlled etching process. It is known that the porosity of a layer is strongly influenced by the local current density, which can be a sensitive function of the applied potential. Lateral voltage drops in nonuniform backside contacts thus may lead to large variations in porosity or other properties of the PSi. Uniform electrical contacts to silicon substrates with low doping levels, however, are not easy to achieve. The formation of a $p^+$ or $n^+$ layer on $n^-$ or $p^-$ substrates, respectively, using ion-implantation with the following proper annealing is the best way to do that.

Research has shown that the position and the geometry of CE electrode also have great influence on PSi layer uniformity. As it is known, the planar uniformity of the PSi surface depends on the distribution of applied electric field lines (i.e., uniformity of the distribution of local current density) during PSi formation. In most studies, it is indicated that during Si porosification a thin cylindrical electrode is used. However, the results of Hossain et al. (2002) indicated that in this case the resulting PSi has both a planar and a transverse non-uniform structure. In particular, PSi layers formed had lateral surface nonuniformity in relative reflectance and photoluminescence spectra. This situation is observed because the electric field lines are denser in the central region of the surface than that in the edges. This leads to higher value local current density and as a result higher porosity in the central region. Experiment has shown that the use of counter electrodes realized as a platinum mesh or sheet gives a significant improvement of lateral uniformity of PSi parameters. At that, a homogeneous current distribution, as desired in most applications, is best achieved by using a counter electrode of the same size and in-plane orientation as the working electrode. If a Pt mesh is used, its total surface area must be comparable to that of the Si electrode and the size of the mesh openings are smaller than the distance to the Si electrode (Lehmann 2002). Platinum black coating of the mesh can reduce the required mesh area. However, studies (Hossain et al. 2002) have shown that even in this case for flat electrodes, the inhomogeneity of porous silicon parameters is possible; the porosity at the edges was found to be greater than that at the central region. Increasing the size of the sample and the cathode, much better uniformity can be achieved around the center. Hossain et al. (2002) also found that similar results could be achieved by taking a cathode larger than the sample and introducing a graphite plate below the sample, that being kept at the same potential of the sample. This bottom graphite plate, electrically connected to the sample, behaves like a guard plate effectively extending the surface of the sample and the electrodes. Hossain et al. (2002) established that using this electrode leads to improvement in both lateral and vertical uniformity of PSi layer even for formation current densities as low as 2 mA/cm$^2$.

The size of the reference electrode is not relevant because it carries no significant current (Lehmann 2002). The internationally accepted primary reference is the normal hydrogen electrode (NHE), which consists of a Pt electrode in a stream of hydrogen bubbles at 1 atm in a solution of unit hydrogen ion activity. The most common reference is the saturated calomel electrode (SCE). To reduce measurement errors, the reference electrode should be placed as close as possible to the Si electrode or it can access the Si electrode via a capillary.

Furthermore, it must be borne in mind that the Si dissolution process especially in aqueous electrolytes always produces H$_2$, and hydrogen bubbles may be formed that cling to the surface of the wafer and cause variations in the local potential. Bubbles sticking to the sample surface can also be a reason for pore degeneration. The sticking bubbles hinder the diffusion process inside pores, and pore degeneration is observed earlier than in the bubble-free regions. According to conclusions made by Lehmann and Gruning (1997), the degeneration is being observed in a spatially limited region corresponding to the bubble diameter. In any case, hydrogen bubbles can cause a considerable inhomogeneity of the PSi parameters, either laterally or in the depth of the PSi layer, and must be avoided. A simple remedy is stirring or pumping of the electrolyte. This is particularly important in the double-cell arrangement, where either the gas evolution and bubble formation during the cathodic reaction can also cause local variations in the potential throughout the substrate. It should be mentioned at this point that the amount of H$_2$ produced by the reactions is usually not of a quantity that is safety relevant but that this is completely different for the case of pore etching in III-V compounds, where AsH$_3$ or PH$_3$ may form gases that are deadly even at minute concentrations below ppm. In addition, the construction of the cell should be conducted in such a way that the conditions allowing accumulating bubbles are not created. For this, for example, the walls of the hole, realized for the contact of the electrolyte with the wafer, should not be vertical but tilted to facilitate the renovation of the electrolyte and the removal of the bubbles on the surface of the wafer.

Precise control of the electrolyte temperature, the current density, and the potential are necessary and this calls for optimized electronic control circuitry (Bisi et al. 2000). The volume of the cell is also an important factor. Larger volumes of the electrolyte may contribute to a more homogeneous etching but raise cost and safety issues.

As an example for practical realizations, two cell concepts will be briefly discussed. First, a very simple cell as used by Kleimann et al. (2000) for the purpose of generation of *n*-Si-macro pores. This cell was made from Teflon and had a volume of 400 cm$^3$. The area of the silicon sample exposed to the electrolyte has a circular shape with an area of 2 cm$^2$. Electron-hole pairs were generated by illuminating the backside of the samples with a 300-W halogen lamp through a circular window formed in a metal ring. The metal ring was used for providing the electrical contact between the sample and the power supply. The photocurrent was adjusted quite simply by varying the distance between the halogen lamp and the sample. As a compromise between providing a good contact without evoking complex and costly ion implantation and keeping the illumination uniform, the backside silicon surface was covered with an aluminum grid. The counter electrode was a platinum grid immersed into the electrolyte. A PC controlled power supply was installed for measurement of the current–voltage characteristic of the electrochemical cell and for controlling the current density during the etching.

More sophisticated cells for cm$^2$ sample sizes like the "Kiel cell" designed in Foell's group and commercialized by ET&TE GmbH (http://www.et-te.com) can be build up from basic components as single or double cells with windows on both sides if required (Hejjo Al Rifai et al. 2000). The electrical scheme of this etching apparatus is shown in Figure 3.5. A simple pseudo reference electrode (Pt wire close to the specimen surface) was incorporated to ensure reproducible results and quantitative control of the junction potential. In above-mentioned electrochemical cell optimized electrolyte flow conditions, the use of reference electrodes, and the possibility to contact the sample in the double-cell mode are provided. A fully software-controlled illumination by an LED array allows for high illumination intensities while minimizing heat production. Full system control ($T$, $I$, $V$, $P_{illu}$) via customized hard- and software, including the possibility of *in-situ* monitoring of pore depth, valence, and many other parameters by dual-mode Fast Fourier Transform (FFT) impedance spectrometry is possible with this cell (Leisner et al. 2008).

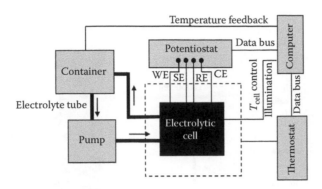

**FIGURE 3.5** The overview of the experimental setup optimal for silicon porosification. The setup has six important components: the electrochemical cell, the electrolyte containers, the peristaltic pumps, the potentiostat, the thermostat, and the computer. The computer controls the thermostat and the potentiostat. (Idea from http://www.porous-35.com.)

A similar approach to designing a chamber for electrochemical etching aimed for application in industrial production was also used in Bosch GmbH (Boehringer et al. 2012). At that, it was shown that PSi application in industrial production requires better controllability and reproducibility of critical PSi properties. In addition, in the restrictive environment of a semiconductor manufacturing facility, additional rigorous constraints apply. Therefore, an anodization tool for use in high-volume production must have the following attributes: safety and environmental protection, compatibility, capability, repeatability, and reliability. In particular, the tool has to be capable of processing hundreds of wafers without a failure demanding assistance or repair. It was concluded that for achievement of such parameters it is necessary to use special configuration of the electrochemical cell, specific materials for the electrodes or seals, fully automated "cassette-to-cassette" processing, continuous circulation of the electrolyte, PC-controlled programmable current source, special control of the HF concentration, and so on (Boehringer et al. 2012).

Several examples of electrolytic cells of O-ring type used for Si porosification in laboratories are shown in Figure 3.6. In these cells, standard black O-rings made of an acrylonitrile-butadiene copolymer (such as Perbunan) have proved to be stable in HF at concentrations up to 50%. If contamination of the silicon sample is an issue, the nitrile O-rings may be replaced by vinylidene fluoride-hexafluoropropylene (Viton) O-rings. One can find description of other variants of electrolytic cells adapted for silicon porosification in Lehmann (2002).

## 3.2 CLASSIFICATION OF ELECTROLYTES USED FOR Si POROSIFICATION

From a general point of view, electrolytes acceptable for silicon porosification must contain chemical species allowing anodic Si dissolution with at least one of the two basic mechanisms such as direct dissolution and oxidation (followed by purely chemical dissolution of the oxide). The chemical reactions for the two cases can be written as (Lehmann 2002)

$$(3.1) \qquad Si + 2HF + 2h^+ \Rightarrow SiF_2 + 2H^+$$

$$(3.2) \qquad Si + 2H_2O + 4h^+ \Rightarrow SiO_2 + 4H^+$$

$$(3.3) \qquad SiO_2 + 2HF_2^- + 2HF \Rightarrow SiF_6^{2-} + 2H_2O$$

One can note that the valence $n$ of the process, that is, the number of elementary charges that need to flow through the external circuit in order to remove one Si atom from the crystal, is implicity given as $n_d = 2$ in the case of the direct dissociation and $n_{ox} = 4$ for dissociation by oxidation. $n$ corresponds to the valency of Si in compound formed during reaction of chemical dissolution. As a rule, micropores and mesopores are formed with direct dissolution being dominant,

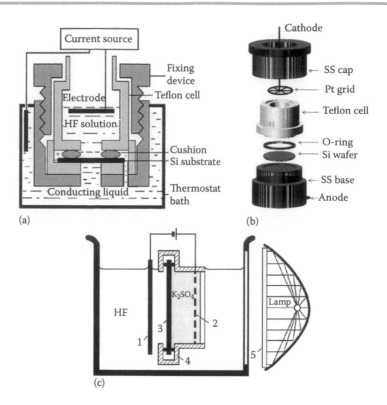

**FIGURE 3.6** (a) A double tank electrochemical etching cell. In this cell, the outside of the etching area was immersed in conducting liquid. A cushion was used to separate HF electrolyte solution and conducting liquid. (Adapted from Zhong F. et al., *Opt. Eng.*, 51(4), 040502, 2012. Published by SPIE as open access.) (b) Electrolytic cell with one-cell configuration. (From Patel P.N. et al., *Adv. Nat. Sci: Nanosci. Nanotechnol.*, 3, 035016, 2012. The Vietnam Academy of Science and Technology (VAST). Copyright 2012: IOP Publishing. With permission.) (c) Schematic of the pore etching cell showing the (1) Pt cathode, (2) Pt grid anode, (3) Si wafer, (4) Si wafer holder, and (5) the opaque scattering window. (From Van den Meerakker, J.E.A.M. et al., *J. Electrochem. Soc.*, 147(7), 2757, 2000. Copyright 2000: Electrochem. Soc. With permission.)

whereas the standard $n$-macropores only form for $n \approx 2.7$, indicating that both direct dissolution and dissolution by oxidation must occur in roughly equal amounts.

Equations 3.1 through 3.3 make clear that any electrolyte must contain some fluoride. This leaves a large number of possibilities. Foell and co-workers (Föll et al. 2002; Christophersen et al. 2003; Frey et al. 2007) introduced the following description for three major types of electrolytes that are being used today for PSi fabrication:

1. Electrolytes based on HF-$H_2O$ mixtures, so called "aqueous electrolytes." The nominal concentration of HF (in weight %) in any form of electrolytes typically ranges from <4% to 40%. This class allows relatively easy oxidation. It includes not only all mixtures of HF (commonly 49%) with water, but also with fluorine bearing salts dissolved in $H_2O$ (e.g., $NH_4F$), additions of ethanol ($C_2H_5OH$), acetic acid ($C_2HOH$), or anything else that serves to reduce the surface tension, adjusts the pH-value or the viscosity, or simply helps to get the desired results.

2. Electrolytes based on HF organic solvent mixtures; so-called "organic electrolytes." This class favors direct dissolution relative to oxidation. Its "oxidation power" is small. Neglecting very special conditions, these organic electrolytes still include some water, usually coming from the HF (49%), and the water is still contained in the solvents. There are a large number of organic solvents besides ethanol that have been used. The most important, perhaps, are acetonitrile (MeCN), dimethylformamide (DMF), and dimethylsulfoxide (DMSO). Alcohols like ethanol or propanol may or may not be included here.

3. Electrolytes for anodic oxidation (called oxidizing electrolytes). These types contain some oxidizing reagents, while they never contain $F^-$ ions. Most common electrolytes without HF addition fall into this category, and one can classify a measurable parameter called "oxidizing power" relating to their ability to form oxide on Si. Pure oxidizing electrolytes have only limited applications, but they could provide a clue for understanding the electrochemistry of Si because their "oxidizing power" also determines to some extent the properties of the HF-based electrolytes with respect to pore etching.

There is a potentially large group of electrolytes left, which one might call "mixed electrolytes" (or "exotic electrolytes") (Föll et al. 2002; Christophersen et al. 2003; Frey et al. 2007). This set contains everything not included in the list above, but thus far, they had little practical significance. Examples, which can be found in the literature, include $H_3PO_4$ with a dash of HF (Parkhutik 1999; Lee and Tu 2007), absolutely water-free organic electrolytes (Propst and Kohl 1994), and diluted HF with some $CrO_4^{-2}$ (Christophersen et al. 2000a). It is important to mention here that all of the above-mentioned electrolytes are extremely dangerous liquids and must be used and disposed of with care.

However, it is necessary to note that the most widespread electrolytes for forming porous silicon would still come under the heading "aqueous electrolytes" even though they often contain large percentages of ethanol. The practical reason for not listing them under "organic electrolytes" is that the ethanol addition often neither changes much the basic properties of aqueous electrolytes, nor allows production of pore types typically observed for "real" organic electrolytes.

## 3.3  MASK TECHNOLOGY

For many applications of PSi, it is essential to obtain locally anodized porous silicon areas, that is, to define by some lithographical technique the areas in which pores are to grow or—in the case of macroporous Si—the location of single pores. For defining the regions where Si is exposed to the electrolyte, different types of etch masks can be used. The peculiarities of using different masks were discussed by Connolly et al. (2002), Splinter et al. (2001), Hedrich et al. (2000), and Defforge et al. (2012). Masking layers have to withstand the aggressive HF electrolyte for durations from ≈30 min up to a 10 h; the time required for etching $n$-Si-macro pores to a depth of >400 μm. It is quite convenient to use the structured photoresist directly for obvious reasons. However, ordinary positive photoresist masks dissolve in standard electrolytes within minutes. Better results are obtained from negative photoresists, but the adherence to silicon substrates is not good enough for this application. The photoresist layer lifts off from the substrate and the result is a lateral underetching (Splinter et al. 2001). A 15-nm layer of chromium between the silicon wafer and the photoresist improves the adhesion and reduces underetching. This mask can withstand the electrolyte for about 30 min with underetch of about 10 μm.

One solution of the problem of mask deterioration would be to use buffered HF (BHF) as an electrolyte, a mixture of $NH_4F$ (ammonium fluoride) and HF, typically in volume ratios 7:1 (Lammel and Renaud 2000). It is known from glass etching that photoresist withstands this mixture better because of the more neutral pH value of about 4.5. However, higher pH values result in some chemical etching of the silicon. Since the wall thickness of the internal sponge-like structure of porous silicon is only in the range of nanometers, this means that it is gradually dissolved at the top while the electrical porosification proceeds into the depth. In addition, the etching conditions in BHF are completely different from those in HF-based solution because of its higher viscosity.

Better results were obtained with metal masks instead of photoresists. Gold can be used with an adhesion layer of chromium, which is normally attacked very slowly by HF. Under the influence of the anodic potential, however, it is also underetched electrochemically. To limit lateral etching, the chromium layer has to be as thin as possible, around 15 nm.

Silicon dioxide may also be employed as a mask (Nassiopoulos et al. 1995). However, this material presents the same issue as the photoresist: the dissolution rate is very high in HF especially at high concentration.

The use of carbides and nitrides is another approach to mask forming (Lang 1996; Cullis et al. 1997; Hedrich et al. 2000; Splinter et al. 2001; Connolly et al. 2002). However, crystalline SiC is not suitable because it is known to become porous in HF-based solution (Shor and Kurtz 1994). Amorphous SiC films deposited by CVD present semi-insulating properties and thus remain inert during anodization. Its etch rate in HF is only several angstroms per hour. This means that amorphous SiC films can be used as masking layers. The main issues of this material are its patterning and removal processes. The SiC is locally removed by $CF_4$ plasma etching. This gas is not selective to silicon and could deteriorate the substrate during the patterning or the porous layer during its removal. In addition, silicon carbide is expensive and has a high intrinsic stress and poor adhesion (Kim et al. 2003). Silicon nitride tends to show better results but is not completely resistant against fluoric solution. It dissolves in HF-ethanol solutions with an etch rate of approximately 15 nm/min. Therefore, silicon nitride can be used only for a short anodization process with duration up to 1 h (Lammel and Renaud 2000). This allows production of over 60 µm thick porous silicon in moderately doped $p$-type wafers. Silicon nitride can be easily deposited using standard low pressure chemical vapor deposition (LPCVD) or plasma enhanced chemical vapor deposition (PECVD) methods, but the thickness of a stoichiometric $Si_3N_4$ layer deposited by those methods is limited to 200 nm as a result of residual stress building up (Splinter et al. 2001). Moreover, stripping this mask to the porous silicon layer obtained after the etching may be difficult without damaging PSi; wet etching of nitride mask necessitates prolonged immersion of nanoporous layers in orthophosphoric acid at 165°C or HF solution (Prochaska et al. 2002). No damage to porous layers was observed during mask removal by dry etching. However, care must be taken to avoid etching the silicon substrate, which would cause a considerable roughness increase and would prevent successful bonding in later stages. The inclusion of an oxide layer under the masking nitride layer allows total nitride removal using dry etching without affecting silicon substrate. After nitride removal, the oxide layer can be etched in 1:10 HF:water solution.

Carbon layers were also studied as a mask (Djenizian et al. 2003). Deposited by electron beam technique, this thin film presents a high resistance in HF depending on the deposition parameters. However, its chemical resistance was only studied for short-time anodization. Moreover, the deposition tool is not widely used even in research laboratories, and it is not suitable for high-throughput production. No postanodization mask removal was performed, but one can assume that $O_2$ plasma may be an efficient film stripping option.

Splinter et al. (2001) stated that the best mask for long-time etching is a sandwich structure of silicon oxide and polycrystalline silicon. In this case, silicon oxide guarantees electrical insulation, and polycrystalline silicon protects the oxide from being dissolved by the HF (except very slowly at the mask edge). Since the polycrystalline silicon is not in direct contact with the silicon substrate, no current flows through it and it is not anodically etched and remains stable. The addition of a relatively thick nitride layer between the poly silicon and silicon dioxide layers should improve masking characteristics. The existence of a nitride layer modifies the net stress of the masking stack, turning it from compressive into tensile, which eliminates the buckling problem. The nitride layer also offers a good mechanical strength to withstand the flushing of evolving bubbles. So, the tri-layer stack can be used as the mask for extremely long duration HF etching. However, one should note that this technology is more complicated.

Defforge et al. (2012) believe that an innovative fluoropolymer (FP)-based layer also is very promising for application in the process of Si porosification. They established that fluoropolymer film presents many advantages compared to conventional materials. This mask owns high chemical resistance into HF-based electrolyte, even under anodic bias and damage-free removal. It is also compatible with the common microelectronic industry processes. This type of masking layer may thus be employed for numerous microelectronic applications such as MEMS, Si/PoSi hybrid substrates, or 3D integration. Thanks to its chemical resistance, the FP may also be employed as a protection for the underlying layers for a backend electrochemical etching process.

A further approach for porous silicon masking material is the possibility to use locally implanted layers (Smith and Collins 1992; Bell et al. 1996; Splinter et al. 2001; Pagonis et al. 2003; Mangaiyarkarasi et al. 2006; Frey et al. 2007). In this case, advantage is taken from the general fact that the porous silicon generation processes for $p$- and $n$-type silicon can be very different because the etching process depends on the concentration of holes in the substrate material. In $p$-doped silicon, the required holes are available as majority carriers, for $n$-type silicon, holes must be generated via illumination or electrical breakdown. Therefore, in the presence of Si areas with both types of conductivity on one substrate (e.g., $n$-implantation in $p$-type substrate), it is easy to find conditions when the etching proceeds only in the $p$-doped areas. Usually the implantation of phosphorous is used for these purposes. The doses of phosphorous implantation can range between $10^{13}$ and $10^{15}$ ions/cm$^2$. A postimplantation treatment includes annealing at temperatures up to 1100°C. According to Pagonis et al. (2003), the most critical parameter in the process is the surface concentration of dopants. The advantage of using protective doped layers instead of thin films is that with doped layers monocrystalline silicon layers in plane with silicon substrate remain after anodization, which provides advantages, for example, for fabrication of membranes, where a step height at the transition between masking layer and porous silicon forms a break point (Bartels et al. 1999; Splinter et al. 2001). Additionally, with localization by doping, the shape of such doped protective regions in depth also can be defined (Mescheder et al. 2001). However, Pagonis et al. (2003) noted that in all cases, some corrosion of the masking area was observed.

Another promising approach to PSi patterning based on ion implantation is the use of the mask formed by hydrogen ion implantation over $p$-type silicon substrate, combined with rapid thermal annealing (RTA) (Galeazzo et al. 2001; Dantas et al. 2003, 2004). Such a procedure can create a buried high resistivity layer due to damaging of silicon lattice and electrical neutralization of boron donor sites by hydrogen (Pankove and Johnson 1991), and thus inhibit electrochemical etching in this area. Moreover, this layer can be "fixed" by formation of H$_2$ bubbles that have low diffusivity, so a stable high resistivity layer can be obtained. In particular, Peres and Ramirez-Fernandez (1997) reported that high resistivity buried layers (above 300 Ω·cm) were obtained in silicon after hydrogen ion implantation with doses between $2 \times 10^{14}$ and $2 \times 10^{16}$ H$^+$/cm$^2$. Thermal annealing is necessary to recover the crystalline structure of Si surface after any implantation process. However, in the present process, it is necessary to do this without losing buried resistivity peak. It was found that the indicated technique has the following advantages (Dantas et al. 2003, 2004): (1) Very homogeneous and isotropic PSi formation even beneath mask borders. (2) There is no time limitation into HF solution. Therefore, thick PSi layers can be obtained (for sensors or optical devices applications, for example) and, additionally, silicon membranes above deep cavities can be fabricated (for MEMS applications). (3) Remaining superficial silicon membranes have good uniformity and no deformation, which is very important to build microstructures. In addition, (4) it is possible to obtain thin silicon membranes (smaller than 1 μm) with good uniformity (Galeazzo et al. 2001). Membrane thickness can be controlled by hydrogen ion implantation and RTA processes parameters. Regarding disadvantages of this technique, they include the following: (1) Ion implantation with the following RTA process is more complicated and requires more processing time than the usual oxidation step. (2) Silicon mask areas are damaged by the ion implantation process. Electronic devices placed on these areas have limited performance, so it is very important to recover silicon crystalline structure (Peres and Ramirez-Fernandez 1997). However, Galeazzo et al. (2001) stated that RTA was adequate to recover the crystalline structure.

Punzon-Quijorna et al. (2012) have shown that implantation of MeV Si ions is also an effective tool for the localized formation of PSi in the micrometer range. It is known that irradiation of materials with MeV heavy ions produces defects in semiconductors along the ion tracks, and the damaged part of the material, depending on the type of material itself, may turn more sensitive or more resistant to chemical etching. Punzon-Quijorna et al. (2012) have found that in the case of heavier ions, its higher damage efficiency allows lower implantation doses to achieve PSi growth inhibition, which allows shorter process times, and at the same time provides good lateral resolution below the micrometric range. As it is seen in Figure 3.7, porous silicon growth is inhibited on irradiated substrates already with Si fluences greater than $10^{12}$ ion/cm$^2$ in the energy range of 5–24 MeV. This effect was explained by strong reduction of the conductivity with increasing irradiation fluence. In addition, Punzon-Quijorna et al. (2012) believe that the usage of ions of the same elementary nature as the target material avoids inconvenient side effects that may be ascribed to the implanted species.

As it is shown in Figure 3.8, the mask, used in the process of Si porosification, strongly affects the shape of the porous silicon layer (Steiner and Lang 1995). For example, when an insulating layer is used for masking, a metal-insulator-semiconductor (MIS)-like sandwich of conductor, insulator, and semiconductor is formed. The potential of the solution is negative with respect to the silicon. This is an important point for the etching behavior of $n$-type

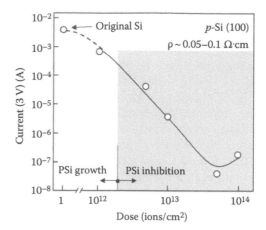

**FIGURE 3.7**  Anodization current through irradiated silicon under voltage bias of 3 V, for irradiation doses from $10^{12}$ to $10^{14}$ ions/cm². Original non-irradiated substrate is shown as having a dose of 1 ion/cm² as a reference. After irradiation, the copper grids were removed and the samples were anodized in an HF-based solution with a current density of 100 mA/cm². (Reprinted from *Nucl. Instrum. Meth. Phys. Res. B*, 282, Punzon-Quijorna, E. et al., 25, Copyright 2012, with permission from Elsevier.)

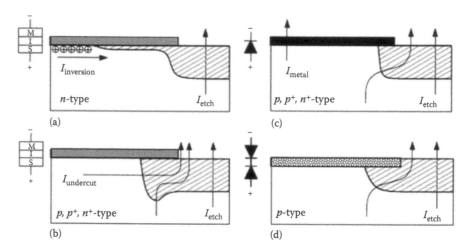

**FIGURE 3.8**  Influence of the masking layer and the substrate doping on the etching behavior: (a) insulating masking for *n*-type silicon; (b) insulating masking for *p*, *p*⁺, and *n*⁺-type silicon; (c) metal masking layer on degenerately doped and *p*-type silicon; (d) *n*-implant (counterdoping) masking on *p*-type substrate. (Reprinted from *Thin Solid Films*, 255, Steiner, P. and Lang, W., 52, Copyright 1995, with permission from Elsevier.)

silicon. Depending on the applied voltage, doping level, and dielectric layer thickness, charge carrier inversion under the wide masking areas may occur. In this case, many holes are available, which can easily drift to the open areas in the masking layers. This causes wide undercutting of the masking (several hundreds of micrometers) and prevents a precise control of the etched areas (see Figure 3.8a).

Veeramachaneni et al. (2011) have found that the formation of porous silicon can also be done selectively by controlling the Fermi level in areas to be etched or not etched, which is typically done by adjusting the level of doping. It was established that implanted fluorine in silicon has demonstrated a donor effect upon annealing at low temperature (600°C), which is reversible as the fluorine outdiffuses during higher temperature annealing (1100°C). Veeramachaneni et al. (2011) have shown that the investigated technique can be used to form crystalline silicon active regions with thickness less than 200 nm completely surrounded by oxidized porous silicon (see Figure 3.9). The described process has the advantages of integration simplicity, with the ability to resolve SOI structures down to the micron and potentially submicron feature sizes.

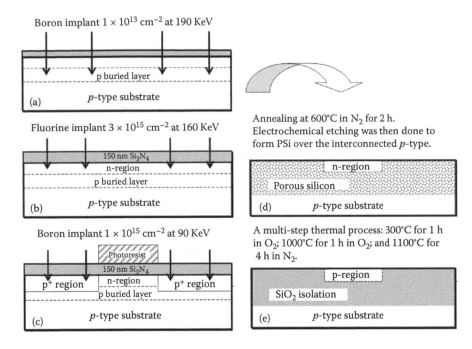

**FIGURE 3.9**    Process sequence for the fabrication of fully isolated crystalline silicon regions, using fluorine for suppression of electrochemical formation of PSi. (Veeramachaneni, B. et al.: *Phys. Stat. Sol. (c)*. 2011. 8(6). 1865. Copyright Wiley-VCH Verlag GmbH & Co. KGaA. Reproduced with permission.)

## 3.4 ETCH-STOP TECHNIQUES

Electrochemical dissolution proceeds exclusively by current flow. Therefore, the etching can be stopped via a switching-off of a power (current) supply or an electrochemical etch-stop. Disconnecting a part of the wafer electrically will passivate that part from etching as well. Dielectric layers or other isolate materials can be used for these purposes.

As it was shown earlier, the etching is mainly determined by the availability of holes in the silicon. Therefore, good selectivity for etching may already be found between $n$-type and $p$-type silicon. One way of using this technically is the generation of a $p$-$n$ junction by implantation as outlined previously. This type of etch stop was used, for example, for silicon on insulator (SOI) structures made by oxidizing porous silicon (Lang 1996). Stopping at a $p$–$n$ junction was also reported by Gupta et al. (1995). Another possibility is the use of $n^-$–$n^+$-type (or $p^-$–$p^+$-type) ISO-junctions (Lang 1996). As it was established by Eijkel et al. (1990), in the absence of light generation the rate of pore formation in $p$- and $n$-type silicon with concentration of free charge carriers smaller than $10^{16}$ cm$^{-3}$ is very low. The relative rates of pore formation in $n$-type and $p$-type silicon of different dopant concentrations are shown in Figure 3.10. In this case, the current density was 0.1 A/cm$^2$ and the HF concentration was 5%.

Results presented in Figure 3.10 testify that on the interface $n^+$–$n^-$ or $p^+$–$p^-$ the change of the rate of Si etching will be great. This effect for etching stop was used by Esashi et al. (1982). The indicated etch-stop techniques may be used for lateral and depth structuring of PSi. Often, for these purposes the buried layer is used. Buried doped regions can be formed by high-energy ion implantation, although the depth is limited in this case (D'Arrigo et al. 2002). Another technique to create buried doped regions is by diffusion or ion implantation with subsequent formation of another silicon layer on top by bonding or epitaxial growth (see Figure 3.11) (Murate et al. 1998; D'Arrigo et al. 2002). The process with epitaxial silicon layer is often called epi-micromachining (Gennissen et al. 1995; French et al. 1997). The same task of etch stop can be resolved through incorporation of a dielectric layer in designed structure.

In this context, one should also note that the pore growth into the depth of the sample always stops by itself eventually because any pore growth mode will switch to some other mode (e.g., electropolishing) because the potential of the pore tip and the concentrations of the relevant

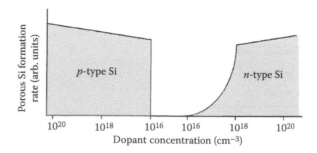

**FIGURE 3.10** Rates of pore formation in *p*-type and *n*-type silicon as a function of dopant concentration with no light generation. *p*-type silicon becomes porous at any doping level, while *n*-type silicon becomes porous only when it is doped highly enough to allow for significant interface tunneling. (Data extracted from Eijkel, C.J.M. et al., *IEEE Electron Device Lett.,* **11**, 588, 1990.)

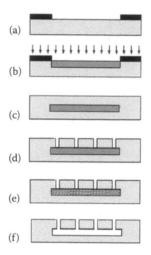

**FIGURE 3.11** Process of a buried *p*-type silicon region formation in *n*-type substrate with following Si anodization and PSi layer removal. (a) *n*-type silicon substrate with mask. (b) Boron ion implantation ($p^+$-type layer). (c) Mask removal and epitaxial growth of *n*-type silicon. (d) Formation of openings for anodization. (e) Porous silicon forming. (f) Porous silicon removal.

chemical species must decrease with increasing pore depth. This has been investigated in some detail for macro pores (Lehmann 2002).

## 3.5 LITHOGRAPHY

When forming the meso- and microporous silicon, lithography is used only for localization of the area of Si porosification (see Figure 3.9). At the same time during the formation of macroporous silicon, this process is important because in contrast to the anodization of metals (e.g., aluminum, titanium) the growth of macropores in silicon is not a self-ordering process. This means that the pores do not form a periodic arrangement naturally, as required by many applications. In this case, a random pore structure is produced. The pore position, however, can be predefined by lithography and subsequent etching. The obtained etch pits work as nucleation sites for the pore growth. The advantage of this process is that there is no restriction to a hexagonal pore arrangement only. Rather, the prepatterning can be varied in the pore arrangement and pore size and hence allows for specifically designed samples. Although the etching process is not a self-ordering process, it is a self-organizing one. For a given doping density, applied voltage and backside illumination an average porosity and pore diameter will arise. Therefore, the lithography has to fit with the intrinsic material parameters.

Experiment has shown that in order to define the starting position of the macropores, standard photolithography and dry (reactive ion etching, RIE) or wet etching usually are used. Figure 3.12 shows a SEM cross-sectional picture of such initial pits used by Grigoras et al. (2001). The main

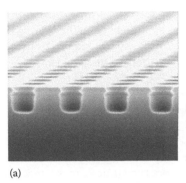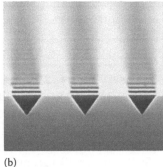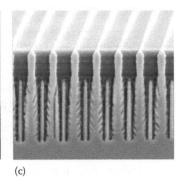

(a)                              (b)                              (c)

**FIGURE 3.12** Initial pits made (a) by reactive ion etching (RIE) through a photoresist (top layer, 250 nm) and amorphous silicon (200 nm) layers (pits of 1.2 μm deep) and (b) by anisotropic alkaline wet etching (pits of 1.7 μm deep); (c) splitting of macropores for alkaline etched inverted pyramid initial pits of 10 μm width. (From Grigoras, K. et al., *J. Micromech. Microeng.*, 11, 371, 2001. Copyright 2001: IOP Publishing. With permission.)

difference between the two types of initial pits is the presence of sharp tips after KOH etching (Figure 3.12b). The electric field, concentrated at those points, attracts the positive charge carriers, initiating the dissolution of silicon. Therefore, a small enough pit is necessary to initiate the formation of only one macropore. In this case, the diameter of the pore and the inter-pore spacing is usually restricted (typically >1 μm with contact lithography). In the case of more flat-bottomed RIE initial pits (Figure 3.12a), the electric field is concentrated along their perimeter. As a result, in this case several pores located along the pit perimeter can start to grow (Ohji et al. 2000). Of course, in some special cases this could even be an advantage; a wider hole would be etched or the density of macropores could be increased in that way. On the other hand, it was found experimentally that four macropores could start to develop even from the initial pit having a form of a sharp-pointed inverted pyramid, when its size exceeds 10 μm and a lightly doped silicon substrate is used (see Figure 3.12c).

Astrova and Vasunkina (2002) have shown that for forming of macropore nucleation centers, ion implantation also can be used. It was found that selective radiation damage or local inversion of the conduction type is sufficient for macropore nucleation in some areas and passivation of others. Unfortunately, this method is expensive and the quality of the structures formed using this method is worse than that of structures obtained by traditional methods. However, Astrova and Vasunkina (2002) believe that the use of ion implantation to generate macropore nucleation centers may be advantageous in cases where a submicrometer pattern is formed by means of a focused ion beam or with substrates having an orientation other than (100).

When choosing the interpore spacing, it is also necessary to keep in mind that the passivation of the pore walls and thus the prevention of the pore walls from being post-etched is a consequence of the presence of space-charge region (SCR). Therefore, for a stabilized pore growth, the remaining silicon between neighboring pores should be completely depleted from charge carriers. From this requirement, a rule of thumb can be derived for the lattice constant of the lithography: the interpore distance $a$ has to be chosen twice as large as the width $W_{SCR}$ (Equation 3.4). It implies that higher doped material is preferentially used for smaller interpore distance and vice versa.

$$(3.4) \qquad W_{SCR} = \sqrt{\frac{2\varepsilon_0\varepsilon_{Si}U_{SCR}}{eN_D}}$$

where $\varepsilon_0$ is the dielectric constant, $\varepsilon_{Si}$ is the dielectric constant of silicon, $U_{SCR}$ is the applied voltage, $e$ is the elementary charge, and $N_D$ is the number of dopants.

With regard to the photolithographic process on the porous silicon, there are two problems that are caused by (1) low chemical resistance of the porous silicon, and (2) difficulty in removing the photoresist from the pores. For solving the first problem, Lai et al. (2011) suggested using PSi passivation by annealing the film in $N_2$ atmosphere at temperatures ranging from 560°C to 800°C, while for the resolving of the second problem, they proposed using an additional simple protection layer (ProLIFTTM). This layer can be used to prevent photoresist seepage. ProLIFTTM is not photosensitive material, but it is easily deposited on the PSi layer by spinning before the coating of photoresist, and then removed even if it seeps into the pores. Therefore, it will be removed from the areas where the photoresist is developed, but will remain under the patterned photoresist.

## 3.6 SPECIFIC PARAMETERS OF POROUS SILICON LAYER FABRICATION

The various applications of PSi depend on the kind of pores and their detailed structure or morphology plus the (chemical) surface properties and the aging behavior in air. This is particularly true for sensors whose design is based on PSi. Parameters like geometric pore type (micro, meso, macro); basic growth mode (crystallographic or current-line orientation); morphology (sponge-like, fractal, heavily branched and interconnected, lightly branched and not interconnected, straight and cylindrical, etc.) as well as secondary or derived parameters like porosity, thickness, pore size distribution, and so on, result from the properties of the bulk Si used (in particular doping type and doping level) and the porosification conditions (Parkhutik 1999; Bisi et al. 2000; Christophersen et al. 2003; Föll et al. 2006; Kochergin and Foell 2006; Frey et al. 2007). Therefore, the manufacture of any devices based on PSi must take into consideration the peculiarities of the etching process that always will be at the core of the designed technology of this device fabrication. In what follows, the basic relations between anodic etching conditions and the parameters of porous layer formed in Si will be considered, first in a more general way and then with regard to specific parameters.

### 3.6.1 GENERAL REMARKS

As it was indicated earlier and illustrated to a small extent in Chapters 1 and 2, there is an amazingly large variety of porous silicon films, ranging from micropores to macropores with diameters in excess of 10 μm and from sponge-like or fractal structures to perfectly cylindrical pores with large aspect ratios (Parkhutik 1999; Lehmann et al. 2000; Föll et al. 2002). Different structures and sizes of pores reflect differences in preparation conditions or different points in a large parameter space. The most important parameters include the following:

- **Semiconductor properties:** Type of conductivity ($n$-type or $p$-type); doping level (heavily doped samples often behave qualitatively differently from lightly doped ones); crystal orientation; surface condition (polished, rough, masked); defect level; minority carrier lifetime or diffusion length; backside conditions; and contact properties (ideally ohmic).
- **Electrolyte parameters:** Basic type (aqueous, organic, or mixed); concentration of the ingredients; additions of surfactants.
- **Experimental settings:** Basic system settings (potentiostatic, galvanostatic, mixed); global and local current densities $j$; external and local potential $U$; temperature $T$; illumination mode (front or backside); intensity of illumination $P$; time dependence of all these parameters (e.g., preset $I(t)$, $U(t)$, and $T(t)$ with $P(t)$ adjusted via a control circuit—the standard case for $n$-Si-macro pores formation); sample preconditioning for pore nucleation (e.g., large potential for a short time).
- **Hardware:** Design of the electrolytic cell; pumping or stirring apparatus; temperature control (e.g., some pores required a $\Delta T < 0.2°C$); electrolyte concentration control; and possibilities for *in-situ* monitoring.

### 3.6.2 COMMONLY USED ELECTROLYTES AND SOME MODIFICATIONS

The electrolyte composition is one of the most important fabrication parameters for well-defined porous layers. Various electrolytes, as classified in Section 3.2, can be used for porous silicon fabrication. With respect to macroporous Si, the situation is especially complex, where just the addition of various alcohols to a basic aqueous electrolyte made a big difference regarding the macropore quality (Leisner et al. 2008).

**TABLE 3.1    Typical Parameters of Silicon Porosification Using Electrochemical Etching**

| Substrate | Electrolytes | Parameters of Oxidation | Porosity | Ref. |
|---|---|---|---|---|
| p-Si (7–13 Ω cm) | HF: $C_2H_5OH$ = 1:1 | $I$ = 200–250 mA/cm$^2$ | 45–60%<br>$d$ ~ 15–40 μm | Barotto et al. 2001 |
| p-Si | HF: $C_2H_5OH$ = 1:1 | $I$ = 20 mA/cm$^2$ | | Racine et al. 1997 |
| n-Si | HF: $C_2H_5OH$ = 1:3 | $I$ = 20 mA/cm$^2$, $t$ = 30 min, illumination (Nd:YAG laser) | 70–80% | Koyama 2006 |
| p-Si (10 Ω cm) | HF: $C_2H_5OH$ = 1:2.5 | $I$ = 5–20 mA/cm$^2$<br>$t$ = 20–120 min | 65–67% | Holec et al. 2002 |
| p-Si (2–9 Ω cm) | HF:$H_2O$: $C_2H_5OH$ = 1:1:2 | $I$ ~ 20–40 mA/cm$^2$ | 57% | Hedrich et al. 2000 |
| n-Si (1 Ω cm) | HF: $C_2H_5OH$ = 7:3 | $I$ ~ 35 mA/cm$^2$<br>$t$ ~ 25 min | $d$ ~ 80 μm | Di Francia et al. 1998 |
| p-Si | HF:$H_2O_2$:ethanol: $H_2O$ = (9–11):1:4:12 | Pt assisted electroless, $t$ = 10 min | 33%<br>65% | Splinter et al. 2000 |
| p-Si (0.01–0.2 Ω cm) | HF: $C_2H_5OH$ = 1:1 | $I$ ~ 14 mA/cm$^2$, $t$ = 275 s<br>$I$ = 172 mA·cm$^{-2}$, $t$ = 42 s | 38%<br>62% | Canham et al. 1996 |
| p-Si (0.02 Ω cm) | HF: $C_2H_5OH$ = 1:2 | $I$ = 20 mA/cm$^2$<br>$I$ = 75 mA/cm$^2$<br>$I$ = 150 mA/cm$^2$ | 74%<br>65% | Lysenko et al. 2002 |
| Si | 30% HF (including Triton X-100) | $I$ = 30 mA/cm$^2$<br>$t$ ~ 60 s | | Connolly et al. 2002 |

However, concerning micro- and mesopores, the most frequently used electrolytes are based on HF–$C_2H_5OH$ mixtures. Table 3.1 gives some details (always for standard 49% HF and absolute ethanol; the substrate orientation is always {100}).

As it follows from Table 3.1, a concentration of 15–40 wt.% of HF is a kind of standard for the fabrication of micro- and mesoporous silicon. The optimal concentration of the HF in an electrolyte depends on what kind of pores are to be etched and which parameter of the pores is to be optimized. Due to the hydrophobic character of the freshly etched (and H covered) Si surface, absolute ethanol ($C_2H_5OH$) is usually added because it might help to overcome surface tension problems (Bisi et al. 2000). In particular, ethyl alcohol (EtOH) added to this solution reduces the surface tension at the silicon-solution interface. As a result, the HF dissolution in ethanol gives better surface wetting in comparison with water solution (Splinter et al. 2001). This is very important for the lateral homogeneity and the uniformity of the PSi layer in depth (Splinter et al. 2001). It was established that in order to be useful, the ethanol concentration should not be less than 15% (Bomchil et al. 1993). In addition, ethanol additions reduce the formation of hydrogen gas bubbles; this is essential for homogeneous layers, as it was pointed out before. In this case, the bubbles of gas produced during Si porosification are smaller and less persistent. Possibly ethanol acts as a surfactant and prevents bubbles sticking to the silicon surface but its action might be more complex than that because when a commercial surfactant was applied in the electrolyte instead of $C_2H_5OH$, the porosity could not exceed 70%.

### 3.6.3 SPECIAL ELECTROLYTES

As we indicated before, standard etching solution for Si porosification includes three components (HF, $C_2H_5O_4$, and water). However, we have to note that in reality the number of components that could be included in the electrolyte for Si etching is much more than 3. For example, in several articles (Christophersen et al. 2000b; Bettotti et al. 2002; Kan and Finstad 2005) it was proposed to use as an electrolyte hydrofluoric acid (HF) diluted with dimethyl sulfoxide (DMSO). The solvent DMSO was chosen because it yields a faster etch rate than the others, such as $H_2O$, ethanol, or dimethylformamide (DMF). DMSO can also produce pores with a larger range of current density and form smaller pores in random (no mask patterning) samples (Christophersen et al. 2000b; Bettotti et al. 2002; Kan and Finstad 2005).

Experiment has shown that surfactant can also be added in the electrolyte. A surfactant is necessary to reduce the surface tension and to avoid hydrogen bubbles sticking to the surface (Sotgiu et al. 1996; Langner et al. 2011). Surfactants are compounds characterized by a high tendency to concentrate at the interfaces. When they are present in an aqueous solution, they are arranged in the solution-air and solution-solid interfaces so that all the surfaces are uniformly covered with surfactant molecules; the covering increases with the surfactant concentration until a critical value, called critical micellar concentration (CMC), is reached. Furthermore, it also has some influence on the pore formation. For example, it was also established that various additives could increase the rate of porous dissolution significantly. In particular, cationic surfactants cause rapid degradation and dissolution of porous silicon in aqueous media, even at acidic pH values (pH < 4). According to Sailor (2012), the origin of this unusual behavior lies in the interaction of the charged head group on the surfactant with the electronic structure of PSi in the presence of water. The role of the cationic surfactant is thought to involve stabilization of negative charge at the porous silicon surface by the positively charged surfactant. Cationic surfactant polarizes negative charges near the surface of the semiconductor, inducing more hydridic character in the Si-H surface species and enhancing the electrophilic nature of the silicon atoms at the surface. The increased reactivity makes the surface more susceptible to attack by water.

There are many different surfactants tested for use during silicon porosification (Sotgiu et al. 1996). SDS and NCW-1002 are examples of such surfactants. SDS is sodium dodecyl sulfate ($C_{12}H_{25}O_4SNa$), a common anionic surfactant normally used in household products as well as the chemical industry. NCW-1002 is a trade name and it is an aqueous solution with 10% polyoxyalkylene alkyl ether and nonionic in nature. It was shown that SDS is well suited for etching of straight pores while NCW is used for diameter-modulated pores (Langner et al. 2011). Some authors also add acetic acid to the electrolyte to further reduce the surface tension.

Replacing $C_2H_5OH$ by other alcohols has been the subject of a systematic study, aiming to decrease the roughness of pore walls in the case of $n$-macro pores (Foca et al. 2007b). Making the electrolyte more viscous by adding glycol or glycerol has emerged more recently as a means to make not only better macropores (e.g., with smoother walls) but also mesopores (Setzu et al. 1999). It has been found that introducing glycerol to the composition of the HF solution decreases the lateral inhomogeneity and surface roughness (Setzu et al. 1999). At present, the mechanism behind this is not clear.

An attempt was also made to optimize the electrolyte through the change of its "oxidation power." Additions of $H_2O_2$ (Yamamoto and Takai 1999), $CrO_3$ (Christophersen et al. 2000a,b), and many other oxidizers including even $HNO_3$ (Yamani et al. 1997) have been tried for various reasons. It was established that $H_2O_2$-based electrolytes have better characteristics in comparison with $HNO_3$-based electrolytes. It was shown that the photochemical etching method with $H_2O_2$ solution did not generate a toxic material unlike the case of $HNO_3$ (Yamamoto and Takai 1999). Moreover, the addition of $H_2O_2$ to the etching mixture raised the pH of the solution and could promote the forming of an ideal Si surface terminated with Si–H bonds thus resulting in a homogeneous PSi surface with a low defect density (Yamani et al. 1997). However, according to the judgment of many researchers, such optimization has not provided the expected result (Bomchil et al. 1993; Sharma et al. 2005). For example, Sharma et al. (2005) concluded that the porous silicon, generated in an ethanol-based electrolyte, has better luminescence properties than the porous silicon, formed by using an $H_2O_2$-based electrolyte at the same current density. This is based on the comparison of the luminescence properties of PSi, formed in HF:ethanol and HF:$H_2O_2$-based electrolytes. Moreover, it was found that PSi layers formed in an ethanol-based electrolyte on textured substrates were relatively mechanically stronger, stable, stress-free, and highly passivated with hydrogen than the PSi layers prepared in an $H_2O_2$-based electrolyte, as it was elucidated by weight loss measurements, PL, SEM, and FTIR studies. Besides that, it was established that in the case of using electrolytes based on $HNO_3$/HF solutions, the process control was more difficult to achieve, and the morphology of the layers differed too much (Bomchil et al. 1993). The last conclusion is understandable because HF:$HNO_3$ mixtures dissolve Si already purely chemically without current flowing.

Therefore, we have to note that in many cases the modification of electrolytes had produced little progress in the optimization of forming micropores and mesopores. However, new types of pores like $n^+$-macro have been obtained in this way, based on predictions from the current burst model (Christophersen et al. 2000).

### 3.6.4 CURRENT-VOLTAGE MODES OF SILICON POROSIFICATION

The I-V curves of Si-electrolyte interface (see Figure 3.13) show some similarities to the normal Schottky diode behavior, but there are some important differences. For instance, while the sign of majority carriers changes between $n$- and $p$-type, the chemical reactions at the interface remain the same. Moreover, the reverse-bias dark currents are at least three orders of magnitude higher than normal Schottky diode expectations. Another anomaly is the open circuit potential for $n$- and $p$-type doped Si, which is not consistent with the difference between bulk Fermi levels.

Under cathodic polarization, for both $n$- and $p$-type materials, Si is stable. The only important cathodic reaction is the reduction of water at the Si/HF interface, with formation of hydrogen gas. This usually occurs only at high cathodic overpotentials, or using Schottky diode terminology, at reverse breakdown. Under anodic polarization, Si dissolves. At high anodic overpotentials, the Si surface electropolishes and the surface retains a smooth and planar morphology. In contrast, with low anodic overpotentials, the surface morphology is dominated by a vast labyrinth of channels that penetrate deep into the bulk of the Si. Just in these conditions PSi is formed.

It was established that the dissolution of Si in the dark could be controlled by either the anodic current (galvanostatic mode) or the potential (potentiostatic mode). Generally, it is preferable to work with constant current because it allows a better control of porosity, thickness, and reproducibility of the PSi layer. Note that we avoid the term "current density" at that point because it is actually the current in Ampères that is the prime quantity externally set and controlled. Of course, giving the total current density $j_G$ with respect to the total area of the sample exposed to the electrolyte is important and should always be reported. However, one should note that the local current density $J_P$ at pore tips by definition is always larger than the global one and not necessarily known. Knowing that the current density at the growing pore tips is the most important parameter for the etching process producing pores, it is essential to keep in mind that this parameter is not really known well.

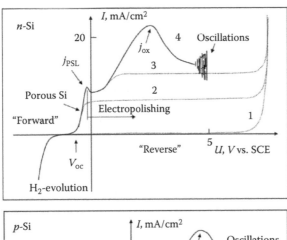

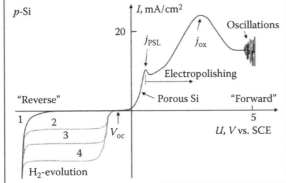

**FIGURE 3.13**   Current-voltage characteristics of Si/HF system for $n$-Si and $p$-Si. $V_{oc}$—equilibrium potential: 1—no illumination; 2–4—illumination ($I_4 > I_3 > I_2$). (Reprinted from *Mater. Sci. Eng. R*, 39, Föll, H. et al., 93, Copyright 2002, with permission from Elsevier.)

Usually, during Si porosification, a current density from 1 to 100 mA/cm² is used. This parameter corresponds to the region of PSi growth on the current-voltage characteristic of *p*-Si/HF system (see Figure 3.14). As it was indicated earlier, micropore or mesopore formation occurs only in the initial rising part of the I-V curve for a potential value below the potential of the small sharp peak. This current peak is called the electropolishing peak, porous silicon layer (PSL) peak, or peak at $J_{PSL}$. According to Lehmann (1993), $J_{PSL}$ can be calculated from the following equation:

$$(3.5) \qquad J_{PSL} = C \cdot c^{1.5} \cdot \exp\left(-\frac{E_A}{kT}\right)$$

where the constant $C$ is 3300 A/cm² (wt.% HF)$^{-1.5}$, $c$ is the HF concentration, the activation energy $E_A$ is 0.345 eV, $k$ is the Boltzmann constant, and $T$ is the temperature.

Experiments carried out by Lehmann (2002) and Zhang (2001) confirmed that the total current density ($J_{PSL}$) for the PSL peak increases with increasing HF concentration. Research has also shown that for current densities above such a value, the reaction is under ionic mass transfer control, which leads to a smoothing of the Si surface (electropolishing). However, as it was shown by Lehmann and Föll (1990), in some cases, in particular for *n*-type Si etched in solution with a relatively small HF concentration (e.g., 5%) under illumination from the backside, for total current densities much smaller than $J_{PSL}$ so-called Si-macropores might be found instead of the usually desired micro- or mesopores. This means that if $J < J_{PSL}$, pores of all kinds can be obtained in Si. However, the pore size and porosity obtainable will not be just a function of $J/J_{PSL}$ but also of $J_{PSL}$ and $J$. It was found that $J$ generally determines the rate of pore growth, while $J/J_{PSL}$ usually controls the porosity scales (Lehmann and Föll 1990).

Some authors explain the transition between the pore formation and electropolishing mode by a change in the chemistry of the silicon dissolution reaction (Zhang et al. 1989; Lehmann and Föll 1990; Smith and Collins 1992; Hamilton 1995). At low current densities, fluoride ions and HF molecules directly remove silicon atoms. At high current density, this reaction competes with the formation of an anodic oxide, which is responsible for the electropolishing (Hamilton 1995). As was shown by Lehmann and Föll (1990), the formation of a homogeneous array of pores by electrochemical etching requires one to have the condition $J_{tip} = J_{PSL}$, where $J_{tip}$ is the current density at the pore tip. In other words, the pore tip is being electropolished during the pore growth.

It is well known that the contact potential between the etching solution and the Si layer strongly depends on both the etching parameters and parameters of the used wafers, such as the nature of the dopant (*p*- or *n*-type doping) and the concentration of the dopant atoms in the Si layer. This produces large differences in the current-voltage (I-V) characteristics of the Si-solution junctions. For example, for *n*-type substrates, an I-V behavior similar to the I-V behavior of a *p*-Si solution junction is observed only under illumination because a hole supply is needed. Figure 3.14 illustrates

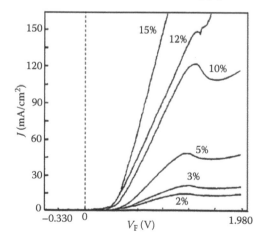

**FIGURE 3.14**    I-V curves illustrating the increase of the critical current density $J_{PSL}$ with HF concentration. (Reprinted from *Mater. Chem. Phys.*, 53, Lee, M.K. et al., 231, Copyright 1998, with permission from Elsevier.)

the above-mentioned discussions for $p$-type and $n$-type Si for various illumination intensities (only the number of photogenerated holes reaching the interface matters) and an HF concentration of about 3%. Between the first and second peak, the reaction shifts increasingly to oxidation, and around the second peak (sometimes called $J_{ox}$ peak) the surface is always completely covered by oxide and electropolishing takes over (Zhang et al. 1989; Lehmann and Föll 1990; Smith and Collins 1992; Hamilton 1995). In the electropolishing region after the $J_{ox}$ peak, a number of complex phenomena are found such as self-induced current (or voltage) oscillations; for details, one can read published articles (Carstensen et al. 1998; Grzanna et al. 2000; Foca et al. 2007a). As it will be shown later, the behavior at high current densities turns out to be useful to produce freestanding µPSi layers. Raising the current density above $J_{PSL}$ by increasing the potential at the end of a micro- or mesopore etching process results in a detachment of these PSi films from the Si substrate.

It is important that the shape of I-V characteristics shown in Figure 3.14 is typical but not mandatory for all types of Si-solution junctions (Smith and Collins 1992). For some cases, for example, for heavily doped Si or for all Si samples in almost all organic electrolytes, the I-V characteristics do not show a PSL peak, meaning that all stated above is not directly applied (Kleimann et al. 2000). The differences from the behavior mentioned above of the I-V characteristics was ascribed by Kleimann et al. (2000) to the use of a water–ethanol solvent mixture, which changed the concentration of fluoride ions as compared with the case of a water solvent. According to Kleimann et al. (2000), these changes influence the kinetics of charge transfer at the interface. It is known that HF is a weak acid, whose dissociation rate depends on the solvent. More basic electrochemical features of the I-V relationship during electrochemical etching can be found in published articles (Beale et al. 1985a,b; Föll 1991; Smith and Collins 1992; Pavesi et al. 1997).

## 3.7 GENERAL GUIDELINES FOR THE IMPROVEMENT OF HOMOGENEITY AND REPRODUCIBILITY OF THE PSi PARAMETERS

Based on their own experience, Föll et al. (2002) and Langner et al. (2011) have formulated the following recommendations, whose implementation could promote the improvement of homogeneity and reproducibility of the porous silicon parameters:

- A key issue is the addition of small amounts of surfactants to increase the "wetability" or to decrease surface tension effects, respectively. Such common surfactants as "dish water detergents," however, are not necessarily active in the acidic environment. The semiconductor industry routinely uses surfactants in its HF baths, which are commercially available (e.g., by Merck) and should be used. Exceptions are aqueous electrolytes diluted with ethanol, which acts as a wetting agent, and some organic electrolytes.
- Electrolyte flow is crucial for many experiments; it should be homogeneous and steady. Uniform pores will only be formed if the supply of reactants and the removal of products are homogeneous and stable. Circulation and/or movement of the electrolyte is mandatory and the flow pattern across the wafer must be either uniform in general or uniform on a time average. This is not easily achieved with the commonly used peristaltic or membrane pumps. In some cases, the results of etching experiments depend sensitively on the pumping speed, which then becomes a parameter to be controlled.
- The chemical reactions always release gases—mostly $H_2$—and for large wafers large amounts are being produced, which may lead to violent bubbling. An efficient means for safe removal of the gases is necessary. Besides, hydrogen bubbles may cling to the surface of the sample producing inactive spots and noise in the controlled parameter (usually the voltage or the light intensity). Bubbles must be avoided by using stirring, pumping, and "sharp" seals at the edge of the sample. In addition, some surfactant should be added in the electrolyte.
- All electrolytes dissolve some $O_2$ from the air, which participates in the dissolution process. Since the oxidizing power of electrolyte is one of its crucial characteristics, displacing dissolved $O_2$ by bubbling of the electrolyte with $N_2$ before use is often necessary for reproducible results.
- If illumination is used, three additional problems arise: homogeneous and controllable illumination with a high intensity is needed, the backside contact must be at the edge of

the sample for backside illumination, and the electrolytic cell must have a light transparent window for frontside illumination. The light source must be controllable because the current in an electrochemical cell depends on the light intensity. Arrays of light-emitting diodes proved to be the best choice.

■ All materials in contact with the electrolyte must be chemically inert in this environment. For some electrolytes, this may limit the choice of materials to Teflon and its derivatives. Transparent windows, heat exchangers, and so on are problematic and should be designed with care. Expensive sapphire windows and more Teflon have to be used.

■ While the electrolyte is primarily chosen from pore etching considerations, it is advisable to pay some attention to its conductivity and its pH value. The conductivity may be improved by adding salts to the solution, for example, $NH_4Cl$ to aqueous electrolytes with low HF concentration or tetrabutylammonium-perclorate for MeCN, but for many organic electrolytes, no suitable salts are known. Conducting salts, of course, could also influence the etching in unpredictable ways. The pH value is mostly determined by the HF concentration, but it can be varied to some extent by the addition of acetic acid or by other means.

■ The use of Si wafers with the set parameters (doping level, *n*- or *p*-type, crystallographic orientation of the Si crystal, etc.) is a prerequisite because inconstancy of the parameters of silicon, used in experiments, is one of the reasons causing nonreproducibility of obtained results.

■ Every task requires the use of silicon with certain parameters. For example, during the formation of macroporous silicon in lowly doped material, large interpore distances and pore diameters can be realized, for example, etching of pore diameters up to 100 μm were reported. While for small interpore distances, highly doped substrates are required.

■ The backside contact is very critical. If the backside illumination is employed, it can be supplied by a contact ring around the perimeter of the wafer, by contact needles (necessarily damaging the Si), or by a light transparent electrolyte or indium tin oxide (ITO) contact. Since the less aggressive electrolyte can be used for the contact, some of the problems encountered on the front side might be less severe for the backside contact in this case.

■ While the current density is not very large (some 10 mA/cm²), the total current easily reaches 10 A. If contact problems exist, this may generate enough local heat to soften the polymers used for the cell, leading to leakage and serious safety problems.

■ For many applications, especially if very perfect and deep pores are to be etched, temperature control is essential. For example, one can find that to avoid inhomogeneities during etching, the temperature has to be kept constant within 0.1°C. In addition, not only is it necessary to keep the temperature constant, but also room temperature is not always the best choice. A heat exchanger system, allowing temperature control between about 0°C and 40°C, should be a part of an etching apparatus. Large pumping and heat exchange equipment is required for removing the heat produced by ohmic losses.

■ The consumption of HF-molecules can be neglected as long as the volume of the electrolyte is large compared to the dissolved volume of silicon. However, things are different at the micrometer scale inside the pores. While a pore is growing, the exchange of reactants with the HF-basin at the front side of the sample becomes more and more affected by diffusion process. As a result, a concentration gradient establishes between the pore bottom and the opening of the pore at the sample surface. Consequently, with a lower concentration, the critical current density $J_{PSL}$ diminishes. This means that the etching speed slows down, porosity increases, the pores grow larger in diameter, and, finally—for $J > J_{PSL}$—electropolishing sets in. To avoid this situation, a correction parameter should be introduced for the illumination intensity independence on the pore depth. This approach is especially important during macroporous silicon forming. In that way, a uniform pore pattern with aspect ratios of several hundreds to one can be obtained.

■ For making prestructured *n*-macro-pores, usually an oxide mask is being used. Pore etching commences at the openings of the mask and by the time the mask has dissolved, the pores are deep enough for stable growth. Etching random pores, or pores through $Si_3N_4$ masks, may require an optimized nucleation step before the conditions for stable pore growth would be established.

■ While polished surfaces are not always required, it would be a good idea to thoroughly clean the wafers if homogeneous random nucleation is desired.

## ACKNOWLEDGMENTS

This work was supported by the Ministry of Science, ICT and Future Planning (MSIP) of the Republic of Korea, and partly by the National Research Foundation (NRF) grants funded by the Korean government (No. 2011-0028736 and No. 2013-K000315).

## REFERENCES

Astrova, E.V. and Vasunkina, T.N. (2002). Formation of macropore nucleation centers in silicon by ion implantation. *Semiconductors,* **36**(5), 564–567.

Barotto, C., Faglia, G., Comini, E. et al. (2001). A novel porous silicon sensor for detection of sub-ppm $NO_2$ concentrations. *Sens. Actuators B,* **77**, 62–66.

Bartels, O., Splinter, A., Storm, U., and Binder, J. (1999). Thick porous silicon layers as sacrificial material for low-power gas sensors. *Proc. SPIE,* **3892**, 184–191.

Beale, M.I.J., Benjamin, J.D., Uren, M.J., Chew, N.G., and Cullis, A.G. (1985a). An experimental and theoretical study of the formation and microstructure of porous silicon. *J. Crystal Growth,* **73**, 622–636.

Beale, M.I.J., Chew, N.G., Uren, M.J., Cullis, A.G., and Benjamin, J.D. (1985b). Microstructure and formation mechanism of porous silicon. *Appl. Phys. Lett.,* **46**(1), 86–88.

Bell, T.E., Gennissen, P.T.J., DeMunter, D., and Kuhl, M. (1996). Porous silicon as a sacrificial material. *J. Micromech. Microeng.,* **6**, 361–369.

Bettotti, P., Dal Negro, L., Gaburro, Z. et al. (2002). P-type macroporous silicon for two-dimensional photonic crystals. *J. Appl. Phys.,* **92**, 6966–6972.

Bisi, O., Ossicini, S., and Pavesi, L. (2000). Porous silicon: A quantum sponge structure for silicon based optoelectronics. *Surf. Sci. Rep.,* **38**, 1–126.

Boehringer, M., Artmann, H., and Witt, K. (2012). Porous silicon in a semiconductor manufacturing environment. *J. Microelectromech. Syst.,* **21**(6), 1375–1381.

Bomchil, G., Halimaoui, A., Sagnes, I. et al. (1993). Porous silicon: Material properties, visible photo and electroluminescence. *Appl. Surf. Sci.,* **65/66**, 394–407.

Canham, L.T., Cox, T.I., Loni, A., and Simons, A.J. (1996). Progress towards silicon optoelectronics using porous silicon technology. *Appl. Surf. Sci.,* **102**, 436–441.

Carstensen, J., Prange, R., Popkirov, G.S., and Föll, H. (1998). A model for current oscillations in the Si-HF system based on a quantitative analysis of current transients. *Appl. Phys. A,* **67**, 459–467.

Carstensen, J., Christophersen, M., Lölkes, S. et al. (2005). Large area etching for porous semiconductors. *Phys. Stat. Sol. (c),* **2**(9), 3339–3343.

Christophersen, M., Carstensen, J., and Föll, H. (2000a). Macropore formation on highly doped *n*-type silicon. *Phys. Stat. Sol. (a),* **182**, 45–50.

Christophersen, M., Carstensen, J., Feuerhake, A., and Föll, H. (2000b). Crystal orientation and electrolyte dependence for macropore nucleation and stable growth on *p*-type Si. *Mater. Sci. Eng. B,* **69–70**, 194–198.

Christophersen, M., Carstensen, J., Voigt, K., and Föll, H. (2003). Organic and aqueous electrolytes used for etching macro- and mesoporous silicon. *Phys. Stat. Sol. (a),* **197**(1), 34–38.

Connolly, E.J., O'Halloran, G.M., Pham, H.T.M., Sarro, P.M., and French, P.J. (2002). Comparison of porous silicon, porous polysilicon and porous silicon carbide as materials for humidity sensing applications. *Sens. Actuators B,* **99**, 25–30.

Cullis, A.G., Canham, L.T., and Calcott, P.D.G. (1997). The structural and luminescence properties of porous silicon. *J. Appl. Phys.,* **82**(3), 909–965.

Dantas, M.O.S., Galeazzo, E., Peres, H.E.M., and Ramirez-Fernandez, F.J. (2003). Silicon micromechanical structures fabricated by electrochemical process. *IEEE Sensors J.,* **3**(6), 722–727.

Dantas, M.O.S., Galeazzo, E., Peres, H.E.M., Ramirez-Fernandez, F.J., and Errachid, A. (2004). HI–PS technique for MEMS fabrication. *Sens. Actuators A,* **115**, 608–616.

D'Arrigo, G., Coffa, S., and Spinella, C. (2002). Advanced micromachining processes for micro-opto-electromechanical components and devices. *Sens. Actuators A,* **99**(1), 112–118.

Defforge, T., Capelle, M., Tran-Van, F., and Gautier, G. (2012). Plasma-deposited fluoropolymer film mask for local porous silicon formation. *Nanoscale Res. Lett.,* **7**(1), 344.

Di Francia, G., De Filippo, F., La Ferrara, V. et al. (1998). Porous silicon layers for the detection at RT of low concentrations of vapoura from organic compounds. In: *Proceeding of European Conference Eurosensors XII,* September 13–16, 1998, Southampton, UK. IOP, Bristol, vol. 1, pp. 544–547.

Djenizian, T., Santinacci, L., Hildebrand, H., and Schmuki, P. (2003). Electron beam induced carbon deposition used as a negative resist for selective porous silicon formation. *Surf. Sci.,* **524**, 40–48.

Eijkel, C.J.M., Branebjerg, J., Elwenspoek, M., and van de Pol, F.C.M.A. (1990). New technology for micromachining of silicon: Dopant selective HF anodic etching for the realization of low-doped monocrystalline silicon structures. *IEEE Electron Device Lett.,* **11**, 588–589.

Esashi, M., Komatsu, H., Matsuo, T. et al. (1982). Fabrication of catheter-tip and sidewall miniature pressure sensor. *IEEE Trans. Electron. Dev.,* **ED-29**, 57–64.

Foca, E., Carstensen, J., and Föll, H. (2007a). Modelling electrochemical current and potential oscillations at the Si electrode. *J. Electroanal. Chem.,* **603**, 175–202.

Foca, E., Carstensen, J., Leisner, M., Ossei-Wusu, E., Riemenschneider, O., and Föll, H. (2007b). Smoothening the pores walls in macroporous *n*-Si. *ECS Transactions,* **6**(2), 367–374.

Föll, H. (1991). Properties of silicon-electrolyte junctions and their application to silicon characterization. *Appl. Phys. A,* **53**, 8–19.

Föll, H., Christophersen, M., Carstensen, J., and Hasse, G. (2002). Formation and application of porous silicon. *Mater. Sci. Eng. R,* **39**, 93–141.

Föll, H., Carstensen, J., and Frey, S. (2006). Porous and nanoporous semiconductors and emerging applications. *J. Nanomater.,* **2006**, 91635 (1–10).

French, P., Gennissen, P., and Sarro, P. (1997). Epi-micromachining. *Microelectronics J.,* **28**(4), 449–464.

Frey, S., Keipert, S., Chazalviel, J.-N., Ozanam, F., Carstensen, J., and Föll, H. (2007). Electrochemical formation of porous silica: Toward an understanding of the mechanisms. *Phys. Stat. Sol. (a),* **204**(5), 1250–1254.

Galeazzo, E., Salcedo, W.J., Peres, H.E.M., and Ramirez-Fernandez, F.J. (2001). Porous silicon patterned by hydrogen ion implantation. *Sens. Actuators B,* **76**, 343–346.

Gennissen, P., French, P., De Munter, D., Bell, T., Kaneko, H., and Sarro, P. (1995). Porous silicon micromachining techniques for accelerometer fabrication. In: *Proceedings of the 25th European Research Conference on Solid State Device, ESSDERC'95,* pp. 593–596.

Grigoras, K., Niskanen, A.J., and Franssila, S. (2001). Plasma etched initial pits for electrochemically etched macroporous silicon structures. *J. Micromech. Microeng.,* **11**, 371–375.

Grzanna, J., Jungblut, H., and Lewerenz, H.J. (2000). Model for electrochemical oscillations at the Si/electrolyte contact: Part II. Simulations and experimental results. *J. Electroanal. Chem.,* **486**(2), 190–203.

Gupta, A., Jain, V.K., Jalwania, C.R. et al. (1995). Technologies for porous silicon devices. *Semicond. Sci. Technol.,* **10**, 698–702.

Halimaoui, A. (1995). Porous silicon: Material processing, properties and applications. In: Vial, J.C. and Derrien, J. (eds.). *Porous Silicon Science and Technology,* Springer-Verlag, Berlin, pp. 33–52.

Halimaoui, A. (1997). Porous silicon formation by anodisation. In: Canham, L.T. (ed.) *Properties of Porous Silicon.* IEE INSPEC, The Institution of Electrical Engineers, London, pp. 12–23.

Hamilton, B. (1995). Porous silicon. *Semicond. Sci. Technol.,* **10**, 1187–1207.

Hedrich, F., Billat, S., and Lang, W. (2000). Structuring of membrane sensors using sacrificial porous silicon. *Sens. Actuators B,* **84**, 315–323.

Hejjo Al Rifai, M., Christophersen, M., Ottow, S., Carstensen, J., and Föll, H. (2000). Dependence of macropore formation in n-Si on potential, temperature, and doping. *J. Electrochem. Soc.,* **147**(2), 627–635.

Holec, H., Chvojka, T., Jelinek, I. et al. (2002). Determination of sensoric parameters of porous silicon in sensing of organic vapours. *Mater. Sci. Eng. C,* **19**, 251–254.

Hossain, S.M., Das, J., Chakraborty, S., Dutta, S.K., and Saha, H. (2002). Electrode design and planar uniformity of anodically etched large area porous silicon. *Semicond. Sci. Technol.,* **17**, 55–59.

Kan, P.Y.Y. and Finstad, T.G. (2005). Oxidation of macroporous silicon for thick thermal insulation. *Mater. Sci. Eng. B,* **118**, 289–292.

Kim, H.-S., Chong, K., and Xie, Y.-H. (2003). Study of the cross-sectional profile in selective formation of porous silicon. *Appl. Phys. Lett.,* **83**(13), 2710–2712.

Kleimann, P., Linnros, J., and Petersson, S. (2000). Formation of wide and deep pores in silicon by electrochemical etching. *Mater. Sci. Eng. B,* **69–70**, 29–33.

Kochergin, V.R. and Foell, H. (2006). Novel optical elements made from porous Si. *Mater. Sci. Eng. R,* **52**, 93–140.

Korotcenkov, G. and Cho, B.K. (2010). Silicon porosification: State of the art. *Crit. Rev. Sol. St. Mater. Sci.,* **35**(3), 153–260.

Koyama, H. (2006). Strong photoluminescence anisotropy in porous silicon layers prepared by polarized-light assisted anodization. *Solid State Commun.,* **138**, 567–570.

Lai, M., Parish, G., Dell, J., Liu, Y., and Keating, A. (2011). Chemical resistance of porous silicon: Photolithographic applications. *Phys. Stat. Sol. (c),* **8**(6), 1847–1850.

Lammel, G. and Renaud, P. (2000). Free-standing, mobile 3D porous silicon microstructures. *Sens. Actuators A,* **85**, 356–360.

Lang, W. (1996). Silicon microstructuring technology. *Mater. Sci. Eng. R,* **17**, 1–54.

Langner, A., Müller, F., and Gösele, U. (2011). Macroporous silicon. In: Hayden, O. and Nielsch, K. (eds.). *Molecular- and Nano-Tubes.* Springer Science+Business Media, New York, pp. 431–460.

Lee, M.-K. and Tu, H.-F. (2007). Stabilizing light emission of porous silicon by *in-situ* treatment. *Jpn. J. Appl. Phys.,* **46**, 2901–2903.

Lee, M.K., Chu, C.H., and Tseng, Y.C. (1998). Mechanism of porous silicon formation. *Mater. Chem. Phys.,* **53**, 231–234.

Lehmann, V. (1993). The physics of macropore formation in low doped *n*-type silicon. *J Electrochem. Soc.,* **140**(10), 2836–2843.

Lehmann, V. (2002). *Electrochemistry of Silicon*. Wiley-VCH, Weinheim, Germany.

Lehmann, V. and Föll, H. (1990). Formation mechanism and properties of electrochemically etched trenches in *n*-type silicon. *J. Electrochem. Soc.,* **137**, 653–659.

Lehmann, V. and Gruning, U. (1997). The limits of macropore array fabrication. *Thin Solid Films,* **297**, 13–17.

Lehmann, V., Stengl, R., and Luigart, A. (2000). On the morphology and the electrochemical formation mechanism of mesoporous silicon. *Mater. Sci. Eng. B,* **69–70**, 11–22.

Leisner, M., Carstensen, J., and Föll, H. (2008). FFT impedance spectroscopy analysis of the growth of anodic oxides on (100) p-Si for various solvents. *J. Electroanal. Chem.,* **615**, 124–134.

Lysenko, V., Perichon, S., Remaki, B., and Barbier, D. (2002). Thermal isolation in microsystems with porous silicon. *Sens. Actuators A,* **99**, 13–24.

Mangaiyarkarasi, D., Breese, M.B.H., Sheng, O.Y., Ansari, K., Vijila, C., and Blackwood, D. (2006). Porous silicon based Bragg reflectors and Fabry-Perot interference filters for photonic applications. In: Kubby, J.A. and Reed, G.T. (eds.). *Silicon Photonics. Proc. SPIE,* 6125, 61250X (1–8).

Mescheder, U., Kovacs, A., Kronast, W., Bársony, I., Ádám, M., and Ducso, C. (2001). Porous silicon as multifunctional material in MEMs. In: *Proceedings of the 1st IEEE Conference on IEEE Nanotechnology,* IEEE-NANO 2001, pp. 483–488.

Murate, M., Iwata, H., and Itoigawa, K. (1998). Micromachining applications of sellective anodization. In: *Proceedings International Symposium on Micromechatronics and Human Science,* November 25–28, Nagoya, Japan, pp. 57–63.

Nassiopoulos, A., Grigoropoulos, S., Canham, L. et al. (1995). Sub-micrometre luminescent porous silicon structures using lithographically patterned substrates. *Thin Solid Films,* **255**, 329–333.

Ohji, H., French, P.J., Izuo, S., and Tsutsumi, K. (2000). Initial pits formation for electrochemical etching in hydrofluoric acid. *Sens. Actuators,* **85**, 390–394.

Pagonis, D., Kaltsas, G., and Nassiopoulou, A.G. (2003). Implantation masking technology for selective porous silicon formation. *Phys. Stat. Sol. (a),* **197**(1), 241–245.

Pankove, J.I. and Johnson, N.M. (eds.). (1991). *Hydrogen in Semiconductors*. Vol. 34: *Hydrogen in Silicon*. Academic Press, New York.

Parkhutik, V. (1999). Porous silicon-mechanisms of growth and applications. *Sol. St. El.,* **43**, 1121–1141.

Patel, P.N., Mishra, V., and Panchal, A.K. (2012). Theoretical and experimental study of nanoporous silicon photonic microcavity optical sensor devices. *Adv. Nat. Sci: Nanosci. Nanotechnol.,* **3**, 035016 (1–7).

Pavesi, L. (1997). Porous silicon dielectric multilayers and microcavities. *La Rivista del Nuovo Cimento,* **20**, 1–76.

Peres, H.E.M. and Ramirez-Fernandez, F.J. (1997). High resistivity silicon layers obtained by hydrogen ion implantation. *Braz. J. Physics A,* **27**(4), 237–239.

Prochaska, A., Mitchell, S.J.N., and Gamble, H.S. (2002). Porous silicon as a sacrificial layer in production of siliocn diaphragms by precision grinding. In: Tay, F.E.H. (ed.), *Materials and Process Integration for MEMS*. Kluwer Academic Publisher, the Netherlands.

Propst, E.K. and Kohl, P.A. (1994). Electrochemical oxidation of silicon and formation of porous silicon in acetonitrile. *J. Electrochem. Soc.,* **141**, 1006–1013.

Punzon-Quijorna, E., Torres-Costa, V., Manso-Silvan, M., Martin-Palma, R.J., and Climent-Font, A. (2012). MeV Si ion beam implantation as an effective patterning tool for the localized formation of porous silicon. *Nucl. Instrum. Meth. Phys. Res. B,* **282**, 25–28.

Racine, G.A., Genolet, G., Clerc, P.A., Despont, M., Vettiger, P., and De Rooij, N.F. (1997). Porous silicon sacrificial layer technique for the fabrication of free standing membrane resonators and cantilever arrays. In: *CD Proceeding of the 11th European Conference on Solid State Transducers Eurosensors XI,* September 21–24, 1997, Warsaw, Poland, Vol. 1, pp. 285–288.

Sailor, M.J. (2012). *Porous Silicon in Practice: Preparation, Characterization and Applications*. Wiley-VCH, Weinheim.

Setzu, S., Letant, S., Solsona, P., Romestain, R., and Vial, J.C. (1999). Improvement of the luminescence in *p*-type as-prepared or dye impregnated porous silicon microcavities. *J. Lumin.,* **80**, 129–132.

Sharma, S.N., Sharma, R.K., and Lakshmikumar, S.T. (2005). Role of an electrolyte and substrate on the stability of porous silicon. *Physica E,* **28**, 264–272.

Shor, J. and Kurtz, A. (1994). Photoelectrochemical etching of 6 H-SiC. *J. Electrochem. Soc.,* **141**, 778–781.

Smith, R.L. and Collins, S.D. (1992). Porous silicon formation mechanisms. *J. Appl. Phys.,* **71**(8), R1–R22.

Sotgiu, G., Schirone, L., and Rallo, F. (1996). Effect of surfactants in the electrochemical preparation of porous silicon. *IL Nuovo Cimento,* **18**D(10), 1179–1186.

Splinter, A., Stürmann, J., and Benecke, W. (2000). New porous silicon formation technology using internal current generation with galvanic elements. In: *CD Proceeding of the 13th European Conference on Solid-State Transducers, EUROSENSORS XIIV,* August 27–30, 2000, Copenhagen, Denmark, Abstract T2P03, pp. 423–426.

Splinter, A., Bartels, O., and Benecke, W. (2001). Thick porous silicon formation using implanted mask technology. *Sens. Actuators B,* **76**, 354–360.

Steiner, P. and Lang, W. (1995). Micromachining applications of porous silicon. *Thin Solid Films,* **255**(1), 52–58.

Van den Meerakker, J.E.A.M., Elfrink, R.J.C., Roozeboom, F., and Verhoeven, J.F.C.M. (2000). Etching of deep macropores in 6 in. Si wafers. *J. Electrochem. Soc.,* **147**(7), 2757–2761.

Veeramachaneni, B., Winans, J.D., Hu, S. et al. (2011). A novel technique for localized formation of SOI active regions. *Phys. Stat. Sol. (c),* **8**(6), 1865–1868.

Yamamoto, N. and Takai, H. (1999). Blue luminescence from photochemically etched silicon. *Jpn. J. Appl. Phys.,* **38**, 5706–5709.

Yamani, Z., Thompson, W.H., AbuHassan, L., and Nayfeh, M.H. (1997). Ideal anodization of silicon. *Appl. Phys. Lett.,* **70**, 3404.

Zhang, X.G. (2001). *Electrochemistry of Silicon and Its Oxide.* Kluwer Academic/Plenum, New York.

Zhang, X.G., Collins, S.D., and Smith, R.L. (1989). Porous silicon formation and electrochemical polishing of silicon by anodic polarization in HF solutions. *J. Electrochem. Soc.,* **136**, 1561–1565.

Zhong, F., Lv, X.-Y., Jia, Z.-H., and Mo, J. (2012). Fabrication of porous silicon-based silicon-on-insulator photonic crystal by electrochemical etching method. *Opt. Eng.,* **51**(4), 040502 (1–3).

# Silicon Porosification

## Approaches to PSi Parameters Control

4

**Ghenadii Korotcenkov**

## CONTENTS

The previous chapter discussed general issues of the technology of Si porosification. However, the analysis conducted in that chapter gave only a general idea about the technology and did not involve issues relating to the control of the parameters of porous layers formed. At the same time, obtaining a porous silicon with certain parameters such as the thickness, porosity, pore size, and distribution in size and thickness is the determining factor for the development of real devices with required performances. Exactly these parameters determine performance of the developed PSi-based devices. Therefore, the present chapter focuses on the consideration of approaches that allow controlling the above-mentioned parameters of porous silicon.

However, we have to note that most studies carried out in the field of silicon porosification usually have been focused on the investigation of one or two preparative parameters' influence on the structures of PSi formed, while the preparative parameters cannot be treated as independent variables because of the strong correlation between them (Salonen et al. 2000). This means that the results of different authors can hardly be compared with each other because of differences in preparation. Therefore, we hope that the reader will be treated with understanding to the fact that his or her results may not coincide with the results presented in this chapter.

## 4.1 THE METHODS OF PSi PARAMETERS CONTROL

For quantitative characterization of porous films, a parameter such as porosity is usually used. However, the porosity is a macroscopic parameter that helps in discussing trends but which does not give microscopic information on the morphology of the PSi layers. A better knowledge of the PSi structure can be achieved if the pore shape, depth, size, and topological distribution are determined. However, no complete understanding of the mechanisms that determine PSi morphology exists because the Si porosification is a very complicated process, which depends on a great number of parameters involved. Nevertheless, some general trends can be derived for different types of starting Si substrates. In particular, in Table 4.1, one can find the information summarized concerning those peculiarities (Bisi et al. 2000).

Regarding detailed analysis of the influence of electrochemical etching conditions on porous semiconductor parameters such as the pore diameter, the pore depth, and the pore growth direction, one can find the answer to those questions in the following sections, as well as in several published articles (Bomchil et al. 1989; Canham 1997; Lehmann and Gruning 1997; Pavesi 1997; Christophersen et al. 2001; Korotcenkov and Cho 2010a,b).

### 4.1.1 INFLUENCE OF THE TYPE OF CONDUCTIVITY

As was shown in the previous chapters and sections, the type of silicon conductivity without a doubt relates to the parameters that control both the conditions of formation and parameters of porous layers. In particular, the need for illumination during silicon anodization or possibility to conduct silicon porosification without lighting is directly dependent on the type of Si conductivity.

**TABLE 4.1    Effect of Anodization Parameters on PSi Formation**

| An Increase of | Porosity | Etching Rate | Critical Current |
|---|---|---|---|
| HF concentration | Decreases | Decreases | Increases |
| Current density | Increases | Increases | – |
| Anodization time | Increases | Almost constant | – |
| Temperature | – | – | Increases |
| Wafer doping (p-type) | Decrease | Increases | Increases |
| Wafer doping (n-type) | Increases | Increases | |

*Source:*  Reprinted from *Surf. Sci. Rep.* 38, Bisi O. et al., 1, Copyright 2000, with permission from Elsevier.

The porosity of the films, pore size, morphology, and other parameters of PSi are also strongly dependent on the type of conductivity (Lehmann and Ronnebeck 1999). However, despite this, in this section we will not dwell on a detailed analysis of this parameter's influence on the features of porous layers formation. This information can be found in the following sections.

## 4.1.2 THE INFLUENCE OF ELECTROLYTE COMPOSITION

As it was established in numerous studies (Bomchil et al. 1989; Splinter et al. 2001; Föll et al. 2006), there is a strong effect of the electrolyte compositions on porous silicon formation (morphology, thickness, porosity) in the anodization procedure. Reviewing the results presented in Figures 4.1 and 4.2, one can see that a change of the HF concentration in the solution is accompanied by a strong change of both the pore diameter (see Figure 4.2) and the film porosity (see Figure 4.2). The increase in the HF concentration results in the decrease of the porosity and pore diameter (Zhang 1991). This means that in order to obtain micropore materials, electrolytes with high HF concentration (often just 49% HF mixed with an equal volume of ethanol) need to be used. In this case, microporous layers grow rather rapidly and uniformly (Lehmann 1993; Van den Meerakker et al. 2000). For preparing macroporous materials, the electrolytes with low HF concentration are more suitable.

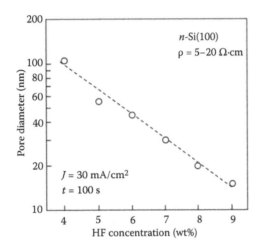

**FIGURE 4.1** Diameter of mesopores on *n*-Si (100) versus HF concentration. (From Föll H. et al., *J. Nanomater.* 2006, 91635, 2006. Published by Hindawi Publishing Corporation as open access.)

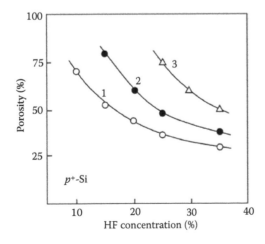

**FIGURE 4.2** Porosity in heavily doped *p*-type mesoporous silicon as a function of HF concentration at (1) *J* = 20, (2) *J* = 80, and (3) *J* = 240 mA/cm². (Reprinted from *Appl. Surf. Sci.* 41/42, Bomchil G. et al., 604, Copyright 1989, with permission Elsevier.)

The same effect was observed by Kordas et al. (2001) and by Nakagawa et al. (1998) (see Figure 4.3). They have shown that in the HF:C₂H₅OH:H₂O system, the porosity was increasing with reduced HF concentration and increased EtOH concentration.

It is clear that such changes in PSi parameters in the absence of proper control of electrolyte composition could lead to catastrophic changes of the parameters of devices whose design is based on such porous films. Examining the results of the influence of porosification on the response of the humidity sensors whose design is based on porous silicon and SiC, presented in Figure 4.4, one can make a conclusion about the importance of monitoring the parameters of the electrolyte used.

However, one should note that the last regularity regarding film porosity is not obligatory for all technological routs. In other works, we can find other regularities. For example, Kan and Finstad (2005) found that in the solution HF-dimethyl sulfoxide the growth of HF concentration is accompanied by the growth of porosity of the formed layer (see Figure 4.5).

Another research has shown that the porosity has a clearly defined maximum in the dependence of the porosity on the HF concentration in the electrolyte used (see Table 4.2).

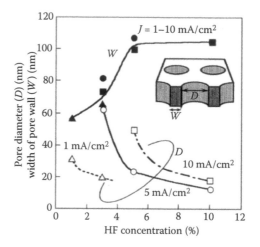

**FIGURE 4.3**  The pore diameter, $D$, and the width of $n^+$-Si wall, $W$, as a function of HF concentration for several samples with various anodization current densities ($\rho = 0.01–003$ $\Omega\cdot$cm, $T = 0°$C). Also shown in the inset is a schematic illustration of the sample structure. (From Nakagawa T. et al., *Jpn. J. Appl. Phys.* 37, 7186, 1998. Copyright 1998: The Japan Society of Applied Physics. With permission.)

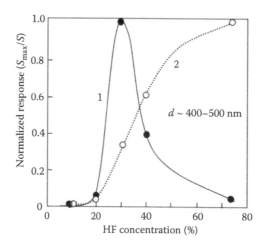

**FIGURE 4.4**  The influence of HF concentration in electrolytes for porosification on the maximum sensitivity of (1) polysilicon and (2) SiC ($d \sim 400–500$ nm) to humidity; the sensitivity was measured for a relative humidity (RH) change from 10 to 90% RH. (Reprinted from *Sens. Actuators B* 99, Connolly E.J. et al., 25, Copyright 2002, with permission from Elsevier.)

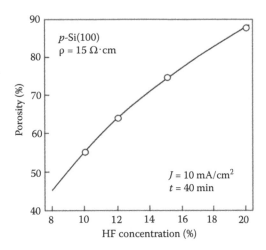

**FIGURE 4.5**  The porosity versus HF concentration (diluted with DMSO) for etching a *p*-type Si(100) wafer with a resistivity of 15 Ω·cm. The etching time was 40 min with the current density of 10 mA/cm². (Reprinted from *Mater. Sci. Eng. B* 118, Kan P.Y.Y. and Finstad T.G., 289, Copyright 2005, with permission from Elsevier.)

---

**TABLE 4.2    Porosity of *p*-Type Poly-Silicon Samples Prepared in HF:EtOH Electrolyte with Different Values of HF Concentrations**

| HF (%) | 8 | 10 | 11.5 | 13 | 14.5 | 16.5 | 17.5 | 19.5 |
|---|---|---|---|---|---|---|---|---|
| Porosity (%) | 1 | 23 | 65 | 73 | 62 | 58 | 55 | 35 |

*Source:*  Reprinted from *Sens. Actuators B* 100, Iraji zad A. et al., 341, Copyright 2004, with permission from Elsevier.

*Note:*  ρ = 0.4–2 Ω·cm, *J* = 32 mA/cm², *t* = 5 min.

---

Such divergence of the results obtained by various teams testifies that porosification, and especially the influence of the solution's composition on that process, requires more intensive study.

Another parameter of PSi, which is controlled by the HF concentration, is the maximum obtainable pore depth. This parameter is especially important for macroporous semiconductors. Lehmann and Gruning (1997) concluded that the maximum obtainable pore depth is limited by the process of degeneration. The degeneration of pore shape is most probably due to the minority carriers, penetrating the region between the pores, which leads to enhanced etching of the pore walls. However, the described effect is not understood in detail yet. The dependence of the pore degeneration depth on the HF concentration in the electrolyte is shown in Figure 4.6.

As follows from Figure 4.6, the maximum obtainable pore depth can be the same order of magnitude as the wafer thickness (300–500 μm); however, this depth requires low electrolyte concentration, low temperature, and etching times as long as several days (Lehmann and Gruning 1997).

At the same time, it was established that the thickness of the porous layer grown during a fixed time was increased for solutions with higher HF concentration and decreased for solutions with higher concentration of EtOH. For example, it was found by Lehmann (1996a) that the growth rate of macro-pores in low doped *n*-type silicon increased from 0.1 to 5 μm/min simply by increasing the HF concentration from 1% to 10%. The same effect was observed for mesopores Si by Föll et al. (2006). This means that the increased HF concentration allows faster pore growth at a nominally constant porosity for thicker porous layers in a given time (see Figure 4.7).

As it was established by Van den Meerakker et al. (2000), the etch rate usually linearly depends on the HF concentration (see Figure 4.8). This situation corresponds to conclusions made by Lehmann (1993). In accordance with Lehmann (1993), the diameter of the macropores was determined by the temperature, the current density, and the HF concentration, while the etch rate was only dependent on the temperature and the HF concentration.

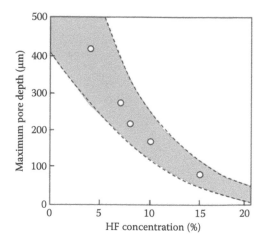

**FIGURE 4.6**    The maximum pore depth, shown by circles, which can be achieved in macroporous Si without degradation at the pore tips as a function of electrolyte concentration (with surfactant) for cylindrical pores. The shaded area shows the region of degradation for the pore geometries differing from cylindrical. (Reprinted from *Thin Solid Films* 297, Lehmann V. and Gruning U., 13, Copyright 1997, with permission from Elsevier.)

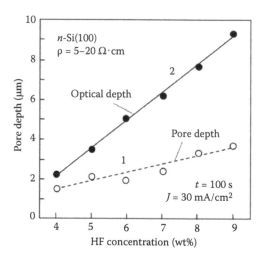

**FIGURE 4.7**    Depth of mesopores on *n*-Si(100). The "optical thickness" (ellipsometer) differs from the geometric thickness because the pore density decreases, while the refractive index increases with increasing HF concentration. (From Föll H. et al., *J. Nanomater.* 2006, 91635, 2006. Published by Hindawi Publishing Corporation as open access.)

The same dependence of etch rate on HF concentration in HF-$H_2O$ and (HF-HCl)-$H_2O$ solutions was determined by Koker and Kolasinski (2000). As the concentration of pure HF(aq) was increased, the etch rate increased, as it is shown in Figure 4.9. The increase in rate was linear over a wide range of molalities.

Experiment has shown that the peak current density $J_{PSL}$ is also strongly dependent on the concentration of electrolyte (see Figure 4.10). Van den Meerakker et al. (2000) established that in temperature range 15–50°C at every temperature a linear dependence of $J_{PSL}$ on HF concentration was observed for electrolyte containing water, ethanol, and HF. Figure 4.10 shows results obtained at 30°C.

Therefore, the results presented testify that the optimal concentration of the HF in electrolyte depends on what kind of pores are to be etched and which parameter of the pores is to be optimized. For example, to make a complicated (and not yet fully understood) issue short: micropores (and "small" mesopores) with very high porosity are the best produced at relatively high HF

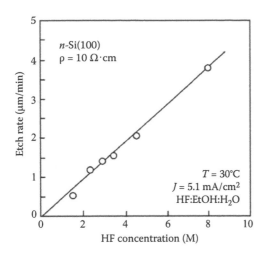

**FIGURE 4.8**  Etch rate ($R$) as a function of HF concentration ($H_2O$/EtOH = 2.8). (From Van den Meerakker J.E.A.M. et al., *J. Electrochem. Soc.* 147(7), 2757, 2000. Copyright 2000: The Electrochemical Society. With permission.)

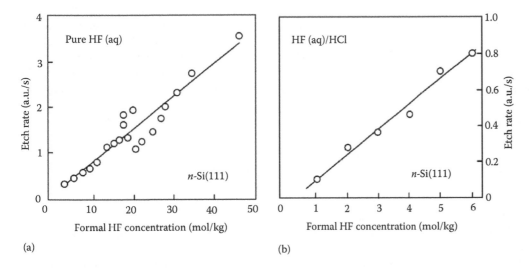

(a)                                                          (b)

**FIGURE 4.9**  Dependence of *n*-type Si(111) etch rate on HF concentration in pure (a) HF-$H_2O$ and (b) (HF-HCl)-$H_2O$ solutions under illumination 15 mW HeNe laser ($\lambda$ = 1.633 nm, $P_{illu}$ = 7 W/cm²). (Reprinted from *Mater. Sci. Eng. B* 69–70, Koker L. and Kolasinski K.W., 132, Copyright 2000, with permission from Elsevier.)

concentrations. On the other hand, pore etching at high HF concentrations usually terminates earlier, that is, at smaller thickness of the porous layer because of diffusion limitations. It means that for the low concentration the maximal length of the macropores could easily exceed 500 µm whereas at the high concentration pore growth terminates around 150 µm or earlier (Halimaoui 1997).

Regarding the influence of concentration of other electrolyte components on the PSi parameters, we do not have such detailed information as we do for HF. However, based on the results presented by Splinter et al. (2001), one can conclude that when a commercial surfactant was applied in the electrolyte instead of EtOH, the porosity could not be increased beyond 70%. However, the thickness values were analogous to the results obtained in that case when the electrolyte contained EtOH.

It was also found that the ethanol concentration is a factor of pore fineness (Splinter et al. 2001). In increasing the ethanol concentration, one achieves finer pores and a better uniformity of the porous layer. As it is known, during the formation of porous silicon, there is hydrogen evolution. When purely aqueous HF solutions are used for the PSi formation, the hydrogen bubbles stick to the surface and induce lateral and in-depth inhomogeneity. To improve the PSi layer

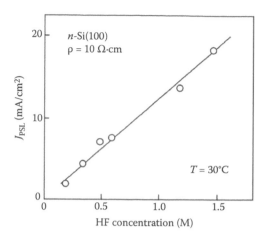

**FIGURE 4.10**   Peak current density ($J_{PSL}$) as a function of HF concentration in electrolyte H$_2$O:EtOH:HF at 30°C. (Reprinted from *Mater. Sci. Eng. B* 69–70, Koker L. and Kolasinski K.W., 132, Copyright 2000, with permission from Elsevier.)

uniformity, these bubbles must be readily eliminated. One of the most appropriate means to overcome this problem is to add a surfactant agent to the HF solution. Commercially available wetting agents, such as "Mirasol" (registered trademark of Tetanal) can be used (Lehmann and Föll 1990). In this case, only a few drops of the wetting agent are necessary for an efficient removal of the bubbles (Halimaoui 1997). However, the most widely used surfactant in the case of PSi formation is absolute ethanol. For efficient bubble elimination, the ethanol concentration should not be less than 15%. Another interesting and efficient surfactant, which can be used, is acetic acid. This surfactant allows a better control of the solution pH (Halimaoui 1997). In fact, low pH solutions can be obtained at any HF concentration. In addition, only a few percent (~5%) of acetic acid is required for efficient bubble removal.

It was also found that ethanoic HF solution completely infiltrates the pores while a purely aqueous solution does not, due to wettability and capillary phenomena (Halimaoui 1993, 1994). For example, Kim and Cho (2012) have shown that when the etchant (isopropyl alcohol, IPA), with the highest solvent ratio (60%) was dropped on the Si surface, the etchant spread out rapidly. In contrast, in the case of electrolytes with an IPA ratio of 0%, 5%, and 33%, the etchant did not spread but rather stayed like a drop of water on a lotus leaf. These phenomena play an important role in the smoothness of the interface between Si and porous Si and thus in the uniformity of the PSi thickness. In fact, when the anodization is performed in a purely aqueous solution, the electrolyte does not completely infiltrate the pores and the propagation of the dissolution reaction, which takes place at the interface, it is not uniform, thus leads to interface roughness and thickness inhomogeneity. Consequently, the role of ethanol (or any other surfactant) is to improve the PSi layer uniformity by elimination of the hydrogen bubbles and to improve the electrolyte penetration in the pores (Halimaoui 1997). This means that the sensitivity of pore diameter to HF concentration strongly depends on the solvent (Levy-Clement et al. 1994; Lehmann and Ronnebeck 1999; Chazalviel et al. 2000; Hasse et al. 2000).

Urata et al. (2012) have shown that the type of alcohol affects the pore structure of PSi layer as well. After analyzing the layer formation in a low-concentration HF solution (8 wt.% HF, $J$ = 14 mA/cm$^2$, $t$ = 1 h) with different solvents such as methanol (MeOH) (CH$_3$OH), ethanol (EtOH) (C$_2$H$_5$OH), 2-propanol (PrOH) ((CH$_3$)$_2$CHOH), t-butanol (BuOH) (C$_4$H$_{10}$O), and diethyl ether (Et2O) (C$_2$H$_5$)$_2$O), Urata et al. (2012) found that the use of alcohol with a large number of carbon atoms improves the conditions of stable macropore structure formation. In particular, they established that when MeOH was used as solvent, pores in *p*-Si(100) were not obtained (see Figure 4.11). With EtOH, pore walls started to be formed but the depth of pores was not deep. When using PrOH or BuOH, pores were obtained comparatively well. Macropores become stable against the surface dissolution. This means that the stability of porous layers increases with the increasing carbon numbers in the alcohol. It is important that when Et2O was added to the HF solutions with alcohol, the dissolution of pore walls was suppressed and stable macroporous

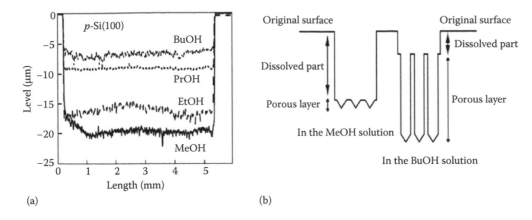

(a)                                    (b)

**FIGURE 4.11**    (a) Cross-sectional profiles of silicon anodized in low-concentration HF solutions with MeOH, EtOH, PrOH, and BuOH; (b) a schematic cross-sectional view illustrating porous layers anodized in the MeOH and BuOH solutions. (With kind permission from Springer Science+Business Media: *Nanoscale Res. Lett.*, 7, 2012, 329, Urata T. et al.)

layers were obtained even with MeOH. Urata et al. (2012) believe that this effect is connected with polarity of the solvent; the polarity of alcohol becomes lower with the increasing number of carbon in alcohol. Et2O also has low polarity.

Regarding the importance of the "organic electrolyte" composition for Si porosification, one can make the same conclusion as we did earlier for HF-$C_2H_5OH$-based electrolytes. Analyzing the influence of organic solvents such as acetonitrile (MeCN), dimethylformamide (DMF), formamide (FA), dimethylsulfoxide (DMSO), hexamethylphosphoric triamide (HMPA), and dimethylacetamide (DMA) on the morphology of PSi obtained in HF-based solutions (~4%) (see Figure 4.12), it was established that electrolytes with little oxidizing power as, for example, MeCN, will hardly produce macropores at all, whereas electrolytes with strong oxidizing power (e.g., FA, comparable to $H_2O$ or stronger) produce micropores instead of macropores (Lehmann et al. 2000; Föll et al. 2002). Of all the electrolytes tried, DMSO and DMA produce the most perfect *p*-macropores. Since the oxidizing power of the organic electrolytes is so important, Föll et al. (2002) proposed to bubble the electrolyte with $N_2$ before use, in order to remove all traces of $O_2$ dissolved from air. There is little doubt that their combination of relatively large oxidizing power and sufficient passivation power produces the best compromise for etching well-formed *p*-macropores. A wider range of pore diameters can be obtained rather in organic solvents, than in aqueous solutions.

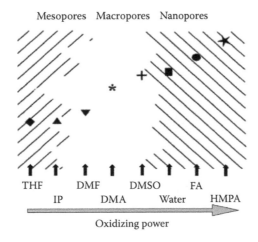

**FIGURE 4.12**    Correlation of the oxidizing power and pore formation for various organic electrolytes (and $H_2O$ for comparison. (Reprinted from *Mater. Sci. Eng. R* 39, Föll H. et al., 93, Copyright 2002, with permission from Elsevier.)

It is interesting that above-mentioned regularities, established for Si electrochemical etching in HF-C$_2$H$_5$OH-H$_2$O electrolytes are also correct for metal-assisted pore etching, which uses the solution containing HF, H$_2$O$_2$, and C$_2$H$_5$OH. The hydrogen peroxide in this solution is added as support for the partial parasitic electrochemical oxidation of silicon. Hydrofluoric acid is used for the dissolution of the silicon, and the parasitic oxide is used to create the pores. Ethanol reduces the surface tension of the solution. As we have shown before, surface wetting is important for good pore uniformity. Splinter et al. (2000) established that the electrolyte composition in metal assisted pore etching is also an important fabrication parameter for well-defined porous layers. The pore dimensions and porosity change with different parts of HF, H$_2$O$_2$, and ethanol. If the hydrofluoric acid concentration in the etching solution decreases, the dimension of the pores increases and, therewith, the porosity of the porous layer increases as well. Pore dimensions refer to the diameter of the pores and the size of the silicon skeleton. On the other hand, if the concentration of hydrogen peroxide increases, the pore dimensions and porosity increase as well. As a further effect, the porous silicon etching rate increases, too.

### 4.1.3 THE INFLUENCE OF CURRENT DENSITY

Research has shown that current density during porosification is another parameter important for porosity control. It was established that the diameter of pores formed on both $p$-Si and $n$-Si generally increases with increasing potential and current density over a wide range of doping concentrations (Bomchil et al. 1983; Herino et al. 1987; Zhang 1991; Pavesi 1997; Lehmann et al. 2000). It was found that the porosity of a PSi layer in $p$-type substrates is increased by increasing the current density (Bomchil et al. 1989; Lehmann 1993). For heavy doped $p$-type Si such dependence is shown in Figure 4.13. As we indicated before, for fixed current density, the porosity decreases with HF concentration. For high current density, the porosity becomes very high and direct electropolishing of the silicon surface occurs. The same result was also obtained by Pavesi (1997) and Halimaoui (1997). At that, it was found that the decrease of the free charge carrier concentration increases the porosity of PSi layer formed at a fixed current density.

However, Herino et al. (1987) studying lightly $p$-doped and heavily $n$-type doped Si have found that the porosity as a function of current density can be quite different from the corresponding curves obtained for $p$-type heavy doped substrates. Herino et al. (1987) established that the porosity exhibited a minimum around 20–50 mA/cm$^2$ (see Figure 4.14).

At that, if for lightly $p$-doped substrates there is only a slight change of porosity, for heavily doped $n$-silicon there is a sharp minimum. For all electrolyte concentrations, a minimum in the average porosity curve for $n$-Si was found for about 40 mA/cm$^2$. For higher current densities, the porosity increases with the forming current density, as with heavily $p$-doped samples. The same

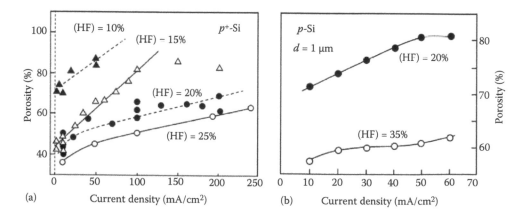

**FIGURE 4.13**    Porosity of $p$-PSi as a function of current densities for different HF concentrations. (a) Highly doped and (b) lightly doped $p$-type silicon substrate. (Data extracted from Bomchil G. et al., *Appl. Surf. Sci.* 41/42, 604, 1989; Halimaoui A., *Properties of Porous Silicon,* Canham L.T. (ed.). IEE INSPEC, London, p. 12, 1997; and Pavesi L., *La Rivista del Nuovo Cimento* 20, 1, 1997.)

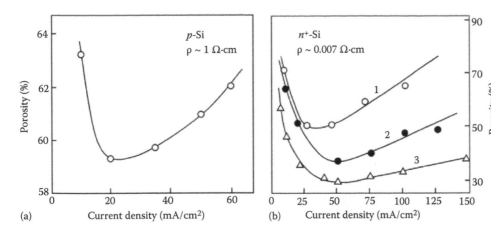

**FIGURE 4.14**   Gravimetric porosity as a function of the forming current density for porous layers with thickness 10 μm prepared on (a) lightly doped p-Si in 35% HF and (b) heavily doped n-Si in (1) 10% HF, (2) 15% HF, (3) 25% HF. Silicon was anodized in ethanoic electrolytes, composed of 50% ethanol. (From Herino R. et al., *J. Electrochem. Soc.* 134, 1994, 1987. Copyright 1987: The Electrochemical Society. With permission.)

regularity for this range of current was also obtained by Rumpf et al. (2009) (see Figure 4.15). As it is seen in Figure 4.15, the increase in current density is accompanied by a significant increase in the pore diameter, while a distance between the mesopores is not changed so greatly. As it was shown in Figure 4.16, for macroporous silicon the pore size also increases with increasing current density. According to Rumpf et al. (2009), this relationship between pore diameter and current density is well described by the equation proposed by Lehmann (1993):

$$(4.1) \qquad d_{pore} = S_{pore} \left( \frac{J}{J_{PSL}} \right)^{1/2}$$

where $S_{pore}$ is the center-to-center pore spacing, and $J_{PSL}$ is the critical current density.

According to Herino et al. (1987), the minimum in the porosity curve which appears in the lower current density range (see Figure 4.15) can be attributed to a chemical dissolution of the porous layer in the electrolyte during its formation (with given thickness, the lower the current density is, the longer is the anodization time). However, it seems that this explanation is

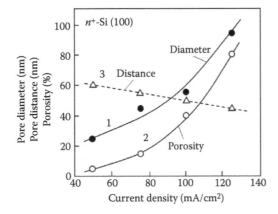

**FIGURE 4.15**   Relation between current density and pore-diameter (1), porosity (2) as well as pore-distance (3) of the PSi-formation is achieved by control of the electrochemical parameters during anodization of a (100) n+-Si-wafer. The fabrication of the porous silicon template was performed in a 10 wt% aqueous HF-solution. (Rumpf K. et al.: *Phys. Status Solidi* (c). 2009. 6(7). 1592. Copyright Wiley-VCH Verlag GmbH & Co. KGaA. Reproduced with permission.)

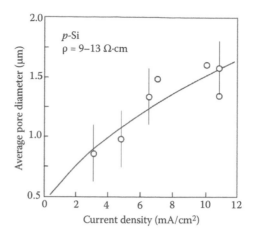

**FIGURE 4.16**  Average pore size as a function of current density for macroporous silicon samples fabricated from $p^-$-type bulk silicon ($\rho$ = 9–13 $\Omega$·cm) in an electrolyte containing 1(49%) HF:14 Acetonitrile (CAN). The error bars represent one standard deviation. (From Peckham J. and Andrews G.T., *Semicond. Sci. Technol.* 28, 105027, 2013. Copyright 2013: IOP. With permission.)

too simple because the difference in microstructure is observed. In $n$-type doped Si, the layers obtained at low current density have a finer structure.

The comparison of the porosity measured for layers formed under the same conditions on $p^+$- and $n^+$-substrates is shown in Figure 4.17. One can see that there is large difference. The values of porosity obtained from $p^+$ substrates are much higher, pointing to a difference in the charge exchange mechanism for the two kinds of substrates. This indicates that caution must be taken when comparing porous layers from $p$- and $n$-type substrates.

Lehmann and co-workers (Lehmann and Föll 1990; Lehmann et al. 1995; Lehmann and Gruning 1997) found that the porosity $P$ of an array of bsi-macropores is detemined by the etching current density according to equation

(4.2) $$P = A_{\mathrm{pore}}/A_{\mathrm{gen}} = J/J_{\mathrm{PSL}},$$

where $J_{\mathrm{PSL}}$ is the critical current density at the pore tips; $A_{\mathrm{pore}}$ is pore cross-sectional area $A_{\mathrm{pore}} = \pi d^2/4$ (with $d$ = pore diameter of an individual pore); and $A_{\mathrm{gen}}$ is a pore generation area. Equation 4.2 holds if the two areas are interpreted as the averages over many individual pores. For an

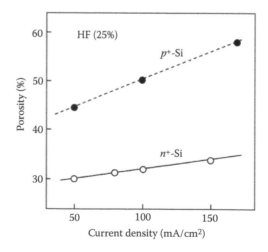

**FIGURE 4.17**  Comparison of porosity values of layers formed on heavily doped $p$-Si (100) and $n$-Si, as a function of forming current density. (Reprinted from *Mater. Sci. Eng. B* 118, Kan P.Y.Y. and Finstad T.G., 289, Copyright 2005, with permission from Elsevier.)

orthogonal pattern of pores, the pore diameter $d$ via etching current $J$ can be described by simple equation $d = J \cdot P^{1/2}$ if the cross-section of a pore is approximated by a square. The validity of Lehmann's formula was demonstrated in a series of experiments where the distance between pores was varied between 4 and 64 mm (Föll et al. 2002). However, it is necessary to take into account that for heavily $n$-type doped Si, the porosity as a function of current density is quite different from the corresponding curves obtained for $p$-type doped substrates.

At that, it is important to note that microporous Si is usually produced at current densities close to the $J_{PSL}$ value of the system, while $p$-macropores are found at current densities much lower than $J_{PSL}$ (Föll et al. 2002). While there is a general agreement in the literature that [HF] determines $J_{PSL}$, there are few and seemingly contradictive quantitative data. Lehmann (1993) reported an exponential relationship of $J_{PSL}$ on [HF], while Van den Meerakker et al. (2000) found a linear dependence. The actual data, however, if plotted in the same diagram, are reconcilable as shown in Figure 4.18.

The periodic variation of the current density during the etching process allows the production of multilayers formed by layers with different optical thicknesses. Another advantage of the formation process of porous silicon is that once a porous layer has been formed, no more electrochemical etching occurs for this area during the following current density variations. Hence, the porosity can be modulated in depth. This fact allows fabricating any refractive index profile.

The etch rate is another parameter of the porosification process strongly dependent on the current density. The growth rate of a PSi layer can vary over a wide range depending on formation conditions. It can be as low as a few Å/s and as high as 400 nm/s (Arita and Sunahama 1977; Parkhutik et al. 1983). Typical etch rates are 100 nm/s at a porosity of 50%. According to Lammel and Renaud (2000), the etch speed $dz/dt$ can, in general, be calculated as

$$(4.3) \qquad \frac{dz}{dt} = \frac{J}{Pv} \frac{M}{\rho \cdot eN_A},$$

where $J$ is the current density, $P$ is the porosity (ratio of voids per total volume, 1.0 for electropolishing), $v$ is the electrovalence (2.0–2.8 for porosification, 4.0 for electropolishing), $M$ is the atomic mass per mole, $\rho$ is the mass density, $e$ is the unity charge, and $N_A$ is Avogadro's constant. For silicon, the second factor on the right side is 1.249 μm cm²/(A·s). Experimental dependences of PSi layer growth rate are shown in Figure 4.19.

It is seen that for $p$-Si, the growth rate of PSi at a given silicon specific resistance appears to increase linearly with respect to current density (Arita and Sunahama 1977; Koker and Kolasinski 2000). For $n$-Si the growth rate increases also with current density but this relation is not linear

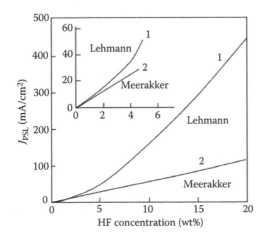

**FIGURE 4.18**  The influence of HF concentration in aqueous-based etching electrolyte on the value of $J_{PSL}$ determined for p-micropores and p-macropores. Curve (1) shows $J_{PSL}$ as described by the Lehmann model (Lehmann 1993) and curve (2) as described by Van den Meerakker et al. (2000). The inset shows measured data taken for low HF concentrations. (Reprinted from *Mater. Sci. Eng. R* 39, Foell H. et al., 93, Copyright 2002, with permission from Elsevier.)

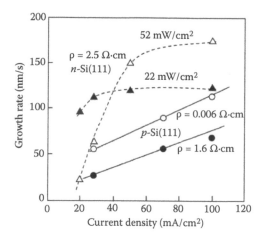

**FIGURE 4.19**  Effect of current density on PSi formation rate. (From Arita Y. and Sunahama Y., *J. Electrochem. Soc.* 124, 285, 1977. Copyright 1977: The Electrochemical Society. With permission.)

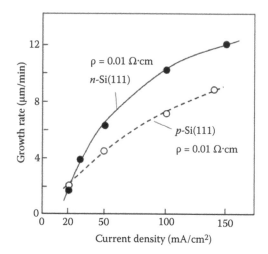

**FIGURE 4.20**  Relation between growth rote and current density. Porosification in 48% HF. The light intensities (Xe lamp) were $J_p$ = ~10–60 mW/cm². *P*-type silicon was anodized without illumination. (From Watanabe Y. et al., *J. Electrochem. Soc.* 122, 1351, 1975. Copyright 1975: The Electrochemical Society. With permission.)

(see Figure 4.19) (Arita and Sunahama 1977). This means that additional factors must be taken into account during *n*-Si porosification. Similar results were also obtained by Watanabe et al. (1975) (see Figure 4.20). At that, the growth rates for *n*-type silicon were larger than those for *p*-type silicon for equal current densities.

### 4.1.4 INFLUENCE OF ETCHING TIME

Usually the etching time is considered as a parameter controlling the thickness of a porous silicon layer. As is seen in Figures 4.21 through 4.23 for obtaining a thicker layer, a longer anodization time is required regardless of the electrolyte used. As anodic time increases, porous silicon thickness increases linearly under each anodic current density (Figures 4.21 and 4.22) and HF concentration (Figure 4.23). At that, Hedrich et al. (2000) established that doubling the current density from 20 to 40 mA/cm² increases the etching rate by a factor of 2. Unno et al. (1987) believe that these results support the fact that (1) the anodic reaction is limited by the interface reaction between HF solution and silicon and (2) the HF solution with a constant concentration is provided at the reaction interface during anodization.

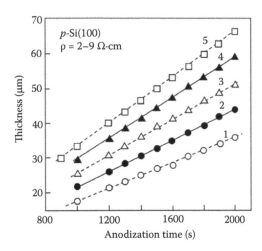

**FIGURE 4.21** Porous silicon layer thickness as a function of etching parameters in HF 25%: (1) $J = 20$ mA/cm$^2$, $P = 65.9\%$; (2) $J = 25$ mA/cm$^2$, $P = 66.5\%$; (3) $J = 30$ mA/cm$^2$, $P = 67.1\%$; (4) $J = 350$ mA/cm$^2$, $P = 67.7\%$; (5) $J = 40$ mA/cm$^2$, $P = 68.2\%$. (Reprinted from *Sens. Actuators B* 84, Hedrich F. et al., 315, Copyright 2000, with permission from Elsevier.)

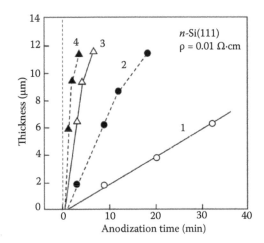

**FIGURE 4.22** Kinetic dependence of the PSi thickness on anodization time in HF (48%): (1) $J = 0.1$ mA/cm$^2$; (2) $J = 1$ mA/cm$^2$; (3) $J = 5$ mA/cm$^2$; (4) $J = 30$ mA/cm$^2$. (Reprinted from *Surf. Technol.* 20, Parkhutik V.P. et al., 265, Copyright 1983, with permission from Elsevier.)

Chao et al. (2000) have shown that the presence of surfactant in the electrolyte also does not change the linear dependence of the thickness on the etching time (see Figure 4.24).

According to Parkhutik et al. (1983), such constant growth rate at a constant current density means that the PSi formed is uniform in thickness. Effective surface area remains constant assuming reaction kinetics are the same. At a large thickness the growth may deviate from linearity due to the effect of diffusion in the electrolyte within the pores. It has been found that for a very thick PSi layer (150 μm), there is about 20% difference in HF concentration between that at the tips of the pores and that in the bulk solution.

Studies carried out by Yaakob et al. (2012) have shown that at higher current densities, previously mentioned dependencies may vary from linear relations especially in the initial time of anodization when the maximum rate of pore growth is observed (see Figure 4.25).

However, conducted research has found that in addition to the thickness, the anodization time influences the porosity of the forming layer (see Figure 4.26). The results obtained by Torres et al. (2008) during Si porosification in a mixture of hydrofluoric acid (HF) at 10% (p/v) and isopropyl alcohol (IPA) have found that an almost linear increase of the porosity is being observed as the anodization time increases. The increase of the etching time is also accompanied by the increase

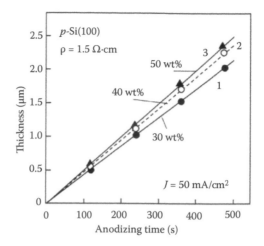

**FIGURE 4.23** Relationship between porous silicon thickness and anodic time with HF concentration as a parameter. (Adapted from Unno H. et al., *J. Electrochem. Soc.* 134, 645, 1987. Copyright 1987: The Electrochemical Society. With permission.)

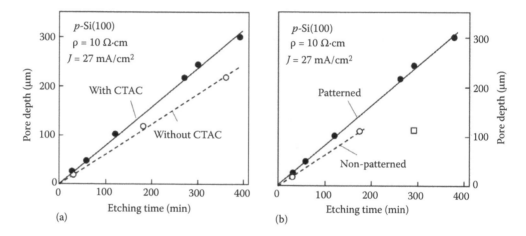

**FIGURE 4.24** Effect of (a) surfactant (10 mM CTAC) and (b) patterning on the growth rate of macropores in *p*-Si (100) in 8% HF: CTAC-surfactant etyltrimethylammonium chloride. □ in (b) indicates that the bare *p*-Si sample was electrochemically polished. (From Chao K.J. et al., *Sol.-St. Lett.* 3(10), 489, 2000. Copyright 2000: The Electrochemical Society. With permission.)

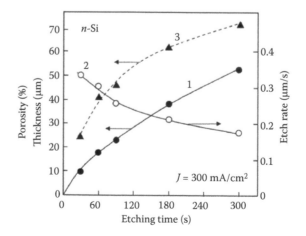

**FIGURE 4.25** Thickness (1) and porosity (3) of PSi layer, and etch rate (2) of the *n*-Si (100) ($n \sim 1.49 - 6.33 \times 10^{18}$ cm$^{-3}$) etched at 300 mA/cm$^2$ using electrolyte HF:ethanol = 1:1, as a function of etching time. (Data extracted from Yaakob S. et al., *J. Phys. Sci.* 23(2), 17, 2012.)

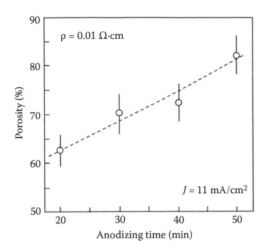

**FIGURE 4.26**  Porosity as a function of the *p*-Si (100) anodization time: electrolyte – HF(10%) in isopropyl alcohol (IPA). (Reprinted from *Microelectronics J.* 39(3–4), Torres J. et al., 482, Copyright 2008, with permission from Elsevier.)

of the pore size. This means that while using long etching time, one can obtain only macropous layers (Jakubowicz 2007).

One should also take into account that with fixed HF concentration and current density, the porosity increases with thickness, and porosity gradients occur in depth (Iraji zad et al. 2004). Iraji zad et al. (2004) believe that this happens because of the extra chemical dissolution of the porous silicon layer in HF. The thicker the layer, the longer the anodization time, and the longer the residence of Si in the HF solutions are, the higher is the mass of chemically dissolved PSi. This effect is much more important for lightly doped Si, while it is almost negligible for heavily doped Si because of the lower specific surface area (Bisi et al. 2000). The amount of chemical dissolution increases also with decreasing HF concentration (Unno et al. 1987). One can find confirmation of the above-mentioned statement in results presented in Figure 4.27. The data in Figure 4.27 can also be used for estimation of the chemical dissolution rate on the surface of pore walls. For PSi with density of 50% and an average pore diameter of 3 nm, the chemical dissolution rate is estimated to be about 6 × $10^{-5}$ nm/s, assuming that PSi consists of straight cylindrical pores of an equal diameter. The order of magnitude is in agreement with the planar etching rate of silicon in concentrated HF solutions.

However, we have to note that the conclusion about the influence of the thickness on the formed layer porosity is not generally accepted. For example, Hedrich et al. (2000) found that the porosity was almost independent of the layer thickness (15–70 μm) and, therefore, the porosity was nearly

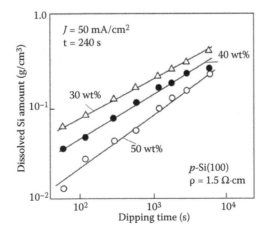

**FIGURE 4.27**  Relationship between amount of silicon dissolved from porous silicon and dipping time in HF solution for various HF concentration. (From Unno H. et al., *J. Electrochem. Soc.* 134, 645, 1987. Copyright 1987: The Electrochemical Society. With permission.)

homogeneous over the layer. Besides the conclusions mentioned above, it is also important to note that the growth of pores passes through several phases. For example, Foell and co-workers (Hejjo Al Rifai et al. 2000a,b) analyzing the growth of random macropores, established that it is necessary to distinguish between the nucleation phase and the phase of stable pore growth. It was found that the formation of randomly nucleated macropores involves a prolonged nucleation phase. Starting from a polished surface, the first macropores occur after a certain amount of Si has been homogeneously dissolved. Moreover, it was found that shallow pores generated during the nucleation phase often had higher density than the pores in the stable phase, that is, a number of these pores vanish before stable growth conditions are obtained. During this redistribution process, some pores stop growing and terminate, whereas other pores continue to grow with increased diameters. Therefore, only when this redistribution process is finished, the phase of stable pore growth starts yielding an arrangement of pores that is stable. At that, it was established that in this nucleation phase, the thickness of the homogeneously dissolved Si strongly depends on the doping level and the temperature but just weakly depends on the applied bias. Some results obtained by Föll and co-workers (Hejjo Al Rifai et al. 2000a,b) are shown in Figures 4.28 and 4.29.

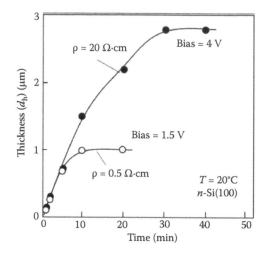

**FIGURE 4.28** The thickness $d_h$ of a homogeneously dissolved Si-layer is plotted as a function of time and doping level. When the thickness reaches its plateau value, the nucleation of pores is completed. (With kind permission from Springer Science+Business Media: *J. Porous Mater.* 7(1–3), 2000, 33, Hejjo Al Rifai M. et al.)

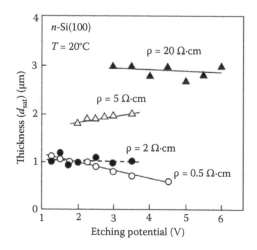

**FIGURE 4.29** The saturation thickness ($d_h$) of a homogeneously dissolved Si-layer as a function of the doping level and etching potential $U$ (temperature ~ 20°C). (From Hejjo Al Rifai M. et al., *J. Electrochem. Soc.* 147(2), 627, 2000. Copyright 2000: The Electrochemical Society. With permission.)

It was established that when the thickness ($d$) reaches its plateau value ($d_h$), the nucleation of pores is completed. From results presented in Figures 4.28 and 4.29, one can see that the duration of the nucleation phase decreases strongly with doping. Foell and co-workers (Hejjo Al Rifai et al. 2000a,b) believe that because both curves start with nearly the same slope, the mechanism for the homogeneous anodic dissolution of silicon seems to be almost independent of the doping. It was also shown that the thickness $d_{sat}$ does not depend strongly on the applied potential (see Figure 4.29).

It is necessary to note that the above-mentioned regularities are being observed for organic electrolytes as well. At that, as it is shown in Figure 4.30, time of the nucleation phase depends on both organic solvent and surface orientation.

Temporal dependence of mass transfer during Si etching is another aspect of porosification, which should be taken into account (Föll et al. 2002). It is know that the vertical homogeneity of PSi layers depends mainly on two different effects: chemical dissolution of silicon in the electrolyte and changes in the HF concentration during the anodization process. According to Canham (1991) and Lehmann (Lehmann and Gosele 1991) the dissolution is restricted by the carrier's transport in the PSi structure and by mass transport of the reactants through the pores, which follows Fick's law

(4.4) $$F_{HF} = D_{HF}(C_{top} - C_{bot})/d,$$

where $F_{HF}$ is the flux of the HF molecules, $D_{HF}$ is the diffusion coefficient of the molecules into the pores, $C_{top}$ and $C_{bot}$ are the HF concentrations at the top and the bottom of the PSi layer, correspondingly, and $d$ is the layer thickness. Therefore, without any doubt, for deep pores, that is, long time etching, diffusion effects should be taken into account. The diffusion of molecules into and out of the pores becomes more difficult as the pore depth increases, while the diffusion of holes to the pore tip becomes easier as the distance between the pore tip and the illuminated backside decreases. For example, Lehmann (1993) found that for a very thick PSi layer ($d \sim 150\ \mu m$) there is about a 20% difference in HF concentration between the tips of pores and the bulk solution.

Both effects mentioned above change $J$ and $J_{PSL}$, so simply keeping $J$ constant during an etching experiment will not keep the pore diameters constant. At that, it is necessary to take into account that for a fixed layer thickness, the decrease of the HF concentration from the top to the bottom of the layer will be larger for higher applied current densities. This means that for long time etching, a porosity gradient from high to low porosities in depth is expected. Therefore, for very deep pores, the diameter may increase with a decreased growth rate due to the effect of the diffusion process inside pores. Of course, the depth at which this occurs will be dependent on current density and HF concentration (Allongue et al. 1997). However, without any doubt, low temperature and low growth rate will favor formation of deep uniform pores.

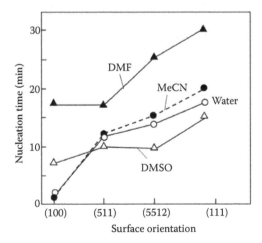

**FIGURE 4.30**  Nucleation time during p-macropores forming for various surface orientations and organic electrolytes (and H$_2$O for comparison); DMF—dimethylformamide; DMSO—dimethylsulfoxide; MeCN—acetonitrile. (Reprinted from *Mater. Sci. Eng. R* 39, Föll H. et al., 93, Copyright 2001, with permission from Elsevier.)

Lehmann (1993) analyzed the above-mentioned effects in detail and developed software that compensates the chemical diffusion effects during etching taking into account the temperature dependence of the processes as well. Several scientific teams used this software and thus automatically obtained the geometric parameters independent of temperature and diffusion in principle. There are, however, still pronounced effects, particularly with respect to the etching temperature (Carstensen et al. 2000b), that are not included in the software (or not even understood).

Another approach was proposed by Billat et al. (1997). For improvement of the PSi layer homogeneity, the authors of this paper proposed to interrupt the etch process several times during the formation of a layer. Billat et al. (1997) believed that during etch stops the HF concentration could recover and the overall depth homogeneity of the layer could be improved. Experiments related to Si anodization in the solution HF:C$_2$H$_5$OH=1:2:1 confirmed this assumption.

### 4.1.5 INFLUENCE OF pH

Unfortunately, a systematic study regarding the influence of pH solution on peculiarities of porosification has not been conducted. At present, while considering this issue, one can only rely on the results found in a few works (Lehmann 1993). Based on those results, one can state that the increasing of pH values promotes the increase of porosity of a formed PSi layer. The correlation between porosity and pH values is caused by chemical dissolution of the porous silicon branches by OH$^-$ ions present in the electrolyte. The dissolution rate increases with increasing levels of the OH$^-$ ions in the electrolyte and therefore with increasing pH values. This chemical dissolution continues for as long as the porous silicon remains in contact with electrolyte, increasing the porosity of a layer even after the anodization process is completed. The dissolution rate is partially dependent on the surface area available for reaction. The effect of chemical dissolution on a porous silicon skeleton should reduce the thickness of individual silicon branches. At higher porosities, already thin branches may disappear, weakening the remaining structure.

### 4.1.6 INFLUENCE OF THE TEMPERATURE

The temperature of electrolyte is an important parameter of the porosification process. However, there are not enough detailed studies published in the literature regarding the influence of the temperature on this process.

There have been some previous investigations regarding the influence of the etching temperature on the quality of the porous silicon (Setzu et al. 1988; Ono et al. 1993; Lehmann and Gruning 1997). However, some conclusions have been contradictory. Based on that data, one can make general conclusions that the electrolyte's temperature influences the rate of the pore growth, their diameter, and the maximum achievable thickness. In particular, the temperature rise promoted the increase of the rate of the pore's growth. Influence of temperature on the formation rate is shown in Figure 4.31. It is seen that the formation rate is not being changed much with increasing temperature in the range from 5°C to 65°C.

It was also established that the increase of electrolyte's temperature is accompanied by an increase in the pore's diameter. At that, the maximum obtainable pore depth without degeneration is being increased (Lehmann and Gruning 1997).

Only results presented in later works (Blackwood and Zhang 2003) allowed specifying some dependencies. It was experimentally shown that the temperature decrease was accompanied by a large increase of the number of pores. However, the pores had smaller diameters (see Figure 4.32). For example, in PSi formed at −10°C the structures with pore size even less than 10 nm have been observed, while for a temperature of 37°C, the typical pore size was no less than 500 nm. It was supposed that at lower temperature smaller nanocrystals were stabilized due to a combination of their reduced solubility and the increased viscosity of the diffusion layer that led to a higher localized concentration of silicon ions, allowing smaller nanocrystals to be in equilibrium with their surroundings (Blackwood and Zhang 2003).

The trend of decreasing crystal size with etching temperature decrease was also observed in atomic force microscope (AFM) and scanning electron microscope (SEM) images. The analysis of AFM images shows that the average surface roughness is decreased with a decrease of the etching

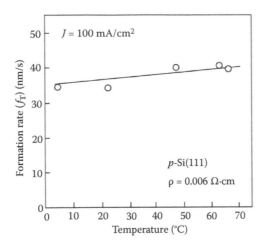

**FIGURE 4.31**  HF temperature ($T$) effect on the PSi formation rate (8). Anodic reaction was performed by xenon lamp (300 and 500 W) illumination to generate holes in the $n$-type silicon but without illumination for the $p$-type silicon. (From Arita Y. and Sunahama Y., *J. Electrochem. Soc.* 124, 285, 1977. Copyright 1977: The Electrochemical Society. With permission.)

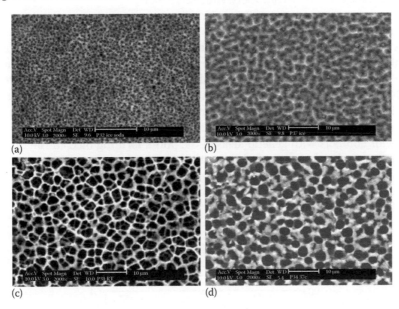

**FIGURE 4.32**  SEM photographs of porous silicon films etched at temperatures of (a) –10°C, (b) 0°C, (c) 25°C, and (d) 37°C. Specimens were fabricated from a $p$-type (100) silicon wafer with a resistivity of 5 Ω·cm. Gallium-indium eutectic was used to make ohmic contacts to the backside of the wafers while acid-resistant epoxy resin was applied to reduce the exposed surface area on the front face to 10 mm × 8 mm. The specimens were weighed (without the epoxy resin) prior to and after the etching process. The porous silicon was obtained by electrochemical etching in a solution of 1:1:2 HF:$H_2O$:ethanol held in a two-electrode PTFE closed-cell, which was in turn placed inside a temperature controlled bath. Etching was performed at a constant current of 20 mA/cm² applied for a period of 20 min. After etching, the porous silicon films were rinsed with ethanol and dried with pentane. (From Blackwood D.J. and Zhang Y., *Electrochim. Acta* 48, 623, 2003. Copyright 2003: The Electrochemical Society. With permission.)

temperature. At that, it was found that the average depths of the porous silicon layers, determined from the profilometer measurements over different areas of the specimens' surfaces, did not show any consistent variation with etching temperature change. Figure 4.33 shows that the average porosity was increasing as the etching temperature was lowered. However, Blackwood and Zhang (2003) anticipate that while the porous silicon is being formed at low temperatures, it is necessary to be careful because the films are inclined to cracking due to thermal expansion in the resultant films.

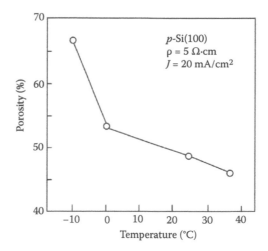

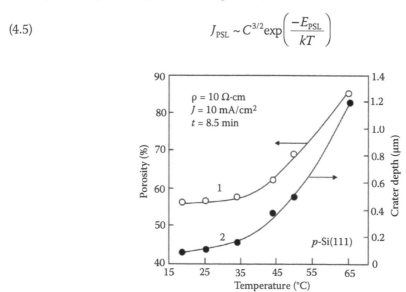

**FIGURE 4.33**  Dependence of the average porosity of the porous silicon films on the temperature of the HF:H$_2$O:ethanol = 1:1:2 etching solution. For porosofication, a p-type (100) Si wafer with a resistivity of 5 Ω·cm was used. (From Blackwood D.J. and Zhang Y., *Electrochim. Acta* 48, 623, 2003. Copyright 2003: The Electrochemical Society. With permission.)

It is necessary to note that a general trend in the temperature dependence of the porosity of the PSi films determined by Blackwood and Zhang (2003) is similar to that previously reported by Setzu et al. (1988). However, research made by Balagurov et al. (2006) did not confirm results presented by Blackwood and Zhang (2003). Results obtained by Balagurov et al. (2006) are shown in Figure 4.34. It can be seen the porosity increases from 56% to 85% with increased temperature. At that, AFM measurements showed that the surface morphology of PSi layers prepared at different temperatures between 18°C and 65°C was almost the same.

Such behavior conflicts with general ideas of the pore's forming mechanism during the process of electrochemical etching. As it is known, the porosity decreases with increased difference between the etching current density and the critical current density. The critical current density ($I_{PSL}$) separates the electropolishing mode from the porous silicon formation mode. It is given by the empirical expression (John and Singh 1995):

$$(4.5) \qquad J_{PSL} \sim C^{3/2} \exp\left( \frac{-E_{PSL}}{kT} \right)$$

**FIGURE 4.34**  Temperature dependence of the porosity and the crater depth obtained after anodization of 10 Ω·cm Si wafer in HF:C$_2$H$_5$OH = 5:1 electrolyte at current density of 10 mA/cm$^2$ for 8.5 min. (Reprinted from *Electrochim. Acta* 51, Balagurov L.A. et al., 2938, Copyright 2006, with permission from Elsevier.)

where $C$ is the HF concentration in electrolyte, $E_{PSi}$ is an activation energy equal to 0.345 eV, $T$ is the absolute temperature, and $k$ is the Boltzmann constant. As the critical current density increases with temperature (John and Singh 1995) and the etching current density is constant, the difference between the etching current density and the critical current density increases. Therefore, the porosity should decrease with temperature.

Balagurov et al. (2006) concluded that the increase of porosity with anodization temperature is being observed due to the chemical etching process, which takes place simultaneously with the electrochemical process of PSi formation (see Figure 4.35). This dependence was found to be an exponential one with activation energy of 0.55 eV. The analysis of obtained results allowed concluding that the chemical etching rate of PSi is determined by the concentration of OH⁻ ions in the solution, supporting the assumption on the central role of hydroxyl ions in the chemical etching process of PSi (Balagurov et al. 2006).

Hejjo Al Rifai et al. (2000b) also established that the increase of the electrolyte temperature was accompanied by a decrease of the thickness of the macroporous layer and by an increase of the density of pores (see Figure 4.36a and b). Such behavior was explained by the temperature influence on the dissolution rate: a high electrolyte temperature increases the electrochemical dissolution rate.

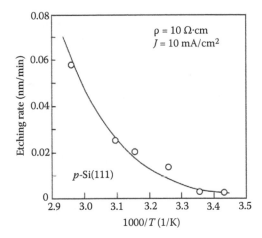

**FIGURE 4.35**  Temperature dependence of PSi chemical etching rate for HF:C₂H₅OH = 5:1 electrolyte. (Reprinted from *Electrochim. Acta* 51, Balagurov L.A. et al., 2938, Copyright 2006, with permission from Elsevier.)

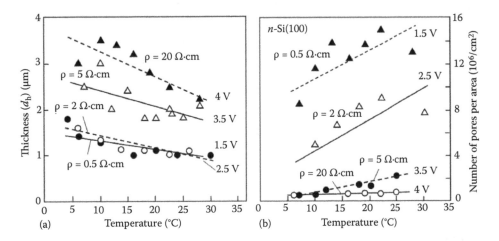

**FIGURE 4.36**  The influence of the temperature on (a) the thickness $d_h$ and (b) the density of macropores of a homogeneously dissolved *n*-Si (100)-layer for different doping levels: 4 wt% HF-solution; $J$ = 5 mA/cm²; backside illumination. (With kind permission from Springer Science+Business Media: *J. Porous Mater.* 7(1–3), 2000, 33, Hejjo Al Rifai M. et al.)

### 4.1.7 INFLUENCE OF WAFER'S DOPING

As it was indicated earlier, depending on etching conditions, the average pore diameter and the pore wall thickness can range from nanometers to tens of micrometers and therefore can cover four orders of magnitude. As it follows from conclusions made by Lehmann et al. (1997), no parameter other than the current density determines the pore diameter. However, research has shown that the doping concentration is also crucial for the pore diameter. Moreover, many researchers believe that the doping of the substrate is the most important parameter in the process of porosification. The effect of Si doping on pore diameter in PSi layers is shown in Figure 4.37.

Figure 4.37 shows that for *n*-Si, the diameter of pores decreases with doping concentration at different current densities (Lehmann et al. 2000). In contrast, the pore diameter of *p*-Si of moderate or high doping concentrations decreases with increasing doping concentration.

The influence of doping density on the pore wall thickness is shown in Figure 4.38. It was found that the pore wall thickness was always smaller than the pore diameter. At that, a square root of the pore wall thickness on the substrate doping density was observed (Lehmann and Ronnebeck 1999). Simple calculations have shown that the average pore-wall thickness observed

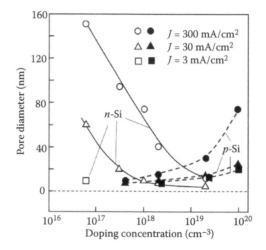

**FIGURE 4.37**   Pore diameter of the PSi formed in 50% HF + ethanol (1:1) as a function of doping concentration and current density. (Reprinted from *Mater. Sci. Eng. B* 69–70, Lehmann V. et al., 11, Copyright 2000, with permission from Elsevier.)

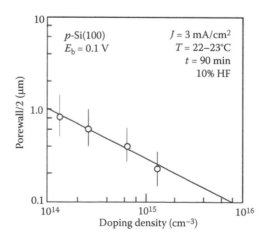

**FIGURE 4.38**   Values of pore wall thickness in macroporous Si as a function of doping density (error bars) together with the space-charge region width for a potential of about 0.1 V (line). Porosification of the *p*-type Si was carried out in 2 M HF solution in acetonitrile prepared from 68 mL of 50% HF and 932 mL acetonitrile. (From Lehmann V. and Ronnebeck S., *J. Electrochem. Soc.* 146, 2968, 1999. Copyright 1999: The Electrochemical Society. With permission.)

for random macropore formation on $p$-type Si corresponds to two times the space-charge region width for an energy barrier of about 0.1 eV.

It was established that the doping concentration in addition to the pore diameter also affects the etch rate, the pore density, and the pore wall thickness. It is generally assumed the pore walls are in full depletion. This means that the pore wall thickness is roughly equal to twice the width of the space charge layer ($W_{SCL}$). Since the $W_{SCL}$ is inversely proportional to the free charge carrier concentration, the doping level will determine the pore wall thickness and the pore spacing (i.e., pore density). According to Zhang (2005), the PSi layers formed on different substrates can be roughly grouped into four main categories according to doping concentration: (1) moderately doped $p$-Si ($10^{15}$–$10^{18}$ cm$^{-3}$); (2) heavily doped materials, $p^+$-Si and $n^+$-Si ($>10^{19}$ cm$^{-3}$); (3) non-heavily doped $n$-Si ($<10^{18}$ cm$^{-3}$); and (4) lowly doped $p$-Si ($<10^{15}$ cm$^{-3}$). The PSi formed on moderately doped $p$-Si has extremely small pores typically ranging from 1 to 10 nm. For heavily doped $p$- and $n$-types, the pores have diameters typically ranging from 10 to 100 nm. For non-heavily doped $n$-Si, the pores have a wide range of possible diameters from 10 nm to 10 μm. For lowly doped $p$-Si, the PSi can have two distinct distributions of pore diameters: large pores with a distribution of diameters on the order of μm and small pores on the order of nm.

Lehmann et al. (2000) proposed a detailed listing of the morphological features as a function of doping for $p$- and $n$-type Si (see Figures 4.39 and 4.40). According to Lehmann (1996a), the microporous silicon with the pore diameter and the pore wall thickness <2 μm is an isotropical, sponge-like material forming on the low doped $n$-type silicon in the dark. The thickness of the wall between pores is equal to double the space charge region (SCR). Therefore, in $n$-type doped Si, the pore size and interpore spacing decreases with increasing dopant concentrations. The pores follow the electrical current lines and, therefore, are mainly perpendicular to the wafer surface. The pores show side branches, which are reduced with low current density. Usually lightly doped $n$-type substrates anodized in the dark have low porosity (1–10%).

When the silicon wafer is $p$-doped, the front end is a forward biased Schottky contact. Plenty of holes are present and no space charge region develops (Lang 1996). The etching goes on and the structures get smaller until they reach a size around 3 nm. According to the results of research carried out by Smith and Collins (1992), for $p$-type doped Si both pore size and interpore spacing are very small, typically between 1 and 5 nm, and the pore network looks very homogeneous and interconnected. Sometimes this material is called nanoporous as well. In this case, for explanation of pore wall passivation from dissolution, the mechanism based on quantum confinement usually is used. It is necessary to note that the pore size and wall thickness in $p$-Si are much smaller than in $n$-type doped Si. The porosity of $p$-PSi material depends on the etching condition

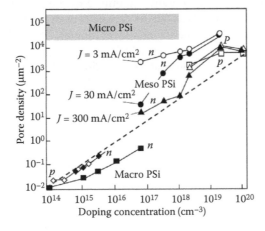

**FIGURE 4.39**  Pore density versus silicon electrode doping concentration for PSi layers of different size regimes. The mesopore formation current (in mA/cm²) is indicated in the legend. The dashed line shows the pore density of a triangular pore pattern with a pore pitch equal to double the space charge region (2 × SCR) width at 3 V. Note that only macropores on $n$-type substrates may show a pore spacing significantly exceeding the SCR width. The regime of stable macropore array formation is indicated by a dot pattern. (Reprinted from *Mater. Sci. Eng. B* 69–70, Lehmann V. et al., 11, Copyright 2000, with permission from Elsevier.)

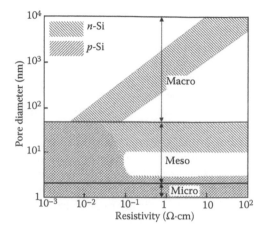

**FIGURE 4.40** The different pore sizes that form on *n*-type and *p*-type silicon substrates during anodization in hydrofluoric acid are shown as a function of the substrate resistivity. (Data extracted from Lehmann V., *Proceedings of IEEE Conference MEMS'96*, Feb. 11–15, 1996b, San Diego, CA, p. 1.)

and it can be varied between 20% and 70%. Pores, as a rule, display cylindrical shapes and they are parallel to each other. As the dopant concentration increases, pore size and inter-pore spacing increase, while the specific surface area decreases. The structure becomes anisotropic, with long voids running perpendicular to the surface, very evident in highly *p*-type doped Si ($p^+$).

Mesoporous silicon can be made from *n*-type wafers as well, if the wafer is illuminated while being etched. It was established that under illumination, higher values of porosity can be achieved, and mesopores are being formed together with macropores. The adsorbed photons generate electron-hole pairs within the pore walls and the passivation mechanism does not work anymore. For this reason, illuminated *n*-type silicon has morphology similar to that of the *p*-type material. However, usually if the low-doped *n*-type substrate is illuminated during anodization, the pore dimension increases, leading to macroporous structures with average dimensions >50 nm. The final structure depends strongly on anodization conditions, especially on the light intensity and current density. In the dark, the etching of moderately or low doped *n*-type Si produces only "breakdown" (bd) mesopores. In this case, the voltages might be as large as 100 V (Föll et al. 2002).

Porous silicon layers, formed on degenerately doped $p^+$ or $n^+$-silicon, exhibit structures in the mesoporous size (2–50 nm) region. No light is required in the case of $n^+$ Si because avalanche breakthrough is easy and enough current will be produced even at small voltages. Lehmann et al. (2000) conducted a detailed study of these kinds of mesopores. The etching of highly doped Si in organic electrolytes also produces mesopores, which are quite similar to mesopores fabricated in HF-$C_2H_5OH$-$H_2O$ solutions (Carstensen et al. 2000a; Christophersen et al. 2000a,b, 2001).

According to Bomchil et al. (1989), for heavily doped silicon similarly to low doped material, the PSi network presents pores with a preferential orientation mainly along the current lines. However, there are branches emerging from the main pores. The pore diameter is in the range of 4–12 nm. In the case of doped $n^+$-substrates, either the microstructure or the porosity critically depend on the exact value of the doping level. Further analysis of the wafer parameter influence on the porosity of the PSi layer established that the porosity of a layer increases with the decrease of the substrate doping (Lehmann 1993). This is understandable because the pore growth depends sensitively on the extension of the depletion region around the pore tips. It was also established that for *p*-Si the growth rate increases linearly with logarithmic dopant concentration, but for *n*-Si the dependence of growth rate on dopand concentration is more complicated (Guendouz et al. 2000).

The region of stable pore diameters in macroporous PSi as a function of the substrate resistivity, determined by Lehmann and Gruning (1997), is shown in Figure 4.41. These results indicate that the increase of Si resistivity promotes the growing of the pore size. Note that either branching or dying of pores does not change the overall porosity, which solely depends on the current density. It is necessary to note that the doping inhomogeneities in the substrate could lead to the changing of a perfect pore pattern to an imperfect one. It is important that in many cases mesopores are found inside macropores, that is, the macropores are filled with mesopores (Föll et al. 2002).

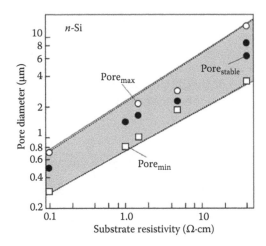

**FIGURE 4.41**    Influence of substrate resistivity on diameter of macropores formed in *n*-Si wafers. (Reprinted from *Thin Solid Films* 297, Lehmann V. and Gruning U., 13, Copyright 1997, with permission from Elsevier.)

The stable formation of macropore arrays requires a certain relation between the pore diameter and the *n*-type doping density of the silicon substrate. The upper and lower limits of stable pore formation are shown as a function of substrate resisitivity (dashed lines) (Lehmann and Gruning 1997). According to Lehmann and Gruning (1997), a good rule of thumb for selection of the substrate material is to take the square of the desired pore diameter (in μm), which gives the appropriate substrate resistivity (in ohms). For example, 2 μm pores can be better etched by using a 4 Ω·cm *n*-type substrate. A misadjustment of pitch and doping density leads to branching of pores for high doping densities or to dying of pores for low doping densities (Lehmann and Gruning 1997).

Thus, as it follows from the conducted analysis, the resistivity and the type of the starting silicon substrate should be determined precisely because the properties of porous samples, prepared on substrates with different resistivities, might be very different. More detailed information about the peculiarities of Si porosification and possible pore structure may be found in the recent books and reviews (Feng and Tsu 1994; Vial and Derrien 1995; Canham 1997; Parkhutik 1999; Bisi et al. 2000; Zhang 2001, 2005; Föll et al. 2002; Lehmann 2002).

It is also necessary to note that there are no clear boundaries between etching parameters used for preparing micro-, meso-, or macropore Si. For example, for heavily doped Si, mesopores will result under conditions that would otherwise produce micropores. Therefore, we have a possibility to state only that there is some transition from micropores to mesopores if the doping level is increased and another transition from micropores to macropores will take place if the current and/or the HF concentration is lowered.

## 4.1.8 INFLUENCE OF WAFER'S ORIENTATION

Using the results of the analysis presented by Lehmann and Gruning (1997), one can make a conclusion that the remarkable straightness of the pores is a prominent feature of the etching process. Most models are silent on this point. The expectation at best would be that pores grow directly toward the source of the holes, that is, perpendicular to any surface. However, in a real situation we do not observe such a simple case (Föll et al. 2002). Moreover, the influence of wafer orientation depends on both the type of Si conductivity (Meek 1971a; Levy-Clement et al. 1993; Yamani et al. 1997; Christophersen et al. 2000b) and parameters of technological process. Detailed investigations carried out by Ronnebeck et al. (1999) showed that during backside illumination of the substrate, the pores grow mainly toward the source of minority carriers and along the (100) crystal directions. It is known that etching in the (100) direction is about 50 times faster than in the (111) direction (Jakubowicz 2007). The morphology of these macropores is always describable as a main pore in one of these two directions and side pores or branches in some of the others. Only in case of an Si(100) substrate do these directions coincide. Christophersen et al. (2000b) established that macropores in *p*-Si can grow in the (113) direction as well. Earlier, the same effect was observed for *n*-Si (Ronnebeck et al. 1999). In addition, Christophersen et al. (2000b) have shown that the anisotropy is strongly enhanced by reducing the oxidizing species. Figures 4.42 and 4.43 show some examples of macropore morphology in the dependence on the Si orientation and the type of conductivity used.

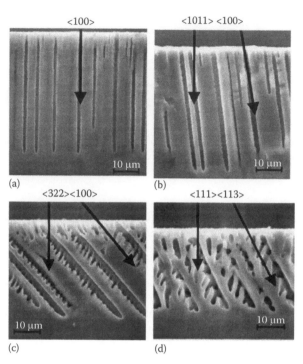

**FIGURE 4.42**   Orientation dependence of (random) *n*-macropores (aqu/bsi). The substrate orientation is (a) (100), (b) (111), (c) (322), and (d) (111). In (d) (113) oriented pore tripods result. (Reprinted from *Mater. Sci. Eng. R* 39, Föll H. et al., 93, Copyright 2002, with permission from Elsevier.)

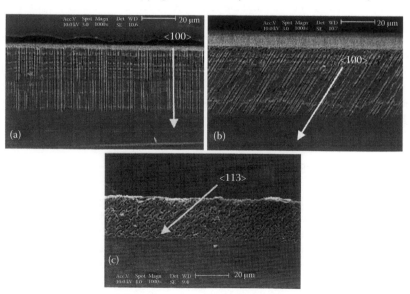

**FIGURE 4.43**   SEM images of anisotropic macropore-formation on *p*-type Si in the (100)- and (113)-directions. According to Christophersen et al. (2000b), organic electrolytes especially DMSO (diemethylsulfoxide) are beneficial for pore formation. (Reprinted from *Mater. Sci. Eng. B* 69–70, Christophersen M. et al., 194, Copyright 2000, with permission from Elsevier.)

For substrates with other orientations, we do not have such auspicious conditions for the pore growth. It was found that on (111) and (110) substrates, the pores grow toward the source of minority carriers, with a strong tendency to branching. For low current density and higher bias breakdown, the pores are developed along the (100) direction. Thus far it is not possible to obtain well-ordered arrays of macropores on (111) or (110) substrate orientations. A slight misalignment between the (100) crystal direction and the source of carriers of 10 to 15 degrees seems to be acceptable. In this case, the pores are titled to the surface and follow the (100) direction (Lehmann and Gruning 1997). However, misorientation enhances the tendency to branching.

For other kinds of pores, the dependence of the pore morphology on sample orientation is complex as well (Föll et al. 2002). For example, Christophersen et al. (2000a) have shown that mesopores only grow in the (100) direction and since they are usually branched, the branches are at right angles to the main pores. In instances where this seems not to be the case, the branches do not have a well-defined geometric shape (Lehmann et al. 2000). Whenever mesopores assume some defined geometric shape (which seems to be tied to low current densities), it consists of a connected octahedral. At higher current densities, the mesopores are still growing in the (100) directions, but the geometric shape of the branches is lost; they appear "cloudy" (Lehmann et al. 2000).

## 4.1.9 INFLUENCE OF LIGHTING

As we indicated earlier, silicon dissolution requires holes. It means that for $n$-type silicon, where holes are the minority carriers, the electrochemical dissolution of Si should be strongly dependent on the hole/electron pair generation by illumination. As it was shown by Halimaoui (1997), when lightly doped (majority carrier concentration below $\sim 10^{18}$ cm$^{-3}$) $n$-type silicon is anodized in the dark, the formation of porous silicon is observed only at high voltages (>5 V). Breakdown and impact ionization processes are supposed to provide the holes required for the dissolution step. The layer obtained in such case is macroporous, that is, tubular pores are running perpendicularly to the surface with a diameter greater than 0.2 μm. If the anodization is performed under illumination, porous silicon is formed at lower potentials (<1 V). The resulting material consists of two parts. The top surface layer is nanoporous (pore diameter of about 3 nm) and its thickness lies in the 0.2–1 μm range, depending on the substrate doping. The underlying part is macroporous. For a heavily doped ($\sim 10^{19}$ cm$^{-3}$) substrate, porous silicon formation is observed even in the dark. The charge exchange mechanism involved, the most probably, is the electron injection into the conduction band by tunneling and via the thin (<10 nm) space charge (Meek 1971b; Halimaoui 1997).

Therefore, the illumination of a semiconductor during electrochemical etching really is one of the important methods of the porosity control. By changing the light intensity, the generation rate of minority carriers within the silicon and, in consequence, the etching current, may be controlled independently of the anodic potential (Hejjo Al Rifai et al. 2000b). The influence of light intensity is shown in Figure 4.44.

At high concentration of HF-H$_2$O solution (48%), as the photon flux is increased, the etch rate increases up to a certain point, after which a little increase is being observed (Figure 4.44a). However, at lower concentration (25% HF), the rate increases linearly up to the maximum flux available, with no suggestion of a saturation region (Figure 4.44b). As would be expected from the investigation above, the rate was lower in less concentrated HF(aq) for any given laser flux. This suggests that the rate limiting process is diffusion of reactive species to, or products away from, the surface, and not the transport of holes to the surface. Regarding the effect of saturation, Koker and Kolasinski (2000) believe that the rate exhibits saturation effects at moderate laser intensities due to significant temperature rise.

As it was established, the influence of lighting also depends on the method of illumination. For example, if the $n$-type silicon is illuminated through the electrolyte during anodization at high anodic bias (>2 V), macropore formation is being observed. As it is known, the holes, necessary for Si porosification, could be generated in an $n$-type silicon electrode by illumination with photon energy exceeding the bandgap of silicon (1.1 eV). However, the macropores, formed using such illumination, will be conical in shape (Föll 1991). According to Lehmann (1996a), this is due to the fact that holes are generated in the pore walls as well as in the bulk of the Si electrode. Consequently, the pore walls are etched too. If the Si electrode is illuminated from the backside,

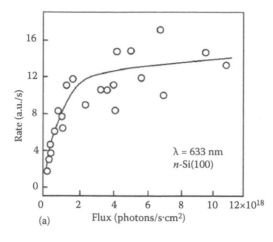
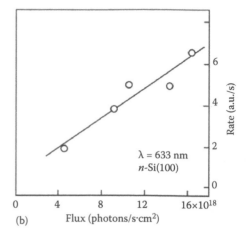

**FIGURE 4.44**    Dependence of etch rate of *n*-Si (111) on photon flux at λ = 633 nm in (a) 48% and (b) 25% HF-H$_2$O solutions. Illumination 15 mW HeNe laser (7 W cm$^{-2}$) (HF-H$_2$O solution). (Reprinted from *Mater. Sci. Eng. B* 69–70, Koker L. and Kolasinski K.W., 132, Copyright 2000, with permission from Elsevier.)

holes are only collected by the pore tips due to the field of the space charge region existing in an *n*-type substrate under anodic bias. Consequently, the pore walls become depleted of holes, which is the reason for their passivation. A prerequisite for this backside illumination technique is a diffusion length for minority carriers in silicon of several hundred micrometers, which allows generating the required holes by illuminating the backside of the wafer, while the front side is etched. It is known that the diffusion length for holes in silicon is up to several hundred micrometers and the wafer has the width of only 525 μm (Jakubowicz 2007). Unfortunately, the diffusion length of semiconductors with a direct bandgap like GaAs is much smaller. Therefore, this technique cannot be applied to those materials.

Jakubowicz (2007) also studied the influence of illumination on the process of Si porosification. They established that the lowest current density was being observed for etching without illumination. That backside illumination, in comparison with frontside illumination, resulted in a lower current density. They also found out that the smallest diameters of pores (~25 nm) were achieved for highly doped *n*-Si(111), etched without illumination. It was also shown that application of illumination gives an additional possibility for influence on PSi layer structures. For example, when the light source is sinusoidally modulated, the pore diameter evolves periodically, and photonic crystals can be designed (Schilling et al. 2001a).

### 4.1.10  INFLUENCE OF MAGNETIC FIELD

First, the effect of magnetic field on porous silicon formation has been reported by Nakagawa et al. (1996). Nakagawa et al. (1996) have shown that a high magnetic field (>1 T), applied perpendicularly to the silicon surface, increases both the porosity and the optical isotropy of the PSi layer. Hippo et al. (2008) fabricated ordered mesopores on a 100-nm scale with highly flat pore walls and a high aspect ratio of approximately 160 using magnetic field-assisted electrochemical etching. They have shown that conditions for etching highly flat pore walls have a very narrow window. It was established that a small change in pore distances causes a significant change in pore wall etching properties.

The SEM images of PSi layers fabricated using conventional and magnetic field assisted electrochemical etching are shown in Figure 4.45. It is seen that in the conventionally prepared PSi layer the shape of the pore tip is nonuniform. In contrast, the PSi layer prepared by magnetic field-assisted anodization has a highly uniform structure in the entire region of the sample.

We need to note that the above-mentioned highly ordered structures cannot be prepared using conventional methods of silicon anodic etching. To date, patterned structures with high aspect ratios and highly uniform trenches produced by electrochemical etching methods are only being used for macroporous silicon with pore diameters above 300 nm (Lehmann and Gruning

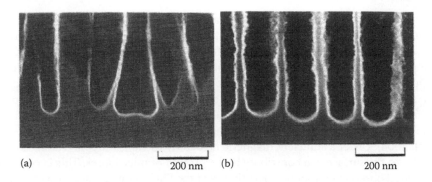

(a)  200 nm  (b)  200 nm

**FIGURE 4.45** The SEM micrographs of two PSi layers formed in $n^+$-Si ($\rho = 0.01$–$003$ $\Omega$·cm) under the same conditions (HF (5%):ethanol = 1:1, $J = 10$ mA/cm² for 5 min, $T = 0$°C) at a magnetic field of (a) H = 0 T, and (b) H = 1.9 T. (From Nakagawa T. et al., *Jpn. J. Appl. Phys.* 37, 7186, 1998. Copyright 1998: The Japan Society of Applied Physics. With permission.)

1997; Schilling et al. 2005). Only a few reports have appeared on studies of uniform pores below 200 nm in diameter (Letant et al. 2004; Nishio et al. 2005).

Koshida and co-workers (Gelloz et al. 2008) also studied the influence of magnetic field (MF) (0–4 T) on the process of *p*- and $n^+$-Si porosification in HF:H$_2$O:C$_2$H$_5$OH=80:140:220 under 40 mA/cm² and observed that at the highest MF (4 T), the pores exhibit no percolation and their diameter remains rather uniform through the PSi depth. In contrast, as the MF is lowered, a deterioration of the in-depth uniformity of the pore diameter is noticeable, and some pore walls have been etched resulting in straight pore propagation with smooth inside walls. They also established that the mean diameter of pores was not strongly affected by the magnetic field in the range of 2–4 T, while the pore's diameter tended to decrease with increasing magnetic field in the range of 0–2 T. Another effect of a high magnetic field application was to suppress the outside top surface etching during the pore growth. This resulted in an increase of both the pore's aspect ratio and the PSi layer thickness at higher MF intensity. Actually deep pores were uniformly fabricated at 4 T in comparison to the case of 2 T under the same anodization time.

Nakagawa et al. (1996) and Hippo et al. (2008) suggested that this effect is due to the deflection of the holes (majority carrier) by the induced Lorenz force (see Figure 4.46). The electrochemical

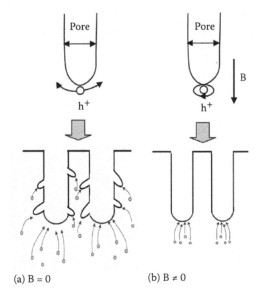

(a) B = 0    (b) B ≠ 0

**FIGURE 4.46** Pore formation in Si wafers by electrochemical anodization without (a) and with (b) an external magnetic field applied perpendicular to the substrate surface. Top figures schematically show the motion of holes generated at the pore tips by tunneling (a) without and (b) with an external magnetic field. The direction of the magnetic field is indicated by the down arrow. (Idea from Hippo D. et al., *Jpn. J. Appl. Phys.* 47(9), 7398, 2008, and Granitzer P. et al., *ECS Transactions* 50(37) 55, 2013.)

dissolution of silicon is initiated by holes, and thus it is critical to control the motion of holes to improve the etching properties of a silicon electrode (Halimaoui 1997). The external magnetic field is considered to play a significant role in this process. In a conventional method, holes diffuse in the pore wall regions, which cause the undesirable etchings, such as spiking or bending of pores. During magnetic field assisted etching, the motion of holes diffusing from pore tips to pore walls is restricted by the Lorentz force induced by the external magnetic field applied parallel to the direction of pore growth. Therefore, the effect of the external magnetic field is to avoid undesirable etchings, such as spiking or bending of pores, which tends to occur at pore walls (Hippo et al. 2008). Under a sufficient magnetic field, only the holes supplied along the direction perpendicular to the surface can contribute to the growth of micropores. This is consistent with the experimental observation that the growth rate of the PSi layer increases with increasing magnetic field (Nakagawa et al. 1998).

### 4.1.11 INFLUENCE OF ULTRASONIC AGITATION

As it is known, during the DC anodic etching process, the reaction product, silicon fluoride (Bisi et al. 2000), tends to deposit at the pore tips. In addition, the $H_2$ bubbles, which adsorb at the surface of silicon pillars because of interfacial tension, can block the silicon pores and lead to a reduction of HF concentration inside the pores (Lehmann 1993). Thus, the etching process will be slowed down. Meanwhile, the dissolved species will increase the resistance of silicon wafer and hence decrease the current density. This factor also slows down the etching rate. In 1996, Hou et al. (1996) proposed a pulsed anodic etching method, which carries out the anodic etching in an intermittent mode. This etching method lets the chemical products diffuse from the silicon pores during the pause period of anodic current and therefore Hou et al. (1996) and Xiong et al. (2002) believe that this method is superior compared to DC anodic etching. However, the speed of natural diffusion of silicon fluoride and $H_2$ bubbles using pulsed anodic etching is still very slow. Besides, there is an assumption that the chemical reaction products deposit at silicon pores, mostly at pore tips, and prevent the dissolution of silicon wafer, consequently enlarging the lateral etching.

Research carried out by Liu et al. (2003) has found that the use of anodic etching with ultrasonic agitation can help in resolving the above-mentioned problems. Due to ultrasonic press

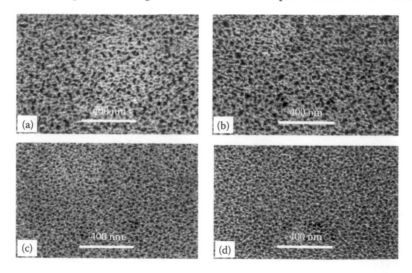

**FIGURE 4.47**   Surface SEM micrographs of porous silicon samples fabricated by four different etching methods: (a) DC etching, (b) pulsed etching, (c) ultrasonic etching, and (d) ultrasonic plus pulsed etching. The substrate used was a (100)-oriented highly doped (0.01 Ω·cm) $p^+$-type silicon single crystal wafer. It was placed in a Teflon etching cell and etched in the dark with an HF (40%): $C_2H_5OH$ (99%):$H_2O$ = 1:1:2 (by volume) electrolyte solution. The ultrasonic wave frequency of the ultrasonic generator was 33.3 kHz. After the etching process, all the samples were immediately rinsed with deionized water and dried. (Reprinted from *Sol. St. Commun.* Liu Y. et al., 127, 583, Copyright 2003, with permission from Elsevier.)

effect and acoustic cavitation (Feng 1999), the diffusion of the dissolved species and $H_2$ bubbles from silicon pores can be accelerated. In addition, the other ultrasonic effects, such as vibration, should also speed up the diffusion of chemical products. Liu et al. (2003) established that PSi layers prepared by ultrasonic anodic etching really had improved qualities in surface morphology, layer interface smoothness, etching efficiency, and optical characteristics compared with the sample prepared by conventional DC etching.

Figure 4.47 shows the surface SEM images of samples prepared using different methods of electrochemical etching. The pores in the PSi samples are seen as dark dots. One can see that the pores of sample (a) in Figure 4.47 are distributed randomly and are irregularly shaped. Most pores have large dimensions that seem to be joined by two or more smaller pores. The uniformity of pores in sample (b) has been slightly improved. Samples (c) and (d), fabricated using ultrasonic wave, have pores with much smaller diameters and more circular shape than in samples (a) and (b) and more uniform distributions of homogeneous pores peculiar to samples (c) and (d). Obtained results indicate that a sample fabricated using ultrasonic anodic etching has a more uniform PSi layer with smaller silicon pores and the etching efficiency is also higher than those prepared by conventional methods. The studies of both PSi single layer and PSi microcavity carried out by Liu et al. (2003) have also shown that ultrasonic etching optimizes the sample's optical characteristic.

## 4.2 INFLUENCE OF ANODIZATION PARAMETERS ON STRUCTURAL PARAMETERS OF PSi LAYERS

### 4.2.1 INFLUENCE OF ANODIZATION PARAMETERS ON MICROSTRUCTURE OF PSi LAYERS

Figure 4.48 shows how big the variation in microstructure of PSi layers formed at different conditions may be. Schematically possible structures of porous Si are shown in Figure 4.49.

As is shown in Figures 4.48 and 4.49, the microstructure of porous Si layers has great variety. Zhang (2005) selected the following peculiarities of PSi layer microstructures:

1. Pores can be either straight with smooth walls (Figure 4.49a[i]) or can be branched (Figure 4.49a[ii–vi]). Research shows that straight large pores with smooth walls can be formed by backside illumination of *n*-Si of the (100) orientation (Lehmann and Föll 1990; Lehmann 1993; Lehmann and Gruning 1997). According to Lehmann and co-workers (Lehmann and Föll 1990; Lehmann 1993; Lehmann and Gruning 1997), the smallest possible pore diameter for a regular pore array formed with surface patterning and backside illumination is about 0.3 μm below which branching at the pore bottom occurs; the largest pores are found to be about 20 μm, above which formation of straight and smooth pores becomes a problem due to hydrogen bubble formation. According to Lehmann et al. (2000), the lower limit for the pore diameter is established by breakdown, which leads to light-independent pore growth and spiking.

   However, experiments carried out by Fukami et al. (2008) and Nakagawa et al. (1998) have shown that ordered structures of PSi with smooth walls could have pores with diameter about 100 nm. They established that highly doped silicon substrates are required to achieve pore geometries below 100 nm in scale. It was established that the space charge regions around pores experience greater fluctuations as the doping density increases, and mesopore growth processes are greatly affected by local electric field distributions. These effects result in spiking or bending of pores, and they are the major problems with the existing mesopore formation techniques (Hippo et al. 2008). One can see ordered structures fabricated by Nakagawa et al. (1998) in Figure 4.45. For preparing straight periodic silicon nanostructures, Nakagawa et al. (1998) used the etching of heavy doped *n*-Si in dark in 5% HF:ethanol solution at 10 mA/cm². Nakagawa et al. (1998) found that the required condition for producing uniform periodic Si nanostructures corresponds to the transition regime from ideal anodization to electropolishing. In case of a higher density ($J > 10$ mA/cm²) for a lower HF concentration, electropolishing occurs.

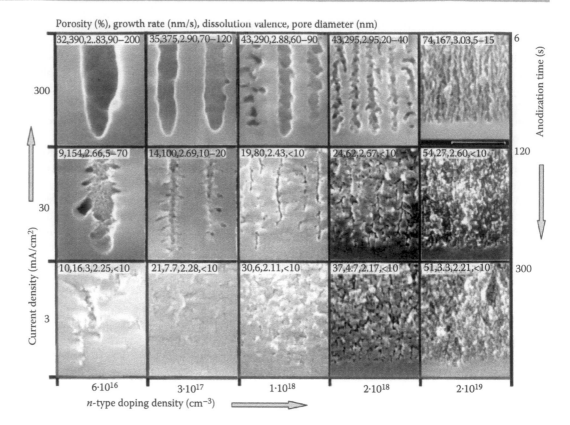

**FIGURE 4.48** Scanning electron micrographs of the interface between bulk and porous silicon for *n*-type doped (100) silicon electrodes anodized galvanostatically in ethanoic HF. (Reprinted from *Mater. Sci. Eng. B* 69–70, Lehmann V. et al., 11, Copyright 2000, with permission from Elsevier.)

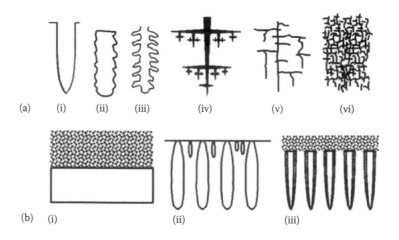

**FIGURE 4.49** Variants of microstructure of the PSi layers fabricated using different parameters of anodization: (a) variants of branching; (i) smooth pore wall; (ii) branches shorter than diameter; (iii) second level branches only; (iv) dendritic branches; (v) main pores with second and third level branches; (vi) dense, random, and short branched; (b) depth variation of PSi layer; (i) single layer of microporous Si; (ii) single layer of macro-porous Si with smaller pores near the surface; (iii) a layer of micropores Si on the top of macropores Si. (With kind permission from Springer Science+Business Media: *Modern Aspects of Electrochemistiy*, Vayenas C. (ed.), 2005, p. 65, Zhang G.X., Number 39.)

Top views and cross-section views of PSi with ordered structures fabricated by Fukami et al. (2008) are shown in Figure 4.50a.

For meso PSi shown in Figure 4.50c, 28 wt% HF/H$_2$O/ethanol solution was used. A *p*-type Si (100) with a resistivity of 0.0045–0.0060 Ω·cm was anodized at 50 mA/cm$^2$ for 30 s. For medium-sized pores, an *n*-type Si (100) with a resistivity of 0.0100–0.0180 Ω·cm was anodized in 6 wt% HF + 8 mM KMnO$_4$ + 3000 ppm NCW-1001 (surfactant) solution. Medium-sized pores with smooth wall and branched wall were formed by electrodissolution at 25 mA/cm$^2$ and 15 mA/cm$^2$, respectively. For ordered macro pores, a *p*-type Si (100) wafer with a resistivity of 10–20 Ω·cm was used. Prior to anodization, ordered etch pits were created on the surface of the Si wafers by a standard photolithographic technique and the subsequent alkaline etching process in 25 wt.% tetramethylammonium hydroxide (TMAH) aqueous solution for 5 min at 363 K. The prepared etch pits were orderly set in array at 5-μm intervals. The solution to produce macropores was 47 wt.% HF/H$_2$O/2-propanol with a composition of 5:6:29 in volume ratio.

More detailed information regarding peculiarities of ordered porous structures fabrication can be found in papers related to this problem (Gruning et al. 1996; Lehmann and Gruning 1997; Lehmann and Ronnebeck 1999; Hejjo Al Rifai et al. 2000a; Kleimann et al. 2000; Christophersen et al. 2000a, 2003; Föll et al. 2002; Badel et al. 2004).

2. PSi can have a surface transition layer. It was found that the pores at the surface usually have smaller diameter than those in the bulk of PSi (Figure 4.49b[ii]) (Watanabe et al. 1975; Unagami 1980; Zhang 1991; Smith and Collins 1992). Such an increase in pore diameter from the surface to bulk is due to the transition from pore initiation to steady growth. The thickness of this transition layer is related to the size of pores; the smaller the pores are, the thinner is the surface transition layer.

3. PSi can have a layer structure (Figure 4.49b[ii, iii]). Two-layer PSi, with a micro PSi layer on the top of a macro PSi layer, is being formed only under certain conditions. For *p*-Si, a two-layer structure was observed only on lowly doped substrates. For moderately or highly doped *p*-Si or for *n*-Si in the dark, the formation of a two-layer PSi has not been observed. For *n*-Si, the formation of a two-layer PSi is associated with front illumination, although it can also be formed with back illumination (Lehmann 1993; Levy-Clement et al. 1994; Ponomarev and Levy-Clement 1998). For the micro PSi layer formed on front illuminated *n*-Si, the pore diameter is less than 2 nm and the thickness of PSi changes

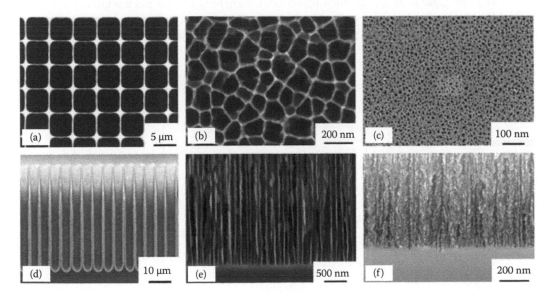

**FIGURE 4.50**   Top views (a–c) and cross-sectional-views (d–f) of porous silicon with different pore diameters: (a, d) ordered macropores, (b, e) medium-sized pores, and (c, f) mesopores. Actual average diameters of the pores, which are determined from the SEM images, are 20 nm (mesopore), 120 nm (medium-sized pore), and 5 μm (macropore), respectively. (Reprinted from *Electrochem. Commun.* 10, Fukami K. et al., 56, Copyright 2008, with permission from Elsevier.)

with illumination intensity and the amount of charge passed. For example, there is a correlation between the occurrence of two-layer PSi and the saturation photo current value (Osaka et al. 1995). It was established that a single micro PSi layer (Figure 4.49b[i]) is being formed at a photo current density below the photo saturation value while a two-layer structure of PSi is being formed at current densities above the saturation current as shown in Figure 4.51. This means that a macro PSi layer is formed only after a certain amount of charge, which is necessary to pass for the initiation of macropores etching (Searson et al. 1992).

4. Individual pores, depending on formation conditions, may propagate straight in the preferred direction with very little branching (Figure 4.49a[ii, iii]) or with the formation of numerous side or branched pores (Figure 4.49a[iv through vi]). The orientation of primary pores is in general in the <100> direction for all the PSi formed on all types of (100) substrates (Smith and Collins 1992; Jager et al. 2000). In general, the conditions that favor the formation of small pores also favor branching. The branched pores can have second level, third level, or further levels of branches. Branched and hierarchical pore structure has been found to form on all types of substrates. Branched pores are generally smaller than the primary pores (Beale et al. 1985a; Chuang et al. 1989; Smith and Collins 1992; Jager et al. 2000). The degree of branching and interpore connection depends strongly on doping concentration. For heavily doped materials, pores are generally branched (Beale et al. 1985a,b). However, the most highly connected branching structure was found in the PSi layers formed in lowly doped $p$-Si and in microporous layers in $n$-Si formed under illumination (Smith and Collins 1992).

On the other hand, well separated and straight pores with smooth walls were generally found in PSi formed in moderately or lowly doped $n$-Si in the dark (Zhang 2005). Also, on $n$-Si, smooth and straight pores without branching (Figure 4.49b[ii, iii]) can be obtained under back illumination using a surface patterned substrate. The macropores formed on lowly doped $p$-Si generally have no side pores longer than the pore diameter (Rieger and Kahl 1995; Lehmann and Ronnebeck 1999).

Like main pores, branched pores are highly directional, propagating preferentially in the <100> direction (Lehmann and Föll 1990; Searson et al. 1992; Smith and Collins 1992). For the PSi with dendritic structure (Figure 4.49a[iv]), pores propagate along the (100) direction even on the (110) and (111) substrates (Smith et al. 1988; Lehmann and Föll 1990; Jager et al. 2000). Such directional branching can produce regularly spaced three-dimensional structures (Smith and Collins 1992). The tendency to branch is stronger on the (110) and (111) substrates than on the (100) because for substrates of non-(100) orientations, the <100> direction does not coincide with the direction of the hole source.

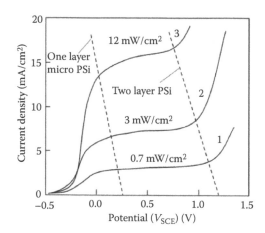

**FIGURE 4.51**   Voltammograms of $n$-Si electrodes in aqueous solution containing 10 wt.% HF and 35 wt.% $C_2H_5OH$ at 5 mV under different light intensities. The illumination intensity was adjusted by changing the distance between the light source and the $n$-Si electrode. (Reprinted from *J. Electroanal. Chem.* 396, Osaka T. et al., 69, Copyright 1995, with permission from Elsevier.)

Therefore, due to the large tendency to branch, it is difficult to produce straight perpendicular pores with smooth walls on (110) and (111) samples, even on surface patterned samples (Allongue et al. 1997).

5. Pore diameters of the PSi layer formed under a given set of conditions can have different distributions. Normal, log-normal, bimodal, fractal, and non-uniform distributions have been found for PSi (Herino et al. 1987; Yaron et al. 1993; Houbertz et al. 1994; Frohnhoff et al. 1995; Binder et al. 1996; Billat et al. 1997; Osaka et al. 1997). For the PSi formed on heavily doped silicon, the pore diameters have a narrower distribution at a lower current at a given HF concentration and the distribution is narrower at a lower HF concentration at a given current density. Bimodal distributions of the pore diameters are generally associated with two-layer PSi formed on lowly doped $p$-Si and illuminated $n$-Si. Pores with multiple distributions have been observed for the PSi that has a surface micropore layer and smaller branched pores in addition to the main pores (Binder et al. 1996; Billat et al. 1997). The distribution of pore diameters for highly branched PSi has been found to be fractal-like (Chuang et al. 1989; Smith and Collins 1992). Illumination during formation of porous layers on $p$-Si can also affect the distribution of pore diameters. It was found that illumination increases the amount of the smaller nanoscrystals, while it reduces the amount of larger crystals (Thonissen et al. 1996a). It was also shown that for the PSi formed under illumination, the relative number of small crystals tends to increase with reduction of light wavelength (Thonissen et al. 1996a). Anodization was started after removal of an $SiO_2$ resist film. Applied current density was 13 mA/cm$^2$, and it was electrolyzed for 60 min.

6. The depth of macropores depends on current density and HF concentration. Low HF concentration, low temperature, and low growth rate favor formation of deep uniform pores. Pore size and depth variation of porous layers on $n$-Si are very different for front and back illuminated $n$-Si samples.

More detailed description of possible morphologies and microstructures of PSi layer can be found in published articles (Lehmann et al. 2000; Chazalviel et al. 2002; Zhang 2005).

## 4.2.2 INFLUENCE OF ANODIZATION PARAMETERS ON THE SURFACE AREA OF PSi LAYERS

The surface area of porous silicon is of great importance in the development of many devices, including gas sensors and heterogeneous catalysts because this parameter determines how intense the interaction of porous silicon with surrounding atmosphere will be. For example, to achieve the maximum efficiency of these devices, the surface area should be maximized. However, it should be noted that, unfortunately, in real experiments, relating to the formation of porous silicon, the surface area is controlled very rarely. We found only a few studies where this parameter was controlled. These include the works of Herino et al. (1987), Bomchil et al. (1993), Halimaoui (1994), Salonen et al. (2000), and Du Plessis (2007a,b). Therefore, this section presents mainly the results obtained by these authors.

During these studies, it was concluded that the surface area is not strictly proportional to the porosity as one might expect (Bomchil et al. 1989, 1993). For example, it is often incorrectly supposed that an increase of porosity is associated with an increase of the surface area. In fact, as a rule, the dependence of surface area on the porosity has the maximum, and at a certain value of the porosity the surface area begins to decrease with increasing porosity (see Figure 4.52). According to studies, this reduction, depending on conditions of anodization, begins when the porosity reaches 50–60%. As a result, on curves, which characterize the dependence of surface area on the current density during anodization, the maxima appear as well. For example, Figure 4.53 shows the effect of current density on the surface area of porous silicon formed in the $p^+$-Si (100) wafers using HF:H$_2$O:EtOH-based electrolyte. These dependences were determined by Salonen et al. (2000). The etching times were varied from 10 to 40 min depending on the current density. It was established that in all studied cases, the maximum specific surface area was observed in samples with a porosity of about 60%. This arises from the fact that the specific surface area per weight continuously increases as a function of current density (Figure 4.53b

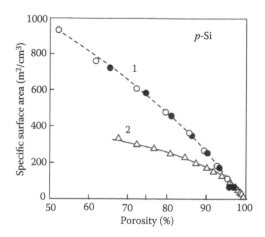

**FIGURE 4.52**    Specific surface area as a function of PSi porosity: (1) Layer thickness ~1 μm. (Data extracted from Herino R. et al., *J. Electrochem. Soc.* 134, 1994, 1987; Halimaoui A., *Surf. Sci.* 306, L550, 1994 (*p*⁺-Si (100), ρ = 1 Ω·cm, 25%HF in EtOH:H₂O = 1:1); Du Plessis M., *Phys. Stat. Sol. (a)* 204(7), 2319, 2007 (*p*-Si, HF:EtOH).)

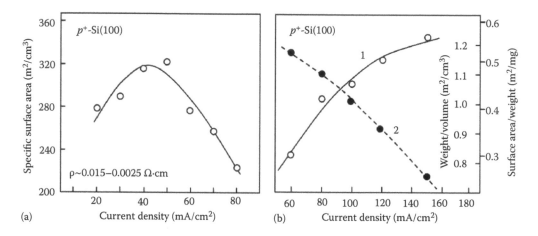

**FIGURE 4.53**    Influence of current density on (a) specific surface area per volume and (b) surface area per weight of *p*⁺-PSi (100): (a) PSi was formed in HF (28%):EtOH electrolyte; (b) PSi was formed in 25% HF in ethanol:water = 1:1 solution. Surface area was measured by the BET method. (Salonen J. et al.: *Phys. Stat. Sol. (a)*. 2000. 182. 249. Copyright Wiley-VCH Verlag GmbH & Co. KGaA. Reproduced with permission.)

[curve 1]), but simultaneously, the weight per volume decreases as the porosity increases (Figure 4.53b [curve 2]), leading to the observed behavior. Using X-ray small-angle scattering of PSi samples with different porosities, Vezin et al. (1992) have also found that the surface area of the *p*⁻ samples decreased with the increase of porosity from 50% to 85%. Similar regularities were obtained by Takemoto et al. (1994) using BET measurements. Bomchil et al. (1993) also observed that the increasing of the porosity up to 85% leads to a complete dissolution and disappearance of the smallest particles, to the increase of pore sizes due to pore coalescence, and to the decrease of surface area. This result testifies that use of a PSi with a maximum porosity is not always the optimal way for the development of specific devices.

Salonen et al. (2000) also showed that the specific surface area depends on the electrolyte composition and concentration of HF (see Figure 4.54). However, the influence of HF concentration was not as significant as the effect of current density. The changes in specific surface areas as a function of HF concentration were quite small: the difference between the highest and lowest values was less than 25 m²/cm³. At that, the largest specific surface areas were observed in the samples anodized in HF–water solutions (water-based electrolyte). However, these samples were less homogeneous than the samples anodized in HF:H₂O:EtOH solutions.

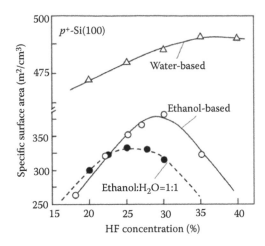

**FIGURE 4.54**   Maximum obtained specific surface areas as a function of HF concentration for three different electrolyte contents. (Salonen J. et al.: *Phys. Stat. Sol. (a)*. 2000. 182. 249. Copyright Wiley-VCH Verlag GmbH & Co. KGaA. Reproduced with permission.)

In addition, Salonen et al. (2000) have found that the achievement of the maximum surface area requires careful selection of the anodization conditions, which depend on both the composition of the electrolyte, and the concentration of HF (see Figure 4.55). In particular, it was shown that an increase of ethanol concentration in the electrolyte causes a reduction in the specific surface area, but on the other hand, it also increases the average pore size, which was found to be linearly decreasing with the HF concentration. In addition, the current density, needed to produce the maximum specific surface areas, behaved as a linear function of HF concentration, but contrary to pore size, it was linearly decreasing. Indicated dependences show that the formation of PSi with morphology required for a specific application is a difficult task, which requires a complete understanding of such a complex relationship between parameters of porous silicon and formation conditions.

It should also be taken into account that the surface area depends on the resistance of silicon wafers used. In particular, as it is seen from the data shown in Figure 4.56, for achievement of the maximum surface area the substrates used should be with low-resistance. Such behavior is to be expected because the decrease in resistance contributes to the formation of porous silicon with a smaller pore diameter, which promotes the growth of the surface area (see Figures 4.40 and 4.41). Therefore, one can conclude that if greater porosities occur by thinning of the silicon particles,

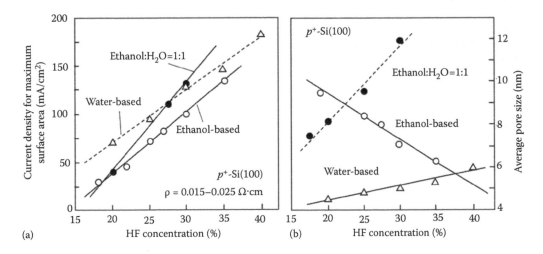

**FIGURE 4.55**   (a) Current densities used to obtain the maximum specific surface areas (plotted in Figure 5.54) for each HF concentration; (b) average pore sizes measured from the samples exhibiting maximum specific surface areas. (Salonen J. et al.: *Phys. Stat. Sol. (a)*. 2000. 182. 249. Copyright Wiley-VCH Verlag GmbH & Co. KGaA. Reproduced with permission.)

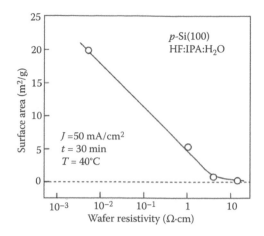

**FIGURE 4.56** The change in the surface area per weight for the PSi as a function of the resistivity of the wafers: IPA-isopropyl alcohol ($C_3H_8O$); surface area was determined by BET method. (With kind permission from Springer Science+Business Media: *Nanoscale Res. Lett.* 7, 2012, 408, Kim H. and Cho N.)

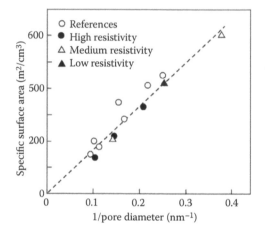

**FIGURE 4.57** Experimental specific surface area $S$ of porous $p$-Si as a function of inverse of pore diameter $d_p$. (Du Plessis M.: *Phys. Stat. Sol. (a)*. 2007. 204(7). 2319. Copyright Wiley-VCH Verlag GmbH & Co. KGaA. Reproduced with permission.)

thus by enlarging the pore size, the surface area should decrease. This means that a large specific surface area can only be achieved with a corresponding small pore diameter. Moreover, the results presented in Figure 4.57 show that the coefficient of proportionality between the specific surface area and $1/d_p$ does not depend on porous growth conditions (different impurity doping densities, HF concentrations, etc.) and is approximately 1725 $m^2 \cdot nm/cm^3$ (Du Plessis 2007b).

It should be noted that in contrast to the specific surface area, between the porosity and the pore size a linear relationship is observed. As it was shown in previous sections, the pore size and porosity have identical dependences on the current density and HF concentration in the electrolyte. However, it is necessary to have in mind that those correlations are valid only for the samples of the same type and doping level (Bomchil et al. 1989).

## 4.3 POROUS MULTILAYERS FORMATION

The opportunity to form multilayers of porous materials during the process of anodization is an important peculiarity of Si porosification. It provides additional great possibilities for elaboration of various structures, prospective for sensor and optoelectronic applications (Bettotti et al. 2002). It was established that the opportunity of forming multilayer structures using PSi of different porosities relies on the basic characteristics of the etching process. The etching is self-limiting and occurs only in the area of the pore tips. At that, the already etched structure is not affected by further electrochemical etching of the wafer.

There are two types of PSi multilayers with controlled variations of the porosity and microstructure, which are classified by the way the porosity is changed from one layer to another (Thonissen et al. 1996b). Schematically these ways are shown in Figure 4.58. In the first type of multilayer, the etching parameters (mostly the current density) are changed during anodization (Osaka et al. 1995; Pavesi and Mulloni 1999), whereas in the second type the change in porosity is determined by changing the depth profile of substrate doping (Berger et al. 1994, 1995). Heavily doped regions are being etched faster than the low-doped regions. In the darkness, $n$-type doped regions embedded into $p$-type doped regions are not attacked, and controlled doping profiles result in controlled PSi formation (Higa and Asano 1996). This last approach can produce very sharp interfaces (Berger et al. 1994), but it turned out to be less convenient than the current variation approach because it requires epitaxial growth of the Si substrate. For this reason, the first type of PSi multilayers, described in the literature, is more common (Pavesi 1997; Trifonov et al. 2005; Quiroga-Gonzalez et al. 2014).

Using the first method, it is possible to vary the porosity and, therefore, the refractive index only at the etch front by changing the current density during the anodization process. In this way, an arbitrary current versus time profile is transferred into porosity versus depth profile. According to Lehmann's model (Lehmann 1993) for a regular pore arrangement, the porosity $P$ of the macropore array is determined by the ratio of the total current density $J$ to the critical current density $J_{PSL}$. For example, in the case of square pore arrangement, the porosity, $P$, is related to the current density ratio according to

$$(4.6) \qquad P = \frac{A_{\text{pore}}}{A_{\text{cell}}} = \frac{\pi}{4}\left(\frac{d}{a}\right)^2 = \frac{J}{J_{PSL}},$$

where $A_{\text{pore}}$ and $A_{\text{cell}}$ are the pore cross-section and the area of unit cell, respectively. Parameters $d$ and $a$ denote the pore diameter and the pitch, respectively. $J_{PSL}$ is the current density for the PSL peak (electropolishing peak). The current density $J_{PSL}$ is constant under the given conditions of temperature, etchant concentration, and depth. Therefore, as the current density at the pore tips is constant, varying the total current $J$ can control the pore diameters $d$ since $\pi(d/2a)^2 J_{PSL} = J$. Experiment has shown that electrochemically etched pores with modulated diameter actually have a circular or slightly faceted shape in cross-section due to anisotropic properties of the electrochemical etching process. The cross-sectional shape of the pores cannot be changed, whereas the pore shape along the growing direction can be modeled by designing an appropriate profile of the applied etching current.

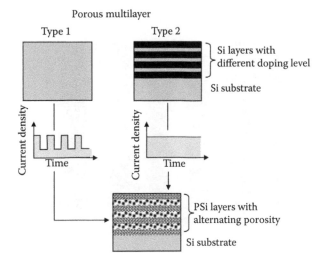

**FIGURE 4.58**  Preparation of Type 1 and Type 2 porous silicon multilayer structures. Type 1 multilayers are prepared from uniformly doped crystalline silicon substrates. Type 2 multilayers are prepared from silicon substrates with differently doped epitaxial layers.

However, one should note that for forming high quality structures, it is necessary to control not only current density (Quiroga-Gonzalez et al. 2014). For example, good *n*-Si macropore formation implies that at the vicinity of the pore tip, the pore walls are almost completely passivated, that is, that a full SCR (space-charge region) is formed, but this is not always assured, however. The dissolution process induces charged surface states, which may reduce the width of the SCR and thus contribute to the thinning or overetching of the pore walls. Since pore diameter modulations also modulate the speed of the pore growth into the bulk, the passivation kinetics should also be considered. Faster pore growth with increasing depth, for example, requires electrolyte additives (surfactants) that reduce dissolution kinetics by improved passivation. HF concentration and substrate doping should be chosen carefully in order to obtain the maximum variations of the refractive index.

It was also established that in *p*-type Si, the modulation of the pore diameter is more complicated than in *n*-type Si (Quiroga-Gonzalez et al. 2014). In this case, the etching process is generally omnidirectional owing to the high hole concentration and faster hole mobility at the interface, and lateral pore growth is severely limited by the SCR between pores (Lehmann and Ronnebeck 1999; Carstensen et al. 2000a; Kochergin and Föll 2009). Since the availability of electronic holes necessary for the etching is much higher in *p*-type silicon than in *n*-type silicon, a selective passivation of pore walls is much more difficult to achieve. This necessary selective passivation is strongly influenced by several factors: (1) the crystal anisotropy, for example, {100} oriented pore walls are more susceptible to overetching than {111} pore walls; (2) overlapping SCR, enforced by prestructured nucleation sites with suitable geometry; (3) strongly surface-passivating electrolytes; (4) a large curvature around the pore tips, leading to a large electrical field at the pore tip. Therefore, Föll and co-workers (Quiroga-Gonzalez et al. 2011, 2014) proposed for pore diameter modulation in *p*-type Si to use current densities much lower than $J_{PSL}$. In this case, it is possible to reduce the pore diameters by decreasing the current during etching. Additionally, it has been reported that electrolyte additives like polyethyleneglycol (PEG) enhance attempted diameter modulations by improving the passivation of the pore walls (Ossei-Wusu et al. 2013).

It is also necessary to take into account that much bigger variations of the refractive index are possible for heavily *p*-type doped substrates, both for the large value of critical current density and the wide range of porosities achievable with a given HF concentration. Moreover, the etch rate is higher and the inner surface is lower for *p*-type doped substrates, and, therefore, the additional chemical dissolution due to the permanence in HF is almost negligible. However, the refractive index and etching rate for a single layer are modified by the presence of the multilayer structure, and values lower than those determined for thick layers obtained with the same current density have been systematically observed for both the refractive index and the etching rate (Mazzoleni and Pavesi 1995). The examples of PSi layers with modulated porosity are shown in Figure 4.59.

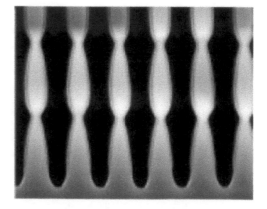

**FIGURE 4.59**   Macropores with modulated pore diameter in depth. The pores were etched by applying sawtooth-like current profile (2.5 wt.% HF; 1.4 V; $t = 280$ min). The fabricated microstructure consists of 10-cycle ratchet-type pores with a period of about 8.2 μm. (Trifonov T. et al.: *Phys. Stat. Sol. (c)*. 2005. 2. 3104. Copyright Wiley-VCH Verlag GmbH & Co. KGaA. Reproduced with permission.)

However, we need to recognize that the variety of attainable pore geometries is limited. Not every pore profile can be realized by modulating the etching current only. Trifonov et al. (2008) have found the following limitations:

1. At first glance, the shown structures look pretty good but there are several imperfections that can be noticed after detailed SEM inspection. For example, the pores are slightly conical with a decreasing diameter in depth. Such carrot-shaped pores are an effect of the pure chemical dissolution of silicon. Dissolution of an *n*-type electrode occurs even though the electrode is kept in the dark, due to the chemical nature of the process. This chemically induced dark current, which is on the order of a few $\mu A/cm^2$, leads to a consumption of HF not only at the pore tips but also at the pore walls. It increases with increasing etching time because the pore surface area becomes higher. In addition, dark current also depends on the temperature and on the concentration of oxygen diluted in the electrolyte (Lehmann 2002) and is therefore difficult to quantify.

2. Another source of imperfections is the limited diffusion in narrower and deeper pores. The HF concentration decreases toward the pore tips because of the diffusion-limited transport through the pores. These diffusional limitations reduce the critical current $J_{PS}$ and, hence, the growth rate. In addition, the dark current discussed above also contributes to the reduction of the growth rate. Consequently, the variations in the pore diameter are not equidistant in depth. Deviations of about 20% in the length of the period from top to bottom for 100 $\mu m$ porous film can be observed (Müller et al. 2000). This reduction of the growth rate is found to be a nonlinear function of the pore depth and can be approximated by a second-order polynomial (Barillaro and Pieri 2005).

3. We can also see from the shown SEM micrographs that the etched pores do not have sharp modulation, even though the used current profile has sharp peaks. These peaks are smoothed out in the resulting structure. As pointed out by Matthias et al. (2005b), the combination of current modulation and constant anodic potential will always restrict the achievable pore shapes to smoothly modulated ones. This means that modulations with sharp edges, like square or trapezoidal, are not possible. As a rule of thumb, the shorter the length of modulation, the lower is the variation of pore diameter.

For improvement of the formed structure parameters, Trifonov et al. (2008) suggest using the following approaches:

- To optimize conditions of electrochemical etching. For example, experiments have shown that etching at reduced temperature helps somewhat to fabricate sharper modulations of the diameter. Bubbling of the electrolyte with nitrogen and using an appropriate surfactant also help to reduce the dark current and to minimize its effect on the structure of fabricated PSi (Trifonov et al. 2005). As it is known, the dark current depends on the temperature and the oxygen concentration in the electrolyte (Kolasinski et al. 2000). However, we have to note that these improvements do not allow the manufacturing of sharp edges.

- To optimize a program of a current changing during the process of electrochemical etching. In particular, the frequency of the current variations must be reduced in order to achieve equidistant changes of pore diameter in depth. A few etching runs usually provide enough experimental data to optimize the process and to get homogeneous pore modulation for more than 100-$\mu m$ deep pores (Müller et al. 2000; Matthias et al. 2004, 2005a). In addition, to obtain a constant pore diameter, we should increase the applied current $J_{applied}$ during the etching in order to keep $J_{real}$ constant. The example of ordered PSi structure with modulated pores fabricated in optimal conditions is shown in Figure 4.60.

The change of backside illumination can also be used for pore modulation (Schilling et al. 2001b; Matthias et al. 2004). Increasing the backside illumination intensity causes a higher hole generation rate, which results in an increased total current density $J$ flowing through the whole sample. Since $J_{PS}$ remains fixed at the pore tips, from Equation 4.5 the increased total current $J$ will give rise to a larger pore cross-section and thus, to wider pores. Therefore, we can modulate

**FIGURE 4.60** SEM images of porous structure with pore diameter modulation. Sharp modulations of pore diameter are achieved by varying the applied anodic potential together with the etching current (5 wt.% HF; 15 °C; $t = 93$ min). The length of a period is almost the same as the pitch of the pattern. (Reprinted from *Sens. Actuators A* 141, Trifonov T. et al., 662, Copyright 2008, wth permission from Elsevier.)

the pore diameter in the growth direction by varying the backside illumination intensity. To achieve periodic modulation of the pore diameter, the light intensity must be varied periodically while the pores grow into the substrate. Modulation of the illumination intensity with a triangular profile leads to a sinusoidal pore diameter variation. With an asymmetric current density profile, a ratchet-type pore shape is possible as well. An example of a PSi structure fabricated using the indicated approach is shown in Figure 4.61.

Matthias et al. (2005a) also analyzed conditions of macroporous silicon forming and have shown that the conventional regime when an applied voltage is constant and kept below the breakdown potential really has a limitation in pore diameter modulation. In this regime, all the holes are supplied by backside illumination. They also found that applying a constant voltage close to the breakdown potential allows for the fabrication of strong modulations in diameter. Increasing the voltage above the breakdown potential results in the generation of holes near the pore tips by interband tunneling of electronic holes. However, it was established that this growth regime is rather unstable. Some pores branch and others die. The etch front becomes non-uniform and neighboring pores no longer look the same. In order to avoid this, Matthias et al. (2005a) proposed to combine the advantages of stable macroporous silicon growth with the growth near breakdown. The illustration of pore growth using the proposed regime is shown in Figure 4.62. Figure 4.62b presents the applied voltage and current profile. At the beginning of each modulation, a low etching current combined with a high-voltage period is applied. The voltage is approximately the breakdown potential. As shown by Matthias et al. (2005a), an increase of the voltage in the beginning of

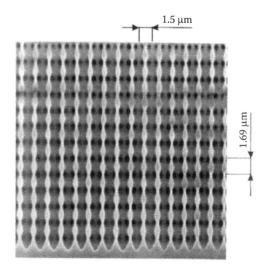

**FIGURE 4.61** SEM image showing a longitudinal section of the modulated pore structure. The variation in the pore diameter with depth can be modulated by a sinusoidal light intensity modulation. (Reprinted with permission from Schilling J. et al., *Appl. Phys. Lett.* 78, 1180, 2001. Copyright 2001. American Institute of Physics.)

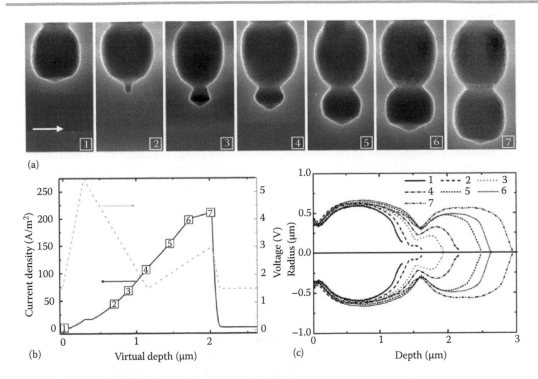

**FIGURE 4.62**   Sequence showing the growth of one diameter modulation. (a) Seven stages (indicated by numbers) of the growth of one modulation. (b) Applied current–voltage profile for the growth of the modulation. The points where the etching was interrupted are numbered. (c) Measured pore shape. Experiments were performed on float-zone silicon wafers with (100) orientation and a resistivity of 0.5 Ω·cm. The hydrofluoric acid has a concentration $C_{HF}$ = 5wt.% at $T$ = 10°C. (Reprinted with permission from Matthias S. et al., *J. Appl. Phys.* 98, 023524, 2005b. Copyright 2005. American Institute of Physics.)

a new modulation can be used to form pores with very small diameters. During this step, a certain amount of charge carriers is generated by breakdown and only a few carriers are generated by the illumination. After a tiny pore has formed, this unstable regime is left by decreasing the voltage again and the pore is widened in its diameter by an increased backside illumination intensity. Figure 4.62a shows SEM cross-sectional images of a series of seven samples, which were etched with the same current–voltage profile and stopped at different phases of one modulation. The different stages for the images are indicated by numbers. The resulting graph is depicted in Figure 4.62c and emphasizes the details of pore growth during the different phases.

It was found that the modulation of both current and voltage allows for strong diameter variations on a length scale close to the in-plane periodicity and the fabrication of sharp edges. Matthias et al. (2005a) established that the most important step to achieve the strong modulations using the proposed regime is the fabrication of the tiny pore (Figure 4.62a [image 2]). Growing longer pores under breakdown conditions leads to additional breakdown pores along the other (100) directions, mostly pronounced at the neck of the previous modulation. This limits the length of the tiny pore in our experiments to values around 300 nm. A careful adaptation of the applied potential to the varying pore-tip radius could perhaps increase this further.

## 4.4 THREE-DIMENSIONAL STRUCTURING

The structures, shown in Section 4.3, do not actually possess a real 3D geometry. To realize a real 3D structure, we must connect the neighboring pores in a direction perpendicular to the pores. Trifonov et al. (2008) consider two possibilities for this task's realization. First, we can use, for example, higher currents so that the ratio $J_{max}/J_{PSL} = P$ defines porosity $P$ greater than unity. In this way, overlapping pores will be etched only at the positions of current maxima. Etching of overlapping pores is also possible and has been used to fabricate silicon nanowires (Van den Meerakker et al. 2003). However, the etching process becomes rather unstable at such high porosities and the system can easily be drawn out from the steady-state regime of stable pore growth. Second, the good way to achieve truly 3D structures is to widen the pores after the etching. This can be done by several oxidation/oxide-stripping cycles, which leads to a dissolution of the pore walls only at the positions of diameter maxima. The neighboring pores will then be connected along the (011) directions. The transformation of PSi structure after such treatments is illustrated in Figure 4.63.

Trifonov et al. (2008) have shown that the etching by itself is quite fast: it takes less than 5 min to etch one period of modulation. This 3D shaping can be applied on a large scale, that is, on the whole silicon wafer. This can be particularly helpful for the preparation of 3D silicon molds, which can be used to cast 3D structures of unusual materials. Figure 4.64 shows a bird's-eye-view of the resulting structure at different magnifications. It is a real 3D network. It also seems that this 3D structure resembles a simple cubic lattice of overlapping air spheres embedded in silicon.

The obtained geometry, however, is slightly different from that of overlapping spheres. First, since the length of the period was fixed at 5.7 µm and the pitch of the lattice was 4 µm, the structure does not approximate a perfectly spherical shape in a direction along the pore axis. To obtain

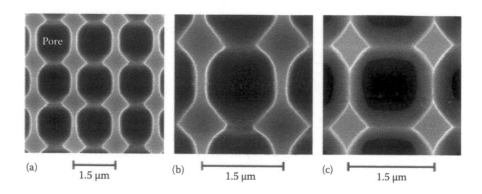

**FIGURE 4.63** SEM images taken of three-dimensionally µ-shaped macroporous silicon. The pores are arranged in a square lattice with a pitch of 1.5 µm. (a) A cut-out of 3 pores with 3 modulations fabricated by the developed etching process. (b) Single modulation. The length of the modulation is equal to the lateral dimensions of the lithographically defined pitch dimensions of 1.5 µm. (c) Widened single modulation. The samples were annealed at 900°C and the grown oxide was removed by hydrofluoric acid. The resulting structure is interconnected along the Cartesian axes. (Matthias S. et al.: *Adv. Mater.* 2004. 16. 2166. Copyright Wiley-VCH Verlag GmbH & Co. KGaA. Reproduced with permission.)

**FIGURE 4.64** Bird's-eye-views of the 3D structure produced by etching symmetrically modulated pores and by subsequent widening of the pores. The structure is a fully 3D network of interconnected voids in silicon. (Reprinted from *Sens. Actuators A* 141, Trifonov T. et al., 662, Copyright 2008, with permission from Elsevier.)

this, the period of modulation must be equal to the pitch of the square lattice. Trifonov et al. (2008) have shown that the quality of these 3D networks mainly depends on the etching process and on the ability to etch homogeneously modulated pores. For example, the desired sphericity will hardly be achieved because any sharp features along the pore axis will be smoothed out due to over-etching. To achieve an ideal spherical geometry, intersections between the overlapping air spheres must exhibit extremely sharp cusps. While the cusp between adjacent pores is still preserved, the cusp along the pore axis is smoothed out due to enhanced etching. This problem has been circumvented to some extent by carefully adjusting the current–voltage profile and, specifically, the stage where low etching current is combined with a high-applied voltage (Matthias et al. 2005a). The interplay between both etching parameters is a key factor in achieving the smallest radius of curvature (relative to the pore radius) that can be etched.

Other versions of 3D structures for photonic crystal applications fabricated using photoelectrochemical etching and anisotropic post treatment (Matthias et al. 2004) is shown in Figure 4.65. Potassium hydroxide (KOH)—the alkaline solution used here—etches in the <100> and <110> directions, two orders of magnitude faster than in the <111> direction (Kolasinski et al. 2000) resulting in cubic structures with squared cross-section. Such a cross-section of the pores breaks the cylindrical symmetry, and from this moment, the relative orientation with respect to the axes of the square lattice arrangement of the pores becomes important.

One can see that this approach allows fabricating a columnar structure of the air pores arranged in a simple cubic lattice in (001)-oriented $n$-type silicon wafers. The diameter of these cylindrical pores is strongly modulated and varies between 0.8 and 1.75 μm with a period that equals the lateral lithographically defined lattice constant (Figure 4.65a). Isotropic erosion, by several oxidization and subsequent etching steps, transfers this columnar structure into the simple cubic lattice of overlapping air spheres.

However, it is necessary to note that there are also other approaches to fabrication of 3D structures based on PSi. For example, Lourtioz and co-workers (Chelnokov et al. 2000; Wang et al. 2003) and Schilling et al. (2005) used a combination of photoelectrochemical macropore etching in silicon and subsequent drilling of pore set (see Figure 4.66). In the first step, a conventional two-dimensional array of straight pores is photoelectrochemically etched. Afterward, additional pores are drilled under oblique angles from the top using a focused ion beam (FIB). In this way, a set of three different pore directions is established which cross each other depth-wise. The FIB drilling of the etched macroporous silicon is faster than bulk material, and the problem of redeposition of milled material is minimized at the same time. This technique was successfully applied to fabricate an alternative 3D photonic crystal structure.

However, a complete three-dimensional band gap is not yet shown since the angles between the three different pore sets have not been properly aligned. Besides, the depth of the structure is limited due to the FIB process (Schilling et al. 2001a,b). In structures fabricated by Schilling et

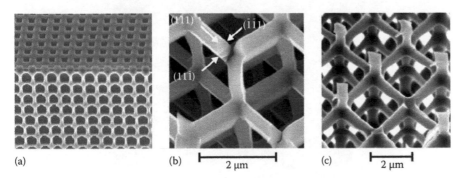

(a)    (b)    2 μm    (c)    2 μm

**FIGURE 4.65**    (a) SEM picture of strongly modulated pores in macroporous silicon arranged in a cubic primitive lattice with a lattice constant of 2 μm. Initial structure before a further postprocessing. (b–c) SEM pictures of cubic three-dimensional photonic crystals in macroporous silicon: (b) Side view of the scaffoldlike structure. The surfaces of the bars in the $x$ and $y$ directions are formed by the (111) planes of the silicon crystal. (c) Bird's-eye-view of the scaffoldlike structure obtained by anisotropic widening of macroporous silicon. (Reprinted with permission from Matthias S. et al., *J. Appl. Phys.* 98, 023524, 2005. Copyright 2005b. American Institute of Physics.)

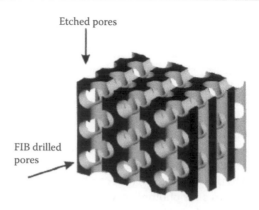

Etched pores

FIB drilled pores

**FIGURE 4.66**    Model of the orthorhombic structure consisting of two hexagonal pore sets: 3D view revealing the interpenetrating network of pores. (Reprinted with permission from Schilling J. et al., *Appl. Phys. Lett.* 86, 011101, 2005. Copyright 2005. American Institute of Physics.)

al. (2005), a 100-μm-long trench was milled out of the porous region applying a dual beam FIB (model DB235 from FEI).

The technique described by Christophersen et al. (2000a,b) also can be considered as a method of 3D structure fabrication. In contrast to the pore growth on a (100)Si surface in case of a (111) Si surface, the pores grow into <113> directions. As there are three equivalent <113> directions available from the (111) surface, three pores start to grow from each nucleation point at the surface. Band structure calculations for a corresponding structure show that the pores along the three <113> directions grow at suitable angles such that the structure should exhibit a three-dimensional complete photonic bandgap. Figure 4.67 shows an image of such a structure where the nucleation spots of the pores at the (111) surface are still randomly distributed. Therefore, this structure does not yet have the described lon-range periodicity of crossing pores and exhibits no photonic bandgap. However, the crossing pores are clearly visible and the intended structure can be imagined (Schilling et al. 2001a,b).

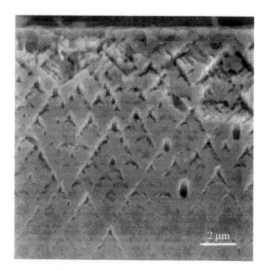

2 μm

**FIGURE 4.67**    Crossing pores caused by the photoelectrochemical etching of a (111) *n*-type silicon surface. The nucleation spots of the pores at the (111) surface are random so that a strict periodic arrangement cannot yet be obtained. (Reprinted with permission from Schilling J. et al., *Appl. Phys. Lett.* 78, 1180, 2001a. Copyright 2001. American Institute of Physics.)

## ACKNOWLEDGMENTS

This work was supported by the Ministry of Science, ICT and Future Planning (MSIP) of the Republic of Korea.

## REFERENCES

Allongue, P., de Villeneuve, C.H., Bernard, M.C., Peou, J.E., Boutry-Forveille, A., and Levy-Clement, C. (1997). Relationship between porous silicon formation and hydrogen incorporation. *Thin Solid Films* **297**, 1–4.

Arita, Y. and Sunahama, Y. (1977). Formation and properties of porous silicon film. *J. Electrochem. Soc.* **124**, 285–295.

Badel, X., Rajendra Kumar, R.T., Kleimann, P., and Linnros, J. (2004). Formation of ordered pore arrays at the nanoscale by electrochemical etching of n-type silicon. *Superlattice Microstr.* **36**, 245–253.

Balagurov, L.A., Loginov, B.A., Petrova, E.A., Sapelkin, A., Unal, B., and Yarkin, D.G. (2006). Formation of porous silicon at elevated temperatures. *Electrochim. Acta* **51**, 2938–2941.

Barillaro, G. and Pieri, F. (2005). A self-consistent theoretical model for macropore growth in n-type silicon. *J. Appl. Phys.* **97**, 116105 (1–3).

Beale, M.I.J., Benjamin, J.D., Uren, M.J., Chew, N.G., and Cullis, A.G. (1985a). An experimental and theoretical study of the formation and microstructure of porous silicon. *J. Crystal Growth* **73**, 622–636.

Beale, M.I.J., Chew, N.G., Uren, M.J., Cullis, A.G., and Benjamin, J.D. (1985b). Microstructure and formation mechanism of porous silicon. *Appl. Phys. Lett.* **46**(1), 86–88.

Berger, M.G., Dieker, C., Thonissen, M. et al. (1994). Porosity superlattices: A new class of Si heterostructures. *J. Phys. D: Appl. Phys.* **27**, 1333–1336.

Berger, M.G., Thonissen, M., Arens-Fischer, R. et al. (1995). Investigation and design of optical properties of porosity superlattices. *Thin Solid Films* **255**, 313–316.

Bettotti, P., Dal Negro, L., Gaburro, Z. et al. (2002). p-type macroporous silicon for two-dimensional photonic crystals. *J. Appl. Phys.* **92**, 6966–6972.

Billat, S., Thonissen, M., Arens-Fischer, R., Berger, M.G., Kruger, M., and Luth, H. (1997). Influence of etch stops on the microstructure of porous silicon layers. *Thin Solid Films* **297**, 22–25.

Binder, M., Edelmann, T., Metzger, T.H., Mauckner, G., Goerigk, G., and Peisl, J. (1996). Bimodal size distribution in p⁻ porous silicon studied by small angle X-ray scattering. *Thin Solid Film* **276**, 65–68.

Bisi, O., Ossicini, S., and Pavesi, L. (2000). Porous silicon: A quantum sponge structure for silicon based optoelectronics. *Surf. Sci. Rep.* **38**, 1–126.

Blackwood, D.J. and Zhang, Y. (2003). The effect of etching temperature on the photoluminescence emitted from, and the morphology of, p-type porous silicon. *Electrochim. Acta* **48**, 623–630.

Bomchil, G., Herino, R., Barla, K., and Pfister, J.C. (1983). Pore size distribution in porous silicon studied by adsorption isotherms. *J. Electrochem. Soc.* **130**, 1611–1614.

Bomchil, G., Halimaoui, A., and Herino, R. (1989). Porous silicon: The material and its applications in silicon-on-insulator technologies. *Appl. Surf. Sci.* **41/42**, 604–613.

Bomchil, G., Halimaoui, A., Sagnes, I. et al. (1993). Porous silicon: Material properties, visible photo and electroluminescence. *Appl. Surf. Sci.* **65/66**, 394–407.

Canham, L.T. (1991). Silicon quantum wire array fabrication by electrochemical and chemical dissolution of wafers. *Appl. Phys. Lett.* **57**, 1046–1048.

Canham, L. (ed.) (1997). *Properties of Porous Silicon.* INSPEC, London.

Carstensen, J., Christophersen, M., and Föll, H. (2000a). Pore formation mechanisms for the Si-HF system. *Mater. Sci. Eng. B* **69–70**, 23–28.

Carstensen, J., Christophersen, M., Hasse, G., and Föll, H. (2000b). Parameter dependence of pore formation in silicon with a model of local current bursts. *Phys. Stat. Sol. (a)* **182**, 63–69.

Chao, K.J., Kao, S.C., Yang, C.M., Hseu, M.S., and Tsai, T.G. (2000). Formation of high aspect ratio macropore array on p-type silicon. *Electrochem. Sol.-St. Lett.* **3**(10), 489–492.

Chazalviel, J.-N., Wehrspohn, R.B., and Ozanam, F. (2000). Electrochemical preparation of porous semiconductors: From phenomenology to understanding. *Mater. Sci. Eng. B* **69–70**, 1–10.

Chazalviel, J.-N., Ozanam, F., Gabouze, N., Fellah, S., and Wehrspohn, R.B. (2002). Quantitative analysis of the morphology of macropores on low-doped p-Si. *J. Electrochem. Soc.* **149**(10), C511–C520.

Chelnokov, A., Wang, K., Rowson, S., Garoche, P., and Lourtioz, J.-M. (2000). Near-infrared Yablonovite-like photonic crystals by focused-ion-beam etching of macroporous silicon. *Appl. Phys. Lett.* **77**, 2943–2945.

Christophersen, M., Carstensen, J., and Föll, H. (2000a). Macropore formation on highly doped n-type silicon. *Phys. Stat. Sol. (a)* **182**, 45–50.

Christophersen, M., Carstensen, J., Feuerhake, A., and Föll, H. (2000b). Crystal orientation and electrolyte dependence for macropore nucleation and stable growth on *p*-type Si. *Mater. Sci. Eng. B* **69–70**, 194–198.

Christophersen, M., Merz, P., Quenzer, J., Carstensen, J., and Föll, H. (2001). Deep electrochemical trench etching with organic hydrofluoric electrolite. *Sens. Actuators A* **88**, 241–245.

Christophersen, M., Carstensen, J., Voigt, K., and Föll, H. (2003). Organic and aqueous electrolytes used for etching macro- and mesoporous silicon. *Phys. Stat. Sol. (a)* **197**(1), 34–38.

Chuang, S.-F., Collins, S.D., and Smith, R.L. (1989). Porous silicon microstructure as studied by transmission electron microscopy. *Appl. Phys. Lett.* **55**(15), 1540.

Connolly, E.J., O'Halloran, G.M., Pham, H.T.M., Sarro, P.M., and French, P.J. (2002). Comparison of porous silicon, porous polysilicon and porous silicon carbide as materials for humidity sensing applications. *Sens. Actuators B* **99**, 25–30.

Du Plessis, M. (2007a). A gravimetric technique to determine the crystallite size distribution in high porosity nanoporous silicon. *ECS Trans.* **9**(1), 133–142.

Du Plessis, M. (2007b). Relationship between specific surface area and pore dimension of high porosity nanoporous silicon—Model and experiment. *Phys. Stat. Sol. (a)* **204**(7), 2319–2328.

Feng, R. (1999). *Ultrasonic Handbook*. Nanjing University Press, Nanjing.

Feng, Z.C. and Tsu, R. (eds.) (1994). *Porous Silicon*. Word Science, Singapore.

Föll, H. (1991). Properties of silicon-electrolyte junctions and their application to silicon characterization. *Appl. Phys. A* **53**, 8–19.

Föll, H., Christophersen, M., Carstensen, J., and Hasse, G. (2002). Formation and application of porous silicon. *Mater. Sci. Eng. R* **39**, 93–141.

Föll, H., Carstensen, J., and Frey, S. (2006). Porous and nanoporous semiconductors and emerging applications. *J. Nanomater.* **2006**, 91635 (1–10).

Frohnhoff, St., Marso, M., Berger, M.G., Thonossen, M., Luth, H., and Munder, H. (1995). An extended quantum model for porous silicon formation. *J. Electrochem. Soc.* **142**, 615–620.

Fukami, K., Harraz, F.A., Yamauchi, T., Sakka, T., and Ogata, Y.H. (2008). Fine-tuning in size and surface morphology of rod-shaped polypyrrole using porous silicon as template. *Electrochem. Commun.* **10**, 56–60.

Gelloz, B., Masunaga, M., Shirasawa, T., Mentek, R., Ohta, T., and Koshida, N. (2008). Enhanced controllability of periodic silicon nanostructures by magnetic field anodization. *ECS Trans.* **16**(3), 195–200.

Granitzer, P., Rumpf, K., Ohta, T., Koshida, N., Poelt, P., and Reissner, M. (2013). Magnetic field assisted etching of porous silicon as a tool to enhance magnetic characteristics. *ECS Trans.* 50(37), 55–59.

Gruning, U., Lehmann, V., Ottow, S., and Busch, K. (1996). Macroporous silicon with a complete two-dimensional photonic band gap centered at 5 μm. *Appl. Phys. Lett.* **68**, 747–749.

Guendouz, M., Joubert, P., and Sarret, M. (2000). Effect of crystallographic directions on porous silicon formation on patterned substrates. *Mater. Sci. Eng. B* **69–70**, 43–47.

Halimaoui, A. (1993). Influence of wettability on anodic bias induced electroluminescence in porous silicon. *Appl. Phys. Lett.* **63**, 1264.

Halimaoui, A. (1994). Determination of the specific surface area of porous silicon from its etch rate in HF solutions. *Surf. Sci.* **306**, L550–L554.

Halimaoui, A. (1997). Porous silicon formation by anodisation. In: Canham L.T. (ed.) *Properties of Porous Silicon*. IEE INSPEC, The Institution of Electrical Engineers, London, pp. 12–23.

Hasse, G., Christophersen, M., Carstensen, J., and Föll, H. (2000). New insights into Si electrochemistry and pore growth by transient measurements and impedance spectroscopy. *Phys. Stat. Sol. (a)* **182**, 23–29.

Hedrich, F., Billat, S., and Lang, W. (2000). Structuring of membrane sensors using sacrificial porous silicon. *Sens. Actuators B* **84**, 315–323.

Hejjo Al Rifai, M., Christophersen, M., Ottow, S., Carstensen, J., and Föll, H. (2000a). Dependence of macropore formation in *n*-Si on potential, temperature, and doping. *J. Electrochem. Soc.* **147**(2), 627–635.

Hejjo Al Rifai, M., Christophersen, M., Ottow, S., Carstensen, J., and Föll, H. (2000b). Potential, temperature and doping dependence for macropore formation on *n*-Si with backside-illumination. *J. Porous Mater.* **7**(1–3), 33–36.

Herino, R., Romchil, G., Boala, K., and Bertrand, C. (1987). Porosity and pore size distributions of porous silicon layers. *J. Electrochem. Soc.* **134**, 1994–2000.

Higa, H. and Asano, T. (1996). Fabrication of single-crystal Si microstructures by anodisation. *Jpn. J. Appl. Phys.* **35**, 6648–6651.

Hippo, D., Nakamine, Y., Urukawa, K. et al. (2008). Formation mechanism of 100-nm-scale periodic structures in silicon using magnetic-field-assisted anodization. *Jpn. J. Appl. Phys.* **47**(9), 7398–7402.

Hou, X.Y., Fan, H.L., Xu, L. et al. (1996). Pulsed anodic etching: An effective method of preparing light-emitting porous silicon. *Appl. Phys. Lett.* **68**, 2323.

Houbertz, R., Memmert, U., and Behm, R.J. (1994). Morphology of anodically etched Si(111) surfaces: A structural comparison of $NH_4F$ versus HF etching. *J. Vac. Sci. Technol. B* **12**(6), 3145–3148.

Iraji zad, A., Rahimi, F., Chavoshi, M., and Ahadian, M.M. (2004). Characterization of porous poly-silicon as a gas sensor. *Sens. Actuators B* **100**, 341–346.

Jager, C., Finkenberger, B., Jager, W., Christophersen, M., Carstensen, J., and Föll, H. (2000). Transmission electron microscopy investigations of the formation of macropores in *n*- and *p*-Si(001)/(111). *Mater. Sci. Eng. B* **69–70**, 199–204.

Jakubowicz, J. (2007). Nanoporous silicon fabricated at different illumination and electrochemical conditions. *Superlattice. Microst.* **41**, 205–215.

John, G.C. and Singh, V.A. (1995). Porous silicon: Theoretical studies. *Phys. Rep.* **263**, 93–151.

Kan, P.Y.Y. and Finstad, T.G. (2005). Oxidation of macroporous silicon for thick thermal insulation. *Mater. Sci. Eng. B* **118**, 289–292.

Kim, H. and Cho, N. (2012). Morphological and nanostructural features of porous silicon prepared by electrochemical etching. *Nanoscale Res. Lett.* **7**, 408.

Kleimann, P., Linnros, J., and Petersson, S. (2000). Formation of wide and deep pores in silicon by electrochemical etching. *Mater. Sci. Eng. B* **69–70**, 29–33.

Kochergin, V. and Föll, H. (2009). *Porous Semiconductors: Optical Properties and Applications*. Springer, London.

Koker, L. and Kolasinski, K.W. (2000). Applications of a novel method for determining the rate of production of photochemical porous silicon. *Mater. Sci. Eng. B* **69–70**, 132–135.

Kolasinski, K.W., Barnard, J.C., Koker, L., Ganguly, S., and Palmer, R.E. (2000). In situ photoluminescence studies of photochemically grown porous silicon. *Mater. Sci. Eng. B* **69–70**, 157–160.

Kordas, K., Remes, J., Beke, S., Hu, T., and Leppavuorim, S. (2001). Manufactoring of porous silicon: Porosity and thickness dependence on electrolyte composition. *Appl. Surf. Sci.* **178**, 190–193.

Korotcenkov, G. and Cho, B.K. (2010a). Porous semiconductors: Advanced material for gas sensor applications. *Crit. Rev. Sol. St. Mater. Sci.* **35**, 1–23.

Korotcenkov, G. and Cho, B.K. (2010b). Silicon porosification: State of the art. *Crit. Rev. Sol. St. Mater. Sci.* **35**(3), 153–260.

Lammel, G. and Renaud, P. (2000). Free-standing, mobile 3D porous silicon microstructures. *Sens. Actuators B* **85**, 356–360.

Lang, W. (1996). Silicon microstructuring technology. *Mater. Sci. Eng. R* **17**, 1–54.

Lehmann, V. (1993). The physics of macropore formation in low doped *n*-type silicon. *J Electrochem. Soc.* **140**(10), 2836–2843.

Lehmann, V. (1996a). Porous silicon formation and other photoelectrochemical effects at silicon electrodes anodized in hydrofluoric acid. *Appl. Sur. Sci.* **106**, 402–405.

Lehmann, V. (1996b). Porous silicon—A new material for MEMS. In: *Proceedings of IEEE Conference MEMS'96*, February 11–15, 1996, San Diego, CA, pp 1–6.

Lehmann, V. (2002). *Electrochemistry of Silicon*. Wiley-VCH, Weinheim, Germany.

Lehmann, V. and Föll, H. (1990). Formation mechanism and properties of electrochemically etched trenches in n-type silicon. *J. Electrochem. Soc.* **137**, 653–659.

Lehmann, V. and Gosele, U. (1991). Porous silicon formation: A quantum wire effect. *Appl. Phys. Lett.* **58**(8), 856–858.

Lehmann, V. and Gruning, U. (1997). The limits of macropore array fabrication. *Thin Solid Films* **297**, 13–17.

Lehmann, V. and Ronnebeck, S. (1999). The physics of macropore formation in low-doped p-type silicon. *J. Electrochem. Soc.* **146**, 2968–2975.

Lehmann, V., Hönlein, W., Reisinger, H., Spitzer, A., Wendt, H., and Willer, J. (1995). A new capacitor technology based on porous silicon. *Solid State Technol.* **38**, 99.

Lehmann, V., Stengl, R., and Luigart, A. (2000). On the morphology and the electrochemical formation mechanism of mesoporous silicon. *Mater. Sci. Eng. B* **69–70**, 11–22.

Letant, S.E., van Buuren, T.W., and Terminello, L.J. (2004). Nanochannel arrays on silicon platforms by electrochemistry. *Nano Lett.* **4**, 1705–1707.

Levy-Clement, C., Lagoubi, A., Ballutaud, D., Ozanam, F., Chazalviel, J.-N., and Neumann Spallart, M. (1993). Porous n-silicon produced by photoelectrochemical etching. *Appl. Surf. Sci.* **65/66**, 408–414.

Levy-Clement, C., Lagoubi, A., and Tomkiewicz, M. (1994). Morphology of porous n-type silicon obtained by photoelectrochemical etching I. Correlations with material and etching parameters. *J. Electrochem. Soc.* **141**, 958–967.

Liu, Y., Xiong, Z.H., Liu, Y. et al. (2003). A novel method of fabricating porous silicon material: Ultrasonically enhanced anodic electrochemical etching. *Sol. St. Commun.* **127**, 583–588.

Matthias, S., Müller, F., Jamois, C., Wehrspohn, R.B., and Gosele, U. (2004). Large-area three-dimensional structuring by electrochemical etching and lithography. *Adv. Mater.* **16**, 2166–2170.

Matthias, S., Müller, F., Schilling, J., and Gosele, U. (2005a). Pushing the limits of macroporous silicon etching. *Appl. Phys. A* **80**, 1391–1396.

Matthias, S., Müller, F., and Gösele, U. (2005b). Simple cubic three-dimensional photonic crystals based on macroporous silicon and anisotropic posttreatment. *J. Appl. Phys.* **98**, 023524.

Mazzoleni, C. and Pavesi, L. (1995). Application to optical components of dielectric porous silicon multilayers. *Appl. Phys. Lett.* **67**, 2983–2985.

Meek, R.L. (1971a). Anodic dissolution of *n*-pluss silicon. *J. Electrochem. Soc.* **118**, 437–442.

Meek, R.L. (1971b). $n^+$ silicon-electrolyte interface capacitance. *Surf. Sci.* **25**, 526–536.

Müller, F., Birner, A., Schilling, J., Gosele, U., Kettner, C., and Hanggi, P. (2000). Membranes for micro-pumps from macroporous silicon. *Phys. Stat. Sol. (a)* **182**, 585–590.

Nakagawa, T., Koyama, H., and Koshida, N. (1996). Control of structure and optical anisotropy in porous Si by magnetic-field assisted anodization. *Appl. Phys. Lett.* **69**, 3206.

Nakagawa, T., Sigiyama, H., and Koshida, N. (1998). Fabrication of periodic Si nanostructure by controlled anodization. *Jpn. J. Appl. Phys.* **37**, 7186–7189.

Nishio, K., Yasui, K., Matsumoto, F., Kanezawa, K., and Masuda, H. (2005). Direct nanoimprinting of Si single crystals using SiC molds for ordered anodic tunnel etching. *Adv. Mater.* **17**, 1293–1295.

Ono, H., Gomyou, H., Morisaki, H. et al. (1993). Effects of anodization temperature on photoluminescence from porous silicon. *J. Electrochem. Soc.* **140**, L180–L182.

Osaka, T., Ogasawara, K., Katsunuma, M., and Momma, T. (1995). Control of the porous structure of n-type silicon and its electroluminescence properties. *J. Electroanal. Chem.* **396**, 69–75.

Osaka, T., Ogasawara, K., and Nakahara, S. (1997). Classification of the pore structure of n-type silicon and its microstructure. *J. Electrochem. Soc.* **144**, 3226–3237.

Ossei-Wusu, E., Carstensen, J., Quiroga-Gonzalez, E., Amirmaleki, M., and Föll, H. (2013). The role of poly-ethylene glycol in pore diameter modulation in depth in p-type silicon. *ECS J. Solid State Sci. Technol.* **2**, 243–247.

Parkhutik, V. (1999). Porous silicon-mechanisms of growth and applications. *Sol. St. El.* **43**, 1121–1141.

Parkhutik, V.P., Glinenko, L.K., and Labunov, V.A. (1983). Kinetics and mechanism of porous layer growth during n-type silicon anodization in HF solution. *Surf. Technol.* **20**, 265–277.

Pavesi, L. (1997). Porous silicon dielectric multilayers and microcavities. *La Rivista del Nuovo Cimento* **20**, 1–76.

Pavesi, L. and Mulloni, V. (1999). All porous silicon microcavities: Growth and physics. *J. Lumin.* **80**, 43–52.

Peckham, J. and Andrews, G.T. (2013). Effect of anodization current density on pore geometry in macro-porous silicon. *Semicond. Sci. Technol.* **28**, 105027.

Ponomarev, E.A. and Levy-Clement, C. (1998). Macropore formation on *p*-type Si in fluoride containing organic electrolytes. *Electrochem. Sol.-St. Lett.* **1**, 42–45.

Quiroga-Gonzalez, E., Ossei-Wusu, E., Carstensen, J., and Föll, H. (2011). How to make optimized arrays of Si wires suitable as superior anode for Li-ion batteries. *J. Electrochem. Soc.* **158**, E119–123.

Quiroga-Gonzalez, E., Carstensen, J., Glynn, C., O'Dwyer, C., and Föll, H. (2014). Pore size modulation in electrochemically etched macroporous *p*-type silicon monitored by FFT impedance spectroscopy and Raman scattering. *Phys. Chem. Chem. Phys.* **16**, 255–263.

Rieger, M.M. and Kahl, P.A. (1995). Mechanism of (111) silicon etching in HF-acetonitrile. *J. Elechochem. Soc.* **142**, 1490–1496.

Ronnebeck, S., Carstensen, J., Ottow, S., and Föll, H. (1999). Crystal orientation dependence of macropore growth in *n*-type silicon. *Electrochem. Solid State Lett.* **2**, 126–128.

Rumpf, K., Granitzer, P., Poelt, P., and Krenn, H. (2009). Transition metals specifically electrodeposited into porous silicon. *Phys. Stat. Sol. (c)* **6**(7), 1592–1595.

Salonen, J., Bjorkqvist, M., Laine, E., and Ninisto, L. (2000). Effects of fabrication parameters on porous *p*⁺-type silicon morphology. *Phys. Stat. Sol. (a)* **182**, 249–254.

Schilling, J., Müller, F., Matthias, S., Wehrspohn, R.B., Gosele, U., and Busch, K. (2001a). Three-dimensional photonic crystals based on macroporous silicon with modulated pore diameter. *Appl. Phys. Lett.* **78**, 1180.

Schilling, J., Wehrspohn, R.B., Birner, A. et al. (2001b). A model system for two-dimensional and three-dimensional photonic crystals: Macroporous silicon. *J. Opt. A: Pure Appl. Opt.* **3**, S121–S132.

Schilling, J., White, J., Scherer, A., Stupian, G., Hillebrand, R., and Gosele, U. (2005). Three-dimensional macroporous silicon photonic crystal with large photonic band gap. *Appl. Phys. Lett.* **86**, 011101.

Searson, P.C., Macaulay, J.M., and Ross, F.M. (1992). Pore morphology and the mechanism of pore formation in *n*-type silicon. *J. Appl. Phys.* **72**(1), 253.

Setzu, S., Lerondel, G., and Romenstain, R. (1988). Temperature effects on photoluminescence properties of porous silicon. *J. Appl. Phys.* **84**, 3129.

Smith, R.L. and Collins, S.D. (1992). Porous silicon formation mechanisms. *J. Appl. Phys.* **71**(8), R1–R22.

Smith, R.L., Chuang, S.-F., and Collins, S.D. (1988). A theoretical model of the formation morphologies of porous silicon. *J. Electron. Mater.* **17**(6), 533–541.

Splinter, A., Stürmann, J., and Benecke, W. (2000). New porous silicon formation technology using internal current generation with galvanic elements. In: *CD Proceeding of the 13th European Conference on Solid-State Transducers, EUROSENSORS XIIV,* August 27–30, 2000, Copenhagen, Denmark, Abstract T2P03, pp. 423–426

Splinter, A., Bartels, O., and Benecke, W. (2001). Thick porous silicon formation using implanted mask technology. *Sens. Actuators B* **76**, 354–360.

Takemoto, K., Sugiyama, I., and Nittono, O. (1994). Microstructure and crystallinity of *n*-type silicon. *Jpn. J. Appl. Phys.* **33**, 6432–6436.

Thonissen, M., Berger, M.G., Arens-Fisher, R., Gluck, O., Kruger, M., and Luth, H. (1996a). Illumination-assisted formation of porous silicon. *Thin Solid Film* **276**, 21–24.

Thonissen, M., Billat, S., Frotscher, U. et al. (1996b). Depth inhomogeneity of porous silicon layers. *J. Appl. Phys.* **80**, 2990.

Torres, J., Martinez, H.M., Alfonso, J.E., and Lopez, L.D. (2008). Optoelectronic study in porous silicon thin films. *Microelectron. J.* **39**(3–4), 482–484.

Trifonov, T., Marsal, L.E., Rodriguez, A., Pallares, J., and Alcubilla, R. (2005). Fabrication of two- and three-dimensional photonic crystals by electrochemical etching of silicon. *Phys. Stat. Sol. (c)* **2**, 3104–3107.

Trifonov, T., Rodriguez, A., Marsal, L.F., Pallares, J., and Alcubilla, R. (2008). Macroporous silicon: A versatile material for 3D structure fabrication. *Sens. Actuators A* **141**, 662–669.

Unagami, T. (1980). Formation mechanism of porous silicon layer by anodization in HF solution. *J. Electrochem. Soc.* **127**, 476–483.

Unno, H., Imai, K., and Muramoto, S. (1987). Dissolution reaction effect on porous-silicon density. *J. Electrochem. Soc.* **134**, 645–648.

Urata, T., Fukami, K., Sakka, T., and Ogata, Y.H. (2012). Pore formation in *p*-type silicon in solutions containing different types of alcohol. *Nanoscale Res. Lett.* **7**, 329.

Van den Meerakker, J.E.A.M., Elfrink, R.J.G., Roozeboom, F., and Verhoeven, J.F.C.M. (2000). Etching of deep macropores in 6 in. Si wafers. *J. Electrochem. Soc.* **147**(7), 2757–2761.

Van den Meerakker, J.E.A.M., Elfrink, R.J.G., Weeda, W.M., and Roozeboom, F. (2003). Anodic silicon etching; the formation of uniform arrays of macropores or nanowires. *Phys. Status Solidi (a)* **197**, 57–60.

Vezin, V., Goudeau, P., Naudon, A., Halimaoui, A., and Bomchil, G. (1992). Characterization of photoluminescent porous Si by small-angle scattering of X-rays. *Appl. Phys. Lett.* **60**, 2625.

Vial, J.C. and Derrien, J. (eds.) (1995). *Porous Silicon: Science and Technology.* Les Editions de Physique, Les Ulis and Springer, Berlin.

Wang, K., Chelnokov, A., Rowson, S., and Lourtioz, J.-M. (2003). Extremely high-aspect-ratio patterns in macroporous substrate by focused-ion-beam etching: The realization of three-dimensional lattices. *Appl. Phys. A: Mater. Sci. Process.* **76**, 1013–1016.

Watanabe, Y., Arita, Y., Yokoyama, T., and Lgarashi, Y. (1975). Formation and properties of porous silicon and its application. *J. Electrochem. Soc.* **122**, 1351–1355.

Xiong, Z.H., Liao, L.S., Ding, X.M. et al. (2002). Flat layered structure and improved photoluminescence emission from porous silicon microcavities formed by pulsed anodic etching. *Appl. Phys. A* **74**, 807–811.

Yaakob, S., Abu Bakar, M., Ismail, J., Abu Bakar, N.H.H., and Ibrahim, K. (2012). The formation and morphology of highly doped n-type porous silicon: Effect of short etching time at high current density and evidence of simultaneous chemical and electrochemical dissolutions. *J. Phys. Sci.* **23**(2), 17–31.

Yamani, Z., Thompson, W.H., AbuHassan, L., and Nayfeh, M.H. (1997). Ideal anodization of silicon. *Appl. Phys. Lett.* **70**, 3404.

Yaron, A.A., Bastide, S., Maurice, J.L., and Clement, C.L. (1993). Morphology of porous n-type silicon obtained by photoelectrochemical etching. 2. Study of the tangled Si wires in the nanoporous layers. *J. Lumin.* **57**, 67–71.

Zhang, X.G. (1991). Mechanism of pore formation on *n*-type silicon. *J. Electrochem. Soc.* **138**, 3750–3756.

Zhang, X.G. (2001). *Electrochemistry of Silicon and Its Oxide.* Kluwer Academic/Plenum, New York.

Zhang, G.X. (2005). Porous silicon: Morphology and formation mechanisms. In: Vayenas C. (ed.) *Modern Aspects of Electrochemistiy*, Number 39. Springer, New York, pp. 65–133.

# Properties and Processing

# Methods of Porous Silicon Parameters Control

## 5

**Mykola Isaiev, Kateryna Voitenko,
Dmitriy Andrusenko, and Roman Burbelo**

CONTENTS

## 5.1 INTRODUCTION

Porous silicon is a promising material for application in various areas of modern technology. Successful use of porous silicon in devices of nano- and optoelectronics, sensorics, MEMS, and so on is not possible without reliable information about its properties. Chapter 1 presented methods for investigation of main characteristics of porous matrix, such as porosity, specific surface, pores size distribution, and chemical composition. In this chapter, we discuss methods for study of optical, luminescence, electrical, mechanical, and thermal properties of porous silicon. Exactly these parameters determine the applicability of porous silicon for the development of devices of various purposes and their performances. In this review, we will try to point out the specifics of application of classical methods for porous silicon study and will not discuss these methods in detail. Readers can get this information from classic books and reviews.

## 5.2 OPTICAL PROPERTIES

Porous silicon is successfully applied in optoelectronics, photonics, sensors, solar cells, and so on. Wide application of this material is based on the peculiarities of porous silicon optical properties, which could be easily tuned in wide range, for example, by changing porosity or by filling pores. Moreover, there is a possibility to set spatial distribution in the optical parameters.

The main parameters that characterize optical properties of porous silicon are refractive index, reflectivity, and absorption coefficients. Such parameters depend strongly on the porosity of porous matrix. For example, bulk Si has a refractive index of 3.47, whereas air has a refractive index of approximately 1.0. As a crude model, one can assume that the porosity increases from 0 to 100%, the refractive index of porous Si would decrease from 3.47 to 1.0 (see Figure 5.1). This approach is somewhat true; however, the exact relationship between silicon porosity and refractive index is more complex. The most frequently used approximation (Equation 5.1) was proposed by Bruggeman (1935).

$$(5.1) \qquad n_{PSi} = \sqrt{\frac{(-3P\varepsilon_{Si} + 3P + 2\varepsilon_{Si} - 1) + \sqrt{(-3P\varepsilon_{Si} + 3P + 2\varepsilon_{Si} - 1)^2 + 8\varepsilon_{Si}}}{4}}$$

where $P$ represents the fraction of void in the layer, $\varepsilon_{Si}$ is the dielectric constant of silicon, and the dielectric constant of air is equal to 1. In addition, one should take into account that a refractive index is also determined by gas surrounding, by the substances, which fill pores, and by the degree of silicon oxidation (Astrova and Tolmachev 2000; Kulathuraan et al. 2012). Thus, methods of porous silicon optical parameters evaluation are absolutely important. It should be noted that optical techniques are mostly non-contact, which is very important in the case of porous silicon study.

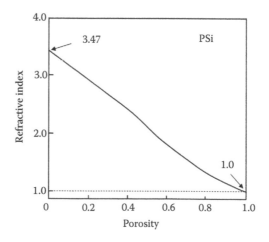

**FIGURE 5.1**  Graph showing relationship between PSi porosity and refractive index according to the Bruggeman approximation.

## 5.2.1 TRANSMISSION AND REFLECTION TECHNIQUES

Transmission and reflection techniques are classical methods for investigation of optical properties of solid state. The use of such methods (see Figure 5.2.) is based on measurement of intensities of reflected ($I_r$) from the sample surface and transmission ($I_{tr}$) through the sample beam relatively to the initial intensity ($I_0$) or to the intensity reflected from etalon with some normalization ($I_e$). In the case of free-standing porous silicon, the treatment of experimentally obtained data give a possibility to evaluate its transmittance ($T$) and reflectance ($R$) (Pap et al. 2006):

$$(5.2) \qquad T = \frac{I_t}{I_0} \quad R = \frac{I_r}{I_e}$$

In the area of transparency of porous silicon film, the transmission specter is oscillatory in nature due to interference of multiple reflected waves from the front and back sample surface. From the enveloping functions of local minimum ($T_{\min}$) and maximum ($T_{\max}$) of transmission spectra, there is a possibility to evaluate refractive index and optical absorption coefficient of porous silicon (Manifacier et al. 1976; Kordás et al. 2004) (see Figure 5.3).

$$(5.3) \qquad n = \sqrt{N + \sqrt{N^2 - n_0^2 n_1^2}}$$

$$(5.4) \qquad \alpha = -\frac{1}{d}\ln\left(\frac{(n+n_0)(n_1+n)\left(1-\sqrt{T_{\max}/T_{\min}}\right)}{(n-n_0)(n_1-n)\left(1+\sqrt{T_{\max}/T_{\min}}\right)}\right)$$

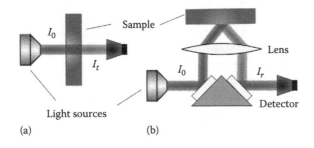

(a)    (b)

**FIGURE 5.2** Schematic sketch of experimental setups used for optical (a) transmission and (b) reflection measurements.

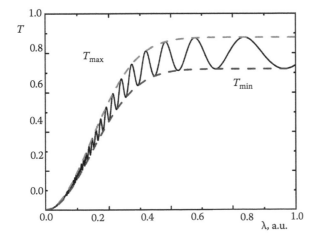

**FIGURE 5.3** Schematic view of transmission spectra of thin film.

where $N = \left(n_0^2 + n_1^2\right)/2 + 2n_0 n_1 \left(T_{max} - T_{min}\right)/(T_{max} T_{min})$, $n_0$ and $n_1$ are refractive indices of surrounding media on the two sides of the sample, and $d$ is sample thickness.

Investigation of spectral characteristics of the reflective index, reflection, and absorption coefficients gives the possibility to obtain information about energetic states in porous silicon (band gap, doping, defect, and impurity levels) (Matsui and Adachi 2012), and dielectric characteristic of this material.

In the case of layered structure "porous silicon on silicon substrate" study, reflectance methods for measuring porous layer refraction are often used (Pérez et al. 2007). Such methods are used to avoid significant influence of silicon substrate in the transmission methods.

One should note that the papers dealing with light absorption in porous silicon are not too numerous. There is an intrinsic difficulty in doing absorption spectroscopy on porous Si, that is, the relatively low absorption coefficient ($10^2$–$10^3$ cm$^{-1}$) (Amato and Rosenbauer 1997). Therefore, the application of the standard transmission technique to this system requires self-supporting layers with thickness ~100 μm. In such films, the homogeneity is expected to be poor, and the absorption spectrum is likely to be an average of different spectra from layers having different porosity and crystallite size. In addition, Pap et al. (2006) have found that when studying the optical properties of porous media of <30 nm pore size, one has to consider photon scattering from the pores in the near infrared spectrum. If scattering takes place, the optical parameters cannot be derived precisely from the transmission spectra using the envelope method because the values of local extrema in the transmission spectra $T_{max}(\lambda)$ and $T_{min}(\lambda)$ used for calculating the index of refraction, absorption, and film thickness are affected by the scattering losses. Surface roughness also has influence on the results of measurements (Theiss 1997). In the case when the characteristic sizes of pore and crystallite are less than wavelength of the light, the assumption of effective media (Garnett 1904; Bruggeman 1935) could be used. According to Canham (1997), from the optical point of view, in the visible and infrared wavelength range, PSi can be specified as an effective medium, whose optical properties depend on the relative volumes of silicon, pore filling medium, and, in some cases, silicon oxide, that is, mainly on the porosity and on the degree of oxidation of the PSi layer. This is an important statement because during aging the porous silicon pore walls oxidize, which leads to a decrease of the refractive index due to the fact that silicon oxide has a much lower refractive index than silicon (Theiss and Hilbrich 1997). This holds even if the volume of the solid matter expands during oxidation by approximately a factor of 2.

Let us note that in the case when porous silicon is produced on monocrystalline substrate, the usage of the transmission method is impossible in visible spectra diapason. Therefore, polarized reflection methods are often used for study of optical properties of porous layer. The main disadvantages of these methods are related to the significant diffusive scattering of light in porous silicon due to its morphology. As a result, there is a significant difficulty in estimating the reflectance and transmission intensities.

## 5.2.2 ELLIPSOMETRY

Ellipsometry is a type of method that makes possible investigation of optical properties of different materials. This method is based on analysis of polarization changing of the light as a result of its interaction with the media. As a result, there are few configurations of ellipsometric setup, the most widely used method based on measurements of polarization state in reflected beam.

In this case, the change in light polarization could be expressed in terms of complex reflection (Riedling 1988)

(5.5)
$$\rho = \frac{E_p^r}{E_p^i} \cdot \frac{E_s^i}{E_s^r}$$

where $E_p^i$, $E_s^i$, $E_p^r$, and $E_s^r$ are components of electrical field intensity. Upper index "$i$" ("$r$") indicate that $E$ corresponds to the incident (reflected) wave, lower index "$s$" ("$p$") indicate that direction of electrical field perpendicular (parallel) to plane of incidence is considered (see Figure 5.4).

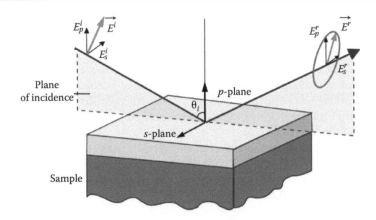

**FIGURE 5.4**   Schematic setup of an ellipsometry experiment.

Usually, ρ is presented as follows:

$$(5.6) \qquad\qquad \rho = \tan(\psi)\exp(i\Delta)$$

where $\tan(\psi)$ is the amplitude of ρ, $\Delta$ is denoted the relative change of phase of the $p$ and $s$ components of the electrical field. The parameters $\psi$ and $\Delta$ could be measured, and usage of Fresnel equations gives the possibility to evaluate from this data real $(\varepsilon_1)$ and complex $(\varepsilon_2)$ parts of dielectric constants. For the thick porous silicon layer, the expression for complex dielectric constant can presented as follows (Pickering et al. 1984):

$$(5.7) \qquad \varepsilon = \varepsilon_1 + i\varepsilon_2 = \sin^2(\varphi) + ((1-\rho)/(1+\rho))^2 \sin^2(\varphi)\tan^2(\varphi)$$

where φ is angle of incidents. The expression presented above is obtained in the framework of two layer media assumption (air and porous silicon), and as a result it could be applied only in defined spectral range of incident light and thicknesses of porous layer where the influence of substrate can be neglected.

For the treatment of experimental data obtained from porous silicon, the inhomogeneity of porous media are often neglected, and the Bruggeman effective media approximation is used (Fried et al. 1996). In the frames of this approximation, the dielectric constant (ε) of complex media with two or more phases could be evaluated from the following expression:

$$(5.8) \qquad f_a\left(\frac{\varepsilon_a - \varepsilon}{\varepsilon_a + 2\varepsilon}\right) + f_b\left(\frac{\varepsilon_b - \varepsilon}{\varepsilon_b + 2\varepsilon}\right) + \ldots = 0$$

where $\varepsilon_a$ and $\varepsilon_b$ are the dielectric constants of phases $a$ and $b$, $f_a$ and $f_b$ its volume fractions (Pickering et al. 1984).

But in general case influence of its rough surface and inhomogeneous morphology (Losurdo and Hingerl 2013) should be taken into account. Therefore, to accurately assess porous silicon optical parameters, non-homogeneity distribution of its optical properties even if the sample is etched under the same current should be taken into account. In this case, several approaches can be applied, the most effective one is based on application of the multilayered system model (Zangooie et al. 2001).

### 5.2.3 NONLINEAR OPTICAL PROPERTIES

Nonlinear optics phenomena are a wide class of effects observed as a result of powerful laser irradiation action on the media. To this class of phenomena frequency mixing phenomena, Kerr effect, multiphoton absorption, optical limiting, and so on (Zheltikov et al. 2012) could be assigned. All mentioned phenomena take place when intensity of radiation is big enough to induce changes of optical parameters of media in which it propagates.

Investigation of nonlinear optics phenomena enables obtaining information about fine features of energetic state, interactions between atoms of substance, and the presence of impurities. Most setups for investigation of nonlinear properties of porous silicon are the same as for other materials; however, some peculiarities were taken into consideration: strongly anisotropy of porous silicon, important portion of excited light scattering effects, significant specific surface state, quantum confinement, and so on.

It should be noted that porous silicon is characterized by the presence of strongly third-order optical non-linearity (Sheik-Bahae et al. 1990; Henari et al. 1995; Lettieri et al. 2001; Mathews et al. 2007) that give the possibility to develop methods of porous silicon characterization using relatively simple Z-techniques (Sheik-Bahae et al. 1990).

## 5.3 LUMINESCENT PROPERTIES

After the discovery of luminescence of porous silicon, this material became active. This is because increasing of luminescence quantum efficient enables the practical application of porous silicon in different kinds of optoelectronic devices. In addition, PL spectra (Cullis et al. 1997) and time-resolved PL characteristics (Chen et al. 2013) are important sources of information about sample properties, such as band structure, electron-hole lifetime, and energetic parameters of defects, impurity centers, and surface states (Stolz 1994; Dao and Wen 2007). For example, Table 5.1 shows the porous silicon bandgap for a variety of samples determined using PL measurements.

The measurement of luminescence properties of porous silicon does not have features in comparison with other materials, so in this section they will not be discussed. However, it is important to note that in contrast to bulk silicon, porous silicon has a strong luminescence in visible and even in near UV range. Therefore, the excited source has to have compliance with such spectral characteristics. For efficiently luminescence excitation, the best way is to use light with high energy

**TABLE 5.1  Measures of the Porous Silicon Bandgap for a Variety of Samples**

| Porosity (%) | PL or EL Peak (eV) | Average Bandgap (eV) | Temperature | Ref. |
|---|---|---|---|---|
| 58 | 1.33 (4.2 K) | 1.61 (RT) | | Binder et al. 1996 |
| 66 | 1.36 (4.2 K) | 1.64 (RT) | | |
| 74 | 1.44 (4.2 K) | 1.93 (RT) | | |
| 53–78 | 1.05–1.19 | 1.26–1.43 | 4.2 K | Calcott 1997 |
| 80 | 1.4 | 1.63 | RT | |
| – | 1.55 | 1.78–1.82 | RT | |
| 80 | 1.6 | 1.95 | RT | |
| – | 1.62 | 1.8 | RT | |
| 80 | 1.81 | – | RT | |
| – | 2.0 | 2.2 | RT | |
| – | 2.27 | – | RT | |
| – | 2.34 | – | RT | |
| | 3.1 | – | RT | |
| 60–70 | ~1.57 | | RT | Von Behren et al. 2000 |
| 82 | 1.61 | | RT | |
| 90 | 1.82 | | RT | |
| 92 | 1.91 | | RT | |
| 62 | 1.88 | 1.87 | RT | Behzad et al. 2012b |
| 72 | 1.92 | 1.89 | RT | |
| 77 | 1.97 | 1.94 | RT | |
| 79 | 2.01 | 1.98 | RT | |
| 80 | 2.05–2.11 | 2.01–2.09 | RT | |

photons (in blue and near infrared diapason). Electron subsystem excitation by beams of electron (cathodoluminescence) and voltage (electroluminescence) can also be used for luminescence properties study of PSi (Canham 1997; Dovrat et al. 2007; Biaggi-Labiosa et al. 2008; Das and Chini 2011).

## 5.4 ELECTRICAL PROPERTIES

During a long time period, the investigation of porous silicon was focused only on the study of its optical properties. However, electrical properties and mechanism of charge carrier transport in this material are also important for fabrication of porous silicon-based devices. It should be emphasized that due to specificity and variety of structural and morphological peculiarities of the matrix there is the possibility of realization of porous silicon with defined electrical properties in a wide range: it could manifest itself as a conductor as well as a dielectric. Therefore, depending on the kind of properties that are appropriate to the material, the methods of *alternative* (ac) and *direct* (dc) *current* for investigation of dielectric and conductive properties of porous silicon could be applied.

The main parameters that characterized conductive properties of semiconductors are resistivity, mobility of charge carriers, its concentration, life-time, diffusive length, electrical band gap, and so on. For investigation of such porous silicon properties, the measurement of current-voltage (I-V) characteristics (Anderson et al. 1991; Dimitrov 1995; Ray et al. 1998) are conveniently applied. In this method, the system in the configuration metal/crystalline silicon/porous silicon/metal is often studied. There is a possibility for various metals (aluminum, silver, gold, antimony, calcium, magnesium, etc.) to be used as an electrical contact. Typical forward I-V characteristics are shown in Figure 5.5. They are similar to those obtained for the Schottky contacts. Therefore, to describe the passage of current through the layered structure, a thermionic emission theory can be used, according to which

$$(5.9) \qquad I = I_s \exp\left(\frac{qV}{nkT}\right)\left(1 - \exp\left(-\frac{qV}{kT}\right)\right)$$

where $I_s$ is the current saturation given by

$$(5.10) \qquad I_s = A^* T^2 \exp\left(-\frac{q\Phi_b}{kT}\right)$$

where $A^*$ is Richardson constant, $n$ is the ideality factor, and $\Phi_b$ is the barrier height.

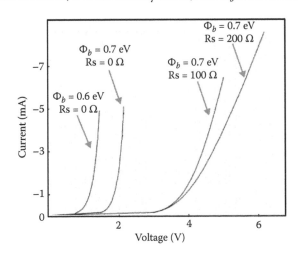

**FIGURE 5.5** Typical forward current-voltage characteristics of metal-porous silicon layer-silicon structures. (Reprinted with permission from Dimitrov D.B., *Phys. Rev. B*, 51, 1562, 1995. Copyright 1995 by the American Physical Society.)

In dc as well as in ac methods, significant difficulties in porous silicon study are related to ensuring reliability of electric contact to porous silicon layer (Diligenti et al. 1996; Parkhutik 1999). Shunting effect of the substrate, on which the porous layer is formed, also has a significant impact on the measurement results. Therefore, the most efficient way to measure real conductive properties of porous silicon is the detaching of the porous silicon from the substrate and the use of four-point probes methods, for example, the Van der Pauw method (Diligenti et al. 1996). It should be noted that the environment media could significantly influence electric properties of porous silicon as well. In particular, Figure 5.6 shows influence of air humidity on the conductivity of porous silicon (Mares et al. 1995). Therefore, the measurement should be provided in well-controlled conditions, for example, in vacuum, in air with constant humidity, or using samples with passivated surfaces.

For the investigation of porous silicon ac characteristics, on both sides of free-standing porous silicon or on a porous surface of "porous silicon–Si substrate," structure contacts (in most cases, aluminum) are evaporated (Ben-Chorin et al. 1995; Axelrod et al. 2002; Theodoropoulou et al. 2004; Urbach et al. 2007). The measurements of capacity of such porous silicon-based capacitor, for example, by means of broadband dielectric spectrometer, enable to obtain its dielectric properties in a wide frequency range (from Hz to MHz). Let us note that this method could be applied for measurements also ac conductance of porous silicon, but it should be taken into account that its dc conductivity is field dependent. Thus, nonlinear effects may occur, so ac current on different ac voltages have to be checked. Also, in this case, the influence of contacted resistance between porous silicon and electrodes should be taken into account (Ben-Chorin et al. 1995). There is a possibility to expand frequency diapason (up to 210 GHz) of dielectric constant of porous silicon measurement when using the broadband extraction method, based on the measurement of the S-parameters of coplanar waveguide transmission lines (CPWT lines) integrated on the porous Si substrate (Sarafis and Nassiopoulou 2014).

Just as porous silicon could be successfully applied as an active element in photovoltaic devices, the photoconductivity methods are also very important. In general, an investigation of photoconductivity can be in steady-state and transient regimes.

The convenient photoconductivity technique is based on the application of a fixed potential difference across the layered structure shown in Figure 5.7, and the measurement of the current under illumination and without illumination. Experimental study of the photoconductivity spectra provides information on the band gap of porous silicon and features of the band structure (Yeh et al. 1993; Kaifeng et al. 1994; Frello et al. 1996; Mehra et al. 1998; Torres et al. 2008). In the experimental geometry presented above, the energy barrier between porous silicon and substrate significantly

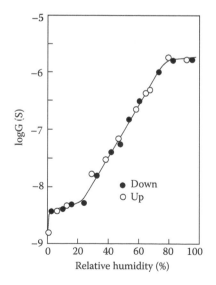

**FIGURE 5.6** Humidity dependence of PSi conductance at room temperature taken with decreasing and increasing humidity. (Reprinted from *Thin Solid Films*, 255, Mares J., Kristofik J., and Hulicius E., 272, Copyright 1995, with permission from Elsevier.)

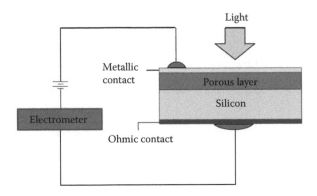

**FIGURE 5.7**    Schematic diagram of the measurement of photoconductivity of porous silicon.

influences the obtained results. To avoid this influence, the experimental measurements on the free-standing porous silicon layer are often provided. When separation of porous layers is not possible (e.g., when it is thin), the steady-state photocarrier grating method for investigation of electrical properties of porous layers can be applied (Schwarz et al. 1995; Lubianiker et al. 1996).

Mobility and a lifetime of photoexcited carriers in porous silicon could be determined from analysis of the photoconductivity dynamic response. Thus, classical photoconductivity decay methods are often applied for such parameters of porous silicon investigation. Also, time-of-flight technique allows estimation of the mobility of electrons and holes in the porous silicon layers in a simple way (Rao et al. 2002; Kojima and Koshida 2005; Fejfar et al. 2013). Drift mobility of holes and electrons can be evaluated from the following equation:

$$\mu = d^2/t_{tr}U$$
(5.11)

where $d$ is the thickness of the porous layer, $t_{tr}$ is the transit time of carriers, and $U$ is the applied voltage.

## 5.5 MECHANICAL PROPERTIES

Ensuring mechanical quality is an essential step in the development process of porous silicon-based devices. The main parameters that define mechanical properties are lattice parameter, crystallites sizes, elastic constants, Poisson's ratio, and Young's modulus. Methods for the porous silicon mechanical properties study are mainly standard as well as for other solid states. The comparison of these parameters for Si, $SiO_2$, and PSi is presented in Table 5.2 (Nassiopoulou and Kaltsas 2000).

*X-ray diffraction* is a classical method (Barla et al. 1984), which provides information about the changes in the lattice parameters. Rocking curves are composed of two distinct Bragg peaks (see

**TABLE 5.2    Comparison of Si, SiO₂, and PSi Parameters**

| Material Parameter | Si | SiO₂ – TEOS | Poly-Si | Porous Si |
|---|---|---|---|---|
| Density (kg/m³) | 2330 | 2200 | 2330 | 466 |
| Young's modulus (N/m²) | 190·10⁹ | 70·10⁹ | 160·10⁹ | 2.4·10⁹ |
| Thermal expansion coefficient (K⁻¹) | 2.5·10⁻⁶ | 0.5·10⁻⁶ | 2.3·10⁻⁶ | 2·10⁻⁶ |
| Poisson ratio | 0.278 | 0.17 | 0.22 | 0.09 |
| Thermal conductivity (W/mK) | 145 | 1.1 | 28 | Variable parameter |
| $C_p$ (J/kg K) | 702.24 | 840 | 702 | 850 |
| Film coefficient (h) (W/m²K) | 30 | 30 | 30 | 30 |
| Emissivity | 0.5 | 0.6 | 0.5 | 0.6 |

*Source:*  Nassiopoulou A.G. and Kaltsas G.: *Phys. Stat. Sol. (a)* 2000. 182. 307. Copyright Wiley-VCH Verlag GmbH & Co. KGaA. Reproduced with permission.

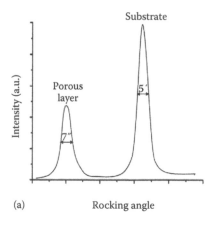

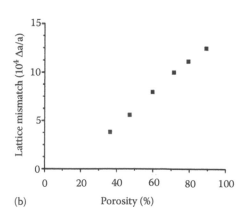

(a)   Rocking angle                     (b)   Porosity (%)

**FIGURE 5.8**   Double crystal rocking curve of porous silicon layer on a silicon substrate. 400 reflection. Porosity: 34% (a). (From Barla K. et al., *J. Cryst. Growth*, 68, 727, 1984. Copyright 1984: Elsevier Science. With permission.) Lattice mismatch parameter Δa/a, between the PSi layer and the substrate, versus the porosity for a series of p⁺-type samples (b). (Reprinted from *Thin Solid Films* 276, Bellet D. and Dolino G., 1, Copyright 1996, with permission from Elsevier.)

Figure 5.8a) that allow unambiguous identification of the lines obtained by porous silicon and the substrate. This demonstrates that porous silicon lattice expands relative to the substrate and this expansion increases when porosity increases (see Figure 5.8b). It was demonstrated that the porous silicon behaves like a nearly perfect monocrystal. Studies of the lattice parameter dependencies on temperature could be used for evaluation of the porous silicon expansion coefficient (Bellet and Dolino 1996).

To investigate the surface of porous silicon layers, atomic force microscopy (AFM) is often used, especially to analyze its surface roughness at the nanoscale range. AFM does not allow demonstrating deep pores. The main method for visualization of the structure features, pores shape, and size distribution is scanning electron microscopy (SEM) (Dian et al. 2004; Khokhlov et al. 2008).

For the investigation of size distribution of porous silicon crystallites, Raman spectroscopy can be applied. This method is based on the detection of the inelastic scattering spectra of optical radiation induced by optical high-frequency phonons. Due to mismatch of the substrate and the porous silicon, the deformation of the porous layer occurs. These deformations could be registered with Raman spectroscopy; Raman shift also indicates change in the interatomic distances. Combining experimental data obtained by X-ray diffraction and Raman spectroscopy, there is the possibility to evaluate the Poisson's ratio and elasticity modulus of porous silicon (Zhenkun et al. 2005). Possibilities of Raman spectroscopy for porous silicon characterization will be discussed in more detail in Chapter 6 of the present book.

Unlike Raman spectroscopy, Brillouin scattering technique is applied for the study of the atom groups vibrations and occurs as a result of the photons' interaction with acoustical phonons. Different types of acoustic modes (Rayleigh, longitudinal, and two transverse) can be investigated by this technique. Such investigation gives the possibility to evaluate velocities of such modes (Andrews et al. 1996; Beghi et al. 1997; Fan et al. 2002; Andrews et al. 2004), which allows estimating the elastic constants and Young's modulus of the porous material.

*Acoustic methods* are based on the measurement of spectral dependences of the acoustic wave's velocity in porous silicon. In the classical method, a sample is placed between two transducers (Figure 5.9a). The first transducer is used for signal excitation and the second for its registration. Detected experimental dependences of voltage on time (see Figure 5.9b) allow estimating the velocity of longitudinal acoustic waves (Aliev et al. 2011):

(5.12)
$$v_{L,psj} = \frac{2jd}{t_j / t_o}$$

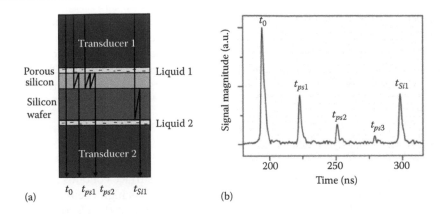

(a)    (b)

**FIGURE 5.9** Scheme for the implementation of acoustic techniques for the study of mechanical properties of porous silicon (a). Example of transmitted signal amplitude in time domain for the specimen (b). (Reprinted with permission from Aliev G.N. et al., *J. Appl. Phys.* 110, 4, 2011. Copyright 2011. American Institute of Physics.)

where $d$ is the thickness of the porous layer, $t_o$ and $t_j$ ($j = 1, 2, 3...$) are the arrival times with no and $j$ roundtrips within the porous layer. Velocity of longitudinal waves in bulk silicon substrate can also be calculated:

(5.13) 
$$v_{L,Si} = \frac{2D}{t_{Si1} - t_o}$$

where $D$ is the thickness of substrate, and $t_{Si1}$ is the arrival time after a single round trip within the substrate.

As a source of the acoustic vibrations, excitation laser radiation can also be used (Zharki et al. 2003). Such techniques enable exciting broadband acoustic signals, and there is no need to make mechanical contact with porous rough surfaces, which is significant in the case of porous silicon study.

*Nanoindentation investigation* may also be useful for the characterization of porous silicon. Schematic diagrams of the setup for porous material investigation with nanoindentation techniques are shown in Figure 5.10a. The method was developed to measure hardness and modulus of elasticity of materials, and it is based on the load-displacement parameters measurement obtained during one loading and unloading cycle (Bellet et al. 1996; Oliver and Pharr 2004; Fang

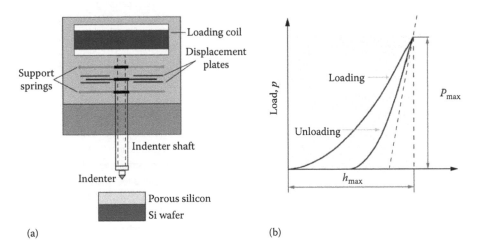

(a)    (b)

**FIGURE 5.10** The experimental apparatus used to perform the indentation experiments (a). A schematic representation of load versus indenter displacement data for an indentation experiment (b). (From Oliver, W.C. and Pharr, G.M., *J. Mater. Res.* 19, 1, 2004. Copyright 2004: Materials Research Society. With permission.)

et al. 2009). There are three quantities that must be measured from the P–h diagram: the maximum load ($P_{max}$), the maximum displacement ($h_{max}$), and the contact stiffness ($S = dP/dh$) (see Figure 5.10b). Then the projected area of the indent at maximum load is calculated from the diamond shape and the elasto-plastic model proposed by Oliver and Pharr (1992) $A = 24.5(h_{max} - 0.75P_{max}/S)$. Material Young's modulus *Ep* of porous silicon is estimated from the experimental values of *A* and *S*: $E_p = ((1 - v^2)/2)S\sqrt{\pi/A}$.

## 5.6 CHEMICAL PROPERTIES

Effective surface area of porous silicon is very large; the pore surface contains high density dangling bonds that remain after etching, and thus this material has high chemical activity. Let us emphasize that chemical composition is the determining factor of the electronic, optical, and electrical properties of porous silicon, so the information about chemical properties is significant for the practical applications of the material.

*Infrared (IR) spectroscopy* is widely used in chemistry and material science methods for substance characterization. Such wide usage of this method is due to its sensitivity to twisting,

**TABLE 5.3    Peaks Observed in FTIR Spectra of Porous Silicon and Species Associated with These Peaks**

| Peak Position, cm$^{-1}$ | Associated Species |
|---|---|
| 465–480 | SiO-Si bending |
| 611–619 | Si-Si (bonds) |
| 624–627 | Si-H bending |
| 661–665 | SiH wagging |
| 827 | SiO bending (O-Si-O) |
| 856 | SiH$_2$ wagging |
| 878–880 | Si-O or Si-O-H bending |
| 905–915 | SiH$_2$ scissor |
| 948 | SiH bending (Si$_2$-H-SiH) |
| 979 | SiH bending (Si$_2$-H-SiH) |
| 1157–1170 | surface oxide species |
| 1061–1110 | Si-O-Si asymmetric stretching |
| 1230 | SiCH$_3$ bending |
| 1463 | CH$_3$ asymmetric deformed |
| 1720 | CO |
| 2073–2090 | SiH stretching (Si$_3$-SiH) |
| 2109–2120 | SiH stretching (Si$_2$-SiH$_2$) |
| 2139–2141 | SiH stretching (Si-SiH$_3$) |
| 2136–2160 | SiH stretching (Si$_2$O-SiH) |
| 2193–2197 | SiH stretching (SiO$_2$-SiH) |
| 2238–2258 | SiH stretching (O$_3$-SiH) |
| 2856 | CH stretching (CH) |
| 2860 | CH sym. stretching (CH$_2$) |
| 2921–2927 | CH assym. stretching (CH$_2$) |
| 2958–2960 | CH assym. stretching (CH$_3$) |
| 3450–3452 | OH stretching (H$_2$O) |
| 3610 | OH stretching (SiOH) |

*Source:* Data extracted from Dietrich G. et al., *J. Mol. Struct.* 349, 109–112, 1995; Mawhinney D.B. et al., *J. Phys. Chem. B* 5647(96), 1202–1206, 1997; Vásquez-A M.A. et al., *Rev. Mexicana Fisica* 53(6), 431–435, 2007.

bending, rotating, and vibrational motions of atoms in a molecule, and that is why the IR spectrum is a specific and unique passport of the individual atoms and interatomic bonds.

Mostly transparency of porous silicon in IR diapason makes this method a powerful and easy tool for surface states of porous silicon characterization, which could be based on convenient transmissions measurements (Ogata 2014). Moreover, *Fourier transform infrared spectroscopy* (FTIR) enables rapidly obtaining IR spectra of porous silicon (Tsao et al. 1990; Xie et al. 1992). IR spectra are sensitive for the presence of substance inside pores, porosities, thickness, impurities level, free charge carriers, surface states of the porous silicon (Theiss 1995). Additionally, this technique can be applied for the study of the fine structural features of porous silicon matrix (Coyopol et al. 2012). FTIR method can be used to study PSi surface state because porous silicon's high surface area allows the incorporation of large amounts of a surface modifier, and because low doped silicon (resistivities > 1 $\Omega$·cm) allows transmission of the IR beam. Highly doped silicon (resistivities < 1 $\Omega$·cm) has higher IR absorption, which prevents transmission IR measurements. Such samples can be studied by using reflectance IR techniques such as attenuated total reflectance (ATR) and diffuse reflectance spectroscopy. The IR frequencies commonly observed in PSi and their attributions are presented in Table 5.3.

*Secondary ion mass spectrometry* (SIMS) method is the most sensitive method capable of detecting impurity elements present in the surface layer. SIMS analysis (Canham et al. 1991; Lee et al. 2007) was first proposed for measurements of main impurities on the surface of microporous silicon.

Also the *thermogravimetric analysis* (TGA) can be used to study the chemical properties of porous silicon. This method is based on the study of material mass changing as a result of heating. Thermogravimetric analysis provides information on physical and chemical phenomena, such as absorption, adsorption, desorption, oxidation, or reduction of porous silicon.

## 5.7 THERMAL PROPERTIES

The information about thermal properties of porous silicon is important for reliability and stability of the devices based on it. Porous silicon is characterized by significant thermal conductivity reduction and its possible usage as a thermal isolator in silicon-based technology. As far as methods of porous silicon, thermal properties are not widely discussed in the literature. Therefore, we will try to provide a more detailed overview on this issue.

It should be emphasized that thermal conductivity of porous silicon strongly depends on its porosity, but it is also very sensitive to its morphology. Thus, porous silicon prepared under different etching regimes and methods could have significantly different thermal conductivity, and values of this parameter could change over time. So, thermal conductivity is another important parameter of porous silicon. But the numbers of overviews describing methods to evaluate thermal parameters of porous silicon are limited. Therefore, in this section, we will analyze them in more detail compared with other methods.

Let us note that the heat capacity is also an important thermal parameter of porous silicon, but due to its significant chemical activity, the measurement of this parameter by classic calorimetry techniques is problematic. Methods for assessing the capacity of porous silicon heat based on the comparison of thermal conductivity and thermal diffusivity values are discussed next. The measurements of thermal expansion coefficient are based mainly on the analysis of dependence of lattice parameters on temperature by X-ray diffraction (Faivre et al. 2000) and have been discussed in the preceding sections.

In a general case, there are huge amounts of techniques to study thermal conductivity of solids. Several measurement techniques used for these purposes are shown in Figure 5.11 (Bernini et al. 2001; Wolf and Brendel 2006; Gomès et al. 2007; Isaiev et al. 2014b). These techniques could vary depending on (1) source of thermal perturbation excitation (e.g., electrical-, photoinduced); (2) method of informative response registration (e.g., measurement of electrical resistivity, acoustic signal, light refraction); and (3) temporal peculiarities (e.g., steady state, transient). We will not consider the classification of such methods in detail; this information can be obtained from monographs and books on experimental study of thermophysical parameters. In this section, we will briefly review main techniques that were successfully applied for thermal property study of porous silicon.

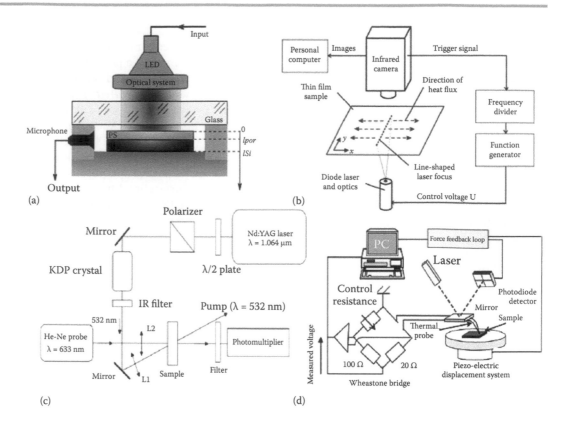

**FIGURE 5.11**   Diagrams illustrating measurement techniques used for characterization of thermal properties of porous silicon: (a) photo-acoustic method; (b) lock-in thermography; (c) pump-probe method; (d) a scanning thermal microscope (SThM) method. (Reprinted from *Mater. Lett.* 128, Isaiev M. et al., 71, Copyright 2014, with permission from Elsevier; Reprinted from *Thin Solid Films* 513, Wolf A. and Brendel R., 385, Copyright 2006, with permission from Elsevier; Bernini U. et al., *J. Phys.: Condens. Mater.* 13, 1141, 2001. Copyright 2001: IOP Publishing; and Gomès S. et al., *J. Phys. D* 40, 6677, 2007. Copyright 2007: IOP Publishing. With permission.)

### 5.7.1 ELECTROTHERMAL METHODS. 3-ω TECHNIQUES

Electrothermal methods are based on the usage of the metal strips as a source of thermal perturbation due to Joule heating and as a resistance thermometer. The most conventional electrothermal methods are 3-ω techniques. In this section, we will focus on these techniques. Other electrothermal methods will be only briefly overviewed.

The 3-ω techniques are based on the presence of the signal on the third harmonic component (3ω) of the alternating voltage applied along metal wire. For the first time, the application of 3-ω techniques for thermal conductivity evaluation of solids was proposed by Cahill (1990), and Cahill et al. (1989). Later, this technique was successfully applied for thermal conductivity study of mesoporous silicon over a wide range of temperature by Gesele et al. (1997).

Let us analyze the physical origin of the presence of such components. In the case of application of alternative current with amplitude $I_0$ and frequency ω, on the electrodes of the metal wire (see Figure 5.12) will be generated thermal power

(5.14)
$$P \sim I_0^2 \exp(2i\omega t)$$

due to Joule's law. As the result time-depend (with the harmonic law) temperature rise (ΔT) in the wire will be induced at the same frequency (2ω) as the applied power. It is clear that the temperature of the wire will depend on the thermal properties of adjacent materials in the case of ideal thermal contact between them.

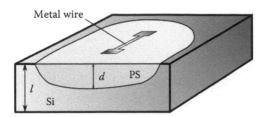

**FIGURE 5.12** Schematic sketch view of the setup for thermal conductivity study by 3-ω technique.

On the other hand, in the frameworks of approximation of linear dependence of resistivity ($R$) of metal wire on temperature ($T$)

$$(5.15) \qquad R = R_0(1 + \alpha(T - T_0))$$

where $T_0$ is some starting temperature (in most cases, room or another equilibrium/quasi-equilibrium temperature), $R_0$ is resistivity at temperature $T_0$, and $\alpha$ is proportional coefficient that is precisely measured for most metals. Therefore, the voltage ($V = IR$) on the third harmonic frequency will depend on the thermal properties of the sample.

In the experimental conditions, the dependence of amplitude voltage on the third harmonic component on frequency of applied alternative current is measured. The analysis of experimental results is based mainly on solution of equations for thermal perturbation diffusion. In the case of the harmonic source, it could be presented in the terms of thermal wave formalism (Carslaw and Jaeger 1959). In the frames of this formalism, it is easy to show that thermal perturbation is a rapidly damped wave. The wavelength of such waves ($\lambda_{th}$) and the characteristic length of its damping ($l_{th}$) are the same:

$$(5.16) \qquad \lambda_{th} = l_{th} = \sqrt{\frac{D_{th}}{2\Omega}}$$

where $D_{th}$ is the thermal diffusivity of the sample, and $\Omega$ is the frequency of the heat source modulation (in the case of 3-ω techniques $\Omega = 2 \cdot \omega$). By $\Omega$ varying it is easy to vary spatial distribution of thermal perturbation.

In the general case, the evaluation of thermal perturbation in the two- or even three-dimensional case is a challenging task, especially in the inhomogeneous media. Therefore, often some particular cases are considered. For example, when the wavelength of the thermal wave is more than the wire width and smaller than the layer thickness, the thermal perturbation could be presented as follows (Gesele et al. 1997):

$$(5.17) \qquad \Delta T = \frac{P}{\pi l D_{th}}\left[-\frac{1}{2}\log\left(\omega\frac{1}{Hz}\right) + \frac{1}{2}\log\left(\frac{a}{b^2}\frac{1}{Hz}\right) + \text{const} - i\frac{\pi}{4}\right].$$

Thus, in the frequency region of this equation reliability the dependence of $\mathrm{Re}(\Delta T)$ on the frequency in the logarithmic scale has a linear character. It is clear that the slope of this function depends on the thermal diffusivity of the porous silicon layer.

In Figure 5.13, the typical experimental dependence of real part of $\Delta T$ on frequency in a logarithmic scale obtained from the porous silicon on monocrystalline substrate is presented. One can see that in the frequency region (full dots) defining conditions of Equation 5.16 reliability such dependence is linear.

This could be the basis for in-plane measurements of thermal conductivity of porous silicon on silicon substrates in cases when wire width is bigger than $d$. In cases when the wire width is less than such width, the 3-ω component will be sensitive for cross-sectional conductivity of the porous silicon layer (Borca-Tasciuc et al. 2001).

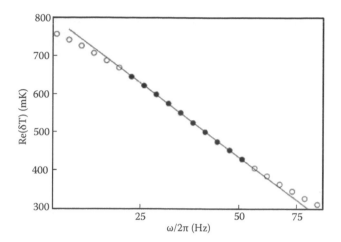

**FIGURE 5.13**    Real part Re(δT) of the amplitude δT of the temperature oscillation within the metal heater versus the logarithm of the frequency ω of the heating current. The thickness of the PS layer is 31 μm, the width of the metal line is 22 μm. The full dots within ω/2π = 22–50 Hz correspond to a penetration depth $l_0$ = 16–25 μm. They were used for the calculation of $λ_{PS}$. (From Gesele G. et al., *J. Phys. D: Appl. Phys.* 30, 2911, 1997. Copyright 1997: IOP Publishing. With permission.)

Besides ac, electrothermal method with dc is effective as a source of heating, for thermal conductivity study of porous silicon (Siegert et al. 2012). In the case when the cross-sectional thermal conductivity of porous silicon is under investigation, the electrothermal setups with two parallel metal wires are also often used (Feng et al. 2009). In such setups, one wire is a heater and another is a thermometer. Thermal properties of thins layers can be evaluated based on time delay (or phase shift) of thermal perturbation achieved from heater to thermometer. Various materials with well-known thermal conductivity can be used as the calibration standards (Table 5.4) (Gomès et al. 2007).

Therefore, electrothermal methods are powerful tools for thermal conductivity study of porous silicon. The main disadvantage of these methods is that its realization is based on contacted measurement. As a result, the information about thermal contact in such methods is crucially important. Moreover, in the case of porous materials study, the material of the electrodes could penetrate with time in the pores and consequently change the experimentally obtained value (Behzad et al. 2012a).

**TABLE 5.4    Thermal Conductivity of the Materials That Can Be Used as Calibration Standards**

| Materials | Thermal Conductivity (W/m·K) | Roughness (nm) |
|---|---|---|
| SiO$_2$ | 1.2 | 3.25 |
| ZnS | 1.73 | 6.27 |
| ZnO | 2.5 | 2.19 |
| CdSe | 3.49 | 0.86 |
| CdTe | 6.28 | 3.7 |
| ZnTe | 12.39 | 5.08 |
| CdS | 15.9 | 1.71 |
| V | 31 | 2.35 |
| Steel40 | 40 | 1.49 |
| Ge | 58.6 | 0.604 |
| Dural | 147 | 1.29 |

*Source:* Gomès S. et al., *J. Phys. D* 40, 6677, 2007. Copyright 2007: IOP Publishing. With permission.

## 5.7.2 RAMAN SHIFT METHOD

In this section, we will discuss application of Raman scattering techniques for thermal conductivity evolution of porous silicon thermal conductivity, focusing on the Raman shift method. This method is fully contactless—information about thermal properties of studied material could be found in the scattering from the sample surface light. First, the Raman shift method was applied to study thermal properties of porous silicon (Perichon et al. 1999, 2000), and then became popular after its successful application by Balandin et al (2008) to study thermal conductivity of graphene.

The physical principles of the Raman shift method usage are based on: (1) photothermal effect—heating of the media by absorbed irradiation and (2) dependence of the Raman peak shift on the sample temperature. Therefore, in the most experimental setups (see Figure 5.14), absorbed laser irradiation has formed at the thermal source, as a result depending on thermal properties of investigated samples, the temperature distribution in the material will occur.

When the porous silicon layer thickness is much bigger than the source radius and in the case of surface radiation absorption, the simple equation for describing local rise of surface temperature ($\Delta T$) could be used (Nonnenmacher and Wickramasinghe 1992).

$$(5.18) \qquad \Delta T = \frac{2P}{\pi a \chi}$$

where $P$ and $a$ are power and radius of heat source, respectively, and $\chi$ is the sample thermal conductivity.

In Figure 5.15, the numerous theoretical calculations and experimental measurements of Raman shift ($\Delta k$) dependence on temperature are presented (Stoib et al. 2014a,b). From these results, the linear dependence of the Raman shift on temperature in the diapason 300–700 K can be stated as

$$(5.19) \qquad \Delta k = A + B \cdot T$$

where $A$ and $B$ are constants that could be precisely evaluated from numerical simulation or experimental data, and $T$ is sample temperature. Taking this into consideration, the simple linear dependence between Raman shift and source power could be achieved, the slope of this dependence will be defined by thermal conductivity of the porous silicon layer.

Let us notice that in the case of relatively thin porous silicon layers, the simple relation (Equation 5.18) is not applicable. In this case, the stationary thermal conductivity equation has been solved with consideration of volume heat source distribution (Newby et al. 2013). It enables improving spatial resolution of in-plane thermal conductivity evaluation of porous silicon.

Therefore, the Raman techniques are fully contactless techniques for thermal conductivity evaluation of porous materials. Realization of such methods enables mapping thermal conductivity at the submicron scale. But to achieve tangible shift of Raman peak, the investigated materials

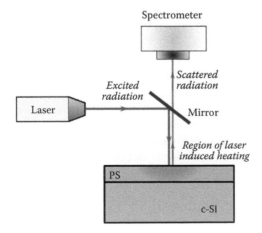

**FIGURE 5.14**   Schematic sketch view of the setup for thermal conductivity study by Raman shift method.

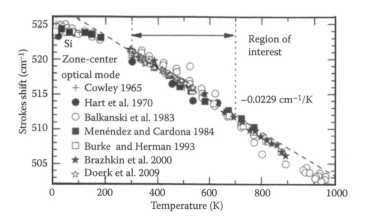

**FIGURE 5.15**    T-dependence of the Stokes shift Δ𝑘 of the zone-center optical mode for crystalline Ge and crystalline Si. In the region of interest for this work, a linear dependence is assumed. (Reprinted with permission from Stoib B. et al., *Appl. Phys. Lett.* 104, 161907, 2014. Copyright 2014. American Institute of Physics.)

have to be substantially heated, and as a result the thermal stresses that occurred in the sample could significantly distort experimentally obtained data. Partial solution of this problem can be found using the Raman technique, which is based on measurement relation between Stockes/anti-Stockes components of scattered light. However, this method is not widely used, mainly due to significant difficulties with stabilization of experimental conditions.

### 5.7.3 PHOTOTHERMAL AND PHOTOACOUSTIC TECHNIQUES

Photothermal effects arise also as a result of nonradiation photoexcitations relaxation, which leads to media heating. Usually, it is considered that in this case the source of excitation has time-depended intensity (modulated or pulse), in comparison with, for example, the Raman technique. Application of photothermal techniques are based on measurement of photoinduced temperature distribution directly (i.e., by IR detector [Ould-abbas et al. 2012]) or caused by them changing media parameters (e.g., reflectivity [Bernini et al. 2001]).

Photoacoustic effect is traditionally a photothermal method, based on measurements of the pressure oscillation caused by the media heating.* There are a huge number of different types of experimental realization of photoacoustic techniques. In this section, we discuss the most widely used for thermal conductivity study of porous silicon: gas-microphone and piezoelectric.

For the first time, gas-microphone photoacoustic techniques for thermal conductivity evaluation of porous silicon were applied by Benedetto et al. (1997a,b). This technique also allows evaluating thermal properties of porous silicon with important oxide (Lysenko et al. 1999) and amorphous fraction (Isaiev et al. 2014b). Measurement of the photoacoustic signal characteristics at different wavelengths allows precise evaluation of thermal conductivity of microporous silicon (Obraztsov et al. 1997; Maliński et al. 2005; Tytarenko et al. 2014) and porous germanium (Isaiev et al. 2014c). Let us note that the usage of photoacoustic technique also enables obtaining thermal properties of the porous silicon samples with different kinds of fillers (Behzad et al. 2012a; Andrusenko et al. 2014).

Application of gas-microphone technique for thermal conductivity determination of solid was first described by Rosencwaig and Gersho (1975, 1976). These authors considered formation of photoacoustic signals in the photoacoustic cell shown in Figure 5.16.

In the case when harmonically modulated incident radiation is absorbed in the investigated sample, the volume density of thermal power can be presented as follows:

$$(5.20) \qquad \rho_P(x,t) = \frac{1}{2} I_0 \alpha \exp(\alpha x)(1 + \cos(\omega t))$$

---

* Let us note that in most general cases, photoacoustic effect is the conversion of absorbed light by the media light with the non-stationary intensity distribution into elasticity distribution. Besides photothermal, there are other mechanisms of photoacoustic effect.

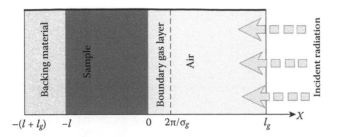

**FIGURE 5.16** Schematic sketch view of the typical cylindrical photoacoustic cell.

where $I_0$ is intensity of incident radiation, $\alpha$ is absorbed coefficient of the media. As results in the sample will excite thermal perturbation, spatial distribution of which can be described in terms of thermal wave formalism (see Section 5.7.1). As a result, the gas adjusted to sample surface will be also be heated and its expansion will create the "gas piston," which will lead to pressure oscillation in the cell.

In their paper, Rosencwaig and Gersho (1976) had shown that such pressure oscillation depended on the sample surface temperature oscillations and as a result, on its thermal properties. These pressure oscillations could be easily detected by microphone. The most widely used photoacoustic techniques are also based on the presence of peculiarities on amplitude-/phase-frequency dependencies of photoacoustic signals, which are determined by the relation between wavelengths of thermal wave and thicknesses of the sample.

The gas-microphone photoacoustic method is an easy tool for thermal conductivity evaluation of porous silicon layer on monocrystalline substrate. Figure 5.17 presents the amplitude-frequency dependences (in double logarithmic scale) of gas-microphone registered photoacoustic signal obtained from porous silicon on monocrystalline substrate for different porosity of porous silicon layers. All samples have the same thickness of porous layer. From Figure 5.17 one can observe the bends that are presented on these dependences. The frequency of this bending ($\omega_b$) increases with porosity decrease. This could be explained in the frames of thermal wave formalism. Therefore, when wavelength of thermal wave is less than the thickness of the top layer ($d$), behavior of thermal perturbation is determined only by thermal properties of this layer. When the value of the wavelength increases and becomes more than $d$, the thermal perturbation will depend also on thermal properties of the bottom layer, and the influence of this layer will become more and more significant with the increase of the wavelength. Thus, the bending frequency

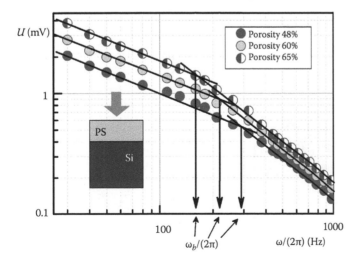

**FIGURE 5.17** The amplitude-frequency dependence of photoacoustic signal experimentally obtained on the two-layer system "porous silicon on monocrystalline substrate" with different porosities (48, 60, and 65%) of porous layer.

corresponds to the wavelength of thermal wave approximately equal to the thickness of the porous silicon layer

$$(5.21) \qquad \lambda_{th} = d = \sqrt{\frac{D_{th}}{2\omega_b}}$$

Therefore, the thermal diffusivity of porous silicon layer from the Equation 5.19 could be easy evaluated.

Let us note that gas-microphone techniques are non-contact, and for efficient signal excitation it is enough to heat the sample only into a part of a Celsius degree. But due to specification of these methods, there are no possibilities to provide measurement without gas surrounding. Another important limitation of the usage of these methods is related to shorter frequency diapason of providing measurements due to resonance oscillation of gas in the photoacoustic cell. Therefore, as far as resolution of thermal wave methods increases with the increase of modulation frequency, there is no possibility to obtain thermal conductivity of samples with a size of layer less than a few microns.

As it was already mentioned, gas-microphone photoacoustic signal is sensitive to the sample's surface temperature; in the case of piezoelectric photoacoustic configuration signal, it depends on the temperature spatial distribution.

When compared with gas-microphone, these techniques have advantages, such as the possibility of providing measurements in vacuum and under low temperature, better sensitivity, and the possibility to provide measurement at much wider frequency diapason.

For the first time, piezoelectric photoacoustic technique for investigation of solid state was proposed by Jackson and Amer (1980), who introduced a mathematical model of photoacoustic signal formation. Later, Blonskij et al. (1996) introduced a simplified mathematic model for piezoelectric photoacoustic signal formation, which stipulated application of piezoelectric technique for different kinds of material.

Alekseev et al. (2011) present an experimental study of piezoelectric photoacoustic signal formation in the multilayered system (see Figure 5.18a) with porous silicon under its irradiation by square-wave modulate irradiation. A typical shape of experimentally obtained oscillogram (the first half-period) from investigated system "porous silicon on monocrystalline Si substrate" at the frequency of light modulation 400 Hz is presented on Figure 5.18b. As one can see, there is a bend on the presented curve at the time $t_b$, which is related to the transition of the heat perturbation from the porous silicon to the wafer and is due to significant difference of elastic properties of these layers. From the value $t_b$ the thermal diffusivity of porous silicon layer can be evaluated

$$(5.22) \qquad a \sim d^2/t_b$$

where $d$ is thickness of the porous silicon layer.

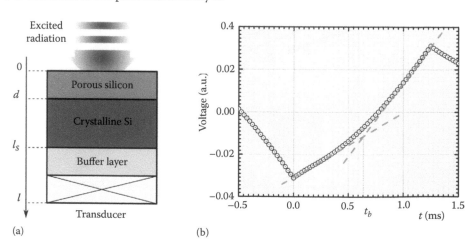

(a)  (b)

**FIGURE 5.18** (a) Schematic sketch view of piezoelectric photoacoustic layered system and (b) experimentally obtained piezoelectric photoacoustic signal shape.

Isaiev et al. (2014a) proposed the mathematical model of photoacoustic response formation in these systems, and as a result significant influence of thermal properties of porous silicon layer on the photoacoustic signal shape was demonstrated.

Let us note that the shape of photoacoustic signal in this configuration is sensitive also to the moving liquids filling the pores of porous matrix (Burbelo et al. 2011; Andrusenko et al. 2012), and therefore can be successfully applied to obtain information about the matrix percolation as well as liquid properties in small volume.

## ACKNOWLEDGMENTS

This work was supported by the Ministry of Science, ICT and Future Planning (MSIP) of the Republic of Korea, and partly by the Moldova Government under grant 15.817.02.29F and ASM-STCU project #5937.

## REFERENCES

Alekseev, S., Andrusenko, D., Burbelo, R., Isaiev, M., and Kuzmich, A. (2011). Photoacoustic thermal conductivity determination of layered structures PS-Si: Piezoelectric detection. *JPCS* 278, 012003.

Aliev, G.N., Goller, B., and Snow, P.A. (2011). Elastic properties of porous silicon studied by acoustic transmission spectroscopy. *J. Appl. Phys.* 110, 043534.

Amato, G. and Rosenbauer, M. (1997). Absorption and photoluminescence in porous silicon. In Amato, G., Delerue, C., and Bardeleben, H.J. (eds.), *Optical and Structural Properties of Porous Silicon Nanostructures*. Gordon and Breach, Newark, NJ, 5, pp. 3–52.

Anderson, R.C., Muller, R.S., and Tobias, C.W. (1991). Investigations of the electrical properties of porous silicon. *J. Electrochem. Soc.* 138, 3406–3411.

Andrews, G.T., Zuk, J., Kiefte H., Clouter, M.J., and Nossarzewskaorlowska, E. (1996). Elastic characterization of a supported porous silicon layer by Brillouin scattering. *Appl. Phys. Lett.* 69(9), 1216–1219.

Andrews, G.T., Clouter, M.J., and Mroz, B. (2004). Brillouin scattering studies of surface acoustic waves in SiC. *Mater. Sci. Forum* 457–460, 653–656.

Andrusenko, D., Isaiev, M., Kuzmich, A., Lysenko, V., and Burbelo, R. (2012). Photoacoustic effects in nanocomposite structure "porous silicon-liquid." *Nanoscale Res. Lett.* 7(1), 411.

Andrusenko, D., Isaiev, M., Tytarenko, A., Lysenko, V., and Burbelo, R. (2014). Size evaluation of the fine morphological features of porous nanostructures from the perturbation of heat transfer by a pore filling agent. *Microporous Mesoporous Mater.* 194, 79–82.

Astrova, E.V. and Tolmachev, V.A. (2000). Effective refractive index and composition of oxidized porous silicon films. *Mater. Sci. Eng. B* 69–70, 142–148.

Axelrod, E., Givant, A., Shappir, J., Feldman, Y., and Sa'ar, A. (2002). Dielectric relaxation and transport in porous silicon. *Phys. Rev. B* 65, 1–7.

Balandin, A.A., Ghosh, S., Bao, W. et al. (2008). Superior thermal conductivity of single-layer graphene. *Nano Lett.* 8(3), 902–907.

Balkanski, M., Wallis, R.F., and Haro, E. (1983). Anharmonic effects in light scattering due to optical phonons in silicon. *Phys. Rev. B* 28(4), 1928–1934. doi:10.1103/PhysRevB.28.1928.

Barla, K., Herino, R., Bomchik, G., Pfister, J.C., and Freund, A. (1984). Determination of lattice parameter and elastic properties of porous silicon by X-ray diffraction. *J. Cryst. Growth* 68, 727–732.

Beghi, M.G., Bottani, C.E., Ghislotti, G., Amato, G., and Boarino, L. (1997). Brillouin scattering of porous silicon. *Thin Solid Films* 297, 110–113.

Behzad, K., Yunus, W.M.M., Talib, Z.A., Zakaria, A., and Bahrami, A. (2012a). Preparation and thermal characterization of annealed gold coated porous silicon. *Materials* 5(12), 157–168.

Behzad, K., Yunus, W.M.M., Talib, Z.A., Zakaria, A., Bahrami, A., and Shahriari, E. (2012b). Effect of etching time on optical and thermal properties of p-type porous silicon prepared by electrical anodisation method. *Adv. Opt. Technol.* 2012, 581743.

Bellet, D. and Dolino, G. (1996). X-ray diffraction studies of porous silicon. *Thin Solid Films* 276, 1–6.

Bellet, D., Lamagnère, P., Vincent, A., and Bréchet, Y. (1996). Nanoindentation investigation of the Young's modulus of porous silicon. *J. Appl. Phys.* 80(7), 3372–3376.

Ben-Chorin, M., Moller, F., and Koch, F. (1995). Hopping transport on a fractal: Ac conductivity of porous silicon. *Phys. Rev. B* 51(4), 2199–2213.

Benedetto, G., Boarino, L., Brunetto, N., Rossi, A., Spagnolo, R., and Amato, G. (1997a). Thermal properties of porous silicon layers. *Philos. Mag. B* 76(3), 383–393.

Benedetto, G., Boarino, L., and Spagnolo, R. (1997b). Evaluation of thermal conductivity of porous silicon layers by a photoacoustic method. *Appl. Phys. A: Mater. Sci. Process.* 64(2), 155–159.

Bernini, U., Lettieri, S., Maddalena, P., Vitiello, R., and Di Francia, G. (2001). Evaluation of the thermal conductivity of porous silicon layers by an optical pump-probe method. *J. Phys.: Condens. Matter.* 13(5), 1141–1150.

Biaggi-Labiosa, A., Fonseca, L.F., Resto, O., and Balberg, I. (2008). Tuning the cathodoluminescence of porous silicon films. *J. Lumin.* 128(3), 321–327.

Binder, M., Edelmann, T., Metzger, T.H., Mauckner, G., Goerigk, G., and Peisl, J. (1996). Bimodal size distribution in p– porous silicon studied by small angle X-ray scattering. *Thin Solid Film* 276, 65–68.

Blonskij, I.V., Tkhoryk, V.A., and Shendeleva, M.L. (1996). Thermal diffusivity of solids determination by photoacoustic piezoelectric technique. *J. Appl. Phys.* 79(7), 3512–3516.

Borca-Tasciuc, T., Kumar, R., and Chen, G. (2001). Data reduction in 3ω method for thin-film thermal conductivity determination. *Rev. Sci. Instrum.* 72(4), 2139.

Brazhkin, V.V., Lyapin, S.G., Trojan, I.A., Voloshin, R.N., Lyapin, A.G., and Mel'nik, N.N. (2000). Anharmonicity of short-wavelength acoustic phonons in silicon at high temperatures. *J. Exper. Theoretical Phys. Lett.* 72(4), 195–198. doi:10.1134/1.1320111.

Bruggeman, D.A.G. (1935). Calculation of various physics constants in heterogeneous substances. I. Dielectricity constants and conductivity of mixed bodies from isotropic substances. *Ann. Phys.* 24, 636–664.

Burbelo, R., Andrusenko, D., Isaiev, M., and Kuzmich, A. (2011). Laser photoacoustic diagnostics of advanced materials with different structure and dimensions. *Arch. Metall. Mater.* 56(4), 1157–1162.

Burke, H.H. and Herman, I.P. (1993). Temperature dependence of Raman scattering in Ge 1 – x Si x alloys. *Phys. Rev. B* 48(20), 15016–15024. doi:10.1103/PhysRevB.48.15016.

Cahill, D.G. (1990). Thermal conductivity measurement from 30 to 750 K: The 3ω method. *Rev. Sci. Instrum.* 61(2), 802–808.

Cahill, D.G., Fischer, H.E., Swartz, E.T., and Pohl, O. (1989). Thermal conductivity of thin films: Measurements and understanding. *J. Vac. Sci. Technol. A* 7, 1259–1266.

Calcott, P.D.J. (1997). Experimental estimates of porous silicon bandgap. In: *Properties of Porous Silicon*, Canham L. (ed.). INSPEC, London, UK, pp. 202–205.

Canham, L.T. (ed.) (1997). *Properties of Porous Silicon.* INSPEC, London.

Canham, L.T., Houlton, M.R., Leong, W.Y., Pickering, C., and Keen, J.M. (1991). Atmospheric impregnation of porous silicon at room temperature. *J. Appl. Phys.* 70(1), 422.

Carslaw, H.S. and Jaeger, J.C. (1959). *Conduction of Heat in Solids,* 2nd ed. Clarendor Press, Oxford.

Chen, Z., Wu, Q., Yang, M. et al. (2013). Time-resolved photoluminescence of silicon microstructures fabricated by femtosecond laser in air. *Opt. Express* 21(18), 21329–21336.

Cowley, R.A. (1965). Raman scattering from crystals of the diamond structure. *J. Phys.* 26(11), 659–667. doi:10.1051/jphys:019650026011065900.

Coyopol, A., Diaz-Becerril, T., Garsia-Salgado, G. et al. (2012). Morphological and optical properties of porous silicon annealed in atomic hydrogen. *Superficies y Vacio* 25(4), 226–230.

Cullis, A.G., Canham, L.T., and Calcott, P.D.J. (1997). The structural and luminescence properties of porous silicon. *J. Appl. Phys.* 82(3), 909–965.

Dao, L.V., Wen, X., and Hannaford, P. (2007). Temperature dependence of photoluminescence in silicon quantum dots. *J. Phys. D: Appl. Phys.* 40(12).

Das, P. and Chini, T.K. (2011). An advanced cathodoluminescence facility in a high-resolution scanning electron microscope for nanostructure characterization. *Curr. Sci.* 101(7), 849–854.

Dian, J., Macek, A., Niznansky, D. et al. (2004). SEM and HRTEM study of porous silicon-relationship between fabrication, morphology and optical properties. *Appl. Surf. Sci.* 238(1–4), 169–174.

Dietrich, G., Grobe, J., and Feld, H. (1995). In-situ FT-IR studies of porous silicon surface reactions. *J. Mol. Struct.* 349, 109–112.

Diligenti, A., Nannini, A., Pennelli, G., and Pieri, F. (1996). Current transport in free-standing porous silicon. *Appl. Phys. Lett.* 68(5), 687.

Dimitrov, D.B. (1995). Current-voltage characteristics of porous-silicon layers. *Phys. Rev. B* 51(3), 1562–1566.

Doerk, G.S., Carraro, C. and Maboudian, R. (2009). Temperature dependence of Raman spectra for individual silicon nanowires. *Phys. Rev. B* 80(7), 073306. doi:10.1103/PhysRevB.80.073306.

Dovrat, M., Arad, N., Zhang, X.-H., Lee, S.-T., and Sa'ar, A. (2007). Optical properties of silicon nanowires from cathodoluminescence imaging and time-resolved photoluminescence spectroscopy. *Phys. Rev. B* 75(20), 205343.

Faivre, C., Bellet, D., and Dolino, G. (2000). X-ray diffraction investigation of the low temperature thermal expansion of porous silicon. *J. Appl. Phys.* 87(5), 2131.

Fan, H.J., Kuok, M.H., Ng, S.C. et al. (2002). Brillouin spectroscopy of acoustic modes in porous silicon films. *Phys. Rev. B* 65(16), 1–8.

Fang, Z., Hu, M., Zhang, W., Zhang, X., and Yang, H. (2009). Mechanical properties of porous silicon by depth-sensing nanoindentation techniques. *Thin Solid Films* 517(9), 2930–2935.

Fejfar, A., Pelant, I., Šípek, E. et al. (2013). Transport study of self-supporting porous silicon Transport study of self-supporting porous silicon. *Appl. Phys. Lett.* 66, 1098.

Feng, B., Ma, W., Li, Z., Zhang, X. et al. (2009). Simultaneous measurements of the specific heat and thermal conductivity of suspended thin samples by transient electrothermal method. *Rev. Sci. Instrum.* 80(6), 064901.

Frello, T., Veje, E., and Leistiko, O. (1996). Observation of time-varying photoconductivity and persistent photoconductivity in porous silicon. *J. Appl. Phys.* 79(2), 1027.

Fried, M., Lohner, T., Polg, O., Petrik, P., Piel, P., and Stehle, L. (1996). Characterization of different porous silicon strucutres by spectroscopic ellipsometry. *Thin Solid Films* 276, 223–227.

Garnett, M.J.C. (1904). Colours in metal glasses and in metallic films. *Phil. Trans. R. Soc. Lond.* 203, 385–420.

Gesele, G., Linsmeier J., Drach, V., Fricke J., and Arens-Fischer, R. (1997). Temperature-dependent thermal conductivity of porous silicon. *J. Phys. D: Appl. Phys.* 30(21), 2911–2916.

Gomès, S., David, L., Lysenko, V., Descamps, A., Nychyporuk, T., Raynaud, M. et al. (2007). Application of scanning thermal microscopy for thermal conductivity measurements on meso-porous silicon thin films. *J. Phys. D: Appl. Phys.* 40(21), 6677–6683.

Hart, T.R., Aggarwal, R.L., and Lax, B. (1970). Temperature dependence of Raman scattering in silicon. *Phys. Rev. B* 1(2), 638–642. doi:10.1103/PhysRevB.1.638.

Henari, F.Z., Morgenstern, K., Blau, W.J., Karavanskii, V., and Dneprovskii, V.S. (1995). Third-order optical nonlinearity and all-optical switching in porous silicon. *Appl. Phys. Lett.* 67(3), 323.

Isaiev, M., Andrusenko, D., Tytarenko, A., Kuzmich, A., Lysenko, V., and Burbelo, R. (2014a). Photoacoustic signal formation in heterogeneous multilayer systems with piezoelectric detection. *Int. J. Thermophys.* 35(12), 2341–2351.

Isaiev, M., Newby, P.J., Canut, B. et al. (2014b). Thermal conductivity of partially amorphous porous silicon by photoacoustic technique. *Mater. Lett.* 128, 71–74.

Isaiev, M., Tutashkonko, S., Jean, V. et al. (2014c). Thermal conductivity of meso-porous germanium. *Appl. Phys. Lett.* 105(3), 031912 (1–5).

Jackson, W. and Amer, N.M. (1980). Piezoelectric photoacoustic detection: Theory and experiment. *J. Appl. Phys.* 51(6), 3343–3353.

Kaifeng, L., Yumin, W., Lei, Z., Shenyi, W., and Xiangfu, Z. (1994). Photoconductivity characteristics of porous silicon. *Chin. Phys. Lett.* 11(5), 289–292.

Khokhlov, A.G., Valiullin, R.R., Kärger, J. et al. (2008). Estimation of pore sizes in porous silicon by scanning electron microscopy and NMR cryoporometry. *J. Surf. Inv. X-ray Synchr. Neutr. Techn.* 2(6), 919–922.

Kojima, A. and Koshida, N. (2005). Ballistic transport mode detected by picosecond time-of-flight measurements for nanocrystalline porous silicon layer. *Appl. Phys. Lett.* 86(2), 022102.

Kordás, K., Pap, A.E., Beke, S., and Leppävuori, S. (2004). Optical properties of porous silicon. Part I: Fabrication and investigation of single layers. *Opt. Mater.* 25(3), 251–255.

Kulathuraan, K., Jeyakumar, P., Prithivikumaran, N., and Natarajan, B. (2012). Effect of polymer fraction on refractive index of nanocrystalline porous silicon. *J. Appl. Sci.* 12(6), 1671–1675.

Lee, J.-H., Grossman, J.C., Reed, J., and Galli, G. (2007). Lattice thermal conductivity of nanoporous Si: Molecular dynamics study. *Appl. Phys. Lett.* 91(22), 223110.

Lettieri, S., Maddalenat, P., Odierna, L.P., Ninno, D., La Ferrara, V., and Di Francia, G. (2001). Measurements of the nonlinear refractive index of free-standing porous silicon layers at different wavelengths. *Philos. Mag. B* 81(2), 133–139.

Losurdo, M. and Hingerl, K. (2013). *Ellipsometry at the Nanoscale.* Springer, Berlin, Heidelberg.

Lubianiker, Y., Balberg, I., Partee, J., and Shinar, J. (1996). Porous silicon as a near-ideal disordered semiconductor. *J. Non-Crystall. Solids* 198–200, 949–952.

Lysenko, V., Boarino, L., Bertola, M. et al. (1999). Theoretical and experimental study of heat conduction in as-prepared and oxidized meso-porous silicon. *Microelectron. J.* 30(11), 1141–1147.

Maliński, M., Bychto, L., Patryn, A. et al. (2005). Investigations of the optical and thermal parameters of porous silicon layers with the two wavelength photoacoustic method. *J. Phys. IV France* 129, 241–243.

Mangolini, L., Jurbergs, D., Rogojina, E., and Kortshagen, U. (2006). High efficiency photoluminescence from silicon nanocrystals prepared by plasma synthesis and organic surface passivation. *Phys. Stat. Sol. (c)* 3(11), 3975–3978.

Manifacier, J.C., Gasiot, J., and Fillard, J.P. (1976). A simple method for the determination of the optical constants n, k and the thickness of a weakly absorbing thin film. *J. Phys. E: Sci. Instrum.* 9, 1002–1004.

Mares, J., Kristofik, J., and Hulicius, E. (1995). Influence of humidity on transport. *Thin Solid Films* 255, 272–275.

Mathews, S.J., Chaitanya Kumar, S., Giribabu, L., and Venugopal Rao, S. (2007). Large third-order optical nonlinearity and optical limiting in symmetric and unsymmetrical phthalocyanines studied using Z-scan. *Opt. Commun.* 280(1), 206–212.

Matsui, Y. and Adachi, S. (2012). Optical properties of porous silicon layers formed by electroless photovoltaic etching. *J. Solid State Sci. Tech.* 1(2), 80–85.

Mawhinney, D.B., Glass, J.A., and Yates, J.T. (1997). FTIR Study of the Oxidation of Porous Silicon. *J. Phys. Chem. B* 5647(96), 1202–1206.

Mehra R., Agarwal V., Jain V.K., and Mathur, P.C. (1998). Influence of anodisation time, current density and electrolyte concentration on the photoconductivity spectra of porous silicon. *Thin Solid Films* 315(1–2), 281–285.

Menéndez, J. and Cardona, M. (1984). Temperature dependence of the first-order Raman scattering by phonons in Si, Ge, and α – S n: Anharmonic effects. *Phys. Rev. B* 29(4), 2051–2059. doi:10.1103/PhysRevB.29.2051.

Nassiopoulou, A.G. and Kaltsas, G. (2000). Porous silicon as an effective material for thermal isolation on bulk crystalline silicon. *Phys. Stat. Sol. (a)* 182, 307–311.

Newby, P.J., Canut, B., Bluet, J.-M. et al. (2013). Amorphization and reduction of thermal conductivity in porous silicon by irradiation with swift heavy ions. *J. Appl. Phys.* 114(1), 014903.

Nonnenmacher, M. and Wickramasinghe, H.K. (1992). Scanning probe microscopy of thermal conductivity and subsurface properties. *Appl. Phys. Lett.* 61(2), 168.

Obraztsov, A.N., Timoshenko, V.Y., Okushi, H., and Watanabe, H. (1997). Photoacoustic spectroscopy of porous silicon. *Semiconductors* 31, 534–536.

Ogata, Y.H. (2014). Characterization of porous silicon by infrared spectroscopy. In L. Canham, ed. *Handbook of Porous Silicon.* Springer International Publishing.

Oliver, W.C. and Pharr, G.M. (1992). An improved technique for determining hardness and elastic modulus using load and displacement sensing indentation experiments. *J. Mater. Res.* 7(6), 1564–1583.

Oliver, W.C. and Pharr, G.M. (2004). Measurement of hardness and elastic modulus by instrumented indentation: Advances in understanding and refinements to methodology. *J. Mater. Res.* 19(1), 3–20.

Ould-abbas, A., Bouchaour, M., and Sari, N.C. (2012). Study of thermal conductivity of porous silicon using the micro-Raman method. *OJPC* 2, 1–6.

Pap, A.E., Kordás, K., Vähäkangas, J. et al. (2006). Optical properties of porous silicon. Part III: Comparison of experimental and theoretical results. *Opt. Mater.* 28(5), 506–513.

Parkhutik, V. (1999). Porous silicon-mechanisms of growth and applications. *Solid-State Electron.* 43, 1121–1141.

Pérez, X.E., Pallares, J., Ferre-Borrull, J., Trifonov, T., and Marsal, L.F. (2007). Low refractive index porous silicon multilayer with a high reflection band. *Phys. Stat. Sol. (c)* 4(6), 2034–2038.

Perichon, S., Lysenko, V., Remarki, B., Barbier, D., and Champagnon, B. (1999). Measurement of porous silicon thermal conductivity by micro-Raman scattering. *J. Appl. Phys.* 86(8), 4700–4702.

Perichon, S., Lysenko, V., Remaki, B., Champagnon, B., Barbier, D., and Pinard, P. (2000). Technology and micro-Raman characterization of thick meso-porous silicon layers for thermal effect microsystems. *Sens. Actuators A* 85, 335–339.

Pickering, C., Beale, M.I.J., Robbins, D.J., Pearson, P.J., and Greef, R. (1984). Optical studies of the structure of porous silicon films formed in p-type degenerate and non-degenerate silicon. *J. Phys. C: Sol. State Phys.* 17, 6535–6552.

Rao, P., Schiff, E.A., Tsybeskov, L., and Fauchet, P. (2002). Photocarrier drift-mobility measurements and electron localization in nanoporous silicon. *Chem. Phys.* 284(1–2), 129–138.

Ray, A.K., Mabrook, M.F., Nabok, A.V., and Brown, S. (1998). Transport mechanisms in porous silicon. *J. Appl. Phys.* 84(6), 3232–3235.

Riedling, K. (1988). *Ellipsometry for Industrial Applications.* Springer-Verlag, New York.

Rosencwaig, A. and Gersho, A. (1975). Photoacoustic effect with solids: A theoretical treatment. *Science* 190(4214), 556–557.

Rosencwaig, A. and Gersho, A. (1976). Theory of the photoacoustic effect with solids. *J. Appl. Phys.* 47(1), 64–69.

Sarafis, P. and Nassiopoulou, A.G. (2014). Dielectric properties of porous silicon for use as a substrate for the on-chip integration of millimeter-wave devices in the frequency range 140 to 210 GHz. *Nanoscale Res. Lett.* 9(1), 418.

Schwarz, R., Wang, F., Ben-Chorin, M., Grebner, S., Nikolov, A., and Foch, K. (1995). Photocarrier grating technique in mesoporous silicon. *Thin Solid Films* 255, 23–26.

Sheik-Bahae, M., Said, A.A., Wei, T.-H., Hagan, D.J., and Van Stryland, E.W. (1990). Sensitive measurement of optical nonlinearities using a single beam. *IEEE J. Quantum Electron.* 26(4), 760–769.

Siegert, L., Capelle, M., Roqueta, F., Lysenko, V., and Gautier, G. (2012). Evaluation of mesoporous silicon thermal conductivity by electrothermal finite element simulation. *Nanoscale Res. Lett.* 7(1), 427(1–7).

Stoib, B., Filser, S., Stötzel, J. et al. (2014a). Spatially resolved determination of thermal conductivity by Raman spectroscopy. *Semicond. Sci. Technol.* 29, 124005(1–13).

Stoib, B., Filser, S., Petermann, N., Wiggers, H., Stutzmann, M., and Brandt, M.S. (2014b). Thermal conductivity of mesoporous films measured by Raman spectroscopy. *Appl. Phys. Lett.* 104(16), 161907(1–4).

Stolz, H. (ed.) (1994). *Time-Resolved Light Scattering from Excitons.* (Springer Tracts in Modern Physics Vol. 130). Springer-Verlag, Berlin.

Theiss, W. (1995). IR spectroscopy of porous silicon. In: Vial J.-C. and Derrien J. (eds.) *Porous Silicon Science and Technology.* Springer-Verlag, Berlin, pp. 189–205.

Theiss, W. (1997). Optical properties of porous silicon. *Surf. Sci. Rep.* 29, 91–192.

Theiss, W. and Hilbrich S. (1997). Refractive index of porous silicon. In: Canham L.T. (ed.) *Properties of Porous Silicon*. INSPEC, London, pp. 221–228.

Theodoropoulou, M., Karahaliou, P.K., and Krontiras, C.A. (2004). Transient and ac electrical transport under forward and reverse bias conditions in aluminum/porous silicon/p-cSi structures. *J. Appl. Phys.* 96(12), 7637.

Torres, J., Martinez, H.M., Alfonso, J.E., and López, L.D. (2008). Optoelectronic study in porous silicon thin films. *Microelectron. J.* 39(3–4), 482–484.

Tsao, S.S., Guilinger, T.R., Kelly, M.J., Stein, H.J., Barbour, J.C., and Knapp, J.A. (1990). Porous silicon oxynitrides formed by ammonia heat treatment. *J. Appl. Phys.* 67(8), 3842.

Tytarenko, A.I., Andrusenko, D.A., Kuzmich, A.G. et al. (2014). Features of photoacoustic transformation in microporous nanocrystalline silicon. *Tech. Phys. Lett.* 40(3), 188–191.

Urbach, B., Axelrod, E., and Sa'ar, A. (2007). Correlation between transport, dielectric, and optical properties of oxidized and nonoxidized porous silicon. *Phys. Rev. B* 75(20), 205330.

Vásquez-A, M.A., Aguila Rodriguez, G., Garsia-Salgado, G., Romero-Paredes, G., and Pena-Sierra, R. (2007). FTIR and photoluminescence studies of porous silicon layers oxidized in controlled water vapor conditions. *Rev. Mexicana Fisica* 53(6), 431–435.

Von Behren, J., Wolkin-Vakrat, M., Jorne, J., and Fauchet, P.M. (2000). Correlation of photoluminescence and bandgap energies with nanocrystal sizes in porous silicon. *J. Porous Mater.* 7, 81–84.

Wolf, A. and Brendel, R. (2006). Thermal conductivity of sintered porous silicon films. *Thin Solid Films* 513(1–2), 385–390.

Xie, Y.H., Wilson, W.L., Ross, F.M. et al. (1992). Luminescence and structural study of porous silicon films. *J. Appl. Phys.* 71(5), 2403.

Yeh, C.C., Hsu Klaus, Y.J., Samanta, L.K., Chen,, P.C., and Hwang, H.L. (1993). Study on the photoconductivity characteristics of porous Si. *Appl. Phys. Lett.* 62(14), 1617.

Zangooie, S., Schubert, M., Trimble, C., Thompson, D.W., and Woollam, J.A. (2001). Infrared ellipsometry characterization of porous silicon Bragg reflectors. *Appl. Opt.* 40(6), 906.

Zharki, S.M., Karabutov, A.A., Pelivanov, I.M., Podymova, N.B., and Timoshenko, V.Yu. (2003). Laser ultrasonic study of porous silicon layers. *Semiconductors* 37(4), 468–472.

Zheltikov, A., L'Huillier, A., and Krausz, F. (2012). Nonlinear optics. In: Träger, F. (ed.) *Springer Handbook of Lasers and Optics*. Springer, Dordrecht, pp. 161–251.

Zhenkun, L., Yilan, K., Hao, C., Yu, Q., and Ming, H. (2005). Experimental study on mechanical properties of micro-structured porous silicon film. *Trans. Tianjin Univ.* 11(2), 85–88.

# Structural and Electrophysical Properties of Porous Silicon

**Giampiero Amato**

## CONTENTS

## 6.1 INTRODUCTION

This deals with aspects strongly related to several applications of porous silicon (PSi) conceived in the 30 years spanning from approximately 1980 to 2010. Few materials found such a wide range of applications like PSi, from epitaxial growth (Sato et al. 1995) to optics (Amato et al. 2000a), sensors (Korotcenkov 2010), explosives (Kovalev et al. 2001), and biological and medical applications (Low et al. 2009). Nowadays, in the fields of solid-state physics and nanotechnology, some topics are still open, for example, in state-of-the-art research on semiconductor nanowires. Molecular doping as well as thermal properties of those systems show interesting resemblances with the same phenomena deeply investigated in PSi.

The first, and simplest structural parameter to be measured about PSi is porosity $P$, that is, the volume fraction of voids with respect to the bulk material. It can be extracted by combining three mass measurements, $m_0$ (mass of pristine sample), $m_1$ (sample mass after porosization), and $m_2$ (mass after PSi removal) in the following way: $P = (m_0 - m_1)/(m_0 - m_2)$. Using a commercial balance with $10^{-5}$ g sensitivity and samples with a porous volume of $10^{-4}$ cm$^3$, the relative error on $P$ can be calculated as greater than 20% in a 70% porosity sample. To keep the uncertainty below 10%, the porous volume must be increased by a factor of 3, either by increasing PSi thickness, or, better, by anodizing larger wafer areas.

The situation is actually more complicated. The measurement of $m_1$ must be done when the sample is completely dry (Berger et al. 1998), but waiting too long can give rise to overestimated mass, due to PSi uptake of foreign atoms (typically $O_2$), or ambient dust. Then, it is advisable to carry out the measurement in a clean-room environment.

Unfortunately, only in a few cases, authors provide details on this obvious but delicate measurement. This hinders the comprehension of several trends of physical parameters on porosity: in fact, results from different groups are often difficult to compare. Silicon wafers from different vendors, moreover, can give rise to a variety of microstructures, even if the macroscopic parameter, that is, porosity, is the same (the effect of different surface chemo-mechanical polishing treatments on the vertical porosity gradient is an example).

Techniques that are more sophisticated are then needed to yield insight on the structural properties of PSi. Most of them are related to the measurement of microscopic quantities, that is, the sizes of pores and Si instances.

## 6.2 CRYSTAL STRUCTURE

After almost 20 years after its discovery, PSi found its first application in silicon-on-insulator technology (Imai 1981). Then, for one decade, many studies focused on the possibility of employing PSi as a substrate for reducing lattice mismatch in epitaxial growth of crystalline semiconductors (Luryi and Suhir 1986). A great deal of work was then done in those years to probe the crystalline quality of PSi and the presence of strain or lattice distortions. Those structural studies were intensified after the first report on the visible light emission from PSi (Canham 1990) because some explanations for this spectacular effect were based on previously known systems, like hydrogenated amorphous silicon (a-Si:H) in which light emission had been clearly observed and studied (Street 1984). Therefore, for some time, PSi structure has been deeply investigated in order to ascertain if its emission could be in some way related to the presence of an amorphous phase. In general, such a strong debate was mainly related to the structure of nano-PSi (an Si sponge system in which both Si structures and pores are in the range of a few nanometers), whereas, for non-emitting versions (meso- and macro-PSi), the consensus about the crystallinity of the structure was almost general. Some silicon-on-insulator applications of meso-PSi, in fact (Bomchil et al. 1988), were indirect evidences that the PSi substrate is a nearly perfect monocrystal. In one of the different recipes to achieve c-Si growth onto PSi, molecular beam epitaxy (MBE) had been used (Konaka et al. 1982).

## 6.2.1 TRANSMISSION ELECTRON MICROSCOPY

It is not surprising how transmission electron microscopy (TEM) has been used from the beginning to understand the structure of PSi. Being a direct technique, even if destructive and requiring preparation, it allows direct observation of the PSi skeleton and gain of information about its crystallinity. Previous studies (Theunissen 1972; Beale et al. 1985a) have shown that the microstructure of porous materials strongly depends on the anodization condition and on the density of impurities within the pristine crystal. PSi layers obtained by anodizing $p$-type Si crystals with acceptor density of $10^{16}$–$10^{17}$ at/cm$^3$ are sponge-like, with a random distribution of nanometric-sized pores. For heavily doped ($10^{18}$–$10^{19}$ at/cm$^3$), no matter if $p$- or $n$-type, porous layers present a highly anisotropic morphology, which consists of many long pores parellel to the <100> direction. Before the discovery of the strong PL at RT, most of the electron microscopy studies were aimed to validate the different models proposed to explain the formation mechanism (Theunissen 1972; Beale et al. 1985a,b). Since the discovery of PL (Canham 1990), the interest shifted to the understanding of this phenomenon, in particular to evidence of the presence of an amorphous phase or the presence of small nanocrystallites, an indirect confirm of the quantum confinement model. Bright field studies (Cullis and Canham 1991; Lehmann et al. 1991; Berbezier and Halimaoui 1993) on highly porous (>80%) $p$-type PSi evidenced the presence of small nanocrystallites (2–3 nm sizes) in which one could expect quantum confinement effects. On the contrary, in lower porosity (<70%) samples, which show a very weak luminescence indeed, it was impossible to show the presence of individual crystallites. In that case, a clear correlation between PL and nanocrystals was not possible.

In the latter case, additional information came from transmission electron diffraction (TED) studies (Beale et al. 1985a; Berbezier and Halimaoui 1993). Figure 6.1 shows sharply defined diffraction spots as in the case of c-Si, thus confirming the single crystal character of low and medium porosity PSi. In case of highly porous layers, however, the situation is not so simple: detailed HRTEM studies (Berbezier and Halimaoui 1993) show the coexistence of two phases: a crystalline one and another phase originating broad diffusion rings (Figure 6.2). These rings could originate either from misoriented nanocrystalline clusters or from an amorphous phase (a layer of native oxide which rapidly grows in these systems cannot be excluded, however).

A more robust system to be investigated by TEM and TED is the $p^+$-PSi because nanostructures are larger and more regular. Pores extend along the <100> direction, perpendicular to the surface in (100)-oriented samples. This led researchers to suggest that the pores propagate along the current flow lines during anodization (Beale et al. 1985a), but well before the discovery of PL from PSi, it was shown that in <111> Si, pores do not follow the current lines but rather "zigzags"

**FIGURE 6.1** TEM image of a 65% porosity nanoporous sample. The corresponding TED pattern shows the single crystal character of the material. (Reprinted with permission from Berbezier I. and Halimaoui A., *J. Appl. Phys.* 74, 5421, 1993. Copyright 1993. American Institute of Physics.)

**FIGURE 6.2** TEM image of an 85% porosity nanoporous sample. The corresponding TED pattern shows spots from a crystalline phase with broad diffusion rings attributed to misoriented nanocrystalline clusters or from an amorphous phase. (Reprinted with permission from Berbezier I. and Halimaoui A., *J. Appl. Phys.* 74, 5421, 1993. Copyright 1993. American Institute of Physics.)

(Unagami and Seki 1978). As a matter of fact, studies on differently oriented substrates indicate that pores propagate preferentially along the component of the current vector parallel to the <100> directions, but secondary dissolution directions are non-negligible (Parisini et al. 1996).

Cross-sectional images on <100> substrates allow envisaging a thread-like structure for the Si skeleton with some dendritic character (Figure 6.3). This picture contrasts with the simple observation that a $p^+$-PSi membrane does not disintegrate when detached from the substrate, that is, a sort of lateral interconnection must be present. It has to be noted that a honeycomb structure with pores separated by Si walls is observed as thread-like in the absence of in-depth plane views, but can account for the mechanical stiffness of the self-supporting sample. As it will be mentioned in the following, electrical transport is strongly anisotropic in PSi from <100>

**FIGURE 6.3** TEM image in cross-section of $p^+$ <100> porous silicon sample with 65% porosity showing the apparent thread-like structure for the Si skeleton. (From Grosman A. et al. In: *Structural and Optical Properties of Porous Silicon Nanostructures*, 1998. Copyright 1998: CRC Press. With permission.)

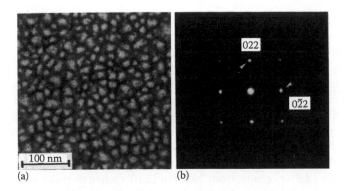

**FIGURE 6.4**   TEM images of (a) 80% porosity sample from p⁺ substrate suggesting its mosaic structure, and (b) 66% porosity sample from *p*⁺ substrate, showing its monocrystalline character. (From Grosman A. et al. In: *Structural and Optical Properties of Porous Silicon Nanostructures,* 1998. Copyright 1998: CRC Press. With permission.)

crystals, suggesting smaller lateral interconnections. A microstructure, intermediate between the two described above, is then likely for this kind of material. PSi from differently oriented substrates has been less investigated, but a more interconnected structure has been reported (Parisini et al. 1996).

TED studies on $p^+$-PS indicate a more precise crystalline character of this material, if compared to $p$-type (Berbezier and Halimaoui 1993). In 66% porosity layers, a well-defined pattern, composed of fine spots (Figure 6.4a), typical of a single crystalline structure, is visualized. This confirms the early report of Barla et al. (1984), who observed by X-ray diffraction the monocrystalline character of this material. On the other hand, when porosity is raised up to 80%, the diffraction pattern consists of small arcs of circles centred on the original spots (Figure 6.4b). This suggests the presence of crystallites on the Si walls having a distribution of orientations. The absence of diffused rings, however, allows us to rule out the presence of an amorphous phase in this material, neither native nor induced by the ion milling process employed in the sample preparation (Grosman et al. 1998).

The first electron microscopy studies of the *varied wildlife* of PSi deriving from anodization of $n$-type and $n^+$-type substrates, in dark or under different illumination conditions, has been rewieved elsewhere (Levy Clement 1995). In the case of macroporous material, most of the electron microscopy studies were carried out with the aim of understanding the mechanism of macropore formation and propagation (Rönnebeck et al. 1999; Christophersen et al. 2001). The mesoporous material deriving from dark anodization of $n^+$ substrates received some attention because of its morphology similar to $p^+$-PSi, in spite of the largely different formation mechanism. The possibility of anodizing $n^+$-PSi in dark makes this material very interesting for several applications, like the realization of $p$-$n$ nano-junctions (Chen et al. 1993), the opening of point surface sites for macropore formation (Parisini et al. 2000), and the realization of buried PSi structures (Amato et al. 2001). In the last case, TEM investigations visualized peculiar bended pore and wall structures, suggesting that in this material, not only the crystallographic directions but also the current flow can steer the pore propagation.

## 6.2.2  X-RAY DIFFRACTION

Another technique, widely employed since the early 1970s (Theunissen 1972) is the X-ray diffraction (XRD). The reason for its success relies on the fact it is a direct technique, non-destructive, and does not require any preparation step. Differently from TEM, uncertainties due to, for example, ion-milling induced amorphization, are avoided in XRD, although less relevant information about the PSi crystallites size and shape can be obtained.

The first observations (Theunissen 1972) pointed out that PSi is a crystalline porous material, but an important result came in evidence a few years later: using X-ray topography and high resolution double-crystal diffraction, it was discovered that PSi behaves as a single crystal and gives diffraction peaks nearly as narrow as those of the perfect Si substrate (Barla et al. 1984). The situation is depicted in Figure 6.5. The rocking curve $I(\omega)$, which gives the variation of the diffracted X-ray intensity as a function of the rotation angle $\omega$ of the sample, shows a characteristic double peak shape: the peak "S" is due to the diffraction from the substrate, whereas the peak "P" corresponds to the diffraction from the porous layer. Commonly, "P" peak is slightly broader indicating a coherence length in the micrometer range. The broad hump "D" below the two peaks is due to diffraction from the small silicon crystallites, as observed by Bensaid et al. (1991).

The angular splitting $\Delta\omega$ between the two narrow peaks "P" and "S" is ascribed to the increase $\Delta a/a$ of the lattice parameter of the porous layer relative to the Si one, along the <001> direction, perpendicular to the sample surface ($\Delta a/a = -\Delta\omega/\mathrm{tg}\theta$). $P^+$-type PSi samples generally show the best quality X-ray diffraction patterns. For as-prepared, non-oxidized samples, the value of $\Delta a/a$ increases quite linearly with porosity, from $3\cdot10^{-4}$ to $12\cdot10^{-4}$ when porosity varies from 30% to 90% (Barla et al. 1984; Bellet and Dolino 1996; Dolino and Bellet 1997). Results of these studies are summarized in Figure 6.6. Broader "P" peaks are generally observed in $p$-type PSi, together with larger values for the expansion of the lattice parameter ($\Delta a/a = 2\cdot10^{-3}$) (Manotas et al. 2001). The situation is more complicated in $n$-type PSi because the in-depth inhomogeneity of the samples, no matter if macroporous, mesoporous, or nanoporous, hinders the interpretation of the curves. In the case of $n^+$ PSi on <111> substrates, rocking curves similar to $p^+$ (Figure 6.5) have been obtained (Labunov et al. 1986).

The interesting question regards the origin of such lattice dilatation. One should consider that in the case of small size particles with a clean surface, a decrease of $\Delta a/a$ is expected (Buttard et al. 1996), being $\Delta a/a = -2\gamma K/3r$, where $\gamma$ is the surface energy, $K$ is the bulk compliance, and $r$ is

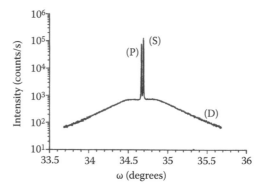

**FIGURE 6.5** X-ray rocking curve I($\omega$) showing the variation of the diffracted X-ray intensity as a function of the rotation angle $\omega$ of the sample. (From Bellet D. and Dolino G. In: *Structural and Optical Properties of Porous Silicon Nanostructures*, 1998. Copyright 1998: CRC Press. With permission.)

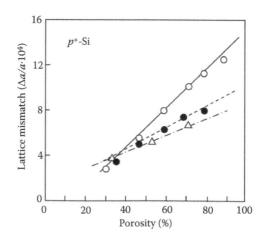

**FIGURE 6.6** Lattice mismatch parameter $\Delta a/a$, between the PSi layer and the substrate, versus the porosity for a series of $p^+$-type samples. (Reprinted from *Surf. Sci. Rep.* 38, Bisi O. et al., 1, Copyright 2000, with permission from Elsevier.)

the radius of the nanoparticle. The difference here is that the surface of the Si nanoparticles composing PSi can never be considered "clean." A freshly prepared sample presents H-termination or traces of native oxide. Both those chemical terminations have been suggested as possible sources for the lattice dilation (Sugiyama et al. 1990; Young et al. 1985; Kim et al. 1991). This viewpoint is confirmed by the observation that, upon controlled oxidation, the lattice parameter further increases (Ito et al. 1991; Gardelis et al. 1995; Bellet and Dolino 1998), as expected because of the well-known volume expansion of 2.3 occurring during Si oxidation. Another confirm for the surface origin of the lattice expansion comes from H desorption experiments, carried out either at 1000°C (Labunov et al. 1986) or at 350°C (Sugiyama et al. 1990). As a consequence for H desorption, a negative lattice variation of a few $10^{-4}$ was observed in both cases.

## 6.2.3  RAMAN SPECTROSCOPY

There are three reasons for the big success of Raman spectroscopy as a characterization tool for PSi: first, it is a rather inexpensive equipment because the good lateral uniformity of PSi does not, in principle, require any micro-Raman tool. Second, most of the studies related to confinement of optical phonons focus on the very intense first order Raman emission, located at 500–520 cm$^{-1}$. The third, fundamental reason, is that the properties of nanosized structures that strongly depend on their morphology (i.e., symmetry, structural geometry, pore diameter, skeleton size, etc.) are reflected in their Raman spectra.

### 6.2.3.1  CONFINEMENT OF OPTICAL PHONONS

In the Raman spectra of porous materials with characteristic structure sizes of a few nanometers, the Raman peak progressively red-shifts and the lineshape gets progressively broader and asymmetric (on the low-energy side) as the structure size gets smaller. This effect can be used to determine the particle size. Typical Raman spectra of the optical phonons of nanocrystals in PSi are shown in Figure 6.7, together with the Raman spectrum of a single crystal. Different confinement models have been used in order to determine particle sizes and size distributions from the measured spectra.

A very widely used phenomenological model, originally proposed for microcrystalline silicon (Richter et al. 1981), was later extended (Campbell and Fauchet 1986) and intensively applied to PSi. In this case, the localization of the phonon in a finite crystal is taken into account by introducing a weighting function; the phonon amplitude is reduced when moving in the direct space from the center ($r = 0$) to the microcrystal boundary ($r = L$). Then, the Raman scattering intensity will be a continuous superposition of Lorentzian curves, centered at different wavenumbers (the phonon dispersion) and with different weights.

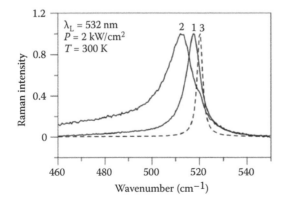

**FIGURE 6.7**    A comparison of Raman spectra from $p^+$-type, 75% porosity, mesoporous (1), $p$-type 65% porosity, nanoporous (2) samples, and a crystalline (3) sample. (Irmer G.: *J. Raman Spectrosc.* 2007. 38. 634. Copyright Wiley-VCH Verlag GmbH & Co. KGaA. Reproduced with permission.)

Some kinds of confinement functions are reported in the literature (Richter et al. 1981; Iqbal and Veprek 1982; Campbell and Fauchet 1986). In the traditional approach to line shape analysis (LSA), the weighting function $W(r,L) = \exp(-\alpha r^2/L^2)$ is mostly used because of the better agreement between experimental spectra and theoretical fits obtained in this way. This is because a gaussian confinement form takes account, in some way, for a distribution of differently sized crystallites. There is no physical meaning, however, in the assumption that the phonon wavefunction amplitude differs from zero at the crystallite boundary. Only one of the confinement models proposed in the past assumes the phonon amplitude wavefunction as going to zero at the boundary of the crystallite, the weighting function being $W(r,L) = \sin(\alpha r)/\alpha r$, by analogy with the ground state of an electron in a hard sphere (Münder et al. 1992). The relative Raman shift versus FWHM is plotted in Figure 6.8a together with the axis on the righthand side indicating the size dimension. The confinement of optical phonons in PSi layers has been investigated by means of this phenomenological model for Raman scattering by numerous authors, by considering different weighting functions (see, e.g., Irmer 2007). The results obtained in these ways have to be taken with some care because the nanocrystal shapes are often not well defined and the nanocrystals have a size distribution, as described next. Influence of particle shape on the results of theoretical simulation is shown in Figure 6.8b. Moreover, strain-induced changes in the phonon wavenumbers have to be taken into account (Kozlowski and Lang 1992; Münder et al. 1992; Xia et al. 1995). It is known that in nanoporous Si, the Raman peak can be shifted by a dilation strain on the order of $10^{-3}$ to lower wavenumbers by about 2 cm$^{-1}$ (Englert et al. 1980). Stress inhomogeneities on a nanometer scale can be expected to broaden and shift the Raman signal, as it will be described in the following. Last, the presence of an amorphous component cannot be neglected.

Brunetto and Amato (1997) proposed a different method, based on the TEM observation of a log-normal distribution (Figure 6.9) of particle diameters in PSi (Kanemitsu 1994). A log-normal distribution is the result of a process in which a variate obeys the law of proportionate effect, say, the change of the variate at any step of the process is a random proportion of the previous value of the variate (Crow and Shimizu 1987). Many examples of the log-normal distribution have been noted in nature like the sedimentary petrology or the erosion phenomena. The electrochemical etching of Si can be included among those. The approach consists of a Monte Carlo based method, where $L$ is no longer considered a deterministic fitting parameter, but as a log–normal stochastic variable (Gaussian distribution of the logarithm of the particle diameter, $x$).

In this framework, the most probable $x$ value ($x_{mp}$) defined by $\mu = ln(x_{mp})$, and $\sigma$ (the shape parameter) are taken as the fitting parameters. A stochastic generation algorithm provides size

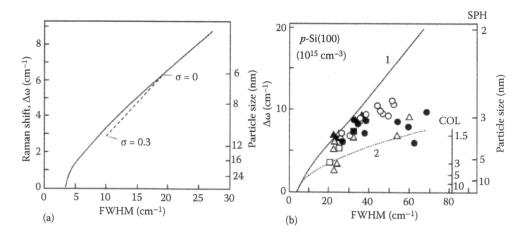

**FIGURE 6.8**   (a) Relation between FWHM and Raman shift with respect to the c-Si case as expected for the hard sphere confinement model in the deterministic (full line) and stochastic (dashes) models. The bifurcation is related to the assumption of a distribution of crystallite sizes in the stochastic case. See text for details. (Reprinted from *Thin Solid Films* 297, Brunetto N. and Amato G., 122, Copyright 1997, with permission from Elsevier.) (b) Experimental values of Raman shift for various films and theoretical values predicted by the LSA model for (1) spherical, SPH, and (2) columnar, COL, particles. (Reprinted from *Thin Solid Films* 349, Papadimitriou D. et al., 293, Copyright 1999, with permission from Elsevier.)

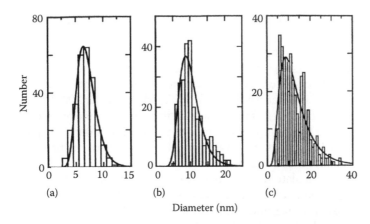

**FIGURE 6.9**    The size distribution of Si nanoparticles ensembles with different average diameter: 3.8 nm (a), 6.8 nm (b), 10.0 nm (c), with the optimum log-normal functions (solid curves). (From Kanemitsu Y., *J. Phys. Soc. Jpn.* 63(suppl. B), 107, 1994. Copyright 1994: The Physical Society of Japan. With permission.)

dimensions, governed by a log–normal $(x, x_{mp}, \sigma)$, to the hard sphere confinement model in order to generate an expected Raman line shape. In this way, a chosen statistical distribution is imposed and the number of fitting parameters is only one more than in the traditional LSA methods. The results report (Brunetto and Amato 1997) both a very good reproduction of the experimental curves (Figure 6.10) and size ranges in agreement with TEM observations. The bifurcation

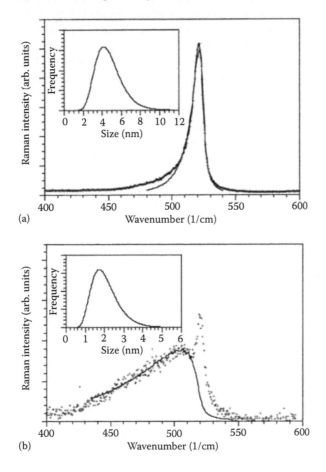

**FIGURE 6.10**    Fits of the experimental Raman spectra for a $p^+$-PSi (a) and a $p$-PSi sample (b). In the inserts, the corresponding size distributions are shown. (Reprinted from *Thin Solid Films* 297, Brunetto N. and Amato G., 122, Copyright 1997, with permission from Elsevier.)

observed in Figure 6.10a between the curves, from the deterministic and stochastic model, can induce errors when inferring the most probable size value.

Other models use the structure factor that is well established in the analysis of small angle X-ray scattering experiments on small particles (Guinier and Fournet 1955) and can be derived by evaluating the susceptibility function over a limited spatial extent (Nemanich et al. 1981). The description of the method is exhaustively reported elsewhere (Irmer 2007); however, it is interesting to note that the final derived formula is similar in both phenomenological and the structure factor models, but with a different physical interpretation. After taking the phonon confinement into account, by considering that the largest possible wavelength of the phonons in the grain is restricted by its diameter and approximating the phonon dispersion curve by a linear chain model, it is possible to calculate the nanoparticles sizes, without involving any arbitrary fitting parameter. Figure 6.11 compares the calculated Raman shift and broadening obtained by both models for different particle size, and compares them with the experimental data.

Being an indirect method to yield information about the structural properties of PSi, Raman spectroscopy of optical phonons requires great care in data manipulation and analysis. The results summarized in Figure 6.11 indicate that the agreement between the phenomenological and the structure factor models is fair, as far as the nanostructure sizes are concerned. Moreover, the arbitrary choice of the nanostructure shape (dots instead of wires) affects the results considerably. Finally yet importantly, both methods give a single value for the sizes of Si instances, whereas a log-normal distribution is observed by TEM.

From the log-normal distribution given by $F(x) = \exp[-(\ln x - \chi)^2/2\sigma^2]$, for typical nano-PSi shape and scale parameters, $\sigma = 0.5$ nm and $x_{mp} = 3$ nm, respectively, one can infer that 80% of the Si nanostructures have a diameter less than $2x_{mp}$. These particles are easily calculated to occupy a volume of the Si skeleton ranging from 45% (assuming the particles as spheres) to 65% (assuming them as wires). This means that a large fraction of the sample volume is occupied by a few large particles, and substantiates the consideration, often invoked by investigators, that PSi is an inhomogeneous material. In other words, while some macroscopic observables (photoluminescence above all) provide information on the lefthand side of the log-normal curve, other ones (e.g., electrical or thermal conductivity) often reflect the properties of the much broader righthand tail. When trying to correlate porosity with other material properties, one must always consider that one is averaging among instances having very different volume occupancies.

In this framework, Raman spectroscopy represents a very powerful tool for the structural characterization of the whole material because from the distortion of the Si optical phonon mode it is possible to extract information about the entire size distribution, instead of a single size value of Si nanoparticles.

As mentioned above, it is also possible to determine the stress in the film, the distance of relaxation (Manotas et al. 2000), stress profile in the porous layer (Papadimitriou et al. 1999), and

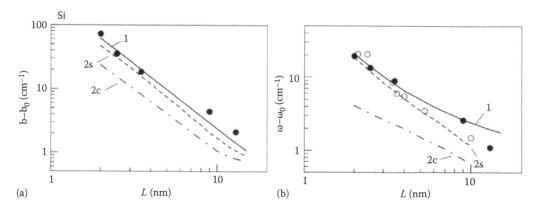

**FIGURE 6.11** Wavenumber broadening (a) and shift (b) of the Raman band of the optical phonon in PSi versus the nanocrystal size $L$. The curves (1, 2s, 2c) show the theoretical results for confinement models discussed in the text (deterministic, structure factor with spheres, structure factor with long cylinders, respectively). (Irmer G.: *J. Raman Spectrosc.* 2007. 38. 634. Copyright Wiley-VCH Verlag GmbH & Co. KGaA. Reproduced with permission.)

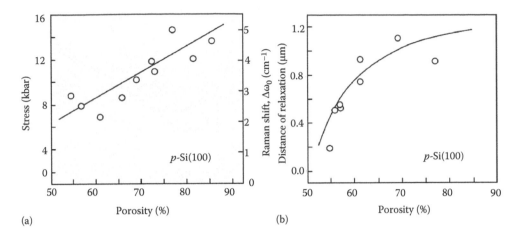

**FIGURE 6.12**    (a) Variation of $\omega_0$ at the interface and calculated compressive stress versus porosity. (b) Distance of relaxation of $\omega_0$ versus porosity. Porous silicon layers were etched in the dark on $p$-Si:B(001) wafers ($\rho = 0.1 - 0.5 \ \Omega\cdot$cm) with a 1:1 HF (48%):ethanol (98%) solution. The current density was varied from 10 to 150 mA/cm$^2$ to obtain samples with different porosities from 55% to 86%. (Manotas S. et al.: *Phys. Stat. Sol. (a)*. 2000. 182. 245. Copyright Wiley-VCH Verlag GmbH & Co. KGaA. Reproduced with permission.)

so on. In particular, Raman study has shown that a compressive stress, attributed to the lattice mismatch between PSi and c-Si, is maximum close to the interface and relaxes approximately within one micrometer (see Figure 6.12). In addition, the stress value and the distance of relaxation increase with porosity (Manotas et al. 2000). At the same time, Papadimitriou et al. (1999) established that a compressive stress relaxes faster in samples of higher porosity.

### 6.2.3.2 CONFINEMENT OF ACOUSTIC PHONONS

Acoustic phonons with vanishing energy at the $\Gamma$ point are not observed in first-order Raman scattering. However, acoustic phonons confined in nanocrystals can be observed (Duval et al. 1986). They appear at low energies in the Raman spectra, with wavenumbers inversely proportional to the diameter of the nanoparticles. The difficulty of approaching the energy of the excitation light (more monochromating stages are needed) is the reason why few observations of confined acoustic phonons in porous semiconductors have been reported, despite the advantage of directly determining the nanocrystal sizes (Gregora et al. 1994; Liu et al. 1996).

## 6.3 CHEMICAL COMPOSITION

In spite of the great efforts carried out by different techniques, only PSi samples with nanostructure sizes above 5 nm can be considered as fully crystalline. When dealing with nanoporous, luminescent PSi, an additional amorphous tissue wrapping the crystalline cores is often invoked to explain halos in TED and XRD patterns or peak broadening in Raman spectra. This amorphous layer, with a varied chemical composition, can affect the optical properties of the material. As an example, the optical absorption spectra of PSi and a-Si:H show common features. The a-Si:H absorption shows a rather broad absorption edge in the 1.5–1.8 eV range, below the energy gap (Amato and Fizzotti 1992), related to shallow electronic traps formed at the elongated bonds due to the inevitable lack of long-range order (Figure 6.13a). The difference is evident when comparing with the similar spectrum of c-Si (Figure 6.13b): the absorption edge is narrow in c-Si, and mostly related to dopant states energetically close to band edges (Amato et al. 1989). When the same investigation is done with PSi (Figure 6.13b), either by PL excitation spectroscopy or photothermal deflection spectroscopy (Amato and Rosenbauer 1998), in both nano- or meso-PSi, the absorption edge shows the same exponential tail with inverse slope values of the order of 200 meV in luminescent samples, whereas at lower porosity, tail slopes comparable to a-Si are found.

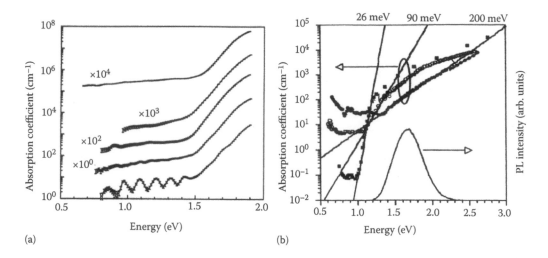

**FIGURE 6.13** A comparison of the band absorption edge of (a) several amorphous silicon samples, and (b) 50% porosity PSi (open circlets), 70% porosity, luminescent PSi (closed circlets), and c-Si (squares). The luminescent spectrum falls at energies in which the exponential tail is observed. ([a] Reprinted with permission from Amato G. and Fizzotti F., *Phys. Rev. B* 45, 14108, 1992. Copyright 1992 by the American Physical Society; [b] from Amato G. and Rosenbauer M., In: *Structural and Optical Properties of Porous Silicon Nanostructures*, 1998. Copyright 1998: CRC Press. With permission.)

The nature of such dramatic smearing of the absorption edge is still unclear, but, if related to the presence of distorted bonds located at the surface of the Si skeleton composing PSi (Petit et al. 1997), they should play a role in the PL process (Koch et al. 1992). Several different models based on PL mechanisms related to the disordered silicon compound tissue that wraps the crystallites, for example, a-Si:H, $SiO_x$, (Prokes 1993), or siloxene (Fuchs et al. 1993) have been proposed. The chemical composition of the external tissue of nano-PSi, then, deserves a careful analysis.

Apart from Si, which is the main constitutant of the crystalline cores, many different chemical elements can be incorporated:

- Dopant atoms are incorporated in the pristine substrate and are likely to affect the characteristics of the PSi itself after anodization.
- The anodization reaction can incorporate other elements, H and F above all.
- The high specific surface of PSi allows it to incorporate a large variety of other elements, either spontaneously ($O_2$ uptake during air storage) or purposely introduced by the investigators.

Concerning the last point, which is beyond the scope of the present chapter, there is extensive literature devoted to incorporating other elements in PSi, to add different functionalities, for example, improvement of light emission at different energies (Setzu et al. 1998; Reece et al. 2004), magnetic (Granitzer and Rumpf 2010) and sensor (Parisini et al. 2000; Rocchia et al. 2003) functionalities, passivation (Song and Sailor 1998), and nanocomposite approaches (Halimaoui et al. 1995; Amato et al. 2005). The chemical composition of PSi without any post-anodization treatments will be discussed in more detail hereafter.

### 6.3.1 DOPANT ATOMS IN THE PRISTINE SUBSTRATE

After porozisation, the electrical conductivity decreases by several orders of magnitude independently on the original doping type and concentration. The first explanation for this effect could be the preferential etching at the dopant sites, that is, the concentration of dopant atoms [D] should decrease dramatically because of electrochemical etching. As it will be discussed in the following, this is not our case. SIMS depth profile investigations on PSi (Lehmann et al. 1995) show that the original [D]/[Si] ratio remains unchanged (at least within a factor of 2).

## 6.3.2 ATOMS INCORPORATED DURING ETCHING

Si dangling bonds created during porosification tend to bond in a relatively stable way with H atoms provided by the HF etchant. Freshly prepared PSi can then be studied by Fourier transform infra-red (FTIR) spectroscopy to assign the different bond groups (Ogata et al. 1995a). The full IR spectrum of the as-prepared sample outgassed in dynamic vacuum at room temperature is reported in Figure 6.14 (thick full solid line). The enlargement of the SiH stretching region (2200–2100 cm$^{-1}$) is reported as the thicker full solid curve: peaks due to $-SiH_3$ (2137 cm$^{-1}$), $-SiH_2$ (2113 cm$^{-1}$), and $-SiH$ (2087 cm$^{-1}$) groups (Ogata et al. 1995a), superimposed on that belonging to the $-Si_2H_2$ species (at 2100 cm$^{-1}$) are seen. The band at 910 cm$^{-1}$ is assigned to the SiH$_2$ scissors mode. Bands related to Si–O–Si groups are very weak (modes centered at 1100 cm$^{-1}$) and absorptions due to O$_y$SiH$_x$ groups are absent (Ogata et al. 1995b) because of the low oxidation state of the surface. Below 700 cm$^{-1}$, intense deformation modes of SiH$_x$ species (Ogata et al. 1998) are present.

In B-doped Si, the formation of the Si-H-B complexes has been reported in the past (Johnson et al. 1991). This compound can effectively reduce the doping efficiency of B up to 99%, as experimentally (Pankove et al. 1984) and theoretically (DeLeo 1991) demonstrated. If the same held for PSi, this should explain the drop of electrical conductivity mentioned in Section 6.3.1. Grosman and Ortega (1997) did not report on the presence of the IR mode of the Si-H-B compound at the expected frequency of 1875 cm$^{-1}$ on freestanding PSi layers. In contrast, they observed this band on samples still lying on the original substrate: this led them to conclude that Si-H-B compounds are present only in the bulk Si region close to the PSi interface. Detailed joint FTIR and Raman spectroscopies together with TPD (thermal-programmed desorption) studies (Boarino et al. 2001) observed no signature of Si-H-B compounds. Instead, B-related Raman modes are clearly observed at 620 cm$^{-1}$ and 640 cm$^{-1}$.

Once stated that H preferentially bonds with the Si dangling bonds, the TPD experiment can be used in conjunction with FTIR to elucidate the nature of the bonds, and the amount of H incorporated (Rivolo et al. 2003). By comparing the TPD curve with the spectrum from a quadrupole mass analyzer, the same authors concluded that the desorbed phase is entirely constituted by H$_2$ with a contribution, 2 orders of magnitude lower, from desorbed silanes. The main peaks from the TPD curve at 380°C (with a shoulder at 240°C), 500°C, and 700°C are assigned, through comparison with FTIR spectra, to different SiH$_x$ species at the skeleton surface. Rivolo et al. (2003) also estimated the amount of incorporated H atoms by the integration of the TPD curve, concluding that the [H]/[Si] ratio at the wall surface is close to 1. Few Si bare atoms are then present on the surface. Lysenko et al. (2005) then proposed PSi as a material for H$_2$ storage. The experiments carried out by Rivolo et al. (2003) indeed demonstrate that the principal impurity introduced by the anodization step is H$_2$.

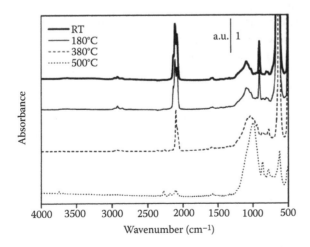

**FIGURE 6.14**  IR transmission spectra collected in a vacuum cell at RT (thick full line), after outgassing at 180°C (thin full line), 380°C (dashed line), and 500°C (dotted line). (Rivolo P. et al.: *Phys. Stat. Sol. (a)*. 2003. 197. 217. Copyright Wiley-VCH Verlag GmbH & Co. KGaA. Reproduced with permission.)

F was supposed as an additional contaminant of PSi in early works (Arita and Sunohara 1977; Beale et al. 1985a,b). Desorption experiments (Zoubir et al. 1994) demonstrated that $SiF_3$ groups desorb at the same temperature as $SiH_3$, indicating that $SiF_3$ groups are present on the pore walls. It has been shown by IR spectroscopy (Banerjee et al. 1994) that SiF and $SiF_2$ groups are present in freshly prepared PSi and it has been proposed that $SiF_x$ bonds are progressively replaced by Si-OH bonds through hydrolysis reaction with water vapor in air. On the other hand, only HF and $SiF_6^{2-}$ were found using $^{19}F$ NMR (nuclear magnetic resonance) (Petit et al. 1997). According to these results, F in the pores is considered a residual product of the electrochemical process.

### 6.3.3 ATOMS UPTAKE AFTER FORMATION

Just after formation, PSi starts to uptake $O_2$ from air. This leads to a progressive substitution of $H_2$ bound on its surface with $O_2$. The trend is described by the FTIR spectra displayed in Figure 6.14 (even if in that case oxidation is accelerated through $H_2$ thermal desorption) with a progressive smearing out of the Si-H peaks and the appearance of a feature around 1000 cm$^{-1}$ due to Si-O bonds. This band is initially centered at 1100 cm$^{-1}$, assigned to Si-O-Si bonds, and progressively shifts to 1000 cm$^{-1}$ with increasing oxidation. The mechanism involves oxidation of the outermost backbonded Si in the form $H_2Si–O_2$, $HSi–O_3$ bonding (Canham et al. 1991, O'Keeffe et al. 1995, Ogata et al. 2000, Petrova et al. 2000). Gupta et al. (1991) and Ogata et al. (2000) also reported a similar mechanism for PSi immersed in water at room temperature showing that during rinsing, the material already starts to oxidize.

## 6.4 ELECTRICAL CONDUCTIVITY

Electrical properties of PSi were intensively studied with the aim to achieve efficient carrier injection and electroluminescence. Those studies continued later on, when PSi was proposed as a possible platform for a fully Si-integrated sensor technology. In that framework, it was demonstrated how the electrical properties of PSi can be affected by the exposure to various molecular species. The chemical composition had been studied as well, to ascertain the role of dopants and other incorporated species on the electrical transport mechanism.

From early works on PSi (Unagami 1980), it was almost clear that the original electrical conductivity of the pristine Si wafer was decreased by several orders of magnitude after porosification. Beale et al. (1985a,b) reported about a resistivity value of 10$^5$ Ω·cm, about an intrinsic difficulty of making ohmic contact on PSi, about the interference of the substrate when performing electrical measurements in planar configuration, and about the increase of the PSi conductivity when exposed to ammonia. It is interesting to note, in passing, that the last two issues were studied in detail only two decades later (Baratto et al. 2001; Chiesa et al. 2003). Except for macroporous Si, which, to some extent, keeps the electrical quality of the original wafer, in all other types of PSi, a dramatic carrier depletion occurs. As mentioned in Sections 6.3.1 and 6.3.2, both the preferential etching of dopants and their passivation with $H_2$ must be ruled out as possible causes, as shown by Lehmann et al. (1995) and Boarino et al. (2001). If B-doped PSi is considered, another possibility could be that the tri-coordinated B atom on the surface of a nanocrystal bonds with the tetra-coordinated Si atoms beneath (without any dangling bond to be formed), losing its doping character this way. Apart from the experimental evidences contradicting this viewpoint, this naïve explanation does not solve the problem with phosphorus-doped PSi.

### 6.4.1 CONDUCTIVITY OF NANOPOROUS Si

Studies on this class of material were the first being pursued after the discovery of PL. The reason is obvious: PSi was a promising material to achieve electroluminescence from Si and the goal was to find a way to efficiently inject electrons and holes into the light emitting nanostructures (Richter et al. 1991; Koshida and Koyama 1992). For this reason, and to get rid of the problems reported by Beale et al. (1985b) about the role of substrate in planar configuration, most of those measurements were carried out in sandwich configuration.

Figure 6.15 depicts the temperature dependence of the dark conductivity $\sigma$ of $p$-type PSi films with different porosities (Lee et al. 2000). In this figure, as the porosity increases, the activation energy $E_a$ of the Arrhenius relation $\sigma = \sigma_0 \cdot \exp(-E_a/k_B T)$ increases (see Figures 6.15 and 6.16), indicating that apparently the porosity has an influence on the charge transport mechanism, that is, on the thermal excitation of carriers in the conduction process.

Anderson et al. (1991) measured resistivity values in $p$-type PSi of the order of $10^7$ $\Omega\cdot$cm and evidenced that, differently from c-Si, the Al/PSi junction shows a blocking behavior, whereas Ben-Chorin et al. (1995) investigated the role played by band alignment at the PSi/c-Si interface on carrier injection. In order to reduce the effects of contacts, which hinder the intrinsic conduction mechanism, Ben-Chorin et al. (1994a) carried out measurements on thick nanoporous samples, suggesting a Poole-Frenkel effect to account for the observed electric field-enhanced conduction. The influence of post-treatments (thermal and aging) was also investigated (Möller et al. 1995). A detailed review on those early studies has been given by Pèlant et al. (1998).

An interesting model proposed by Balberg (2000) assumes the nanoporous Si as a composite material, that is, a "pea-pot" where the disordered tissue that wraps the silicon nanocrystals dominates for nonoxidized PSi. The second route is dominated by tunneling through the crystallites according to their charging energy. This dual transport route model is consistent with the presence of two Meyer-Neldel rules (MNR). The conductivity prefactor $\sigma_0$ has been reported by Lubianiker and Balberg (1997) to follow two distinct energy activation regimes, one that is very similar to what is found in amorphous silicon (see Figure 6.16).

Urbach et al. (2007), using state-of-the-art oxidation of PSi, observed the other route as appearing when the disordered tissue is oxidized, and ascribed it to tunneling and hopping in between the nanocrystals. They suggested that dc conduction is limited by narrow geometrical constrictions along the transport path, giving rise to a Coulomb blockade (Ali and Ahmed 1994) type activation process. The evolution between Coulomb blockade and hopping mechanisms has also been observed in other systems, when the interdot coupling is varied (Romero and Drndic 2005).

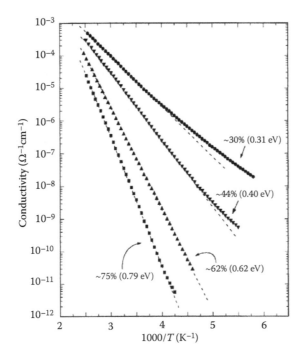

**FIGURE 6.15**  Temperature dependence of the dark conductivity of PSi with different porosities. (Reprinted from *Solid State Commun.*, 113, Lee et al., 519, Copyright 2000, with permission from Elsevier.)

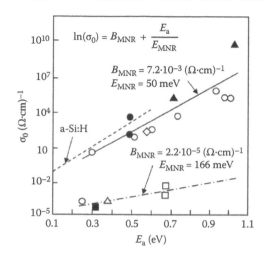

**FIGURE 6.16** The experimentally determined values of the DC conductivity prefactors as a function of the corresponding activation energies in a variety of PSi samples fabricated in different labs. (Data extracted from Lubianiker Y. and Balberg I., *Phys. Rev. Lett.* 78, 2433, 1997.)

Obviating the problem of stable electrical contacts by using photoemission stimulated by hard-ultraviolet/X-ray synchrotron radiation, Hamilton et al. (1998) found a temperature dependence of conductivity of nanoporous silicon. They suggested a percolation process occurring through sites in the porous network and postulated that this activation may be the consequence of a Coulomb blockade effect in the nanoscale channels of the film. This conclusion implies that conduction occurs in weakly confined nanostructures; in other words, the better electrically active part of the material is the worse optically active part, and vice versa.

This explanation has many common points with the "undulating wire" model introduced by Read et al. (1992). It was an approach to understand the low conductivity based on the quantum size of the porous structures. An undulating thickness in the porous skeleton will produce an undulation of the energy gap. Free carriers will be trapped in the larger crystallites, which show a narrower gap than the connecting bottlenecks. Then, in this framework, the activation of σ reported in Figure 6.15 could be related to the opening of different percolation pathways with temperature instead of the presence of trapping levels whose energy depends on porosity. With increasing σ, the availability of those pathways progressively decreases, consistently with a steeper variation of σ with *T*.

An alternative explanation was given by Tsu and Babić (1994), who calculated that a decrease in the dielectric constant can be expected for quantum size Si structures (Tsu et al. 1997). This results in an increase in dopant binding energy and the porous material becomes intrinsic at room temperature.

Quite surprisingly, experiments on nanoporous Si from lightly doped substrates cannot evidence the strong and weak points of these models. Mesoporous Si from $p^+$-substrates show similarities (carrier freeze-out) and differences with respect to this system that allow more insight on this complicated problem.

### 6.4.2 CARRIER MOBILITY IN NANOPOROUS Si

The drift mobility ($\mu_D$) of electrons and holes in nanoporous Si is smaller than in bulk material. In different papers one can find data, almost measured at room *T*, varying from 30 to $10^{-5}$ cm²/V·s. In crystalline semiconductor, the main limitations to the value of μ are acoustic phonon and ionized impurity scattering; in PSi, these mechanisms are also expected to be present, but the very low drift mobility values in comparison to bulk silicon suggest that typical features of PSi, such as the greater surface area, surface chemistry, skeletal structure, or quantum confinement, must play a dominant role. In other words, the different mechanisms listed above to explain the dramatic increase of the material resistivity play a role not only in reducing the concentration of free carriers, but also in the reduction of their mobility. Unfortunately, the scarcity of free carriers inside the Si nanostructures makes the measurement of $\mu$ very difficult, and this could be the reason for the large variation of the values reported by different authors. As it will be described in the following, a better evaluation of carrier mobility can be achieved if an efficient mechanism to free a large number of carriers is found. This mechanism, which recently has been labeled as "molecular doping," is particularly efficient in mesoporous $p^+$-PSi.

### 6.4.3 CONDUCTIVITY OF MESOPOROUS Si

Carrier freeze-out is observed also in mesoporous Si from $p^+$ substrates. A loss of free-carrier absorption in the IR region of the transmission spectrum of a highly doped surface layer, which was transformed into mesoporous silicon by anodization, was first mentioned by Beale et al. (1985b). Such monotonic, featureless absorption from free carriers was investigated later by FTIR on $p^+$-PSi membranes detached from the original substrate and transferred onto wafers transparent in the investigated spectral range (Boarino et al. 2000). A weak IR absorption due to free carriers was observed, but it was also reported that when exposing the PSi to $NO_2$ molecules, such absorption increases, turning back to its original value when vacuum is restored. Apart from the sensoristic applications, discussed in Chapters 1–5 in Porous Silicon: Biomedical and Sensor Applications, such a result states that

■ Original dopant atoms are present in PSi.
■ They are passivated in a reversible way.
■ The amount of background free carriers is a few percent of the concentration of dopant atoms.
■ Such background charge does not participate in conduction.

The result also confirms that other possible explanations for the carrier freeze-out like the preferential etching of dopants, their location at the wall surfaces, and their passivation with hydrogen must be ruled out.

Before going into the discussion on the passivation mechanism, a first analysis about the presence of free carriers in the background is needed. The question is "Why they do not participate to conduction, since a conductivity value of the order of $10^{-5}$–$10^{-9}$ $(\Omega \cdot cm)^{-1}$ (Lehmann et al. 1995) is measured?" The answer is that those background carriers (holes) are confined in bigger nanocrystals (see Figure 6.9). The confinement mechanism, however, cannot originate from quantum effects (Read et al. 1992) because of the larger sizes in $p^+$-PSi. The IR beam, however, "sees" them as free because the oscillation amplitude of such "confined" holes under the electric field is smaller than the dimension of bigger nanocrystals (Timoshenko et al. 2001).

The passivation mechanism of B dopants has been described by Boarino et al. (2001) as the formation of a B-DB (DB = dangling bond), where the extra electron needed by B for its full coordination is provided by the DB in the neighborhoods instead of being extracted by the valence band (Miranda-Durán et al. 2010). At the approach of an electron scavenger like the $NO_2$ molecule, such electron is no longer shared by the DB with the B site, and the last starts behaving as a dopant. The B-DB complex forms again when $NO_2$ is pumped away. The model has been substantiated by nuclear magnetic resonance measurements (Chiesa et al. 2003; Geobaldo et al. 2004), which indicate the subsurface position of the B atoms in the Si nanowires, because of the electrochemical formation process.

When exposed to traces of $NO_2$ (typically in the ppm range), the electrical conductivity of the material raises up by one order of magnitude at least. At such concentrations, the FTIR technique does not reveal any variation of the absorption of free holes in the background. This suggests that such jump in conductivity cannot be related to the reactivation of dopants and consequent release of new carriers, but rather an increase of the mobility of the background free carriers must be considered. In other words, they are no longer confined when $NO_2$ is chemisorbed at the surface of the Si walls.

### 6.4.4 CARRIER MOBILITY IN MESOPOROUS Si

The situation is depicted in Figure 6.17 in which both $\sigma/e = \mu \cdot n$ ($e$ is the electron charge, $n$ the carrier concentration) and $n$ are plotted versus the partial pressure of $NO_2$. The different sets of data have been collected through electrical and FTIR measurements, respectively. For high concentration of $NO_2$, the two sets of data follow approximately the same trend, indicating that the increase in conductivity is almost related to the progressive reactivation of B dopant sites. Nevertheless, the sharp jump of conductivity at low concentrations can be explained by an increase of $\mu$, solely.

Holes "confined" in reservoirs (larger crystallites) can then overcome the potential barrier at bottlenecks. The last ones cannot originate from quantum confinement because of the microstructure of the material, and confinement should not be relaxed by the presence of highly electronegative molecules chemisorbed onto the surface; in other words, a different explanation is needed. Lehmann et al. (1995) proposed a model for such confinement based on a similarity with the "telegraph noise" (Uren et al. 1985) observed in MOS transistors with a sufficiently narrow channel. The basic idea relies on the carrier repulsion due to the presence of surface charged states. If we consider the DB-B complex (Boarino et al. 2001), we notice that instead of having a mobile $h^+$ charge (as expected for the normal doping case), the positive charge is located on the surface dangling bond. Interaction with a single molecule of $NO_2$ creates one $h^+$, but more importantly, wipes out the repulsive potential barrier due to the positive charge at the surface, which confines carriers into reservoirs. Their mobility is then abruptly enhanced from $10^{-3}$ cm²/V·s (Timoshenko et al. 2001) to >$10^0$ cm²/V·s.

Coulomb blockade has been first used to control single electron tunneling in nanostructured devices (Amato and Enrico 2011, and references therein), taking advantage of the repulsive action of an electron confined in a metallic or semiconducting island, which cannot be strictly labeled as a quantum effect.

Borini et al. (2006a) measured the I-V characteristics of mesoporous, $p^+$ samples by means of a three-terminal configuration, to prevent leakages through the substrate (Keithley Instruments Inc. 2000), in vacuum and under $NO_2$ exposure, observing a progressive shrinkage of the Coulomb gap either with $NO_2$ or with increasing temperature. It was shown that this repulsive effect is more pronounced when current flows parallel to the surface, indicating a structural anisotropy (see Section 6.1.2), which reflects on electrical transport (Borini et al. 2006b; Agafonova et al. 2010). Chiesa et al. (2005) performed adsorption of $NH_3$, on $p^+$ meso-PSi generating an isotropic electron spin resonance signal at $g = 1.9984$ independent of temperature. By comparing this value with those concerning shallow donors in bulk silicon, the paramagnetic center was identified with electrons in the conduction band.

The possibility of injecting electron has then been exploited to study the evolution of the conductivity gap, arising from a collective Coulomb blockade phenomenon, under injection of

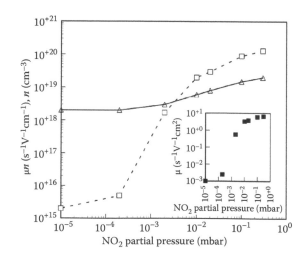

**FIGURE 6.17**   The dependence of $\mu n$ (open squares) and $n$ (triangles) on the $NO_2$ partial pressure, from electrical and spectroscopic measurements, respectively. In the insert, the calculated dependence of $\mu$ is displayed.

negative charges. Borini (2007) found that the conductivity gap is absent (instead of shrunk) with $NH_3$ molecules, as expected for negative charges not subjected to Coulomb blockade from fixed positive charges. This confirms that the Coulomb blockade mechanism observed in mesoporous $p^+$ samples originates from charged surface states, rather than quantum confinement, as proposed for the nanoporous case (Read et al. 1992).

### 6.4.5 TOWARD MOLECULAR DOPING OF PSi AND Si NANOWIRES

The results reported thus far indicate that, while $NH_3$ is a shallow donor, $NO_2$ yields $p$-doping only when passive sub-surface B atoms are present (Figure 6.18). In fact, both the HOMO and the LUMO of the $NO_2$ molecule falls well inside the Si bands, failing to provide shallow states as in conventional impurity doping (Miranda-Durán et al. 2010). Then the latter phenomenon, to some extent, can be considered only as a "fortuitous coincidence." Nevertheless, the possibility of reactivating dopant atoms in PSi shaded light on the conduction mechanism in such class of materials.

Some of the results and interpretations can then be extended to nanoporous $p$-type and mesoporous $n^+$-type silicon. The passivation of B atoms through the formation of a complex with a DB, however, is possible in lightly doped PSi, even if the effect is below the sensitivity of the FTIR measurement. In mesoporous $n^+$-type Si, on the contrary, no experimental methods for turning on/off the dopants have been devised thus far. It is also possible that the low conductivity values reported for mesoporous $n^+$-type silicon (Qu et al. 2009) are equally related to some kind of confinement into carrier reservoirs, even if the mechanism is still unclear. In $p$-type PSi, however, Coulomb blockade can equally occur, but its origin can be more likely ascribed to quantum confinement (Read et al. 1992; Hamilton et al. 1998).

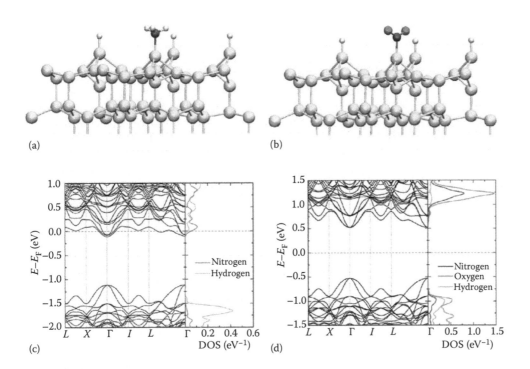

(a)          (b)

(c)          (d)

**FIGURE 6.18**  Adsorption of (a) $NH_3$ and (b) $NO_2$ at an Si dangling bond. $NH_3$ adsorption results in n-type doping, pinning the Fermi level close to the conduction band (c), while the $NO_2$ adsorbed system remains intrinsic, with no half-filled level next to any of the bands (d). The projected DOS of the side panels illustrates the contribution of the atomic species. N and O are shown in black and dark gray spheres, respectively, while gray and (small) white spheres represent Si and H atoms. (Reprinted with permission from Miranda-Durán Á. et al., *Nano Lett.* 10, 3590, 2010. Copyright 2010. American Chemical Society.)

The first and more convincing reports on the $n$-type doping of PSi were related to the adsorption of $NH_3$ (Chiesa et al. 2003; Garrone et al. 2005). The first electrons injected are found to compensate the residual $p$-type character of the material. After that, a systematic increase of the carrier concentration as the $NH_3$ partial pressure is increased is consistently observed (Garrone et al. 2005). First-principles calculations based on density-functional theory (DFT) confirmed that the injected charges are electrons and $n$-type doping is achieved. They also show that the $NH_3$ molecule can easily bind a surface dangling bond via the nitrogen lone pair, as shown in Figure 6.18, then transferring one electron to the Si skeleton (Miranda-Durán et al. 2010; Miranda et al. 2012). The same kind of molecular doping on $n^+$ PSi has been reported as well (Olenych et al. 2014).

Differently from $NO_2$, whose activity is related to the presence of subsurface B atoms passivated with DBs as a consequence for the electrochemical etching, the doping action of $NH_3$ and other Lewis bases (like pyridine, see below) can be transferred to similar systems like Si nanowires produced in a variety of methods (Law et al. 2004; Rurali 2014).

### 6.4.6 EFFECT OF DIELECTRIC CONFINEMENT

The model originally proposed by Tsu and Babić (1994) for the zero-dimensional case and more recently extended to the one-dimensional (Diarra et al. 2007), considers a confinement effect that occurs for sizes slightly larger than those at which quantum confinement can be observed. In fact, the so-called dielectric confinement or dielectric mismatch regime can occur for Si sizes of 5–10 nm, a value normally obtained with the electrochemical etching process of $p^+$ and $n^+$-Si. At these sizes, quantum confinement cannot effectively deactivate dopants by shifting their electronic levels toward midgap, but dielectric confinement can. Timoshenko et al. (2001) proposed a simple model based on dielectric mismatch, and estimated that the energy of a B acceptor decreases from 105 to 30 meV when the media dielectric constant increases from 1 (vacuum) to 25 (ethanol).

Ben-Chorin et al. (1994b) reported that polar vapors (methanol, in their case) induce an increase in the conductivity of microporous Si by orders of magnitude. A condensation of the vapor in the pores during this experiment, however, was excluded by those authors because the accompanying change in the index of refraction was not observed.

Recently, a sensitive experiment to measure the carrier concentration and the refractive index of PSi at the same time has been thought up (Amato et al. 2013). Mesoporous membranes ~18 μm thick, transferred onto IR transparent substrates were investigated by FTIR in the presence of ethanol and pyridine. The chosen thickness allows monitoring of the evolution of the interference fringe pattern during exposure to gas: an increase of carrier density is monitored through a reduction of the fringe's amplitude (i.e., loss of transparency), whereas a change of the pattern periodicity is seen as consequent to the variation of the refractive index $n$. It is found that, while pyridine efficiently injects electrons even at low dosages, with a noticeable increase of $n$ only on forced condensation, in the case of ethanol the holes' concentration and the refractive index follow the same trend with dosage. This suggests that condensation of ethanol already occurs in the pores at pressure values of the order of 1 mbar, relaxing the dielectric screening effects. At ethanol pressures one order of magnitude lower, this does not occur, as reported by Cultrera et al. (2014), who observed the same Arrhenius-type behavior of the unexposed sample with decreasing temperature. When reaching the temperature for condensation, however, the sample conductivity jumps up by a factor of 5, confirming the viewpoint that relaxation of dielectric screening occurs upon pore condensation, solely (Figure 6.19).

Summarizing, we can classify B impurities into three groups: the majority of them are passivated by forming the B-DB complex, another part is passivated by the dielectric screening, and the remaining active ones give rise to the background holes confined into reservoirs.

A reduction of phosphorus ionization energy is predicted by Niquet et al. (2010) when considering the presence of a metallic gate or a high-k dielectric around a stand-alone Si nanowire. Unfortunately, experiments on dielectric screening on $n^+$-PSi are missing; the main reason is probably related to the difficulty of efficiently reducing the ionization energy of P atoms.

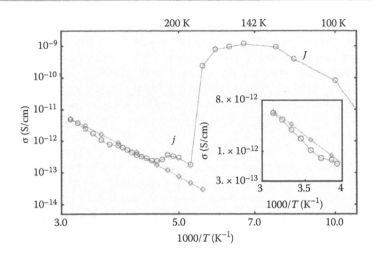

**FIGURE 6.19** Arrhenius plot of conductivity for bare (diamonds) and ethanol exposed (circles) PSi at constant pressure of 0.1 mbar of ethanol vapor. The inset shows the detail around 320 K–280 K. The ethanol-exposed curve features two significant jumps labeled "*j*" and "*J*." The first is assigned to the doping centers reactivation by dielectric screening of ethanol, the second attributed to the saturation of ethanol within the pores with a consequent shunting of the meso-PSi porous network. (Reprinted with permission from Cultrera A. et al., *AIP Adv.* 4, 087134. Copyright 2014. American Institute of Physics.)

### 6.4.7 OTHER NON-LINEAR PHENOMENA

Borini et al. (2007) first reported experimental observations of slow conductivity relaxation, nonergodicity, and simple aging phenomena occurring in mesoporous silicon at room temperature. These effects are ascribed to the strong disorder in the nanocrystalline silicon network constituting the material. The authors suggest that the observed behavior may reflect nonequilibrium glassy dynamics due to Anderson localization and Coulomb interactions. A similar effect has been reported for other stimuli, like the exposure to a molecular dopant like $NH_3$ (Borini 2008). Electrical resistance is found to relax according to a stretched exponential law, independently of the sign of the variation, suggesting a mechanism of rearrangement of trapped charges.

Amir et al. (2011) derived a theoretical curve for relaxation phenomena in PSi and showed that all experimental collapse onto it with a single time scale as a fitting parameter. This time scale is found to be of the order of thousands of seconds at room temperature.

## 6.5 THERMAL PROPERTIES OF PSi

Power dissipation in integrated circuits is favored by the quite high thermal conductivity $\alpha$ of c-Si ($W$ = 150 W/mK). However, the introduction of discontinuities is expected to decrease this value: Song and Chen's (2004) studies on freestanding, single-crystal Si thin films with periodically arranged through-film micropores have evidenced a drastic reduction of $\alpha$ as a size effect.

The interest of thermal transport in PSi grew in the second half of the 1990s, when PSi was proposed as a promising route to integration of sensors on Si chips (Lysenko et al. 2002). As shown in Chapter 9, in complicated nanostructured systems like PSi, however, simple thermal flow measurements are difficult (Lang et al. 1994). For this reason, investigators applied dynamic measurement because the thermal diffusion length $L$, given by $L = (2\alpha/\omega C)^{1/2}$ (where $\omega C$ are the angular frequency of the probing temperature and the heat capacity per unit volume, respectively), can be tuned by varying the frequency, so to restrict investigation to a layer thinner than the film. To introduce the probe signal as a periodic fluctuation of $T$, non-contact methods have been employed. The so-called 3ω method (Cahill 1990; Gesele et al. 1997; Kihara et al. 2005) and the thermal wave techniques have been widely applied to PSi (Amato et al. 1995; Cruz-Orea et al. 1996; Benedetto et al. 1997; Calderón et al. 1997; Bernini et al. 1999; Shen and Toyoda 2003) using different variants and configurations, and in a few cases are able to probe both the thermal and electronic components. Raman spectroscopy has been used by monitoring either the evolution of the Stokes-Antistokes intensity ratio (Amato et al. 2000b) or the red-shift of the Stokes peak (Lysenko et al. 1999b) with temperature. In c-Si isolated microstructures, the two methods are found to provide consistent results (Abel et al. 2007). More recently, with the aim of gaining experimental accuracy, measurements on freestanding membranes have been carried out (Amato et al. 2000b; Bernini et al. 2005).

The remarkable decrease of the thermal conductivity with porosity is reported in Figure 6.20 (Boarino et al. 1999). The big shaded area represents the interval of variation of the results reported by the authors mentioned above on samples with different morphology and pore dimension, consequent to the choice of the substrate doping and anodization conditions. The additional smaller area refers to samples of very high porosity obtained from p-type substrates (crystallite dimension ~2 nm) and dried by supercritical (Canham et al. 1994) or freeze drying (Amato and Brunetto 1996). The thermal conductivity of those *aerocrystals* approaches the air value ($\alpha = 0.026$ Wm$^{-1}$K$^{-1}$).

The same experimental $\alpha$ values, if plotted versus the FWHM of the Raman peak, show a marked dependence on the dimension of the Si instances (compare Figures 6.8 and 6.20). Then, it is expected that the application of effective medium approximations (Hopkins et al. 2009; Fang and Pilon 2011) is valid for a limited range of porosity. When the size of the Si nanocrystals become smaller than the phonon mean free path in c-Si (43 nm at room temperature), the classical Fourier heat conduction theory no longer holds (Chen 1996; Lysenko et al. 1999a; Jean et al. 2014), while disordered phonon scattering at nanocrystallites' boundaries play a major role. Recently, a detailed study of the different mechanisms involved in thermal transport occurring in PSi has been published (Weisse et al. 2012), stating that for Si sizes larger than the phonon mean free path, the thermal conductivity steeply decreases with increasing porosity due to phonon scattering at the pore interfaces, whereas the dependence on the doping concentration and surface roughness is fair. In contrast, when the Si diameter is smaller than the phonon mean free path, the thermal conductivity strongly depends on both the external boundary-phonon scattering and the internal pore interface-phonon scattering, leading to a significant reduction in the thermal conductivity for such systems. Moreover, a further reduction of PSi thermal conductivity has been reported to occur through amorphization achieved by irradiation with swift heavy ions (Newby et al. 2013).

The remarkable thermal isolation capabilities of PSi suggested several applications in the realization of on-chip thermal sensors (Lysenko et al. 1998), radiation detectors (Monticone et al. 1999), gas-sensing devices (Maccagnani et al. 1998; Kaltsas and Nassiopoulou 1999), and the realization of ultrasonic emitters (Shinoda et al. 1999; Koshida et al. 2013).

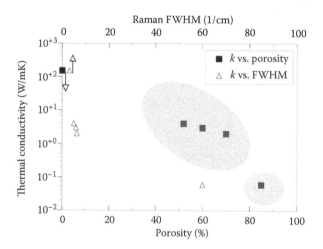

**FIGURE 6.20**   Thermal conductivity as a function of porosity (bottom x-axis) and of FWHM (top x-axis) of Raman spectra. Shaded areas indicate the range of variation of thermal conductivity values versus porosity reported in literature (see text for details) for air dried (bigger area) and specially dried (smaller area), very high porosity samples. (Data extracted from Boarino L. et al., *Microelectron. J.* 30, 1149, 1999.)

Similar values for thermal constants are encountered in the field of Si nanowires (Li et al. 2003) produced in a variety of methods with reduced disorder with respect to electrochemically etched PSi. From a general point of view, nanostructuring is nowadays considered one of the most promising approaches to *on-chip* thermal isolation (Leadley et al. 2014; Lucklum et al. 2014a,b) even for employment at cryogenic temperatures (Valalaki and Nassiopoulou 2014).

More detailed analysis of the thermal properties of porous silicon can be found in Chapter 9, which is devoted to this topic.

## 6.6 MECHANICAL PROPERTIES (STRAINS AND STRENGTH)

As mentioned in Section 6.2, in as-formed PSi the lattice parameter is slightly expanded in the direction perpendicular to the surface, while in the direction parallel to the surface, the lattice parameter of the substrate is generally found to be the same (Young et al. 1985), except in one report (Münder et al. 1992) where a small dilation was reported. The origin of the expansion of as-formed PSi has been explained by invoking the high surface-to-volume ratio of the existing nano-structures, which makes them prone to incorporation of foreign atoms (H, or O), as previously described. According to experimental studies $\Delta a/a$ changes vary in a large range: from a few $10^{-4}$ in the presence of a fluid up to $10^{-2}$ for oxidation (Bellet and Dolino 1996; Dolino and Bellet 1997; Bisi et al. 2000; Manotas et al. 2001). Porosity influence on the lattice mismatch parameter for as prepared PSi layers was shown earlier in Figure 6.6.

It has to be mentioned, in addition, that strain also depends on various parameters, the most important being the porosity of the PSi layer and the doping type and level of the silicon substrate (Korotcenkov and Cho 2010). As it is seen in Figure 6.21, mechanical properties of PSi in general are much worse than those of c-Si (see Figure 6.21). In particular, results presented in Figure 6.21c show that the hardness first decreases gradually with increasing porosity from the hardness of c-Si (11.5 GPa) according to a $(1-P)^{2/3}$ dependence, and then drops rapidly to 10% of this value when the porosity exceeds 75%. Duttagupta et al. (1997) suggested this fast decrease is related to a change in morphology, from a strongly connected network of wire-like objects to a loosely con-nected network of isolated nanocrystallites.

The problem with mechanical strength of PSi is encountered just during air-drying of high porosity PSi with high emission properties. In fact, capillary forces at the liquid-vapor meniscus can easily destroy the Si skeleton. By means of in-situ, time-resolved Raman spectroscopy, stress effects occurring during water evaporation are observed while the Raman peak variations during drying confirm the destruction of smaller nanostructures (Amato et al. 1996b).

The first study of the elastic properties of PSi was reported by Barla et al. (1984) from X-ray measurements of the curvature of a PSi layer supported by a substrate. By this method, they observed a Young's modulus in a $P = 0.54$, $p^+$-PSi layer, one order of magnitude lower than Si sub-strate, and a further reduction with increasing $P$. Those results were later confirmed (Bellet et al. 1996) using nanoindentation techniques in $p^+$-PSi for P ranging from 0.36 to 0.90.

The Young's modulus $E$ had been found to have a parabolic dependence on the relative den-sity $\rho_r = 1 - P$, with a reduction of two orders of magnitude with respect to the bulk velue ($E_B = 162$ GPa) in the case of $P = 0.90$. A further reduction in stiffness has been reported by the same authors in $p$-type PSi with comparable $P$. The observed parabolic dependence of $E$ versus $\rho_r$ can be explained by considering models for the elastic properties of cellular materials: one of these (Gibson and Ashby 1999) considers the mechanical properties of a foam related to both the microstructure and the material composing the cells. In this case, it is found that $E = CE_B\rho_r^2$, where $C$ is a constant accounting for the microstructure and $E_B$ is the Young's modulus of the bulk material. In Figure 6.21a, the good agreement between the experimental data and the men-tioned model is displayed.

The dependence of breaking strength on silicon porosity usually also has nonlinear behavior as described by Equation 6.1 of the empirical nature (see Figure 6.21d):

(6.1) $$\sigma = \sigma_0(1 - P)^m$$

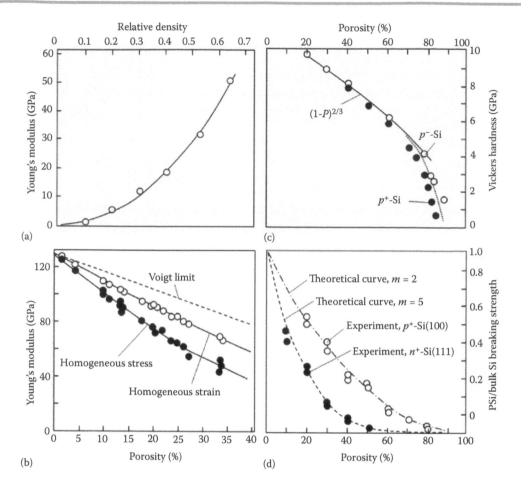

**FIGURE 6.21** Mechanical properties of porous silicon. (a) The parabolic dependence of the Young's modulus of a PSi sample versus the relative density $\rho_r = 1 - P$. (From Bellet D. and Dolino G., In: *Structural and Optical Properties of Porous Silicon Nanostructures,* Copyright 1998: CRC Press. With permission.) (b) Numerical results and interpolations of the homogenized PSi Young's modulus as a function of porosity. Curves are always underneath the Voigt theoretical upper bound (dashed line). (With kind permission from Springer Science+Business Media: Martini R. et al., *Nanoscale Res. Lett.* 7, 2012, 597.) (c) Hardness (H) as a function of porosity (P) for PSi films prepared from $p^-$ and $p^+$-Si wafers. The line represents the expected dependence $H \sim (1-P)^{2/3}$. (Reprinted from *Solid State Commun.* 101(1), Duttagupta S.P. et al., 33, Copyright 1997, with permission from Elsevier.) (d) Relative breaking strength versus porosity dependence. (Reprinted from *Superlatt. Microstruct.* 44, Klyshko A. et al., 374, Copyright 2008, with permission from Elsevier.)

where $m$ is an empirical coefficient that depends on the structure of pores. This is usually explained by the fact that a pore's structure leads to high nonuniform stress distribution along the micro-pores of the porous media during its deformation (Klyshko et al. 2008). As follows from the graphs presented in Figure 6.21d, the results on the $p^+$-PSi are close to the theoretical curve with $m = 2$. Meanwhile, results on $n^+$-PSi samples are much closer to the exponential character with $m = 5$. Klyshko et al. (2008) explained this effect by the different structure of pores of these materials. The anodization on $p^+$-Si(100) gives vertically arranged pores while on the $n^+$-Si(111) wafers, the pores are less regular and have a tree-like structure.

The results mentioned above testify that in a cellular solid like PSi, the mechanical properties have to be considered at different scales: at the microscopic scale, the mechanical quantities are expected to be nearly the same as the bulk material. At the macroscopic level (at which the material can be considered a continuum), the mechanical strength is much smaller than the bulk one. At the intermediate level, where several cells are involved, a marked dependence on the type of nanostructure and their degree of interconnection is expected to occur. For these reasons, different measurement methods could test the mechanical properties of PSi at different scales.

## ACKNOWLEDGMENTS

This work was supported by the Ministry of Science, ICT and Future Planning (MSIP) of the Republic of Korea, and partly by the National Research Foundation (NRF) grants funded by the Korean government (No. 2011-0028736 and No. 2013-K000315).

## REFERENCES

Abel, M.R., Wright, T.L., King, W.P., and Graham, S. (2007). Thermal metrology of silicon microstructures using Raman spectroscopy. *IEEE Trans. Comp. Packag. Technol.* 30, 200–208. doi:10.1109/TCAPT.2007.897993.

Agafonova, E.A., Martyshov, M.N., Forsh, P.A., Timoshenko, V.Y., and Kashkarov, P.K. (2010). Effect of thermal oxidation on charge carrier transport in nanostructured silicon. *Semiconductors* 44, 350–353. doi:10.1134/S1063782610030139.

Ali, D. and Ahmed, H. (1994). Coulomb blockade in a silicon tunnel junction device. *Appl. Phys. Lett.* 64, 2119–2120. doi:10.1063/1.111702.

Amato, G. and Fizzotti, F. (1992). Gap-states distribution in amorphous-silicon films as obtained by photothermal deflection spectroscopy. *Phys. Rev. B* 45, 14108–14113. doi:10.1103/PhysRevB.45.14108.

Amato, G. and Brunetto, N. (1996). Porous silicon via freeze drying. *Mater. Lett.* 26, 295–298. doi:10.1016/0167-577X(95)00244-8.

Amato, G. and Rosenbauer, M. (1998). Absorption and photoluminescence in porous Silicon. In: Amato G., Delerue C., and VonBardeleben H.J. (eds.), *Structural and Optical Properties of Porous Silicon Nanostructures*. CRC Press, Boca Raton, FL, pp. 3–52.

Amato, G. and Enrico, E. (2011). Current status and technological limitations of hybrid superconducting-normal single electron transistors, In: Luiz A. (ed.), *Superconductivity—Theory and Applications*. InTech, Rijeka, pp. 279–300.

Amato, G. Benedetto, G., Spagnolo, R., and Turnaturi, M. (1989). Photoacoustic measurements of doped Silicon wafers. *Phys. Stat. Sol. (a)* 114, 519–523. doi:10.1002/pssa.2211140212.

Amato, G., Boarino, L., Benedetto, G., and Spagnolo, R. (1995). Modulated photothermal reflectance on porous silicon. *Thin Solid Films* 255, 111–114. doi:10.1016/0040-6090(94)05633-O.

Amato, G., Bullara, V., Brunetto, N., and Boarino, L. (1996). Drying of porous silicon: A Raman, electron microscopy, and photoluminescence study. *Thin Solid Films* 276, 204–207. doi:10.1016/0040-6090(95)08053-8.

Amato, G., Boarino, L., Borini, S., and Rossi, A. M. (2000a). Hybrid Approach to Porous Silicon Integrated Waveguides. *Phys. Stat. Sol. (a)* 182, 425–430.

Amato, G., Angelucci, R., Benedetto, G. et al. (2000b). Thermal characterisation of porous Silicon membranes. *J. Porous Mater.* 7, 183–186. doi:10.1023/A:1009630619528.

Amato, G., Boarino, L., Rossi, A.M., Borini, S., Lulli, G., and Parisini, A. (2001). The Belphi-SOI technology. In: Cristoloveanu S. (ed.), *Silicon-on-Insulator Technology and Devices X: Proceedings of the 10th International Symposium on Silicon-on-Insulator Technology*. The Electrochemical Society, Washington, DC, pp. 307–312.

Amato, G., Borini, S., Rossi, A.M., Boarino, L., and Rocchia, M. (2005). Si/SiO$_2$ nanocomposite by CVD infiltration of porous SiO$_2$. *Phys. Stat. Sol. (a)* 202, 1529–1532. doi:10.1002/pssa.200461172.

Amato, G., Cultrera, A., Boarino, L. et al. (2013). Molecular doping and gas sensing in Si nanowires: From charge injection to reduced dielectric mismatch. *J. Appl. Phys.* 114, 204302. doi:10.1063/1.4834576.

Amir, A., Borini, S., Oreg, Y., and Imry, Y. (2011). Huge (but finite) time scales in slow relaxations: Beyond simple aging. *Phys. Rev. Lett.* 107, 186407. doi:10.1103/PhysRevLett.107.186407.

Anderson, R.C., Muller, R.S., and Tobias, C.W. (1991). Investigations of the electrical properties of porous silicon. *J. Electrochem. Soc.* 138, 3406–3411. doi:10.1149/1.2085423.

Arita, Y. and Sunohara, Y. (1977). Formation and properties of porous Silicon Film. *J. Electrochem. Soc.* 124, 285–295. doi:10.1149/1.2133281.

Balberg, I. (2000). Transport in porous silicon: The pea-pod model. *Philos. Mag. Part B* 80, 691–703. doi:10.1080/13642810008209776.

Banerjee, S., Narasimhan, K., and Sardesai, A. (1994). Role of hydrogen- and oxygen-terminated surfaces in the luminescence of porous silicon. *Phys. Rev. B* 49, 2915–2918. doi:10.1103/PhysRevB.49.2915.

Baratto, C., Faglia, G., Sberveglieri, G., Boarino, L., Rossi, A.M., and Amato, G. (2001). Front-side micromachined porous silicon nitrogen dioxide gas sensor. *Thin Solid Films* 391, 261–264. doi:10.1016/S0040-6090(01)00992-0.

Barla, K., Herino, R., Bomchil, G., Pfister, J.C., and Freund, A. (1984). Determination of lattice parameter and elastic properties of porous silicon by X-ray diffraction. *J. Cryst. Growth* 68, 727–732. doi:10.1016/0022-0248(84)90111-8.

Beale, M.I.J., Benjamin, J.D., Uren, M.J., Chew, N.G., and Cullis, A.G. (1985a). An experimental and theoretical study of the formation and microstructure of porous silicon. *J. Cryst. Growth* 73, 622–636. doi:10.1016/0022-0248(85)90029-6.

Beale, M.I.J., Chew, N.G., Uren, M.J., Cullis, A.G., and Benjamin, J.D. (1985b). Microstructure and formation mechanism of porous silicon. *Appl. Phys. Lett.* 46, 86–88. doi:10.1063/1.95807.

Bellet, D. and Dolino, G. (1996). X-ray diffraction studies of porous silicon. *Thin Solid Films* 276(1–2), 1–6.

Bellet, D. and Dolino, G. (1998). X-ray diffraction studies in porous Silicon. In: Amato G., Delerue C., and VonBardeleben H.J. (eds.), *Structural and Optical Properties of Porous Silicon Nanostructures*. CRC Press, Boca Raton, FL, pp. 3–52

Bellet, D., Lamagnère, P., Vincent, A., and Bréchet, Y. (1996). Nanoindentation investigation of the Young's modulus of porous silicon. *J. Appl. Phys.* 80, 3772–3776. doi:10.1063/1.363305.

Ben-Chorin, M., Möller, F., and Koch, F. (1994a). Nonlinear electrical transport in porous silicon. *Phys. Rev. B* 49, 2981–2984. doi:10.1103/PhysRevB.49.2981.

Ben-Chorin, M., Kux, A., and Schechter, I. (1994b). Adsorbate effects on photoluminescence and electrical conductivity of porous silicon. *Appl. Phys. Lett.* 64, 481–483. doi:10.1063/1.111136.

Ben-Chorin, M., Möller, F., and Koch, F. (1995). Band alignment and carrier injection at the porous-silicon-crystalline-silicon interface. *J. Appl. Phys.* 77, 4482–4488. doi:10.1063/1.359443.

Benedetto, G., Boarino, L., Brunetto, N., Rossi, A., Spagnolo, R., and Amato, G. (1997). Thermal properties of porous silicon layers. *Phil. Mag. Part B* 76, 383–393. doi:10.1080/01418639708241101.

Bensaid, A., Patrat, G., Brunel, M., de Bergevin, F., and Hérino, R. (1991). Characterization of porous silicon layers by grazing-incidence X-ray fluorescence and diffraction. *Solid State Commun.* 79, 923–928. doi:10.1016/0038-1098(91)90444-Z.

Berbezier, I. and Halimaoui, A. (1993). A microstructural study of porous silicon. *J. Appl. Phys.* 74, 5421–5425. doi:10.1063/1.354248.

Berger, M.G., Thönissen, M., Theiß, W., and Münder, H. (1998). Microoptical devices based on porous silicon. In: Amato G., Delerue C., and VonBardeleben H.J. (eds.), *Structural and Optical Properties of Porous Silicon Nanostructures*. CRC Press, Boca Raton, FL, pp. 3–52.

Bernini, U., Maddalena, P., Massera, E., and Ramaglia, A. (1999). Photo-acoustic characterization of porous silicon samples. *J. Opt. A: Pure Appl. Opt.* 1, 210. doi:10.1088/1464-4258/1/2/016.

Bernini, U., Bernini, R., Maddalena, P., Massera, E., and Rucco, P. (2005). Determination of thermal diffusivity of suspended porous silicon films by thermal lens technique. *Appl. Phys. A* 81, 399–404. doi:10.1007/s00339-004-2601-6.

Bisi, O., Ossicini, S., and Pavesi, L. (2000). Porous silicon: A quantum sponge structure for silicon based optoelectronics. *Surf. Sci. Rep.* 38, 1–126.

Boarino, L., Monticone, E., Amato, G. et al. (1999). Design and fabrication of metal bolometers on high porosity silicon layers. *Microelectron. J.* 30, 1149–1154. doi:10.1016/S0026-2692(99)00078-6.

Boarino, L., Baratto, C., Geobaldo, F. et al. (2000). $NO_2$ monitoring at room temperature by a porous silicon gas sensor. *Mater. Sci. Eng. B* 69–70, 210–214. doi:10.1016/S0921-5107(99)00267-6.

Boarino, L., Geobaldo, F., Borini, S. et al. (2001). Local environment of Boron impurities in porous silicon and their interaction with $NO_2$ molecules. *Phys. Rev. B* 64, 205308. doi:10.1103/PhysRevB .64.205308.

Bomchil, G., Halimaoui, A., and Herino, R. (1988). Porous Silicon: The material and its applications to SOI technologies. *Microelectron. Eng.* 8, 293–310. doi:10.1016/0167-9317(88)90022-6.

Borini, S. (2007). Effect of ammonia adsorption on the electrical characteristics of mesoporous silicon. *J. Appl. Phys.* 102, 093709. doi:10.1063/1.2805382.

Borini, S. (2008). Experimental observation of glassy dynamics driven by gas adsorption on porous silicon. *J. Phys.: Condens. Matter* 20, 385207. doi:10.1088/0953-8984/20/38/385207.

Borini, S., Boarino, L., and Amato, G. (2006a). Anisotropic resistivity of (100)-oriented mesoporous silicon. *Appl. Phys. Lett.* 89, 132111. doi:10.1063/1.2357882.

Borini, S., Boarino, L., and Amato, G. (2006b). Coulomb blockade Ttuned by $NO_2$ molecules in nanostructured silicon. *Adv. Mater.* 18, 2422–2425. doi:10.1002/adma.200600198.

Borini, S., Boarino, L., and Amato, G. (2007). Slow conductivity relaxation and simple aging in nanostructured mesoporous silicon at room temperature. *Phys. Rev. B* 75, 165205. doi:10.1103/PhysRevB.75.165205.

Brunetto, N. and Amato, G. (1997). A new line shape analysis of Raman emission in porous silicon. *Thin Solid Films* 297, 122–124. doi:10.1016/S0040-6090(96)09363-7.

Buttard, D., Bellet, D., and Dolino, G. (1996). X-ray-diffraction investigation of the anodic oxidation of porous silicon. *J. Appl. Phys.* 79, 8060–8070. doi:10.1063/1.362360.

Cahill, D.G. (1990). Thermal conductivity measurement from 30 to 750 K: The 3ω method. *Rev. Sci. Instrum.* 61, 802–808. doi:10.1063/1.1141498.

Calderón, A., Alvarado-Gil, J.J., Gurevich, Y.G. et al. (1997). Photothermal characterization of electrochemical etching processed *n*-type porous Silicon. *Phys. Rev. Lett.* 79, 5022–5025. doi:10.1103/PhysRevLett. 79.5022.

Campbell, I.H. and Fauchet, P.M. (1986). The effects of microcrystal size and shape on the one phonon Raman spectra of crystalline semiconductors. *Solid State Comm.* 58, 739–741. doi:10.1016/0038-1098(86)90513-2.

Canham, L.T. (1990). Silicon quantum wire array fabrication by electrochemical and chemical dissolution of wafers. *Appl. Phys. Lett.* 57, 1046–1048. doi:10.1063/1.103561.

Canham, L.T., Houlton, M.R., Leong, W.Y., Pickering, C., and Keen, J.M. (1991). Atmospheric impregnation of porous silicon at room temperature. *J. Appl. Phys.* 70, 422–431. doi:10.1063/1.350293.

Canham, L.T., Cullis, A.G., Pickering, C., Dosser, O.D., Cox, T.I., and Lynch, T.P. (1994). Luminescent anodized silicon aerocrystal networks prepared by supercritical drying. *Nature* 368, 133–135. doi:10.1038/368133a0.

Chen, G., (1996). Nonlocal and nonequilibrium heat conduction in the vicinity of nanoparticles. *J. Heat Transfer* 118, 539–545. doi:10.1115/1.2822665.

Chen, Z., Bosman, G., and Ochoa, R. (1993). Visible light emission from heavily doped porous silicon homo-junction pn diodes. *Appl. Phys. Lett.* 62, 708–710. doi:10.1063/1.109603.

Chiesa, M., Amato, G., Boarino, L., Garrone, E., Geobaldo, F., and Giamello, E. (2003). Reversible insulator-to-metal transition in $p^+$-type mesoporous Silicon induced by the adsorption of Ammonia. *Angew. Chem. Int. Ed.* 115, 5186–5189. doi:10.1002/ange.200352114.

Chiesa, M., Amato, G., Boarino, L., Garrone, E., Geobaldo, F., and Giamello, E. (2005). ESR study of conduction electrons in B-doped porous silicon generated by the adsorption of Lewis bases. *J. Electrochem. Soc.* 152, G329–G333. doi:10.1149/1.1872612.

Christophersen, M., Carstensen, J., Rönnebeck, S., Jäger, C., Jäger, W., and Föll, H. (2001). Crystal orientation dependence and anisotropic properties of macropore formation of *p*- and *n*-type Silicon. *J. Electrochem. Soc.* 148, E267–E275. doi:10.1149/1.1369378.

Crow, E.L. and Shimizu, K. (1987). *Lognormal Distributions: Theory and Applications.* CRC Press, Boca Raton, FL.

Cruz-Orea, A., Delgadillo, I., Vargas, H. et al. (1996). Photoacoustic thermal characterization of spark-processed porous silicon. *J. Appl. Phys.* 79, 8951–8954. doi:10.1063/1.362626.

Cullis, A.G. and Canham, L.T. (1991). Visible light emission due to quantum size effects in highly porous crystalline silicon. *Nature* 353, 335–338. doi:10.1038/353335a0.

Cultrera, A., Amato, G., Boarino, L., and Lamberti, C. (2014). A modified cryostat for photo-electrical characterization of porous materials in controlled atmosphere at very low gas dosage. *AIP Adv.* 4, 087134. doi:10.1063/1.4894074.

DeLeo, G.G. (1991). Theory of hydrogen-impurity complexes in semiconductors. *Physica B: Cond. Matter* 170, 295–304. doi:10.1016/0921-4526(91)90141-Z.

Diarra, M., Niquet, Y.-M., Delerue, C., and Allan, G. (2007). Ionization energy of donor and acceptor impurities in semiconductor nanowires: Importance of dielectric confinement. *Phys. Rev. B* 75, 045301. doi:10.1103/PhysRevB.75.045301.

Dolino, G. and Bellet, D. (1997). Strain in porous silicon. In: Canham, L. (ed.), *Properties of Porous Silicon.* INSPEC, London, p. 118.

Duttagupta, S.P., Chenb, X.L., Jenekhe, S.A., and Fauchet, P.M. (1997). Microhardness of porous silicon films and composites. *Solid State Commun.* 101(1), 33–37.

Duval, E., Boukenter, A., and Champagnon, B. (1986). Vibration eigenmodes and size of microcrystal-lites in glass: Observation by very-low-frequency Raman scattering. *Phys. Rev. Lett.* 56, 2052–2055. doi:10.1103/PhysRevLett.56.2052.

Englert, T., Abstreiter, G., and Pontcharra, J. (1980). Determination of existing stress in silicon films on sapphire substrate using Raman spectroscopy. *Solid-State Electron.* 23, 31–33. doi:10.1016/0038-1101(80)90164-1.

Fang, J. and Pilon, L. (2011). Scaling laws for thermal conductivity of crystalline nanoporous silicon based on molecular dynamics simulations. *J. Appl. Phys.* 110, 064305. doi:10.1063/1.3638054.

Fauchet, P.M., Tsybeskov, L., Duttagupta, S.P., and Hirschman, K.D. (1997). Stable photoluminescence and electroluminescence from porous silicon. *Thin Solid Films* 297, 254. doi:10.1016/S0040-6090(96)09438-2.

Fuchs, H.D., Stutzmann, M., Brandt, M.S. et al. (1993). Porous silicon and siloxene: Vibrational and structural properties. *Phys. Rev. B* 48, 8172–8189. doi:10.1103/PhysRevB.48.8172.

Gardelis, S., Bangert, U., Harvey, A.J., and Hamilton, B. (1995). Double-crystal X-ray diffraction, electron diffraction, and high resolution electron microscopy of luminescent porous silicon. *J. Electrochem. Soc.* 142, 2094–2101. doi:10.1149/1.2044247.

Garrone, E., Geobaldo, F., Rivolo, P. et al. (2005). A nanostructured porous Silicon near insulator becomes either a *p*- or an *n*-type semiconductor upon gas adsorption. *Adv. Mater.* 17, 528–531. doi:10.1002/adma.200401200.

Geobaldo, F., Rivolo, P., Borini, S. et al. (2004). Chemisorption of $NO_2$ at Boron sites at the surface of nanostructured mesoporous Silicon. *J. Phys. Chem. B* 108, 18306–18310. doi:10.1021/jp046918k.

Gesele, G., Linsmeier, J., Drach, V., Fricke, J., and Arens-Fischer, R. (1997). Temperature-dependent thermal conductivity of porous silicon. *J. Phys. D: Appl. Phys.* 30, 2911. doi:10.1088/0022-3727/30/21/001.

Gibson, L.J. and Ashby, M.F. (1999). *Cellular Solids: Structure and Properties.* Cambridge University Press, Cambridge.

Granitzer, P. and Rumpf, K. (2010). Porous silicon. A versatile host material. *Materials* 3, 943–998. doi:10.3390/ma3020943.

Gregora, I., Champagnon, B., and Halimaoui, A. (1994). Raman investigation of light-emitting porous silicon layers: estimate of characteristic crystallite dimensions. *J. Appl. Phys.* 75, 3034–3039. doi:10.1063/1.356149.

Grosman, A. and Ortega, C. (1997). Chemical composition of "fresh" porous Silicon. In: Canham, L. (ed.), *Properties of Porous Silicon.* INSPEC, London, p. 118.

Grosman, A., Ortega, C., Wang, Y.S., and Gandais, M. (1998). Morphology and structure of p-type porous Silicon by transmission electron spectroscopy. In: Amato G., Delerue C., and VonBardeleben H.J. (eds.), *Structural and Optical Properties of Porous Silicon Nanostructures.* CRC Press, Boca Raton, FL, pp. 3–52.

Guinier, A. and Fournet, G. (1955). *Small-Angle Scattering of X-rays.* Wiley, New York.

Gupta, P., Dillon, A.C., Bracker, A.S., and George, S.M. (1991). FTIR studies of $H_2O$ and $D_2O$ decomposition on porous silicon surfaces. *Surf. Sci.* 245, 360–372. doi:10.1016/0039-6028(91)90038-T.

Halimaoui, A., Campidelli, Y., Badoz, P.A., and Bensahel, D. (1995). Covering and filling of porous silicon pores with Ge and Si using chemical vapor deposition. *J. Appl. Phys.* 78, 3428–3430. doi:10.1063/1.359972.

Hamilton, B., Jacobs, J., Hill D.A., Pettifer, R.F., Teehan, D., and Canham, L.T. (1998). Size-controlled percolation pathways for electrical conduction in porous silicon. *Nature* 393, 443–445. doi:10.1038/30924.

Hopkins, P.E., Rakich, P.T., Olsson, R.H., El-kady, I.F., and Phinney, L.M. (2009). Origin of reduction in phonon thermal conductivity of microporous solids. *Appl. Phys. Lett.* 95, 161902. doi:10.1063/1.3250166.

Imai, K. (1981). A new dielectric isolation method using porous silicon. *Solid-State Electron.* 24, 159–164. doi:10.1016/0038-1101(81)90012-5.

Iqbal, Z. and Veprek, S. (1982). Raman scattering from hydrogenated microcrystalline and amorphous silicon. *J. Phys. C: Solid State Phys.* 15, 377. doi:10.1088/0022-3719/15/2/019.

Irmer, G. (2007). Raman scattering of nanoporous semiconductors. *J. Raman Spectrosc.* 38, 634–646. doi:10.1002/jrs.1703.

Ito, T., Kiyama, H., Yasumatsu, T., Watabe, H., and Hiraki, A. (1991). Role of hydrogen atoms in anodized porous silicon. *Physica B: Cond. Matter.* 170, 535–539. doi:10.1016/0921-4526(91)90172-B.

Jean, V., Fumeron, S., Termentzidis, K., Tutashkonko, S., and Lacroix, D. (2014). Monte Carlo simulations of phonon transport in nanoporous silicon and germanium. *J. Appl. Phys.* 115, 024304. doi:10.1063/1.4861410.

Johnson, N.M., Doland, C., Ponce, F., Walker, J., and Anderson, G. (1991). Hydrogen in crystalline semiconductors: A review of experimental results. *Physica B: Cond. Matter* 170, 3–20. doi:10.1016/0921-4526(91)90104-M.

Kaltsas, G. and Nassiopoulou, A.G. (1999). Novel C-MOS compatible monolithic silicon gas flow sensor with porous silicon thermal isolation. *Sens. Actuators A* 76, 133–138. doi:10.1016/S0924-4247(98)00370-7.

Kanemitsu, Y. (1994). Visible photoluminescence from nanometer-size Silicon crystallites: Core and surface states. *J. Phys. Soc. Jpn.* 63(suppl. B), 107.

Keithley Instruments Inc. (2000). *WB/259: Low Level Measurements Handbook.* http://www.keithley.it/promo/wb/259.

Kihara, T., Harada, T., and Koshida, N. (2005). Precise thermal characterization of confined nanocrystalline Silicon by a 3ω method. *Jpn. J. Appl. Phys.* 44, 4084–4087. doi:10.1143/JJAP.44.4084.

Kim, K.H., Bai, G., Nicolet, M.-A., and Venezia, A. (1991). Strain in porous Si with and without capping layers. *J. Appl. Phys.* 69, 2201–2205. doi:10.1063/1.348750.

Klyshko, A., Balucani, M., and Ferrari, A. (2008). Mechanical strength of porous silicon and its possible applications. *Superlatt. Microstruct.* 44, 374–377.

Koch, F., Petrova-Koch, V., Muschik, T., Nikolov, A., and Gavrilenko, V. (1992). Some perspectives on the luminescence mechanism via surface-confined states of porous Si. *MRS Online Proc. Library* 283. doi:10.1557/PROC-283-197.

Konaka, S., Tabe, M., and Sakai, T. (1982). A new silicon-on-insulator structure using a silicon molecular beam epitaxial growth on porous silicon. *Appl. Phys. Lett.* 41, 86–88. doi:10.1063/1.93298.

Korotcenkov, G. (2010). *Chemical Sensors: Fundamentals of Sensing Materials*, Vol. 2. Momentum Press, New York.

Korotcenkov, G. and Cho, B.K. (2010). Silicon porosification: State of the art. *Crit. Rev. Solid State Mater. Sci.* 35, 153–260. doi:10.1080/10408436.2010.495446.

Koshida, N. and Koyama, H. (1992). Visible electroluminescence from porous silicon. *Appl. Phys. Lett.* 60, 347–349. doi:10.1063/1.106652.

Koshida, N., Hippo, D., Mori, M., Yanazawa, H., Shinoda, H., and Shimada, T. (2013). Characteristics of thermally induced acoustic emission from nanoporous silicon device under full digital operation. *Appl. Phys. Lett.* 102, 123504. doi:10.1063/1.4798517.

Kovalev, D., Timoshenko, V.Y., Künzner, N., Gross, E., and Koch, F. (2001). Strong explosive interaction of hydrogenated porous silicon with oxygen at cryogenic temperatures. *Phys. Rev. Lett.* 87, 068301.

Kozlowski, F. and Lang, W. (1992). Spatially resolved Raman measurements at electroluminescent porous n-silicon. *J. Appl. Phys.* 72, 5401–5408. doi:10.1063/1.351979.

Labunov, V., Bondarenko, V., Glinenko, I., Dorofeev, A., and Tabulina, L. (1986). Heat treatment effect on porous silicon. *Thin Solid Films* 137, 123–134. doi:10.1016/0040-6090(86)90200-2.

Lang, W., Drost, A., Steiner, P., and Sandmaier, H. (1994). The thermal conductivity of porous Silicon. *MRS Online Proc. Library* 358. doi:10.1557/PROC-358-561.

Law, M., Goldberger, J., and Yang, P. (2004). Semiconductor nanowires and nanotubes. *Annu. Rev. Mater. Res.* 34, 83–122. doi:10.1146/annurev.matsci.34.040203.112300.

Leadley, D., Shah, V., Ahopelto, J. et al. (2014). Thermal isolation through nanostructuring. In: Balestra, F. (ed.), *Beyond-CMOS Nanodevices 1.* John Wiley & Sons, New York, pp. 331–363.

Lee, W.H., Lee, C., Kwon, Y.H., Hong, C.Y., and Cho, H.Y. (2000). Deep level defects in porous silicon, *Solid State Commun.* 113, 519.

Lehmann, V., Cerva, H., and Gösele, U. (1991). Pore formation and propagation mechanism in porous silicon. *MRS Online Proc. Library* 256. doi:10.1557/PROC-256-3.

Lehmann, V., Hofmann, F., Möller, F., and Grüning, U. (1995). Resistivity of porous silicon: A surface effect. *Thin Solid Films* 255, 20–22. doi:10.1016/0040-6090(94)05624-M.

Levy Clement, C. (1995). Characteristics of porous *n*-type silicon obtained by photoelectrochemical etching. In: Vial J.C. and Derrien J. (eds.), *Porous Silicon Science and Technology.* Springer, Berlin, Heidelberg, pp. 329–344.

Li, D., Wu, Y., Kim P., Shi, L., Yang, P., and Majumdar, A. (2003). Thermal conductivity of individual silicon nanowires. *Appl. Phys. Lett.* 83, 2934–2936. doi:10.1063/1.1616981.

Liu, F., Liao, L., Wang, G., Cheng, G., and Bao, X. (1996). Experimental observation of surface modes of quasifree clusters. *Phys. Rev. Lett.* 76, 604–607. doi:10.1103/PhysRevLett.76.604.

Low, S.P., Voelcker, N.H., Canham, L.T., and Williams, K.A. (2009). The biocompatibility of porous silicon in tissues of the eye. *Biomaterials* 30, 2873-2880. doi:10.1016/j.biomaterials.2009.02.008.

Lubianiker, Y. and Balberg, I. (1997). Two Meyer-Neldel rules in porous Silicon. *Phys. Rev. Lett.* 78, 2433.

Lucklum, F., Schwaiger, A., and Jakoby, B. (2014a). Development and investigation of thermal devices on fully porous Silicon substrates. *IEEE Sens. J.* 14, 992–997. doi:10.1109/JSEN.2013.2293541.

Lucklum, F., Schwaiger, A., and Jakoby, B. (2014b). Highly insulating, fully porous silicon substrates for high temperature micro-hotplates. *Sens. Actuators A* 213, 35–42. doi:10.1016/j.sna.2014.04.004.

Luryi, S. and Suhir, E. (1986). New approach to the high quality epitaxial growth of lattice-mismatched materials. *Appl. Phys. Lett.* 49, 140–142. doi:10.1063/1.97204.

Lysenko, V., Roussel, P., Delhomme, G. et al. (1998). Oxidized porous silicon: A new approach in support thermal isolation of thermopile-based biosensors. *Sens. Actuators A* 67, 205–210. doi:10.1016/S0924-4247(97)01777-9.

Lysenko, V., Boarino, L., Bertola, M. et al. (1999a). Theoretical and experimental study of heat conduction in as-prepared and oxidized meso-porous silicon. *Microelectron. J.* 30, 1141–1147. doi:10.1016/S0026-2692(99)00077-4.

Lysenko, V., Perichon, S., Remaki, B., Barbier, D., and Champagnon, B. (1999b). Thermal conductivity of thick meso-porous silicon layers by micro-Raman scattering. *J. Appl. Phys.* 86, 6841–6846. doi:10.1063/1.371760.

Lysenko, V., Périchon, S., Remaki, B., and Barbier, D. (2002). Thermal isolation in microsystems with porous silicon. *Sens. Actuators A* 99, 13–24. doi:10.1016/S0924-4247(01)00881-0.

Lysenko, V., Bidault, F., Alekseev, S. et al. (2005). Study of porous silicon nanostructures as hydrogen reservoirs. *J. Phys. Chem. B* 109, 19711–19718. doi:10.1021/jp053007h.

Maccagnani, P., Angelucci, R., Pozzi, P. et al. (1998). Thick oxidised porous silicon layer as a thermo-insulating membrane for high-temperature operating thin- and thick-film gas sensors. *Sens. Actuators B* 49, 22–29. doi:10.1016/S0925-4005(97)00337-7.

Manotas, S., Agullo-Rueda, F., Moreno, J.D., Ben-Hander, F., Guerrero-Lemus, R., and Martinez-Duart, J.M. (2000). Determination of stress in porous silicon by micro-Raman spectroscopy. *Phys. Stat. Sol. (a)* 182, 245–248.

Manotas, S., Agullo-Rueda, F., Moreno, J.D., Ben-Hander, F., and Martinez-Duart, J.M. (2001). Lattice-mismatch induced-stress in porous silicon films. *Thin Solid Films* 401, 306–309.

Martini, R., Depauw, V., Gonzalez, M., Vanstreels, K. et al. (2012). Mechanical properties of sintered meso-porous silicon: A numerical model. *Nanoscale Res. Lett.* 7, 597.

Miranda, Á., Cartoixà, X., Canadell, E., and Rurali, R. (2012). NH$_3$ molecular doping of silicon nanowires grown along the [112], [110], [001], and [111] orientations. *Nanoscale Res. Lett.* 7, 1–7. doi:10.1186/1556-276X-7-308.

Miranda-Durán, A., Cartoixà, X., Cruz Irisson, M., and Rurali, R. (2010). Molecular doping and subsurface dopant reactivation in Si nanowires. *Nano Lett.* 10, 3590–3595. doi:10.1021/nl101894q.

Möller, F., Ben-Chorin, M., and Koch, F. (1995). Post-treatment effects on electrical conduction in porous silicon. *Thin Solid Films* 255, 16–19. doi:10.1016/0040-6090(94)05623-L.

Monticone, E., Boarino, L., Lérondel, G., Steni, R., Amato, G., and Lacquaniti, V. (1999). Properties of metal bolometers fabricated on porous silicon. *App. Surf. Sci.* 142, 267–271. doi:10.1016/S0169-4332(98)00685-0.

Münder, H., Andrzejak, C., Berger, M.G. et al. (1992). A detailed Raman study of porous silicon. *Thin Solid Films* 221, 27–33. doi:10.1016/0040-6090(92)90791-9.

Nemanich, R.J., Solin, S.A., and Martin, R.M. (1981). Light scattering study of boron nitride microcrystals. *Phys. Rev. B* 23, 6348–6356. doi:10.1103/PhysRevB.23.6348.

Newby, P.J., Canut, B., Bluet, J.-M. et al. (2013). Amorphization and reduction of thermal conductivity in porous silicon by irradiation with swift heavy ions. *J. Appl. Phys.* 114, 014903. doi:10.1063/1.4812280.

Niquet, Y.M., Genovese, L., Delerue, C., and Deutsch, T. (2010). Ab initio calculation of the binding energy of impurities in semiconductors: Application to Si nanowires. *Phys. Rev. B* 81, 161301. doi:10.1103/PhysRevB.81.161301.

O'Keeffe, P., Aoyagi, Y., Komuro, S., Kato, T., and Morikawa, T. (1995). Room-temperature backbond oxidation of the porous silicon surface by oxygen radical irradiation. *Appl. Phys. Lett.* 66, 836–838. doi:10.1063/1.113438.

Ogata, Y., Niki, H., Sakka, T., and Iwasaki, M. (1995a). Hydrogen in porous Silicon: Vibrational analysis of SiH$_x$ species. *J. Electrochem. Soc.* 142, 195–201. doi:10.1149/1.2043865.

Ogata, Y., Niki, H., Sakka, T., and Iwasaki, M. (1995b). Oxidation of porous Silicon under water vapor environment. *J. Electrochem. Soc.* 142, 1595–1601. doi:10.1149/1.2048619.

Ogata, Y.H., Kato, F., Tsuboi, T., and Sakka, T. (1998). Changes in the environment of hydrogen in porous Silicon with thermal annealing. *J. Electrochem. Soc.* 145, 2439–2444. doi:10.1149/1.1838655.

Ogata, Y.H., Tsuboi, T., Sakka, T., and Naito, S. (2000). Oxidation of porous Silicon in dry and wet environments under mild temperature conditions. *J. Porous Mater.* 7, 63–66. doi:10.1023/A:1009694608199.

Olenych, I.B., Monastyrskii, L.S., Aksimentyeva, O.I., and Yarytska, L.I. (2014). Modification of the electrical properties of porous silicon by adsorption of ammonia molecules. *Universal J. Phys. Appl.* 2, 201–205. doi: 10.13189/ujpa.2014.020401.

Pankove, J.I., Wance, R.O., and Berkeyheiser, J.E. (1984). Neutralization of acceptors in Silicon by atomic hydrogen. *Appl. Phys. Lett.* 45, 1100–1102. doi:10.1063/1.95030.

Papadimitriou, D., Bitsakis, J., Lopez-Villegas, J.M., Samitier, J., and Morante, J.R. (1999). Depth dependence of stress and porosity in porous silicon: A micro-Raman study. *Thin Solid Films* 349, 293-297.

Parisini, A., Brunetto, N., and Amato, G. (1996). TEM and photoluminescence characterization of poroussilicon layers from <111>-oriented $p^+$ silicon substrates. *Nuovo Cimento Soc. Ital. Fis., D* 18, 1233–1239. doi:10.1007/BF02464701.

Parisini, A., Angelucci, R., Dori, L. et al. (2000). TEM characterisation of porous silicon. *Micron* 31, 223–230. doi:10.1016/S0968-4328(99)00087-6.

Pèlant, I., Fejifar, A., and Kočka, J. (1998). Transport properties of porous Silicon. In: Amato G., Delerue C., and VonBardeleben H.J. (eds.), *Structural and Optical Properties of Porous Silicon Nanostructures.* CRC Press, Boca Raton, FL, pp. 253–288.

Petit, D., Chazalviel, J.-N., Ozanam, F., and Devreux, F. (1997). Porous silicon structure studied by nuclear magnetic resonance. *Appl. Phys. Lett.* 70, 191–193. doi:10.1063/1.118382.

Petrova, E.A., Bogoslovskaya, K.N., Balagurov, L.A., and Kochoradze, G.I. (2000). Room temperature oxidation of porous silicon in air. *Mater. Sci. Eng., B* 69–70, 152–156. doi:10.1016/S0921-5107(99)00240-8.

Prokes, S.M. (1993). Light emission in thermally oxidized porous silicon: Evidence for oxide-related luminescence. *Appl. Phys. Lett.* 62, 3244–3246. doi:10.1063/1.109087.

Qu, Y., Liao, L., Li, Y., Zhang, H., Huang, Y., and Duan, X. (2009). Electrically conductive and optically active porous Silicon nanowires. *Nano Lett.* 9, 4539–4543. doi:10.1021/nl903030h.

Read, A.J., Needs, R.J., Nash, K.J., Canham, L.T., Calcott, P.D.J., and Qteish, A. (1992). First-principles calculations of the electronic properties of silicon quantum wires. *Phys. Rev. Lett.* 69, 1232–1235. doi:10.1103/PhysRevLett.69.1232.

Reece, P.J., Gal, M., Tan, H.H., and Jagadish, C. (2004). Optical properties of erbium-implanted porous silicon microcavities. *Appl. Phys. Lett.* 85, 3363–3365. doi:10.1063/1.1808235.

Richter, H., Wang, Z.P., and Ley, L. (1981). The one phonon Raman spectrum in microcrystalline silicon. *Solid State Commun.* 39, 625–629. doi:10.1016/0038-1098(81)90337-9.

Richter, A., Steiner, P., Kozlowski, F., and Lang, W. (1991). Current-induced light emission from a porous silicon device. *IEEE Electr. Dev. Lett.* 12, 691–692. doi:10.1109/55.116957.

Rivolo, P., Geobaldo, F., Rocchia, M., Amato, G., Rossi, A.M., and Garrone, E. (2003). Joint FTIR and TPD study of hydrogen desorption from $p^+$-type porous silicon. *Phys. Stat. Sol. (a)* 197, 217–221. doi:10.1002/pssa.200306503.

Rocchia, M., Borini, S., Rossi, A.M., Boarino, L., and Amato, G. (2003). Submicrometer functionalization of porous Silicon by electron beam lithography. *Adv. Mater.* 15, 1465–1469. doi:10.1002/adma.200304919.

Romero, H.E. and Drndic, M. (2005). Coulomb blockade and hopping conduction in PbSe quantum dots. *Phys. Rev. Lett.* 95, 156801. doi:10.1103/PhysRevLett.95.156801.

Rönnebeck, S., Carstensen, J., Ottow, S., and Föll, H. (1999). Crystal orientation dependence of macropore growth in *n*-type Silicon. *Electrochem. Solid-State Lett.* 2, 126–128. doi:10.1149/1.1390756.

Rurali, D.R. (2014). Gas and liquid doping gas and liquid doping of porous Silicon. In: Canham, L. (ed.), *Handbook of Porous Silicon*. Springer International Publishing, Berlin, pp. 1–7.

Sato, N., Sakaguchi, K., Yamagata, K. et al. (1995). High-quality epitaxial layer transfer (ELTRAN) by bond and etch-back of porous Si. In: *Proceedings of International IEEE SOI Conference*, 1995, pp. 176–177. doi:10.1109/SOI.1995.526517.

Setzu, S., Létant, S., Solsona, P., Romestain, R., and Vial, J.C. (1998). Improvement of the luminescence in p-type as-prepared or dye impregnated porous silicon microcavities. *J. Lumin.* 80 129–132. doi:10.1016/S0022-2313(98)00081-7.

Shen, Q. and Toyoda, T. (2003). Dependence of thermal conductivity of porous silicon on porosity characterized by photoacoustic technique. *Rev. Sci. Instrum.* 74, 601–603. doi:10.1063/1.1515897.

Shinoda, H., Nakajima, T., Ueno, K., and Koshida, N. (1999). Thermally induced ultrasonic emission from porous silicon. *Nature* 400, 853–855. doi:10.1038/23664.

Song, J.H. and Sailor, M.J. (1998). Functionalization of nanocrystalline porous Silicon surfaces with Aryllithium reagents: Formation of Silicon–Carbon bonds by cleavage of Silicon–Silicon bonds. *J. Am. Chem. Soc.* 120, 2376–2381. doi:10.1021/ja9734511.

Song, D. and Chen, G. (2004). Thermal conductivity of periodic microporous silicon films. *Appl. Phys. Lett.* 84, 687–689. doi:10.1063/1.1642753.

Street, R.A. (1984). Luminescence in a-Si:H. In: Pankove J.L. (ed.). *Semiconductors & Semimetals*, 21B. Academic Press, New York.

Sugiyama, H. and Nittono, O. (1990). Microstructure and lattice distortion of anodized porous silicon layers. *J. Cryst. Growth* 103, 156–163. doi:10.1016/0022-0248(90)90184-M.

Theunissen, M.J.J. (1972). Etch channel formation during anodic dissolution of n-type Silicon in aqueous hydrofluoric acid. *J. Electrochem. Soc.* 119, 351–360. doi:10.1149/1.2404201.

Timoshenko, V.Y., Dittrich, T., Lysenko, V., Lisachenko, M.G., and Koch, F. (2001). Free charge carriers in mesoporous silicon. *Phys. Rev. B* 64, 085314. doi:10.1103/PhysRevB.64.085314.

Tsu, R. and Babić, D. (1994). Doping of a quantum dot. *Appl. Phys. Lett.* 64, 1806–1808. doi:10.1063/1.111788.

Tsu, R., Babić, D., and Ioriatti, Jr L. (1997). Simple model for the dielectric constant of nanoscale silicon particle. *J. Appl. Phys.* 82, 1327–1329. doi:10.1063/1.365762.

Unagami, T. (1980). Formation mechanism of porous Silicon layer by anodization in HF solution. *J. Electrochem. Soc.* 127, 476–483. doi:10.1149/1.2129690.

Unagami, T. and Seki, M. (1978). Structure of porous Silicon layer and heat-treatment effect. *J. Electrochem. Soc.* 125, 1339–1344. doi:10.1149/1.2131674.

Urbach, B., Axelrod, E., and Sa'ar, A. (2007). Correlation between transport, dielectric, and optical properties of oxidized and nonoxidized porous silicon. *Phys. Rev. B* 75, 205330. doi:10.1103/PhysRevB.75.205330.

Uren, M.J., Day, D.J., and Kirton, M.J. (1985). 1/f and random telegraph noise in silicon metal-oxide-semiconductor field-effect transistors. *Appl. Phys. Lett.* 47, 1195–1197. doi:10.1063/1.96325.

Valalaki, K. and Nassiopoulou, A.G. (2014). Thermal conductivity of highly porous Si in the temperature range 4.2 to 20 K. *Nanoscale Res. Lett.* 9, 1–6. doi:10.1186/1556-276X-9-318.

Weisse, J.M., Marconnet, A.M., Kim, D.R. et al. (2012). Thermal conductivity in porous silicon nanowire arrays. *Nanoscale Res. Lett.* 7, 554.

Xia, H., He, Y.L., Wang, L.C. et al. (1995). Phonon mode study of Si nanocrystals using micro-Raman spectroscopy. *J. Appl. Phys.* 78, 6705–6708. doi:10.1063/1.360494.

Young, I.M., Beale, M.I.J., and Benjamin, J.D. (1985). X-ray double crystal diffraction study of porous silicon. *Appl. Phys. Lett.* 46, 1133–1135. doi:10.1063/1.95733.

Zoubir, N.H., Vergnat, M., Delatour, T., Burneau, A., and de Donato, P. (1994). Interpretation of the luminescence quenching in chemically etched porous silicon by the desorption of $SiH_3$ species. *Appl. Phys. Lett.* 65, 82–84. doi:10.1063/1.113082.

# Luminescent Properties of Porous Silicon

**7**

**Bernard Jacques Gelloz**

CONTENTS

## 7.1 INTRODUCTION

Bulk silicon can emit only infrared light (at about 1.1 eV) and with a very low efficiency due to the indirect nature of its bandgap. The initial report of efficient and visible luminescence from porous silicon (PSi) in 1990 (Canham 1990) created a huge wave of optimism in the field of silicon photonics. The dream of a silicon-based multicolor light emitter and possibly laser seemed within reach in the short term. The discovery also dramatically increased the interest in PSi itself, leading to tremendous amounts of papers on topics such as formation mechanisms, structural properties, as well as various physical and chemical properties.

It was found that the typical luminescence of PSi originates mostly from silicon crystallites smaller than the exciton Bohr radius (~4.5 nm in silicon), due to quantum confinement (bandgap enlargement). The luminescence of porous silicon is broad due to size distribution, with a typical full width at half maximum (FWHM) of about 150 nm. It can be tuned from the near-infrared to the blue, depending on the distribution of sizes. Quantum confinement also increases the efficiency of the emission by enhanced exciton localization, and breaking of the k-conservation rule.

However, the simple picture of size-dependent luminescence due to quantum confinement is too simple. PSi also exhibits a very large surface area due to its porous nature. Thus, the surface also plays a significant role in the luminescence emission energy as well as efficiency and stability. The control of the PSi surface is one of the most difficult challenges in PSi technology. Luminescence quenching may occur because of interaction of PSi with species at the surface, either via dielectric constant change or energy transfer (Lauerhaas et al. 1992; Shimura et al. 1999; Fellah et al. 2000; La Ferrara et al. 2012). Such properties could be used for sensing purposes, which will be discussed in *Porous Silicon: Biomedical and Sensor Applications*.

PSi has been reported to luminescence from the near infrared (even below the Si bandgap energy) to the near UV because of distinct emission bands having different origins (not a result of quantum confinement) and properties. Due to the diversity of the luminescence bands, PSi can be considered for many different types of applications, in optoelectronics, sensing, and imaging. For instance, PSi is non-toxic and biodegradable and may be used in medicine (e.g., tumor imaging [Park et al. 2009]).

This chapter first presents briefly the complexity of the luminescence of PSi and the different mechanisms suggested for its origin. The next section presents the consensus achieved after more than 2 decades of research, showing the generally accepted photoluminescence (PL) bands of PSi and their respective origin. Then, the influence of the characteristics (porosity, thickness, surface termination) of as-formed PSi on the luminescence are discussed. The various treatments that have been proposed as ways to improve the PL efficiency and stability are presented next.

## 7.2  THE NATURE OF POROUS SILICON VISIBLE LUMINESCENCE: APPROACHES TO EXPLANATION

### 7.2.1  ORIGIN OF THE DEBATE OVER THE MODELS OF LIGHT EMISSION

In his first report showing the PL of PSi in 1990 (Canham 1990), Canham suggested quantum size effects in the silicon nanowires in the PSi skeleton as the origin of the luminescence. The same year, at about the same time and independently, a shift in the bulk silicon absorption edge to higher values was reported (Lehmann and Gosele 1991) and attributed to quantum wire formation in PSi. However, the original report by Canham was also received with disbelief by many, in particular those working in the field of amorphous silicon (a-Si), polysilane polymers (e.g., siloxene), and Si/SiO$_x$ systems. Indeed, all these materials can exhibit broad red-orange PL resembling that usually observed in PSi. Consequently, the debate on the origin of the PSi luminescence has generated a tremendous amount of papers in the early 1990s. Many of the interpretations in these early reports are now obsolete, in the light of the generally accepted mechanisms, which are detailed in Section 7.3. Nevertheless, the debate is not completely over.

### 7.2.2  TRICKY NANO-POROUS SILICON

The study of the origins of PSi PL is rendered extremely difficult by the tricky nature of PSi itself. Indeed, conventional PSi consists of interconnected silicon nanostructures with very large surface to volume ratios, and ideally with a hydrogen-terminated surface. However, the real structure of a PSi layer after drying can be quite different, depending on various factors, such as anodization conditions, drying technique, and storage. First, the surface is very reactive. As a result, (1) it could be contaminated by species in the atmosphere (e.g., organic molecules), and (2) it is readily oxidized (formation of native oxide) in a matter of seconds after exposure to air. If the porosity is very high (e.g., when using very long anodization times or unusually high anodization current densities), (1) the immediate oxidation of the surface transforms the original PSi into an oxidized PSi layer, which could contain more oxide than bulk silicon, and (2) stress can lead to a-Si formation at the surface of the PSi skeleton and also to the cracking or partial collapse of the PSi structure. Such layers may still be called PSi, even though they are a mixture of different phases, such as silicon oxide, a-Si, and possibly polysilane polymer fragments. Some of these layers could even hardly be called PSi as per the common definition. Another source of possible wrong interpretation could come from the different structures found in PSi formed from different doping types and level. For example, the PL obtained with PSi formed from *p*-type Si and that formed from heavily doped *n*-type Si in Figure 7.1 (Gelloz et al. 2005) and Figure 7.2 (Gelloz and Koshida 2006a), respectively, are quite different, though they were obtained following similar techniques: anodization followed by high-pressure water vapor annealing (HWA). The details of these spectra will be discussed in Sections 7.4.1 and 7.4.2.4. They are consistent with the generally accepted luminescence mechanism described in Section 7.3, taking into account a 2-peak distribution of size rather than defect state emission for one of the peaks (shorter wavelengths). To summarize, PSi may include different phases (crystalline Si, a-Si, Si oxide, and other species); therefore extreme caution and a good understanding and structural characterization of the layers are paramount when drawing interpretations and extending the conclusions to PSi in general.

### 7.2.3  OVERVIEW OF THE MODELS OF LIGHT EMISSION

The most debated proposed luminescence mechanisms are now briefly discussed. More in-depth discussions of these mechanisms may be found in earlier reviews (Prokes 1996; Cullis et al. 1997; Bisi et al. 2000; Lockwood 2011). The models can be grouped in six different categories, as illustrated in Figure 7.3 (Calcott 1998). As the first three models have been widely ruled out, as thoroughly explained in reviews cited above, only a brief discussion is given here.

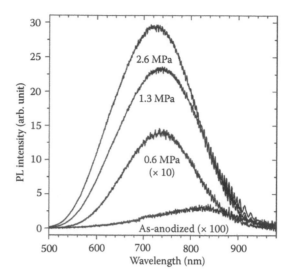

**FIGURE 7.1**    PL spectra of a 20-μm-thick PS layer as a function of the pressure during high-pressure water vapor annealing (HWA). The HWA was done at 260°C for 3 h. Data at zero pressure represents an as-anodized sample with no HWA treatment. (Reprinted with permission from Gelloz B. et al., *Appl. Phys. Lett.* 87, 031107, 2005. Copyright 2005. American Institute of Physics.)

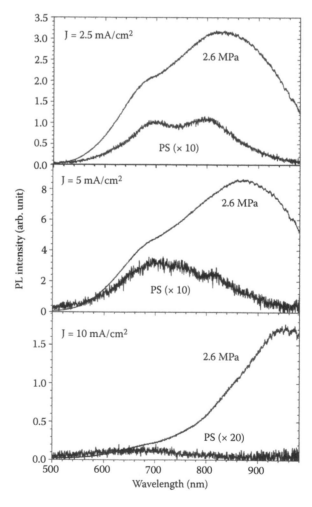

**FIGURE 7.2**    PL spectra of $n^+$-type PS, prepared under different current densities for 10 min, before and after HWA at 2.6 MPa. (From Gelloz B. and Koshida N., *Jpn. J. Appl. Phys.* 45(4B), 3462, 2006. Copyright 2006: The Japan Society of Applied Physics. With permission.)

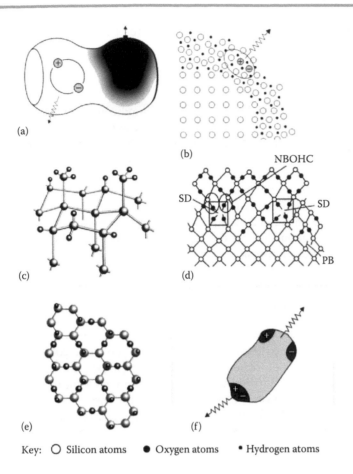

Key:  ○ Silicon atoms   ● Oxygen atoms   • Hydrogen atoms

**FIGURE 7.3**  The six groups of models proposed to explain the PL from porous Si. The model groups are illustrated pictorially under titles referring to the nature of the luminescent material involved. (a) A section of an undulating crystalline Si quantum wire. A surface defect renders one undulation non-radiative, whereas an exciton localized in a neighboring undulation recombines radiatively. (b) A crystalline Si nanostructure surrounded by a surface layer of hydrogenated amorphous Si. Radiative recombination takes place within the surface amorphous layer. (c) A hydride passivated Si surface with mono-, di-, and tri-hydride termination. Radiative recombination occurs from the Si hydride bonds. (d) A partially oxidized Si structure containing various defects that have been proposed as radiative centers in porous Si; the oxygen shallow donor (SD), non-bridging oxygen hole center (NBOHC), and Pb dangling bond. (e) The molecular structure of siloxene, an Si-based polymer proposed to exist on the large internal surface of porous Si and to be responsible for the luminescence from this material. (f) An Si dot with surface states that localize electrons and holes either separately (upper part of figure) or together (lower part of figure); these carriers subsequently recombine radiatively. (Reprinted from *Mat. Sci. Eng. B.* 51, Calcott P.D.J., 132, Copyright 1998, with permission from Elsevier.)

### 7.2.3.1 HYDROGENATED AMORPHOUS SILICON

In favor of this model, a-Si has been observed at the surface of some PSi layers. The PL of a-Si and its alloys, in principle, could emit a broad and tunable PL similar to that of PSi (Zhang et al. 1996). The time resolved dynamics of the PSi PL indicate that disorder is a key factor in its luminescence. However, many arguments can be given against this model, for instance: (1) fine analyses have revealed that conventional PSi (with minimum sample damage) can emit light even without the existence of any a-Si phase, (2) the temperature behavior of a-Si PL and PSi PL follows opposite trends, (3) intense PSi PL can be obtained after oxidation at temperatures high enough to crystallize a-Si:H or change it into oxide, and (4) spectroscopic studies have shown that the luminescent material in PSi has the electronic and vibrational nature of crystalline silicon (some details are shown in Section 7.3.1).

### 7.2.3.2  SURFACE HYDRIDE

In favor of this model is the fact that the surface of as-anodized PSi exhibits a very large area and is terminated by Si-H$_x$ ($x$ = 1, 2, 3) bonds. In addition, the PL is quenched as hydrogen is thermally desorbed and recovered when dipped in HF, which restores the hydrogen coverage of the surface. However, the loss of luminescence upon hydrogen desorption has later been explained by the creation of non-radiative channels (due to dangling bonds). Also against this model are (1) the fact that bright PL can be obtained in partially oxidized PSi in which the surface is passivated by good quality oxide rather than by hydrogen (Petrovakoch et al. 1992), (2) no correlation was found between the PL intensity and the hydrogen coverage, and (3) the requirement of a minimum porosity of about 70% for the appearance of the luminescence in PSi, as shown in Figure 7.4 (such threshold of porosity is not expected in a model involving only species at the PSi surface).

### 7.2.3.3  MOLECULES (SILOXENE)

Siloxene is an Si/H/O-based polymer. Some calculations have suggested it could have a PL similar to that of PSi (Hajnal and Deak 1998). When annealed in the range 200–400°C, it exhibits a PL spectrum very similar to that of the red-orange PL of PSi. However, Prokes (1996) pointed out in her review that, in such cases, the resulting material does not have the ordered siloxene structure any more and consists of an amorphous structure. As in the case of surface hydrides, this model cannot explain the intense PL obtained after high temperature oxidation as siloxene does not survive such treatments.

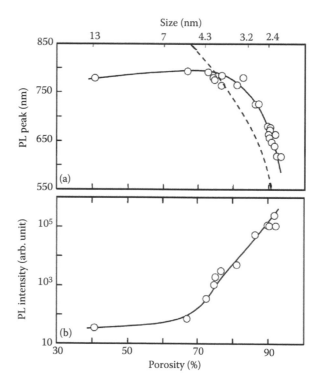

**FIGURE 7.4**    PL peak wavelength (a) and intensity (b) versus porosity (linear scale) and crystal size (nonlinear scale) for PSi films made from 6 Ω·cm $p$-type Si and exposed to air. The thin lines are a guide to the eye and the thick line is the calculated bandgap (see paper for details). (From von Behren J. et al., *J. Porous Mat.* 7, 81, 2000. Copyright 2000: Kluwer Academic Publishers. With permission.)

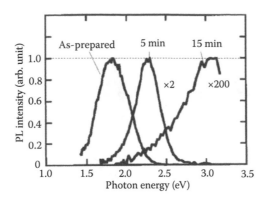

**FIGURE 7.5**   Red, green, and blue PL spectra obtained from PSi by post-anodization illumination with a tungsten lamp. Respective illumination times are also shown. (Reprinted with permission from Mizuno H. et al., *Appl. Phys. Lett. 69*, 3779, 1996. Copyright 1996. American Institute of Physics.)

### 7.2.3.4 DEFECTS

In this model, the luminescence arises from carriers localized at extrinsic centers either in bulk Si or in the oxide located at the surface, or at the Si/SiO$_x$ interface. Indeed, Si oxide is known to have a variety of luminescence centers (Skuja 1998). Actually, blue-emitting oxide centers are considered to contribute significantly to the blue emission observed in oxidized PSi (see Section 7.3.2). The debate is about the red-orange emission, in particular that observed in oxidized PSi. Indeed, continuous and progressive tuning of the PL from red to blue in partially oxidized PSi could not be realized. The PL could only be shifted from near-IR to orange (~2 eV) when oxide was present. Prokes (1996) strongly supports a model in which the emission of partially oxidized PSi is from oxide defects and particularly from non-bridging oxygen hole centers (NBOHC). This center exhibits a typical peak at ~600–670 nm with a full width at half maximum (FWHM) a bit smaller than what is usually observed for PSi. The stress and disorder would allow for a certain degree of PL peak positions in the red-yellow range. However, the energy levels of such defects should be largely insensitive to size of the silicon structure. This model can neither explain oxygen-free PL from red to blue in PSi (Wolkin et al. 1999), nor the large continuous blue-shift of PL from near-IR to orange or blue observed upon leaching in HF (in the dark or under illumination), as illustrated in Figure 7.5 (Mizuno et al. 1996). Nevertheless, in oxidized PSi, NBOHCs may be present in some quantity, depending on samples.

### 7.2.3.5 SURFACE STATES

Since the internal surface of PSi is very large, it was suggested that it should be involved in the luminescence process. Deep surface states that strongly localize carriers have been proposed. A model in which absorption takes place in quantum confined structures and then relaxes to surface states was proposed. In this case, the hole, the electron, both, or none (pure quantum confinement) can be localized (Koch et al. 1993), as illustrated in Figure 7.6. Such a model explains the energy difference between absorption and emission peaks, shown in Figure 7.7. However, spectroscopic analysis (e.g., resonant excitation, polarization memory; these are discussed in Section 7.3.1) have shown the extended nature of the luminescence states. Therefore, deep surface states are unlikely to be involved in the luminescence of PSi. However, shallow defect states, which do not strongly localize carriers, may be involved. This leads us to the next and last model, which is currently the most accepted one.

### 7.2.3.6 QUANTUM CONFINEMENT, WITH EFFECTS OF SHALLOW SURFACE STATES

The generally accepted model for the emission of PSi (red emission of any PSi layer and blue emission of oxygen-free PSi) involves quantum confinement in the silicon nanostructure. However, shallow surface states (in particular, the Si=O surface state) are also necessary for the

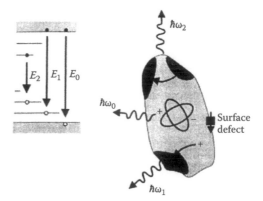

**FIGURE 7.6** Emission processes according to the surface-state mechanism. A photoexcited e-h pair can recombine at the band-edge for extended states ($\hbar\omega_0$), after localization of one partner ($\hbar\omega_1$), or both partners of the geminate pair ($\hbar\omega_2$). The result for an ensemble of differently sized particles is an inhomogeneously broadened line. A luminescence peak results because the radiative efficiency is optimal for processes between $E_0$ and $E_2$. Transitions to the surface defect contribute to the infrared emission band. (Reprinted from *J. Lumin.* 57, Koch et al., 271, Copyright 1993, with permission from Elsevier.)

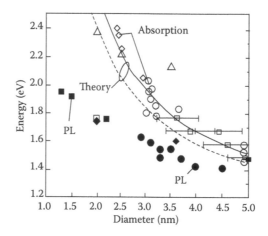

**FIGURE 7.7** Compilation of optical band gaps of silicon crystallites and PSi samples obtained from optical absorption (unfilled symbols) and luminescence (filled symbols). The lines represent calculated values with (dashed line) or without (full line) excitonic correction. (Reprinted from *Surf. Sci. Rep.* 38, Bisi O. et al., 1, Copyright 2000, with permission from Elsevier.)

understanding of the PL of partially oxidized PSi (Wolkin et al. 1999). The basic mechanism is that described in the surface states model, but with quantum confinement and shallow surface states. In this case, because of relaxation of carriers to these states, the PL peak cannot shift toward the blue further than up to ~2 eV, preventing generation of blue emission and inducing the energy difference between absorption and emission peaks shown in Figure 7.7. Figure 7.8 shows the calculation fitting the experimental data (Wolkin et al. 1999). This model could explain many of the results published earlier, which were attributing the PL peak pinned at ~2 eV in oxidized PSi purely to defect luminescence without involvement of quantum confinement. It also explains results that are more recent such as those shown in Figures 7.1 and 7.2 (discussed in Sections 7.4.1 and 7.4.2.4). The publication by Wolkin et al. was a significant milestone in the understanding of PSi luminescence, and made 1999 a kind of transition year in the field of PSi luminescence. This model is discussed in Section 7.3.1. The emission band related to this emission mechanism was named "S" band.

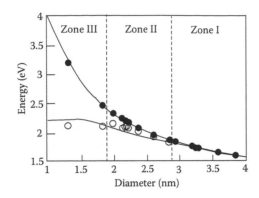

**FIGURE 7.8** Experimental and theoretical PL energies as a function of crystallite size. The upper line is the free exciton band gap and the lower line is the lowest transition energy in the presence of an Si=O bond. The solid and open dots are the peak PL energies of PSi samples kept in Ar and air, respectively. (Reprinted with permission from Wolkin M.V. et al., *Phys. Rev. Lett.* 82, 197, 1999. Copyright 1999 by the American Physical Society.)

### 7.2.4 FINAL REMARKS ON THE DEBATE

It should be noticed that quantum confinement has been proposed as the main mechanism for the luminescence of silicon nanocrystals not only in PSi, but also in a variety of other systems that are more controllable, less disordered, and easier to characterize than PSi, such as single Si nanocrystals (Prokofiev et al. 2009; de Boer et al. 2010; Shirahata 2011; Timmerman et al. 2011; Dohnalova et al. 2013; Sugimoto et al. 2013a,b) and Si nanocrystals in $SiO_2$ matrices or superlattices (Lockwood 2011) prepared by different methods. This suggests that quantum confinement would also be expected in PSi. In particular, the correlation between size and emission energy has been confirmed in several systems.

Nevertheless, it is clear and generally accepted that oxide introduces surface states that do dramatically affect the PL spectrum (Wolkin et al. 1999).

Moreover, as stated in Section 7.2.2, some PSi samples, due to their particular history or fabrication process, may contain different phases of silicon, like a-Si and oxide, and other types of molecular species. For these samples, it is certainly very difficult to distinguish the different PL bands in the observed PL spectrum, as they overlap each other. The various bands that have been clearly identified in PSi are discussed in some detail in Section 7.3.

## 7.3 PHOTOLUMINESCENCE BANDS

Table 7.1 shows a summary of the luminescence bands, which have been observed in as-formed or partially oxidized PSi. The "S" band ("S" for slow) has been the most studied band (Cullis

**TABLE 7.1    Porous Silicon Luminescence Bands**

| Typical Peak Wavelength (nm) | Typical Lifetime at 300 K | Best Efficiency | Band Label |
|---|---|---|---|
| 1100–1500 | 10 ns–10 µs | | IR |
| 400–1300 | ~ ns to ~150 µs | 1–10%; record: 23% | "S" |
| 425–630 | ps range | 0.01% | Hot PL |
| 420–470 | ~10 ns | 0.1% | "F" |
| ~350; 270–290 | ps–ns | | UV |
| 450–540 | ~1 s; 1–8 s in 4– ~200 K range | 2% at 300 K; 8.5% at 4 K | Long-lived |

et al. 1997; Bisi et al. 2000) because of its potential in optoelectronics, photonics, and sensing and because it originates from quantum confinement. It is discussed in detail next. The hot PL band, evidenced rather recently (de Boer et al. 2010), is related to the direct bandgap of silicon. The "F" band has received a rather high attention because of the blue emission and its rather early discovery in the early 1990s. It is observed in oxidized PSi. It often overlaps with some other bands, sometimes making interpretation difficult. Other bands (IR, UV, and long-lived) are related to surface states or oxide-related luminescence centers. They are far less often observed than the "S" and "F" bands and have been much less studied. The IR band was observed in both partially oxidized PSi and oxygen-free samples. It has been related to both quantum-size effect and surface states (Koch et al. 1993).

### 7.3.1 THE "S" BAND

Let us first consider PSi in typical atmospheres, such as air, inert gases, or vacuum. The main characteristics of the "S" band are summarized in Table 7.2. The name comes from the slow ("S") decay of the PL, typically in the microsecond range (much slower than the fast decay rates observed in direct bandgap semiconductors). The emission was shown to originate mostly from exciton recombinations in Si nanocrystals as indicated by polarization memory of PL, PL saturation under high excitation due to Auger recombinations, PL quenching in PSi contact by a liquid electrolyte due to Auger recombinations, as well as resonant excitation and hole-burning experiments (evidencing phonon-mediated recombination and singlet-triplet exciton state splitting). High confinement energy (>0.7 eV) results in the break of k-conservation rules and direct recombination becomes possible (Kovalev et al. 1998). All these experimental and theoretical results have already been thoroughly discussed in review papers (Cullis et al. 1997; Calcott 1998; Bisi et al. 2000).

An example particularly worth mentioning here is the polarization memory exhibited by PSi PL (Koch et al. 1996). A significant fraction of the luminescence of the "S" band is polarized parallel to the exciting light for all the exciting light polarization. The polarization ratio, defined as $\rho = (I_{\parallel} - I_{\perp})/(I_{\parallel} + I_{\perp})$, and the PL intensity as a function of emission energy are shown in Figure 7.9. For the IR band, which has been attributed to surface states, $\rho$ is zero, as expected. For the "S" band, however, it is not zero, and is maximum for emission polarization parallel to the <100> directions. Its energy dependence shows that it tends to zero as the energy approaches that of the silicon band gap. The conclusion is that (1) PSi can be considered an ensemble of randomly oriented dipoles due to aspherical nanocrystals and elongated nanostructures, (2) these elongated nanostructures are preferentially aligned along the <100> direction, and (3) the carrier wave functions extend on the nanoscale nanostructure (bulk-like) rather than on an atomic scale (like for defect emission) because of the sensibility to the shape of the nanostructure.

Another very interesting study deserving special attention is resonant excitation, which is discussed in detail in (Cullis et al. 1997) and summarized in (Bisi et al. 2000) and (Calcott 1998). Figure 7.10 shows that when the excitation energy is within the luminescence band and at low temperature, clear features can be observed in the PL. First, the luminescence onset shows an energy offset $\Delta$ with respect to the excitation energy, which is consistent with the presence

### TABLE 7.2    Main Characteristics of the "S" Band

| Property | Typical Values | Comments |
|---|---|---|
| Peak wavelength | 1300–400 nm | Porosity dependent |
| External quantum efficiency | Up to 23% | At 300 K; depends on porosity and surface passivation |
| FWHM | 150–180 nm | At 300 K; due to size distribution; nanometer range in microcavity |
| Decay time | ns to ~100 µs (from blue to red) | Depends on wavelength, temperature, and surface passivation |
| Degree of polarization | ≤0.2 | Can be anisotropic (effect of crystalline orientations) |
| Fine structure under resonant excitation | Phonon replica at 56 and 19 meV | Consistent with Si phonons |

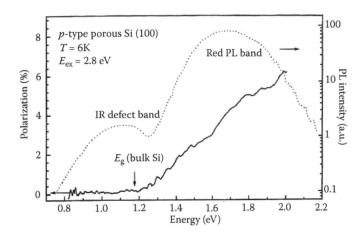

**FIGURE 7.9** Degree of polarization plotted against the log of the PL intensity, both measured at 6 K. The PL spectrum shows twin peaks, the strong peak close to 1.65 eV is "S" band PL, the weaker peak close to 1.1 eV is the IR PL band. The value of the bulk silicon band gap is indicated by $E_g$. (Reprinted from *J. Lumin.* 70, Koch et al., 320, Copyright 1996, with permission from Elsevier.)

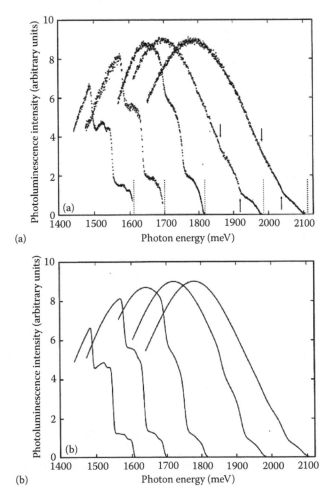

**FIGURE 7.10** (a) The experimentally observed variation of the PL line shape (at 2 K) with laser energy. For each spectrum, the value of the laser energy is indicated by a vertical dotted line. For spectra taken with the highest excitation energies, the position of the first and second phonon replicas of the onset are marked by arrows pointing up and down, respectively. (b) The modeled variation of the PL line shape with laser energy assuming that all the PL is from quantum confined crystalline Si. (Reprinted from *Mat. Sci. and Eng. B.* 51, Calcott P.D.J., 132, Copyright 1998, with permission from Elsevier.)

of a singlet-triplet exciton state splitting. This energy offset $\Delta$ has also been evidenced in the temperature-dependence of the radiative lifetime (discussed later in this section, together with Figure 7.11). Second, replicas of this onset are observed at lower energies, in agreement with phonon-assisted luminescence processes. The energies of the replicas are multiples of 56 and 19 meV, which are the energies of the TO and TA phonons of bulk Si. Thus, these experiments and associated calculations (Figure 7.10b) indicate exciton transitions in an indirect-gap as luminescence mechanism. For high confinement energy (>0.7 eV), quasi-direct recombination actually dominates.

Another interesting observation supporting emission from the nanocrystals was reported recently. Figure 7.12 shows the dependence of the external PL quantum yield on excitation photon energy (Timmerman et al. 2011). A step-like increase in quantum yield was observed for larger photon excitation energies for sample po-Si 1 (and for Si nanocrystals obtained from other

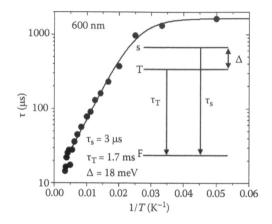

**FIGURE 7.11** PL decay times at 600 nm of oxidized $p$-type PSi at different temperatures. The line is a fit using a thermal equilibrium model with the 3-level system shown in the inset.

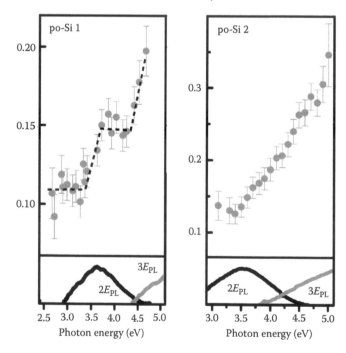

**FIGURE 7.12** Spectral dependence of external quantum yield of photoluminescence for two different PSi layers labeled po-Si 1 and po-Si 2. The lower panels show multiples of the photoluminescence spectra of each sample (i.e., the energy axis is multiplied by either 2 or 3, indicated by $2E_{PL}$ and $3E_{PL}$, respectively). Black dashed lines, indicating the "steps," serve only as a guide to the eye. (Reprinted by permission from Macmillan Publishers Ltd. Timmerman D. et al., *Nat. Nanotechnol.* 6, 710, copyright 2011: NPG.)

routes). This result was attributed to carrier multiplication (more than one electron-hole pair generated per absorbed photon) occurring with high efficiency and close to the energy conservation limit. The occurrence and width of the steps were related to the width of the multiples of photoluminescence spectra (naturally explained by the distribution of nanocrystal sizes). The more or less continuous character of the quantum yield increase for the second sample (po-Si 2) was explained by the broader size distribution and broader PL spectrum in this sample.

In principle, PL from red to blue is achievable due to the quantum size effect. It was indeed confirmed using PSi having its surface terminated by silicon-hydrogen bonds (oxygen-free), just after formation (Wolkin et al. 1999). The surface termination by Si-H$_x$ bonds is naturally obtained from the fact that PSi is in contact with HF just before rinsing and drying.

However, surface states lying inside the bandgap, in particular those introduced by oxygen via Si=O bonds, can prevent the observation of green and blue emission expected from band to band recombinations, leading to red-orange PL instead (Wolkin et al. 1999). This phenomenon is illustrated in Figures 7.8 and 7.13.

In Figure 7.13a, the PL of PSi layers formed in various conditions is shown. The higher the porosity is, the higher the PL peak energy. The PL spans from red to blue. The rather large bandwidth of the PL spectra is a consequence of the nanocrystal size distributions. These samples were kept in Ar to preserve the initial hydrogen surface termination. Figure 7.13b shows the PL spectra for the same samples, but taken after the PSi layers had been exposed to air. The blue emission was not observed in this case. Figure 7.8 shows experimental and theoretical PL energies as a function of crystallite size. It shows that the lowest transition energy is dramatically affected by the presence of an Si=O bond at the surface of the nanocrystal. Experimentally, a few seconds of exposure to air is enough to induce a shift in PL from blue to red, due to the fast formation of native oxide.

The PL lifetime typically ranges from a few nanoseconds for blue emission to 100 μs for red emission at room temperature. It increases with emission wavelength in agreement with the size dependence of quantum confinement strength. It increases at low temperature due to singlet-triplet exciton state splitting (Cullis et al. 1997; Bisi et al. 2000), as illustrated in Figure 7.11. The PL decay follows a stretched exponential curve due to the inhomogeneous nature of the emission. This behavior has been attributed to exciton migration, carrier escape from Si dots, or distribution of dot shape emitting at the same energy (Cullis et al. 1997).

Recently, radiative and nonradiative relaxation processes in freshly prepared and oxidized PSi have been further investigated (Arad-Vosk and Sa'ar 2014), giving more information about the respective roles of silicon cores and surface chemistry. It was shown that radiative processes should be associated with quantum confinement in the core of the Si nanocrystallites and,

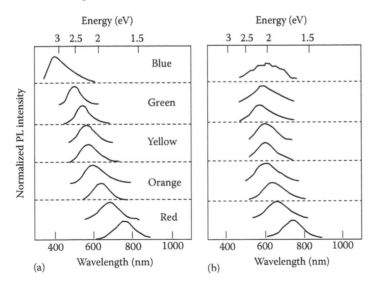

**FIGURE 7.13**    (a) Room temperature PL spectra from PSi samples with different porosities kept under Ar atmosphere. (b) PL of the same samples but after having been exposed to air. (Reprinted with permission from Wolkin M.V. et al., *Phys. Rev. Lett.* 82, 197, 1999. Copyright 1999 by the American Physical Society.)

therefore, are not affected by oxidation. On the other hand, nonradiative relaxation processes are affected by oxidation and by the state of the nanocrystallites surface.

The S-band exhibits a FWHM of about 150 nm, resulting from the size distribution of luminescent nanocrystals. The line width can be drastically decreased by placing the luminescent PSi layer in between two DBRs. Thus, a luminescent microcavity is formed. Figure 7.14 shows the PL of a microcavity as well as the PL of a single PSi layer of the same porosity. The PL FWHM can be decreased down to as low as a few nanometers (Pavesi et al. 1996). Moreover, highly directional tunable PL and EL emission can be obtained (Araki et al. 1996; Pavesi et al. 1996; Chan and Fauchet 1999; Gelloz and Koshida 2010).

The dielectric constant of any material impregnating the PSi pores may influence the PL properties. Although such effect has not been observed when a solid material was incorporated into PSi pores, it was the cause suggested for the different extent of PL quenching observed when PSi was impregnated by different liquids (Fellah et al. 1998, 2000). Figure 7.15 shows the PL intensity at the spectrum maximum, observed in situ in various liquids, as a function of the low frequency

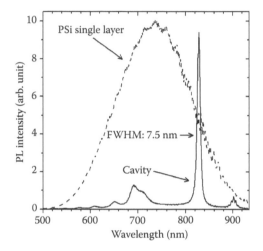

**FIGURE 7.14**    PL spectra of a single 825-nm thick PSi layer and of the same PSi layer sandwiched between an Ag top layer and a PSi Bragg reflector. In both cases, PSi was oxidized electrochemically and then treated by high-pressure water vapor annealing.

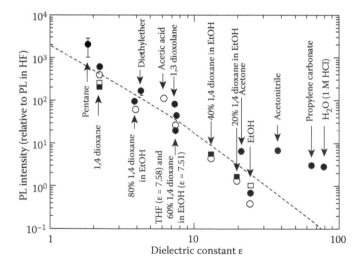

**FIGURE 7.15**    PL intensity at the spectrum maximum of luminescent PSi layers, observed in situ in various liquids, as a function of the low-frequency dielectric constant of the liquids. PL intensity is normalized to that measured in HF in similar excitation conditions (476 nm). The dashed line is the computed variation expected from the model developed in the paper. (Reprinted from *J. Lumin.* 80, Fellah S. et al., 109, Copyright 1999, with permission from Elsevier.)

dielectric constant of the liquids. The difference in PL intensity can be as large as three orders of magnitude. This spectacular result can be quantitatively accounted for in terms of a geminate recombination mechanism and of the variation of the Onsager length with the dielectric constant of the embedding medium, using an effective medium approximation. The model is based on the fact the radiative-recombination probability equals the probability that the electron-hole separation after thermalization is shorter than the Onsager length (which depends on the dielectric constant of the effective medium), at which Coulomb binding energy equals $kT$. For values of the dielectric constant larger than 20, the Onsager length becomes of the order of the nanostructure size, and the effective-medium approximation does not hold any more, and no further quenching is observed.

## 7.3.2 BLUE LUMINESCENCE: "F," HOT PL, AND LONG-LIVED BANDS

This section focuses on the various bands leading to blue PL. They are often present in oxidized samples and their overlapping sometimes makes their interpretation difficult. However, in recent years, a rather good understanding has been reached, as is shown next.

### 7.3.2.1 "F" BAND

Blue PL bands with emission peaks at about 415–500 nm, FWHM of ~0.38–0.5 eV and quantum efficiencies of 0.1% at best, has been reported in different types of oxidized PSi layers (Canham 1997; Cullis et al. 1997; Koyama and Koshida 1997; Qin et al. 1997; Bisi et al. 2000). The oxidation may be carried out thermally or chemically. The lifetimes were usually in the nanosecond range; thus, the naming of the band as "F" band ("F" for fast).

Except for the case of oxygen-free PSi, the blue emission is usually attributed to oxide-related defects (Canham 1997; Cullis et al. 1997; Koyama and Koshida 1997; Qin et al. 1997; Bisi et al. 2000) or carbonyl groups (Loni et al. 1995) resulting from carbon contamination occurring on storage and exposure to air.

Blue-emitting silicon nanocrystals were also prepared by short pulsed laser ablation in liquid media (Svrcek and Kondo 2009; Svrcek et al. 2009). In addition, blue-violet PL was obtained by mechanical milling followed by chemical oxidation (Ray et al. 2009). These emissions were attributed to oxide-related defects. Blue, green, and red light-emitting Si nanocrystals have been fabricated by pulsed laser ablation of crystalline Si in supercritical $CO_2$ (Saitow and Yamamura 2009). The supercritical fluid pressure was found to have a great effect on the resulting PL emission intensity and spectral range.

It was suggested that the "F" band could in fact be the newly identified hot PL band (Prokofiev et al. 2009). This possibility is discussed in Section 7.3.2.4 devoted to the hot PL band.

### 7.3.2.2 BLUE LUMINESCENCE INDUCED BY HIGH-PRESSURE WATER VAPOR ANNEALING

The blue PL component of PSi treated by high-pressure water vapor annealing (HWA) is worth mentioning because of its noticeable clear properties (Gelloz et al. 2008, 2009, 2014). It is most likely the F band, but with clearly identified characteristics due to well-controlled oxide quality. It always exhibits the same spectral shape, regardless of sample history. This indicates the involvement of localized states rather than quantum size effect. Its efficiency can be greater than that generally observed in conventionally oxidized PSi. PL external quantum efficiencies of 2% at room temperature and 8.5% at 4 K have been obtained. In addition, the reproducibility is much better with HWA than with other oxidizing methods.

HWA conducted on as-anodized PSi can produce two co-existing luminescence bands: a red-emitting one (S-band) and a blue emitting one associated with the presence of oxide in the layers. This is shown in Figure 7.16. In order to get only blue-emitting samples, intentional heavy oxidation of PSi was first carried out using thermal oxidation at temperatures above 850°C, and then HWA was conducted (Gelloz et al. 2009).

The mechanism of the blue luminescence of HWA-treated PSi has been attributed to either some kind of oxygen-related centers and/or silicon networks melted in the oxide (Gelloz et al. 2009). Carbon related luminescence centers were ruled out by infrared spectroscopy analysis and also

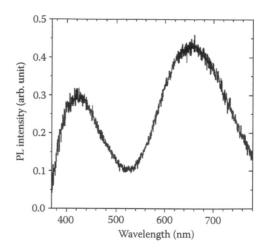

**FIGURE 7.16**    PL spectrum (excitation: 325 nm) of a 500-nm thick PSi layer (porosity 80%) treated by high-pressure water vapor annealing at 260°C for 3 h at 4 MPa.

by cleaning of the surface (Gelloz et al. 2014). The quality of the oxide has a large effect on the blue luminescence intensity. HWA markedly enhances the oxide quality by releasing the stress and decreasing the amount of hydroxyl groups. This is an important difference between HWA and other oxidizing methods. Furthermore, HWA is a relatively low temperature process (<300°C). These beneficial characteristics are definite advantages compared with conventional oxidizing methods.

### 7.3.2.3 LONG-LIVED BAND

The blue emission obtained by HWA includes two main contributions (Gelloz and Koshida 2009, 2012), as shown in Figure 7.17. One is very fast (nanosecond regime) and characterized by a peak wavelength of 410 nm and FWHM of ~50 nm. The other one is very slow (lifetime of several seconds) and is termed here the "long-lived band." Its spectrum shape is independent of sample structural properties but is strongly dependent on the excitation energy (Gelloz and Koshida 2009, 2012), suggesting localized centers as the origin of the luminescence. The best PL external

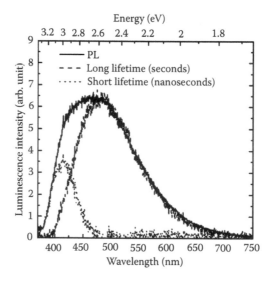

**FIGURE 7.17**    PL spectrum of blue-emitting PSi excited at 325 nm at 4 K. The phosphorescence spectrum (long lifetime) is also shown. It was normalized so that its long wavelength part matched that of the PL spectrum. The spectrum peaked at 3 eV with a FWHM of 0.36 eV (short lifetime), which was obtained by subtracting the phosphorescence from the PL. (Reprinted with permission from Gelloz B. et al., *Appl. Phys. Lett.* 94, 201903, 2009. American Institute of Physics.)

quantum efficiencies of the long-lived band were 2% at room temperature and 8.5% at 4 K (Gelloz and Koshida 2012).

Notice that both the long-lived band and the fast narrow band peaked at 410 nm and could be confused with the F band.

Figure 7.18 shows that the long-lived band can be observed from low temperature up to about 250 K. It is quenched by a thermally activated process. The activation energy of about 0.3 eV matches the peak separation between the long-lived band and the fast band spectra shown in Figure 7.17.

When the excitation energy is greater than ~4.4 eV, all the centers of the long-lived band can be excited via an intersystem crossing mechanism, whereas at lower excitation energies, only a fraction of these centers can be directly excited and a clear luminescence peak energy shift can be obtained upon changing the excitation energy. The broad spectral feature of the luminescence is due to inhomogeneous broadening resulting from local disorder or different chemical environments around the emitting centers (Gelloz and Koshida 2009).

A similar emission band has been observed previously by a few other research groups (Kux et al. 1995; Kovalenko et al. 1999; Wadayama et al. 2002). An emission band peaked at about 540 nm was observed for similar excitation wavelengths (337 and 325 nm) (Kovalenko et al. 1999; Wadayama et al. 2002). Kovalenko et al. (1999) reported a decay time of 0.5 s at 15 K, shorter than that observed by Gelloz and Koshida (2009, 2012), in as prepared but likely slightly aged, as well as in oxidized PSi. This emission was attributed to quantum confinement in very small silicon nanocrystals. Wadayama et al. (2002) reported an emission band decaying very slowly (>1 s) at room temperature in PSi having been subjected to rapid thermal oxidation at 1000°C and subsequent rapid cooling in liquid nitrogen. The last step (fast cooling) was necessary to get the very slow band emission, which remains to be understood.

The long-lived band exhibits several similarities with the emission of the so-called oxygen deficiency center in silicon oxide. However, the lifetime of such defects, in the millisecond range, does not match that of the long-lived band, which is of several seconds. Kux et al. (1995) also observed a similar band in oxidized PSi, using rapid thermal oxidation, and suggested that it could be related to SiOH groups. The same band was recently also observed in pure porous silica

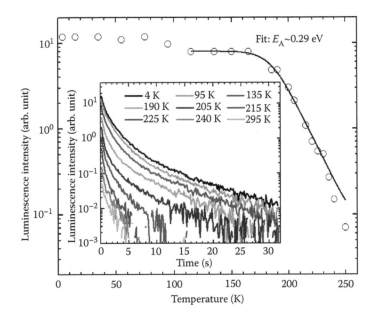

**FIGURE 7.18**    Temperature dependence of the phosphorescence intensity at 2.68 eV (delay: 140 ms; gate time 30 ms). The solid line is a fit using the function $I = I_0/[1 + \beta\exp(-E_A/kT)]$, which assumes thermally activated escape of carriers from the phosphorescent level. The fit leads to an activation energy $E_A$ of 0.29 eV. The inset shows the time evolution of the phosphorescence intensity at different temperatures. (Reprinted with permission from Gelloz B. et al., *Appl. Phys. Lett.* 94, 201903, 2009. Copyright 2009. American Institute of Physics.)

(Gelloz et al. 2014), showing that silicon nanocrystals have no role in the emission process and supporting the attribution of the origin of this band to molecular-like species in or at the surface of the oxide tissue in oxidized PSi. Carbon-related emission was also ruled out (Gelloz et al. 2014). In addition, a similar band has been observed recently in both silica and $Al_2O_3$ nano-particles. The origin of the band was not completely understood but luminescence from photogenerated OH radicals was proposed (Anjiki and Uchino 2012).

The long-lived band is more efficient in partially oxidized PSi than in fully oxidized PSi or pure porous silica, suggesting that silicon nanostructure could act as sensitizer for the excitation of the localized states (Gelloz and Koshida 2009).

### 7.3.2.4 HOT PL BAND

The hot PL band results from direct bandgap core luminescence ($\Gamma$-$\Gamma$ transitions) (Prokofiev et al. 2009). It was referred to as the "hot PL band" because it can be observed only under high excitation. This band was identified using layers of oxide embedded silicon nanocrystals obtained by sputtering (de Boer et al. 2010). In order to get a reasonable probability of radiative $\Gamma$-$\Gamma$ transitions, constant generation of hot carrier at the $\Gamma_{15}$-point (direct gap valley) by Auger recombination of multiple excitons is necessary, to compete with the otherwise very fast (1–10 ps) non-radiative relaxation toward the $\Delta$-valley (indirect gap valley). This mechanism is illustrated in Figure 7.19.

Recently, it was suggested that the "F" band could in fact be the newly identified hot PL band (Prokofiev et al. 2009). Indeed, this hot PL band has only been clearly identified in a partially oxidized silicon nanostructure (de Boer et al. 2010). It could overlap with or be a part of the "F" band, especially the green part. However, it is unlikely to be the "F" band by itself for the following reasons: (1) it is red-shifted as the crystalline core decreases, reaching the yellow range of the visible spectrum, thus not matching the blue emission of the F-band anymore and (2) a fast blue band (lifetime ~10 ns) peaked at about 420 nm (independent of nanocrystal core size) could still be observed as an independent oxide-related band (likely to be part of the F-band) in addition to the hot PL band (de Boer et al. 2010). In addition, the fast blue emission peaked at ~410 nm

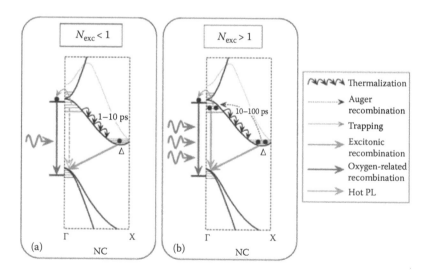

**FIGURE 7.19** Carrier relaxation paths for low ($N_{exc} < 1$) and high ($N_{exc} > 1$) excitation flux. Schematics of the silicon band structure in the $\Gamma$–$X$ direction, showing quantum confinement induced discretization of energy levels in the $\Gamma_{15}$-point (electrons), $\Gamma_{25}$-point (holes), and $\Delta$-valley. (a) When excitation flux is such that $N_{exc} < 1$, carriers can either thermalize within 1–10 ps or trap at a surface-related defect state. Only oxygen-related recombination (~ns) and excitonic recombination (~ms) will be present, and recombination via the direct channel will have low efficiency (dashed line). (b) For $N_{exc} > 1$, multiple carriers are generated by a high flux, allowing for constant filling of the direct channel by Auger recombination of multiple excitons. This will induce enhancement of PL intensity due to radiative recombination via the direct channel. (Reprinted by permission from Macmillan Publishers Ltd. de Boer W.D.A.M. et al., *Nat. Nanotechnol.* 5(12), 878, copyright 2010.)

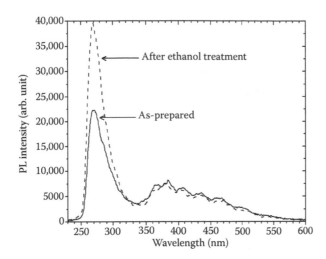

**FIGURE 7.20**    PL of a 2-μm-thick oxidized PSi layer excited at 266 nm at 300 K, in air, as-prepared and after soaking in ethanol for 30 min followed by drying under nitrogen gas.

(FWHM ~50 nm) in heavily oxidized PSi (Gelloz and Koshida 2009, 2012) not containing any significant amount of nanocrystals and not exhibiting any spectral dependence on sample preparation, which would clearly suggest the existence of oxide-related blue emission even without the hot PL band.

### 7.3.3  UV BANDS

The UV band peaked at about 350 nm has been observed only in PSi that was at least partially oxidized. It has been related to oxide luminescence, with the silicon nanocrystals playing a potential role in the photo-excitation process (Qin et al. 1996).

Very recently, another UV band was identified (Gelloz et al. 2014). It exhibits two narrow peaks at ~270 and ~290 nm and has very particular excitation characteristics. It is much stronger when the sample is measured in air than when in vacuum, suggesting that it originates from molecular species present in air and adsorbed at the surface of the porous material pores. This hypothesis was also supported by the fact that the removal of the carbon contaminants at the surface of the samples (using an ethanol treatment; see Figure 7.20) resulted in dramatic enhancement of the emission intensity. The carbon contaminants located at the surface were acting as a barrier preventing further adsorption of molecules.

This narrow UV band is only excited when the excitation energy is slightly higher than the emission energy, identifying a very small Stokes shift in the absorption-emission process and indicating that the emitter is likely a molecular species, and more likely water-related.

## 7.4  FACTORS AFFECTING THE EFFICIENCY AND STABILITY OF THE "S" BANDS

A single Si dot can have high quantum efficiency, typically, 20% (Valenta et al. 2008; Kusova et al. 2010), 35% (Valenta et al. 2002), or even 88% (Mason et al. 1998; Credo et al. 1999). However, the PL quantum efficiency of nc-Si layers is also affected by the dot density, the proportion of luminescent dots, the absorption by non-luminescent structures in the material, and the efficiency of light extraction.

The PL efficiency can be improved, for example, by (1) increasing the number of luminescent nanocrystals, (2) decreasing the light absorption by non-luminescent Si structures, (3) providing good surface passivation to the Si nanocrystals, and (4) increasing the absorption of luminescent nanocrystals, for example, by using plasmonic effects.

**TABLE 7.3    General Proven Effects of Various Treatments on the PL of PSi of Medium Porosity (<75%)**

| Treatment | Effect on PL | | Mechanism of the Changes |
|---|---|---|---|
| | Efficiency | Stability | |
| Thermal oxidation ($T < 600°C$) | Decreases | Decreases | Non-radiative defects at Si/SiO$_2$ interface |
| Rapid thermal oxidation ($T > 800°C$) | Increases | Increases | Good quality of Si/SiO$_2$ interface; enhanced confinement |
| ECO | Increases (~10%) | Increases | Enhanced confinement and surface passivation |
| SCM | Almost no change | Good (months) | Enhanced surface passivation; physical protection of surface |
| HWA | Increases (red: 23%; possible blue emission) | Very good (years) | Relaxed SiO$_2$; good quality of Si/SiO$_2$ interface |
| ECO followed by HWA | Increases (red only) | Very good (years) | Same as ECO and HWA; better preservation of Si dots |
| Combination of HWA and SCM | Increases | Very good | Same as ECO and SCM; SCM before HWA: better preservation of Si dots; good also for high porosity |

*Note:* Oxidation treatments may deteriorate the PL efficiency of high porosity PSi because it oxidizes fully or too much the PSi skeleton. Values given in percentages are record external quantum efficiencies. ECO, SCM, and HWA stand for electrochemical oxidation, surface chemical modification, and high-pressure water vapor annealing, respectively.

Passivation of Si nanocrystals is necessary to prevent competitive non-radiative recombinations (e.g., trapping by defects, charge exchanges with the surface and the environment) and to prevent the PL degradation due to chemical evolution of PSi surface (e.g., oxidation of Si-H bonds in air, contamination by carbon leading to blue emission [Loni et al. 1995]).

Table 7.3 shows the effects of various useful treatments on the PL efficiency and stability.

### 7.4.1  AS-FORMED POROUS SILICON AND EFFECT OF POROSITY

Generally, the PL intensity increases and its peak wavelength decreases when the porosity increases. This is because the higher the porosity, the smaller the average size and the higher the density of luminescent Si nanocrsytals. Figure 7.13a shows the tuning of the emission peak position by changing the porosity. Luminescence cannot be obtained from the whole range of porosities. Typically, a minimum porosity of 70% is necessary to get relatively strong emission from the S band, as shown in Figure 7.4 (Fauchet and vonBehren 1997; Von Behren et al. 2000). The figure also shows that the spectral position also depends on the porosity.

The efficiency usually decreases in the order $n$-type, $p$-type, $n^+$-type, and $p^+$-type PSi due to differences in nanostructures (Cullis et al. 1997). A recent paper again confirmed these trends in $p^+$-type PSi (Li et al. 2014). It is worth noticing that the fabrication of $p$-type and $p^+$-type PSi layers is very well controlled and reproducible. Their structures are typically independent of their thickness. This is true as long as the anodization does not exceed about 20 min because then chemical dissolution of the upper part of the layer may lead to a porosity gradient. However, $n$-type and $n^+$-type PSi are typically not or are only weakly luminescent when the anodization is carried out in the dark. This is the case for layers which are shown in Figure 7.2 (Gelloz and Koshida 2006a). Before HWA was carried out, the PL was very weak. In addition, two peaks are clearly seen in the PL spectrum in the upper panel of the figure. It would be tempting to ascribe the PL peaked at about 650 nm to defect states. However, knowing the structure of these particular types of $n^+$-type PSi layers, these two peaks were attributed to a two-peak size distribution in the complicated silicon structure.

Very bright PL can be obtained from these substrates when the anodization is conducted under illumination. However, in such cases, the resulting PSi layers can exhibit various different structures, which depend on anodization conditions and doping density, and accompanied by porosity gradients through the PSi layer depth. It is then very difficult to identify clear rules for the generation of efficient luminescence for $n$-type Si. As a result, it is mostly $p$-type PSi that is used in PL studies.

Post-anodization dissolution in HF solution of the PSi skeleton either in the dark (Canham 1990) or under illumination using visible light (Mizuno et al. 1996) can be used to induce a further increase in porosity and therefore a blue-shift of the PL peak. Figure 7.5 shows this effect for PSi further dissolved under illumination. The PL could be turned from red to blue. However, the blue emission was rather weak relatively to other emission spectra. In fact, increasing the porosity also means decreasing the amount of material left in the structure. Continuing the dissolution process for a long time eventually leads to a decrease in PL intensity due to disappearance of the porous material. The result of such post-anodization processes designed to decrease the average size (whether it is by dissolution or oxidation of silicon) strongly depends on initial porosity (Gelloz 1997). Applying such a process on an already strongly luminescent PSi layer may actually deteriorate its emission intensity. Obtaining bright blue emission from hydrogen-terminated PSi is currently still a challenge.

In practice, it is difficult to get blue-emitting layers for two other reasons. The first has been presented earlier in this chapter: as-formed PSi is not stable due to progressive oxidation of its large internal surface. The growth of native oxide induces a shift in emission energy due to associated surface states (Wolkin et al. 1999). Within a few seconds, the blue emission shifts to red-orange. The second reason is that, upon drying, high porosity PSi usually suffers from cracking, and the whole layer can be destroyed (Cullis et al. 1997; Bisi et al. 2000). The higher the porosity, the more fragile the nanostructure is. For a given porosity, the thicker the layer, the more prone it is to cracking. At comparable porosity, PSi formed from heavily doped Si is more resistant to cracking than that formed from lightly doped Si due to differences in the nanostructure morphology. Special drying methods have been investigated. The most efficient one is supercritical drying (Cullis et al. 1997; Bisi et al. 2000). However, this technique is quite expensive and difficult to implement. As a result, it is used only when necessary.

Some very particular conditions during anodization have been investigated in a view to enhance the PL intensity of as-formed samples. They remain very isolated cases and not much used. One such technique is conducting the anodization in the presence of a magnetic field (up to 1.9 T). The PL was found more efficient than without the magnetic field. The magnetic field leads to the formation of more spherical nanocrystals (more isotropic nanostructures) (Nakagawa et al. 1996). Luminescent PSi has been formed using $n$-type silicon in the dark, with the assistance of low mechanical pressure during anodization (Naddaf 2012). Under the same equivalent etching condition, pressure-assisted etching can yield PSi layer with stronger room temperature PL intensity than the layer formed by ordinary electrochemical etching. The results are explained by the pressure-induced stress/strain modifying the resistivity of the silicon substrate.

## 7.4.2 OXIDATION OF POROUS SILICON

Oxidation of PSi decreases the nanocrystals size, resulting in enhanced quantum confinement. Size decrease can increase the density of luminescent nanocrystals (increasing the PL intensity) while it induces a blue shift of the emission of already luminescent nanocrystals. Oxidation also affects the nanocrystal passivation via the quality of the surface oxide. It can also modify the carrier localization in Si nanocrystals. The effect of oxidation also depends very much on the initial size of the Si nanocrystals, that is, the porosity in the case of PSi (Gelloz 1997). See Figure 7.21. Indeed, the same oxidation treatment could lead to a PL intensity enhancement in a low porosity PSi layer and to a degradation of PL intensity in a highly porous layer due to full oxidation of most nanocrystals.

### 7.4.2.1 CHEMICAL OXIDATION

Chemical oxidation has been carried out by exposing the opened structure of PSi to various oxidizing gases (e.g., oxygen, ozone, hydrogen peroxide, water) or liquids (mostly boiling water).

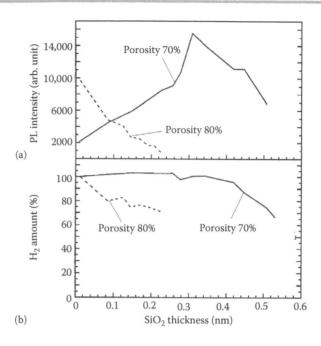

**FIGURE 7.21**    PL peak intensity for a 70% and an 80% porosity porous silicon layer as a function of the oxide thickness (a) and the relative amount of hydrogen on their surface as a function of the oxide thickness (b). (Reprinted from *Appl. Surf. Sci.* 108, Gelloz B., 449, Copyright 1997, with permission from Elsevier.)

These oxidation techniques are not easy to control. Thus, their effects on the PL have not been very reproducible, leading researchers to focus on other techniques like thermal oxidation. Many reports dealing with chemical oxidation were discussed by Gelloz (1997), who showed that their effects are also very dependent on the initial porosity of PSi.

More recently, detailed studies of oxidation by ozone (Caras et al. 2012, 2013) and hydrogen peroxide (Caras et al. 2013) have been carried out during the evolution of the oxidation process, correlating the PL intensity and the surface chemistry (using IR spectroscopy). For both ozone and hydrogen peroxide, the PL first decreased as defects were created in the early stages of oxidation and then increased when oxidation became more complete. The results after complete oxidation were rather the same in both cases, only the path (chemistry evolution of the surface) to get there was found to be different.

In another recent report (Deuro et al. 2012), the apparent oxidation rates of PSi by water vapor was found orders of magnitude faster in the presence of nonaqueous vapor streams that contain just ppm $H_2O$ levels. The nonaqueous analyte vapors served as a vehicle to transport $H_2O$ directly into the hydrophobic, as-formed PSi matrix where the $H_2O$ then oxidizes PSi.

### 7.4.2.2 THERMAL OXIDATION

Typically, thermal oxidation in oxygen at temperatures below 800°C produces an oxide of rather low quality, with a relatively high density of non-radiative centers, whereas higher temperatures lead to better quality oxide, favorable to the luminescence (Petrovakoch et al. 1992; Yamada and Kondo 1992; Meyer et al. 1993; Takazawa et al. 1994). This is illustrated in Figure 7.22 showing the evolution of PL intensity and so-called Pb centers for PSi oxidation at different temperatures.

### 7.4.2.3 ELECTROCHEMICAL OXIDATION

Electrochemical oxidation (ECO) (Bsiesy et al. 1991; Billat 1996) produces a monolayer of oxide at the surface of PSi formed from lightly doped silicon. The oxidation is self-limited by the break of the electrical contact between the substrate and the PSi layer. With PSi formed from highly doped silicon, the oxidation can be heavier due to the coarser PSi structure.

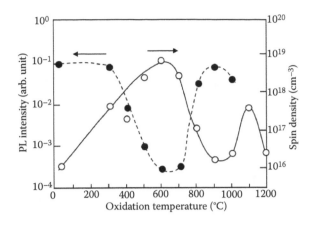

**FIGURE 7.22**  Steady state PL intensity and density of defects as obtained by EPR as a function of the oxidation temperature. (Reprinted with permission from Meyer B.K. et al., *Appl. Phys. Lett.* 63(14), 1930, 1993. Copyright 1993. American Institute of Physics.)

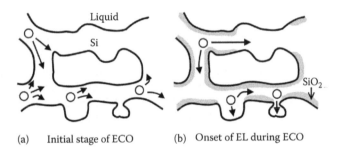

**FIGURE 7.23**  Schematic representation of the anodic oxidation process in porous silicon under galvanostatic conditions. First, holes supplied by the substrate flow through non-confined silicon (a). Later on, because of the oxidation of the surface of non-confined silicon regions, the potential is increased in order to keep the current constant, and carrier injection into luminescent crystallites becomes possible, inducing the electroluminescence (b).

This type of oxidation can result in great efficiency enhancement as well as significant PL blue shift, mainly by improved exciton localization. The originality of this oxidation method is that it can be carried out in such a way that it preferentially oxidizes the coarser parts of PSi without affecting much, and therefore preserving the luminescent Si nanocrystals. This is illustrated in Figure 7.23. About 10% external quantum efficiencies at room temperature have thus been obtained (Vial et al. 1992; Skryshevsky et al. 1996).

ECO has also been taken advantage of to improve the electroluminescence (Gelloz et al. 1998; Gelloz and Koshida 2000) efficiency of PSi (see *Porous Silicon: Opto- and Microelectronic Applications*, Chapter 1, Electroluminescence Devices [LED]).

### 7.4.2.4  HIGH-PRESSURE WATER VAPOR ANNEALING

HWA conducted at 150–300°C under 1–4 MPa (Gelloz and Koshida 2005; Gelloz et al. 2005) is particularly useful for getting highly efficient PL (external quantum efficiency up to 23%). The results depend on initial porosity of PSi and whether ECO or chemical derivatization (Gelloz and Koshida 2007) has been performed before HWA (see Table 7.3). The best results were obtained using lightly doped *p*-type silicon and 68% as initial PSi porosity, as shown in Figures 7.1 and 7.24. The treatment produces a thin oxide layer at the surface of the PSi skeleton.

HWA was also carried out on PSi formed from $n^+$-type silicon, as shown in Figure 7.2 (Gelloz and Koshida 2006a). Also in this case the PL was dramatically increased by HWA. The effect also depended on initial porosity (different anodization current densities). The two peaks in PL spectra

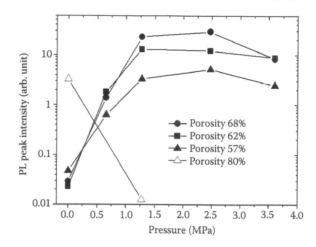

**FIGURE 7.24**   PL peak intensities of HWA-treated PSi samples with different initial porosities as a function of the HWA pressure. The untreated samples are located at the origin of the pressure axis.

were attributed to an initial two-peak size distribution in the complicated silicon structure. The PL band peaked at longer wavelengths was blue-shifted because of size reduction. The other PL band peaked at ~650 nm does not show a blue shift due to its pinning by surface state recombinations.

The key advantage of the treatment compared to conventional oxidation techniques is that the oxide is very stable and of good structural quality because its stress is released and it exhibits a very low density of non-radiative defects (Pb centers). This is well illustrated in Figure 7.25 (Gelloz et al. 2009). PSi (porosity 68%) was first thermally oxidized in $O_2$ at 300°C. The resulting oxide had many oxide defects, and the PL was very weak, due to the large density of non-radiative centers. The oxidized layer was then subjected to HWA at 260°C, and the PL was then very intense. Since the temperature of HWA was even lower than that of the initial thermal oxidation, the increase in PL intensity cannot be ascribed to a size reduction effect, but purely to surface passivation (dramatic decrease of the number of non-radiative centers). Moreover, this technique leads to completely stable layers whereas PSi oxidized by other techniques still suffers from property drifting upon aging.

The most efficient PL stabilization method to date is HWA (Gelloz and Koshida 2005, 2006a,b, 2007; Gelloz et al. 2005, 2006, 2007). It is the only method that offers long-term stabilization. This technique also stabilizes the PSi EL (Gelloz et al. 2006) and PSi photonic structures (Gelloz et al. 2007; Gelloz and Koshida 2010). The stabilization mechanism comes from the oxide grown by HWA, which is relaxed and includes a very low defect density.

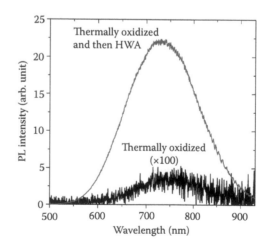

**FIGURE 7.25**   PL of a PSi layer of porosity 68% thermally oxidized at 300°C, before and after HWA conducted at 4 MP and 260°C. (From Gelloz B. et al., *Jpn. J. Appl. Phys.* 48, 04C119, 2009. Copyright 2009: The Japan Society of Applied Physics. With permission.)

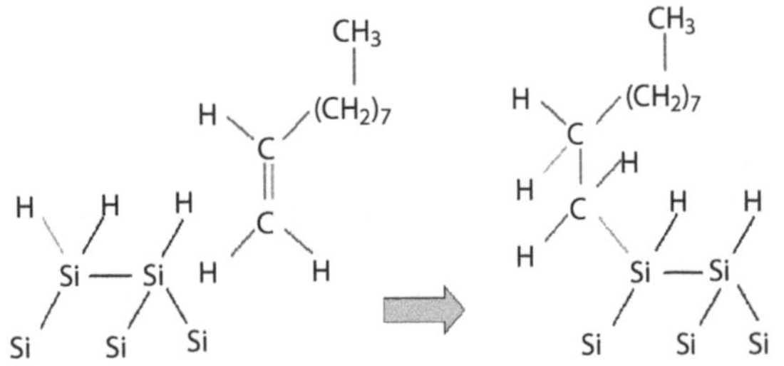

**FIGURE 7.26** Schematic representation of the mechanism of thermally induced hydrosililation at the surface of PSi. The molecule involved here is 1-decene.

### 7.4.3 SURFACE CHEMICAL MODIFICATION WITHOUT OXIDATION

Oxidation changes the properties of the initial PSi film. It may be useful to stabilize PSi without changing much of its original properties. Two approaches can be considered: (1) physical protection of PSi by encapsulation (Giaddui et al. 1996) or pore-filling techniques (Halimaoui et al. 1995) and (2) changing the chemistry of the PSi surface. Derivatization with long alkyl groups led to good stability of PL (Boukherroub et al. 2000, 2001a,b, 2002; Buriak 2002), EL (Gelloz et al. 2003), photonic properties (Ghulinyan et al. 2008), and biosensors (Alvarez et al. 2009). One advantage of this technique is its simplicity. PSi is merely immersed into the organic solution at about 100°C for a few hours. The chemical reaction is that of hydrosilylation involving a carbon-carbon double bond, as shown in Figure 7.26. Typically used organic solutions are 1-decene, 1-dodecene, undecylenate, and undecylenic acid.

### 7.4.4 OTHER POST-ANODIZATION TREATMENTS

A few reports deal with nitridation (Dillon et al. 1991; Anderson et al. 1993; Li et al. 1996; Morazzani et al. 1996) and thermal carbonization (Salonen et al. 2002) of PSi. The latter process was shown to stabilize passive photonic structures without inducing much property shift (Torres-Costa et al. 2008). Silver atoms deposited inside PSi may also help in passivating the Si nanocrystals (Sun et al. 2005; Lu et al. 2006). Also, iron atoms incorporated into PSi have helped stabilized the PL, the exact mechanism remaining unclear (Chen et al. 2001; Mavi et al. 2003; Rahmani et al. 2008, 2012).

Deposition of metals onto PSi has been reported to increase the PL by a relatively small factor (<100%), but whether the improvement is really due to the metal or another factor such as partial oxidation is still not fully clear. An increase of PL intensity by 67% by depositing a thin Al layer (5.2 nm) by sputtering onto PSi has been shown (Kim et al. 2009). The results were explained by the formation of Si-Al bonds. However, the actual mechanism of the enhancement and whether the Al penetrates inside the PSi pores remain unclear. Deposition of Ce onto PSi was done by an electrochemical method (Atyaoui et al. 2012). The small improvement in PL was attributed to a change of Si–H bonds into Si–O–Ce bonds and to a newly formed PS layer during electrochemical Ce coating.

PSi passivated by hydrocarbon groups ($CH_x$) using a plasma source has shown stable PL (Mahmoudi et al. 2007; Benyahia et al. 2013). The extent of the stability remains to be assessed in detail.

A few attempts to introduce Au (de la Mora et al. 2014) or silver nano-particles or nanostructures (Nakamura and Adachi 2012) into PSi have been reported in view to increase the PL by plasmonic effects. Thus far, the effects are not very clear but are still encouraging.

## 7.5 CONCLUSION

PSi exhibits several luminescence bands, spanning from the near-infrared to the UV, making it a potential candidate for applications in optoelectronics, sensing and imaging.

The most studied band is the so-called S-band, which is due to quantum confinement in silicon nanocrystallites. This band is mostly useful for the red emission range. Indeed, blue or even green emission is difficult to achieve due to surface states preventing the band-to-band emission in PSi whose surface is not perfectly free of contamination by foreign atoms or molecules. This band suffers from low stability (due to easy surface changes when exposed to the atmosphere). The techniques used to stabilize it have different effectiveness. Noticeably, for the long term, highly efficient (23%) and stable PL has been obtained by using HWA (Gelloz et al. 2005). Another potential problem with this band is its rather long lifetime. However, there may be some ways to decrease it, for example, by using plasmonic structures.

The bands emitting in the blue-green have also received much attention. They have a range of different origins, oxide-related emitters being one of the most important. One of these bands distinguishes itself by its very long lifetime (several seconds) (Gelloz and Koshida 2009).

# REFERENCES

Alvarez, S.D., Derfus, A.M., Schwartz, M.P., Bhatia, S.N., and Sailor, M.J. (2009). The compatibility of hepatocytes with chemically modified porous silicon with reference to in vitro biosensors. *Biomaterials.* **30**, 26–34.

Anderson, R.C., Muller, R.S., and Tobias, C.W. (1993). Chemical surface modification of porous silicon. *J. Electrochem. Soc.* **140**(5), 1393–1396.

Anjiki, A. and Uchino, T. (2012). Visible photoluminescence from photoinduced molecular species in nanometer-sized oxides: Crystalline $Al_2O_3$ and amorphous $SiO_2$ nanoparticles. *J. Phys. Chem. C* **116**(29), 15747–15755.

Arad-Vosk, N. and Sa'ar, A. (2014). Radiative and nonradiative relaxation phenomena in hydrogen- and oxygen-terminated porous silicon. *Nanoscale Res. Lett.* **9**, 47.

Araki, M., Koyama, H., and Koshida, N. (1996). Controlled electroluminescence spectra of porous silicon diodes with a vertical optical cavity. *Appl. Phys. Lett.* **69**(20), 2956–2958.

Atyaoui, M., Dimassi, W., Monther, G., Chtourou, R., and Ezzaouia, H. (2012). Electrochemical deposition of cerium on porous silicon to improve photoluminescence properties. *J. Lumin.* **132**(2), 277–281.

Benyahia, B., Guerbous, L., Gabouze, N., and Mahmoudi, B. (2013). Photoluminescence, time-resolved emission and photoresponse of plasma-modified porous silicon thin films. *Thin Solid Films* **540**, 155–161.

Billat, S. (1996). Electroluminescence of heavily doped *p*-type porous silicon under electrochemical oxidation in galvanostatic regime. *J. Electrochem. Soc.* **143**(3), 1055–1061.

Bisi, O., Ossicini, S., and Pavesi, L. (2000). Porous silicon: A quantum sponge structure for silicon based optoelectronics. *Surf. Sci. Rep.* **38**(1–3), 5–126.

Boukherroub, R., Morin, S., Wayner, D.D.M., and Lockwood, D.J. (2000). Thermal route for chemical modification and photoluminescence stabilization of porous silicon. *Phys. Stat. Sol. (a)* **182**(1), 117–121.

Boukherroub, R., Morin, S., Wayner, D.D.M. et al. (2001a). Ideal passivation of luminescent porous silicon by thermal, noncatalytic reaction with alkenes and aldehydes. *Chem. Mater.* **13**(6), 2002–2011.

Boukherroub, R., Wayner, D.D.M., Sproule, G.I., Lockwood, D.J., and Canham, L.T. (2001b). Stability enhancement of partially-oxidized porous silicon nanostructures modified with ethyl undecylenate. *Nano Lett.* **1**(12), 713–717.

Boukherroub, R., Wayner, D.D.M., and Lockwood, D.J. (2002). Photoluminescence stabilization of anodically-oxidized porous silicon layers by chemical functionalization. *Appl. Phys. Lett.* **81**(4), 601–603.

Bsiesy, A., Vial, J.C., Gaspard, F. et al. (1991). Photoluminescence of high porosity and of electrochemically oxidized porous silicon layers. *Surf. Sci.* **254**(1–3), 195–200.

Buriak, J.M. (2002). Organometallic chemistry on silicon and germanium surfaces. *Chem. Rev.* **102**(5), 1271–1308.

Calcott, P.D.J. (1998). The mechanism of light emission from porous silicon: Where are we 7 years on? *Mater. Sci. Eng. B* **51**(1–3), 132–140.

Canham, L.T. (1990). Silicon quantum wire array fabrication by electrochemical and chemical dissolution of wafers. *Appl. Phys. Lett.* **57**(10), 1046–1048.

Canham, L.T. (1997). Visible photoluminescence from porous silicon. In: Canham L.T. (ed.) *Properties of Porous Silicon.* INSPEC, The Institution of Electrical Engineers, London, pp. 249–255.

Caras, C.A., Reynard, J.M., Deuro, R.E., and Bright, F.V. (2012). Link between $O_2SiH$ infrared band amplitude and porous silicon photoluminescence during ambient $O_3$ oxidation. *Appl. Spectrosc.* **66**(8), 951–957.

Caras, C.A., Reynard, J.M., and Bright, F.V. (2013). An in-depth study linking the infrared spectroscopy and photoluminescence of porous silicon during ambient hydrogen peroxide oxidation. *Appl. Spectrosc.* **67**(5), 570–577.

Chan, S., and Fauchet, P.M. (1999). Tunable, narrow, and directional luminescence from porous silicon light emitting devices. *Appl. Phys. Lett.* **75**(2), 274–276.

Chen, Q.W., Li, X., and Zhang, Y. (2001). Improvement mechanism of photoluminescence in iron-passivated porous silicon. *Chem. Phys. Lett.* **343**(5–6), 507–512.

Credo, G.M., Mason, M.D., and Buratto, S.K. (1999). External quantum efficiency of single porous silicon nanoparticles. *Appl. Phys. Lett.* **74**(14), 1978–1980.

Cullis, A.G., Canham, L.T., and Calcott, P.D.J. (1997). The structural and luminescence properties of porous silicon. *J. Appl. Phys.* **82**(3), 909–965.

de Boer, W.D.A.M., Timmerman, D., Dohnalova, K. et al. (2010). Red spectral shift and enhanced quantum efficiency in phonon-free photoluminescence from silicon nanocrystals. *Nat. Nanotechnol.* **5**(12), 878–884.

de la Mora, M.B., Bornacelli, J., Nava, R., Zanella, R., and Reyes-Esqueda, J.A. (2014). Porous silicon photoluminescence modification by colloidal gold nanoparticles: Plasmonic, surface and porosity roles. *J. Lumin.* **146**, 247–255.

Deuro, R.E., Richardson, J.P., Reynard, J.M., Caras, C.A., and Bright, F.V. (2012). Parts per million water in gaseous vapor streams dramatically accelerates porous silicon oxidation. *J. Phys. Chem. C* **116**(43), 23168–23174.

Dillon, A.C., Gupta, P., Robinson, M.B., Bracker, A.S., and George, S.M. (1991). Ammonia decomposition on silicon surfaces studied using transmission fourier-transform infrared-spectroscopy. *J. Vac. Sci. Technol. A* **9**(4), 2222–2230.

Dohnalova, K., Poddubny, A.N., Prokofiev, A.A. et al. (2013). Surface brightens up Si quantum dots: Direct bandgap-like size-tunable emission. *Light: Sci. Appl.* **2**, e47.

Fauchet, P.M. and vonBehren, J. (1997). The strong visible luminescence in porous silicon: Quantum confinement, not oxide-related defects. *Phys. Stat. Sol. (b)* **204**(1), R7–R8.

Fellah, S., Wehrspohn, R.B., Gabouze, N., Ozanam, F., and Chazalviel, J.N. (1998). Photoluminescence quenching of porous silicon in organic solvents: Evidence for dielectric effects. *J. Lumin.* **80**(1–4), 109–113.

Fellah, S., Ozanam, F., Gabouze, N., and Chazalviel, J.N. (2000). Porous silicon in solvents: Constant-lifetime PL quenching and confirmation of dielectric effects. *Phys. Stat. Sol. (a)* **182**(1), 367–372.

Gelloz, B. (1997). Possible explanation of the contradictory results on the porous silicon photoluminescence evolution after low temperature treatments. *Appl. Surf. Sci.* **108**(4), 449–454.

Gelloz, B. and Koshida, N. (2000). Electroluminescence with high and stable quantum efficiency and low threshold voltage from anodically oxidized thin porous silicon diode. *J. Appl. Phys.* **88**(7), 4319–4324.

Gelloz, B. and Koshida, N. (2005). Mechanism of a remarkable enhancement in the light emission from nanocrystalline porous silicon annealed in high-pressure water vapor. *J. Appl. Phys.* **98**(1), 123509.

Gelloz, B. and Koshida, N. (2006a). Highly enhanced efficiency and stability of photo- and electro-luminescence of nano-crystalline porous silicon by high-pressure water vapor annealing. *Jpn. J. Appl. Phys., Part 1* **45**(4B), 3462–3465.

Gelloz, B. and Koshida, N. (2006b). Highly enhanced photoluminescence of as-anodized and electrochemically oxidized nanocrystalline *p*-type porous silicon treated by high-pressure water vapor annealing. *Thin Solid Films* **508**(1–2), 406–409.

Gelloz, B. and Koshida, N. (2007). Highly efficient and stable photoluminescence of nanocrystalline porous silicon by combination of chemical modification and oxidation under high pressure. *Jpn. J. Appl. Phys., Part 1* **46**(4B), 2429–2433.

Gelloz, B. and Koshida, N. (2009). Long-lived blue phosphorescence of oxidized and annealed nanocrystalline silicon. *Appl. Phys. Lett.* **94**, 201903.

Gelloz, B. and Koshida, N. (2010). Stabilization and operation of porous silicon photonic structures from near-ultraviolet to near-infrared using high-pressure water vapor annealing. *Thin Solid Films* **518**(12), 3276–3279.

Gelloz, B. and Koshida, N. (2012). Blue phosphorescence in oxidized nano-porous silicon. *ECS J. Solid State Sci. Technol.* **1**(6), R158–R162.

Gelloz, B., Nakagawa, T., and Koshida, N. (1998). Enhancement of the quantum efficiency and stability of electroluminescence from porous silicon by anodic passivation. *Appl. Phys. Lett.* **73**(14), 2021–2023.

Gelloz, B., Sano, H., Boukherroub, R., Wayner, D.D.M., Lockwood, D.J., and Koshida, N. (2003). Stabilization of porous silicon electroluminescence by surface passivation with controlled covalent bonds. *Appl. Phys. Lett.* **83**(12), 2342–2344.

Gelloz, B., Kojima, A., and Koshida, N. (2005). Highly efficient and stable luminescence of nanocrystalline porous silicon treated by high-pressure water vapor annealing. *Appl. Phys. Lett.* **87**(3), 031107.

Gelloz, B., Shibata, T., and Koshida, N. (2006). Stable electroluminescence of nanocrystalline silicon device activated by high pressure water vapor annealing. *Appl. Phys. Lett.* **89**, 191103.

Gelloz, B., Shibata, T., Mentek, R., and Koshida, N. (2007). Pronounced photonic effects of high-pressure water vapor annealing on nanocrystalline porous silicon. *Mater. Res. Soc. Symp. Proc.* **958**, 227–232.

Gelloz, B., Koyama, H., and Koshida, N. (2008). Polarization memory of blue and red luminescence from nanocrystalline porous silicon treated by high-pressure water vapor annealing. *Thin Solid Films* **517**, 376–379.

Gelloz, B., Mentek, R., and Koshida, N. (2009). Specific blue light emission from nanocrystalline porous Si treated by high-pressure water vapor annealing. *Jpn. J. Appl. Phys., Part 1* **48**(4), 04C119.

Gelloz, B., Mentek, R., and Koshida, N. (2014). Ultraviolet and long-lived blue luminescence of oxidized nano-porous silicon and pure nano-porous glass. *ECS J. Sol. St. Sci. Technol.* **3**(5), R83–R88.

Ghulinyan, M., Gelloz, B., Ohta, T., Pavesi, L., Lockwood, D.J., and Koshida, N. (2008). Stabilized porous silicon optical superlattices with controlled surface passivation. *Appl. Phys. Lett.* **93**(6), 061113.

Giaddui, T., Forcey, K.S., Earwaker, L.G., Loni, A., Canham, L.T., and Halimaoui, A. (1996). Reduction of ion beam induced and atmospheric ageing of porous silicon using Al and $SiO_2$ caps. *J. Physics D-Appl. Phys.* **29**(6), 1580–1586.

Hajnal, Z. and Deak, P. (1998). Surface recombination model of visible luminescence in porous silicon. *J. Non-Cryst. Solids* **230**, 1053–1057.

Halimaoui, A., Campidelli, Y., Badoz, P.A., and Bensahel, D. (1995). Covering and filling of porous silicon pores with Ge and Si using chemical-vapor-deposition. *J. Appl. Phys.* **78**(5), 3428–3430.

Kim, H., Hong, C., and Lee, C. (2009). Enhanced photoluminescence from porous silicon passivated with an ultrathin aluminum film. *Mater. Lett.* **63**(3–4), 434–436.

Koch, F., Petrovakoch, V., and Muschik, T. (1993). The luminescence of porous Si—the case for the surface-state mechanism. *J. Lumin.* **57**(1–6), 271–281.

Koch, F., Kovalev, D., Averboukh, B., Polisski, G., and BenChorin, M. (1996). Polarization phenomena in the optical properties of porous silicon. *J. Lumin.* **70**, 320–332.

Kovalenko, N.P., Doycho, I.K., Gevelyuk, S.A., Vorobyeva, V.A., and Roizin, Y.O. (1999). Geminate and distant-pair radiative recombination in porous silicon. *J. Phys.-Condens. Matter* **11**(24), 4783–4800.

Kovalev, D., Heckler, H., Ben-Chorin, M., Polisski, G., Schwartzkopff, M., and Koch, F. (1998). Breakdown of the k-conservation rule in Si nanocrystals. *Phys. Rev. Lett.* **81**(13), 2803–2806.

Koyama, H. and Koshida, N. (1997). Spectroscopic analysis of the blue-green emission from oxidized porous silicon: Possible evidence for Si-nanostructure-based mechanisms. *Solid State Commun.* **103**(1), 37–41.

Kusova, K., Cibulka, O., Dohnalova, K. et al. (2010). Brightly luminescent organically capped silicon nanocrystals fabricated at room temperature and atmospheric pressure. *ACS Nano* **4**(8), 4495–4504.

Kux, A., Kovalev, D., and Koch, F. (1995). Slow luminescence from trapped charges in oxidized porous silicon. *Thin Solid Films* **255**(1–2), 143–145.

La Ferrara, V., Fiorentino, G., Rametta, G., and Di Francia, G. (2012). Luminescence quenching of porous silicon nanoparticles in presence of ascorbic acid. *Phys. Stat. Sol. (a)* **209**(4), 736–740.

Lauerhaas, J.M., Credo, G.M., Heinrich, J.L., and Sailor, M.J. (1992). Reversible luminescence quenching of porous Si by solvents. *J. American Chem. Soc.* **114**(5), 1911–1912.

Lehmann, V. and Gosele, U. (1991). Porous silicon formation—A quantum wire effect. *Appl. Phys. Lett.* **58**(8), 856–858.

Li, G.B., Hou, X.Y., Yuan, S. et al. (1996). Passivation of light-emitting porous silicon by rapid thermal treatment in $NH_3$. *J. Appl. Phys.* **80**(10), 5967–5970.

Li, S., Ma, W., Zhou, Y., Chen, X., Ma, M., Xiao, Y. et al. (2014). Influence of fabrication parameter on the nanostructure and photoluminescence of highly doped *p*-porous silicon. *J. Lumin.* **146**, 76–82.

Lockwood, D.J. (2011). Bringing silicon to light: Luminescence in silicon nanostructures. *ECS Trans.* **33**(33), 17–36.

Loni, A., Simons, A.J., Calcott, P.D.J., and Canham, L.T. (1995). Blue photoluminescence from rapid thermally oxidized porous silicon following storage in ambient air. *J. Appl. Phys.* **77**(7), 3557–3559.

Lu, Y.W., Du, X.W., Sun, J., Han, X., and Kulinich, S.A. (2006). Influence of surface Si-Ag bonds on photoluminescence of porous silicon. *J. Appl. Phys.* **100**(6), 063512.

Mahmoudi, B., Gabouze, N., Guerbous, L., Haddadi, M., and Beldjilali, K. (2007). Long-time stabilization of porous silicon photoluminescence by surface modification. *J. Lumin.* **127**(2), 534–540.

Mason, M.D., Credo, G.M., Weston, K.D., and Buratto, S.K. (1998). Luminescence of individual porous Si chromophores. *Phys. Rev. Lett.* **80**(24), 5405–5408.

Mavi, H.S., Rasheed, B.G., Shukla, A.K., Soni, R.K., and Abbi, S.C. (2003). Photoluminescence and raman study of iron-passivated porous silicon. *Mater. Sci. Eng., B* **97**(3), 239–244.

Meyer, B.K., Petrovakoch, V., Muschik, T., Linke, H., Omling, P., and Lehmann, V. (1993). Electron-spin-resonance investigations of oxidized porous silicon. *Appl. Phys. Lett.* **63**(14), 1930–1932.

Mizuno, H., Koyama, H., and Koshida, N. (1996). Oxide-free blue photoluminescence from photochemically etched porous silicon. *Appl. Phys. Lett.* **69**(25), 3779–3781.

Morazzani, V., Cantin, J.L., Ortega, C. et al. (1996). Thermal nitridation of *p*-type porous silicon in ammonia. *Thin Solid Films* **276**(1–2), 32–35.

Naddaf, M. (2012). Low mechanical pressure during electrochemical etching: Induced modification in optical and structural properties of *n*-type porous silicon. *J. Mater. Sci.: Mater. Electron.* **23**(12), 2173–2180.

Nakagawa, T., Koyama, H., and Koshida, N. (1996). Control of structure and optical anisotropy in porous Si by magnetic-field assisted anodization. *Appl. Phys. Lett.* **69**(21), 3206–3208.

Nakamura, T., and Adachi, S. (2012). Photoluminescence decay dynamics of silver/porous-silicon nano-composites formed by metal-assisted etching. *J. Lumin.* **132**(11), 3019–3026.

Park, J.H., Gu, L., von Maltzahn, G., Ruoslahti, E., Bhatia, S.N., and Sailor, M.J. (2009). Biodegradable luminescent porous silicon nanoparticles for in vivo applications. *Nat. Mater.* **8**(4), 331–336.

Pavesi, L., Guardini, R., and Mazzoleni, C. (1996). Porous silicon resonant cavity light emitting diodes. *Solid State Commun.* **97**(12), 1051–1053.

Petrovakoch, V., Muschik, T., Kux, A., Meyer, B.K., Koch, F., and Lehmann, V. (1992). Rapid-thermal-oxidized porous Si—The superior photoluminescent Si. *Appl. Phys. Lett.* **61**(8), 943–945.

Prokes, S.M. (1996). Surface and optical properties of porous silicon. *J. Mater. Res.* **11**(2), 305–320.

Prokofiev, A.A., Moskalenko, A.S., Yassievich, I.N. et al. (2009). Direct bandgap optical transitions in Si nanocrystals. *JETP Lett.* **90**(12), 758–762.

Qin, G.G., Song, H.Z., Zhang, B.R., Lin, J., Duan, J.Q., and Yao, G.Q. (1996). Experimental evidence for luminescence from silicon oxide layers in oxidized porous silicon. *Phys. Rev. B* **54**(4), 2548–2555.

Qin, G.G., Liu, X.S., Ma, S.Y. et al. (1997). Photoluminescence mechanism for blue-light-emitting porous silicon. *Phys. Rev. B* **55**(19), 12876–12879.

Rahmani, M., Moadhen, A., Zaibi, M.A., Elhouichet, H., and Oueslati, M. (2008). Photoluminescence enhancement and stabilisation of porous silicon passivated by iron. *J. Lumin.* **128**(11), 1763–1766.

Rahmani, M., Moadhen, A., Kamkoum, A.M. et al. (2012). Emission mechanisms in stabilized iron-passivated porous silicon: Temperature and laser power dependences. *Physica B-Cond. Matter* **407**(3), 472–476.

Ray, M., Jana, K., Bandyopadhyay, N.R. et al. (2009). Blue-violet photoluminescence from colloidal suspension of nanocrystalline silicon in silicon oxide matrix. *Solid State Commun.* **149**(9–10), 352–356.

Saitow, K., and Yamamura, T. (2009). Effective cooling generates efficient emission: Blue, green, and red light-emitting si nanocrystals. *J. Phys. Chem. C* **113**(19), 8465–8470.

Salonen, J., Laine, E., and Niinisto, L. (2002). Thermal carbonization of porous silicon surface by acetylene. *J. Appl. Phys.* **91**(1), 456–461.

Shimura, M., Makino, N., Yoshida, K., Hattori, T., and Okumura, T. (1999). Reversible quenching of red photoluminescence of porous silicon with adsorption of alcohol OH-groups I. *Electrochem.* **67**(1), 63–67.

Shirahata, N. (2011). Colloidal si nanocrystals: A controlled organic-inorganic interface and its implications of color-tuning and chemical design toward sophisticated architectures. *Phys. Chem. Chem. Phys.* **13**(16), 7284–7294.

Skryshevsky, V.A., Laugier, A., Strikha, V.I., and Vikulov, V.A. (1996). Evaluation of quantum efficiency of porous silicon photoluminescence. *Mater. Sci. Eng., B* **40**(1), 54–57.

Skuja, L. (1998). Optically active oxygen-deficiency-related centers in amorphous silicon dioxide. *J. Non-Cryst. Solids* **239**(1–3), 16–48.

Sugimoto, H., Fujii, M., Imakita, K., Hayashi, S., and Akamatsu, K. (2013a). Phosphorus and boron codoped colloidal silicon nanocrystals with inorganic atomic ligands. *J. Phys. Chem. C* **117**(13), 6807–6813.

Sugimoto, H., Fujii, M., Imakita, K., Hayashi, S., and Akamatsu, K. (2013b). Codoping n- and p-type impurities in colloidal silicon nanocrystals: Controlling luminescence energy from below bulk band gap to visible range. *J. Phys. Chem. C* **117**(22), 11850–11857.

Sun, J., Lu, Y.W., Du, X.W., and Kulinich, S.A. (2005). Improved visible phiotoluminescence from porous silicon with surface Si-Ag bonds. *Appl. Phys. Lett.* **86**(17), 171905.

Svrcek, V., and Kondo, M. (2009). Blue luminescent silicon nanocrystals prepared by short pulsed laser ablation in liquid media. *Appl. Surf. Sci.* **255**(24), 9643–9646.

Svrcek, V. Mariotti, D., and Kondo, M. (2009). Ambient-stable blue luminescent silicon nanocrystals prepared by nanosecond-pulsed laser ablation in water. *Opt. Express* **17**(2), 520–527.

Takazawa, A., Tamura, T., and Yamada, M. (1994). Photoluminescence mechanisms of porous si oxidized by dry oxygen. *J. Appl. Phys.* **75**(5), 2489–2495.

Timmerman, D., Valenta, J., Dohnalova, K., de Boer, W.D.A.M., and Gregorkiewicz, T. (2011). Step-like enhancement of luminescence quantum yield of silicon nanocrystals. *Nat. Nanotechnol.* **6**(11), 710–713.

Torres-Costa, V., Martin-Palma, R.J., Martinez-Duart, J.M., Salonen, J., and Lehto, V.P. (2008). Effective passivation of porous silicon optical devices by thermal carbonization. *J. Appl. Phys.* **103**(8), 083124.

Valenta, J., Juhasz, R., and Linnros, J. (2002). Photoluminescence spectroscopy of single silicon quantum dots. *Appl. Phys. Lett.* **80**(6), 1070–1072.

Valenta, J., Fucikova, A., Vacha, F. et al. (2008). Light-emission performance of silicon nanocrystals deduced from single quantum dot spectroscopy. *Adv. Funct. Mater.* **18**(18), 2666–2672.

Vial, J.C., Bsiesy, A., Gaspard, F. et al. (1992). Mechanisms of visible-light emission from electrooxidized porous silicon. *Phys. Rev. B* **45**(24), 14171–14176.

Von Behren, J., Wolkin-Vakrat, M., Jorne, J., and Fauchet, P.M. (2000). Correlation of photoluminescence and bandgap energies with nanocrystal sizes in porous silicon. *J. Porous Mater.* **7**(1–3), 81–84.

Wadayama, T., Arigane, T., Hayamizu, K., and Hatta, A. (2002). Unusual photoluminescence decay of porous silicon prepared by rapid thermal oxidation and quenching in liquid nitrogen. *Mater. Trans.* **43**(11), 2832–2837.

Wolkin, M.V., Jorne, J., Fauchet, P.M., Allan, G., and Delerue, C. (1999). Electronic states and luminescence in porous silicon quantum dots: The role of oxygen. *Phys. Rev. Lett.* **82**(1), 197–200.

Yamada, M. and Kondo, K. (1992). Comparing effects of vacuum annealing and dry oxidation on the photoluminescence of porous Si. *Jpn. J. Appl. Phys., Part 2* **31**(8A), L993–L996.

Zhang, R.Q., Costa, J., and Bertran, E. (1996). Role of structural saturation and geometry in the luminescence of silicon-based nanostructured materials. *Phys. Rev. B* **53**(12), 7847–7850.

# Optical Properties of Porous Silicon*

## 8

**Gilles Lérondel**

CONTENTS

---

* This chapter is dedicated to Robert Romestain.

The ability of porous silicon (PSi) to emit visible light at room temperature is not the only property that makes PSi a unique optical material. Another unique property is the ability to tune the effective refractive index of the material from above 3 to almost 1 by simply modulating the amount of air in volume, that is, the porosity of the material. Due to the pore size being much smaller than the diffraction limit, PSi can be seen as a homogenous material, which combined with the ability of making thin films allows for a fully integrated silicon-based alternative photonic technology. This chapter aims at reviewing the optical properties of PSi starting with the "bulk" related properties followed by the thin film related optical properties. We will mainly limit ourselves to nano- and mesoporous silicon, namely microporous silicon as opposed to macroporous silicon.

## 8.1  BULK OPTICAL PROPERTIES

### 8.1.1  DIELECTRIC CONSTANT

First measurements of PSi dielectric constant ($\varepsilon$) were performed in 1984 much before the discovery of its photoluminescence (Pickering et al. 1984). As shown in Figure 8.1, two trends can be observed: first, a decrease of $\varepsilon$ due to the loss of matter and second, a peculiar behavior depending on the doping of the substrate used for anodisation. In the case of p+-type PSi, one clearly observes a reminiscence of the bulk optical resonances attesting to the crystallinity of the material, and this is even more pronounced for low porosity materials. In the case of the p-type PSi, bulk transitions remain but are less pronounced. This effect was explained by a partial oxidation of the material in air, which is expected to be more important for p-type nanoporous silicon than p+-type mesoporous silicon. Under the p+-type silicon, by this we understand a material with resistivity <0.001 Ohm·cm, and under the p-type Si a material with resistivity >1.0 Ohm·cm. Usually, these materials after porosification are mesoporous and nanoporous, correspondingly. Native oxidation of nanoporous silicon was later confirmed by Wolkin et al. (1999) showing that native oxidation prevents the

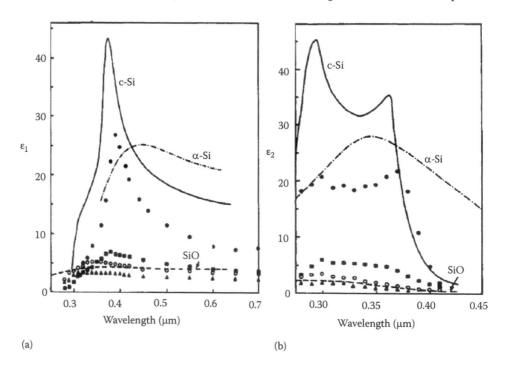

**FIGURE 8.1**  Comparison between the real (a) and imaginary (b) parts of the dielectric constants of bulk silicon (solid line), amorphous silicon (dashes and dots), silicon oxide (dashed line), and porous silicon of different kinds: p+-type of 31% (black dots) and 57% (black squares) of porosity, p-type of 54% (empty dots) and 65% (triangles) of porosity. Measurements were carried out using ellipsometry between 0.3 and 0.7 µm. (From Pickering et al., *J. Phys. C: Solid State Phys.* 17, 6535, 1984. Copyright 1984: IOP publishing. With permission.)

photoluminescence blueshift, as a function of dissolution time in HF, to be observed. Indeed, coming back to Figure 8.1, one can clearly see a clear spectral dependence between the imaginary part of the dielectric constant of a p-type porous silicon sample of 65% porosity and a thin film of oxide.

The bulk nature, however, of the p-type PSi was confirmed later thanks to two optical studies of the reflectivity at high energies. Reflectivity was measured from 1 to 20 eV using either a synchrotron source (Koshida 1993) or high-order laser harmonic radiation both combined with Kramers-Kroening analysis (De Filippo et al. 2002).

The measurements made by Pickering et al. (1984) were also of importance to evidence that no matter the doping, the optical properties of nanoporous silicon, which at the microscopic scale is a heterogenous material, can be described at the macroscopic scale (wavelength scale) by an effective dielectric constant, $\varepsilon_{eff}$.

Within the frame of the effective medium approximation, numerous models can be used to retrieve $\varepsilon_{eff}$ from the dielectric constants of the different elements of the effective medium (air, silicon, oxide, etc.). It is worth mentioning here that such models are usually used to describe isotropic heterogeneous materials with heterogeneities size much smaller than the wavelength of light. Three main models have been intensively used, namely Lorentz-Lorentz (LL), Maxwell Garnet (MG1 and MG2), and Bruggeman (EMA). All of them could be derived from the following generic formula (Aspnes et al. 1979):

$$(8.1) \qquad \frac{\varepsilon_{eff} - \varepsilon_h}{\varepsilon_{eff} + 2\varepsilon_h} = v_1 \frac{\varepsilon_1 - \varepsilon_h}{\varepsilon_1 + 2\varepsilon_h} + v_2 \frac{\varepsilon_2 - \varepsilon_h}{\varepsilon_2 + 2\varepsilon_h} + \cdots \quad \text{with} \sum_i v_i = 1$$

where $\varepsilon_h$, $\varepsilon_1$, $\varepsilon_2$,... are, respectively, the complex dielectric constants of the host medium and the inclusions (type 1, 2,...) with a proportion in volume of $v_1$, $v_2$, ... Relations in the case of the LL (MG1), MG2, and EMA approximations can simply be derived for $\varepsilon_h = 1$, $\varepsilon_h = \varepsilon_{eff}$ and $\varepsilon_h = \varepsilon_i$, respectively. The Sellmeyer model has also been used, and can simply be obtained by removing the denominator and taking $\varepsilon_h = 1$.

While Pickering et al. (1984) have shown that the Bruggeman model could account for the refractive index behavior in the near IR observed for p⁺-type PSi, the model is not sufficient to account for the entire dispersion curve, pointing out the need to account for the morphology as well.

A satisfactory model of the porous silicon dielectric constant as a function of the wavelength was proposed by Theiss et al. (1995) based on the Bergman model (Figure 8.2c′ and d′). As shown by the following expression, this model involves a function called $g$, also called spectral density (Bergman 1978; Theiss et al. 1995).

$$(8.2) \qquad \varepsilon_{eff} = 1 - v \int_0^1 \frac{g(n,f)}{\dfrac{1}{1 - \varepsilon_{Si}} - n} \, dn$$

The expression is given here for the simple case of a medium made of air and a proportion $v$ of bulk silicon. The $g$ function is independent of the material dielectric properties and only depends on its geometry. This theoretical modeling has confirmed that the difference observed for p- and p⁺-type is not related to a native oxidation of the material but rather related to a difference in morphology of the two materials (Smith and Collins 1992). However, it is worth noting here that the dielectric constant deduced from any experimental fitting strongly depends on the $g$ function (Figure 8.2a and b) and therefore requires us to know the morphology of the porous structure, which is not a trivial task. Generally, this model is more suitable for high porosity samples (low dielectric constants) (Theiss et al. 1995). Indeed, one clearly observes that in the case of the p⁺-type sample, the fitting of the fringes pattern is not perfect at low energies due to an overestimation of the theoretical absorption (Figure 8.2a and c′). The Bergman model proposed by Theiss was the most elaborated analytical model proposed for PSi. We will see in the next part on optical thin films properties that the dielectric constant can be retrieved without any modeling assuming that all physical parameters of the porous thin films are known (Lerondel et al. 1997a). Modern simulation softwares based on the

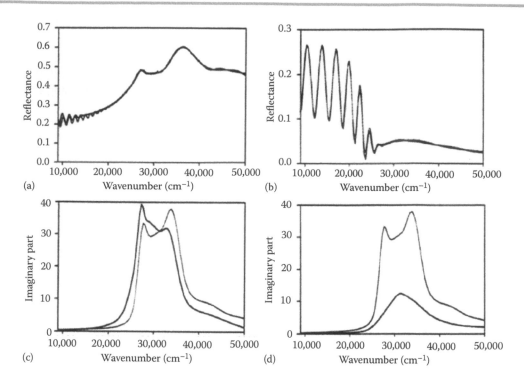

**FIGURE 8.2** Comparison between the experimental reflectivity spectra of 1-μm thick PSi films: (a) p⁺-type (25% of porosity), (b) p⁺-type (75% of porosity) and the theoretical spectra obtained from the Bergman model. (c) and (d) imaginary parts of the dielectric constants of porous silicon as obtained from the fitting of (a) and (b), respectively the light gray line in (c) and (d) corresponds to bulk Si for comparison. (Reprinted from *Surf. Sci. Rep. 29*, Theiss, 91, Copyright 1997, with permission from Elsevier.)

finite difference time domain (FDTD) method, for example, could also be used to some extent to modelize the dielectric constant of PSi assuming, however, a periodic kind of structure as already demonstrated for generic nanoporous films (Yang et al. 2004).

### 8.1.2 OPTICAL ABSORPTION

The observation of PSi photoluminescence has generated numerous studies on the absorption (α) of the material involving different measurement techniques as direct optical transmission or indirect techniques such as photoconduction or photothermal deflection spectroscopy. One of the goals was to observe (or not) a shift toward the high energies to confirm the quantum confinement model. Figure 8.3a shows the dependence of the square root of the absorption multiplied by the energy for self-supported thin films of p- and p⁺-type porous silicon (Sagnes et al. 1993). One observes indeed a shift in the absorption threshold compared with bulk silicon. The shift is even more pronounced for low-doped silicon and high porosity samples.

From the curve one can extrapolate the bandgap for the p⁺-type sample only. These results are in agreement with the results obtained by Pickering et al. (1984) evidencing the crystalline character of p⁺ PSi, compared to p-type PSi. Authors explained the spectrum in the case of p-type silicon using either a distribution of energy bandgap or a high density of states in the bandgap. Similar spectra have been obtained by measuring the photoconductivity of PSi (Koshida 1993; Ookubo 1993).

Interestingly, experimental data obtained by photoluminescent excitation spectra (PLE) led to weaker absorption values, compared to the values obtained with the more classical techniques (Wang et al. 1992; Ookubo 1993; Kux and Ben Chorin 1995). A possible explanation lies in the material itself. PSi can be seen as a heterogeneous material with small scale crystallites absorbing less because of quantum confinement and large scale crystallites absorbing light like bulk silicon.

At low energies (<1.5 eV), the absorption due to the bigger crystallites is prominent. It is worth mentioning that by transmission one probes the entire material. One of the limits of transmission

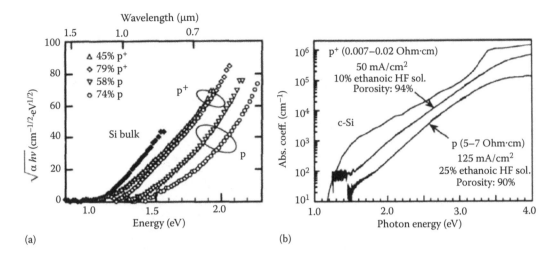

**FIGURE 8.3** Optical absorption of low- (p) and highly doped (p⁺) PSi as measured by direct transmission of self-supporting porous silicon layers. (a) Tauc representation. (Reprinted with permission from Sagnes et al., *Appl. Phys. Lett.* 62(10), 1155, 1993. Copyright 1993. American Institute of Physics.) (b) Absorption coefficient. (From Von Behren et al., *Mater. Res. Soc. Symp. Proc.* 452, 565, 1997. Copyright 1997: Material Research Society. With permission.) Curves have been corrected to account for the porosity.

measurements is that they are performed on suspended thin films. Progress in drying techniques such as hypercritical or supercritical drying has allowed absorption measurements in the high energies range using thin films (Canham et al. 1994; Von Behren et al. 1997).

As shown by Figure 8.3, the absorption spectra are linear over a wide energy range (2 to 3 eV). For a given energy, one finds that the minimum and maximum absorption levels are obtained for bulk and p-type silicon. This kind of behavior, together with the observation of an absorption tail (Urbach tail) near the bandgap edge, are classical characteristics of amorphous silicon (Cody et al. 1981) with a typical energy $E_0$ (Urbach energy). Below $E_0$, the absorption can be described as follows:

$$(8.3) \qquad \alpha(E) = \alpha_0 e^{\left(\frac{E - E_0}{E_A(T,X)}\right)}$$

In the case of hydrogenated amorphous silicon, $E_0$ is equal to 2.2 eV. $E_A$ usually depends on the temperature and the structural disorder as represented by $X$.

A later study on the temperature dependence of the optical absorption in PSi has infirmed this hypothesis as shown in Figure 8.4a (Kovalev et al. 1996). Indeed, the absorption spectra of PSi for different temperatures are parallel to each other whereas in the case of amorphous silicon they all converged to $E_0$ (Figure 8.4b). In addition, authors confirmed that there was no effect of the hydrogen on the absorption spectra. Instead, they interpret this behavior as characteristic of a crystal-like absorption due to phonon-assisted optical transitions, as expected for an indirect semiconductor. First theoretically suggested for silicon nanostructures, the indirect bandgap behavior was later confirmed by phonons replica observations in the excitation spectra (Hybertsen 1994; Kux and Ben Chorin 1995).

Alternatively to the already mentioned techniques, direct transmission and PLE, another interesting technique that does not require the removal of the film and allows for integrated transmission measurements (Lerondel et al. 2000) was introduced. This later point is essential as it makes the measurements insensitive to the interface roughness, which leads to an overestimation of the optical losses as it will be shown later in the second part of this chapter. The technique consists in measuring the transmission using the substrate underneath the PSi thin film (Figure 8.5). The transmitted photons modify the conductivity of the bulk silicon.

Figure 8.5 shows the absorption spectra obtained using this technique. Spectra have been obtained using several samples of different thicknesses ($d$) so that $\alpha \cdot d$ remains close to one on the entire spectral range. This allows the sensitivity to be constant and avoids having to account for multiple reflections (fringes) occurring for small $\alpha \cdot d$ preventing $\alpha$ to be straightforwardly calculated. One observes here

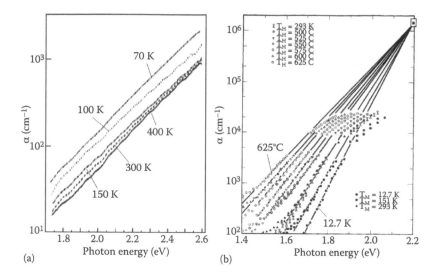

**FIGURE 8.4**  Absorption spectra as a function of the temperature for p-type PSi (70% of porosity) (a) and amorphous silicon (b). ([a] Reprinted with permission from Kovalev et al., *J. Appl. Phys.* 80(10), 5978, 1996. Copyright 1996. American Institute of Physics. [b] Reprinted with permission from Cody et al., *Phys. Rev. Lett.* 47(20), 148081, 1981. Copyright 1981 by the American Physical Society.)

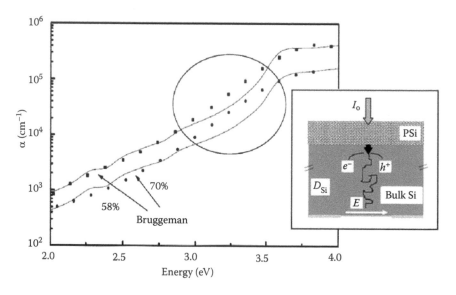

**FIGURE 8.5**  Absorption spectra of p-type porous silicon of 53% (squares) and 70% (dots) of porosity as measured using the photoconduction technique illustrated in the insert. Experimental data are compared with the absorption coefficient calculated using the Bruggeman model for comparable porosities. (Data extracted from Lerondel et al., *Thin Solid Films* 366, 216–224, 2000.)

again an exponential behavior for the lower part of the spectrum and saturation over 3.5 eV, which, however, is much less pronounced than for bulk silicon. Measurements in the low energy range (2 eV) compared rather well with the ones obtained by photoconduction and PDS. However, the measured absorption coefficient is one order of magnitude higher than the one measured by PLE for the same porosity. Again within the frame of the quantum confinement hypothesis this could be explained by the low proportion of luminescent material. In the same energy range, the absorption coefficient is twice as high as the one measured on self-supported films (Sagnes et al. 1993). This could be explained by a porosity gradient in the self-supporting samples as reported by Thönissen et al. (1996). The exponential behavior has already been discussed. The conclusion here is that there is no obvious relation between the amount of matter and the absorption coefficient in nanoporous silicon even if, as shown in Figure 8.5, a qualitative or even quantitative agreement can be obtained using the Bruggeman model. A clear mismatch remains in the 3 to 3.5 eV near the direct bandgap of silicon. It is also not possible to simply explain the behavior by a shift in the absorption threshold expected due to quantum confinement. To summarize, if an accurate measurement of the absorption in nanoporous silicon over a large energy range has been made possible, it seems that the full behavior can still only be accounted for using the usual representation of nanoporous silicon as a nanocrystalline sponge of silicon, meaning a material ordered at short distances (crystal) and disordered at a larger scale. Absorption measurements will again be addressed in the next section on specular reflection.

### 8.1.3 OPTICAL ANISOTROPY

For a highly symmetric material like bulk silicon, almost no optical anisotropy is expected. Neither is it expected from a porous material with a sponge-like morphology. However, the dependence of the refractive index on the morphology became clear when for the same porosity but different morphologies (columnar vs. sponge-like) $p^+$- and p-type anodized porous layers were found to have different refractive indexes. Indeed in 1992, it was reported that the optical birefringence in $p^+$-type samples can be as high as 0.14 (>10%) for highly porous materials (83% of porosity). Weaker but not at all negligible, a birefringence of 0.063 was also reported for p-type samples of comparable porosity. The origin of the optical anisotropy was further investigated through guided light by Mihalcescu et al. (1997). They measured the refractive index of guided modes using the M-line technique (cf. Figure 8.12a) on a PSi planar waveguide obtained using mainly low doped p-type silicon. The porous layers (guide or cladding) appear as a positive uniaxial material ($n_e > n_o$) with the optical axis normal to the layer surface as previously reported. As shown in Figure 8.6a, the highest and lowest optical anisotropies ($\delta = (n_e/n_o)^2 - 1$) are, respectively, obtained for the as-formed and anodically and thermally oxidized samples. In other words, the optical anisotropy increases with the amount of silicon. Using elasto-optic coefficients for bulk silicon, authors also calculated the possible maximum strain-induced anisotropy and found values of 0.009 and 0.003 for p- and $p^+$-type porous materials, respectively, confirming that strain in porous silicon layers is not sufficient to explain the anisotropy. Instead, the authors showed that the results obtained on low-doped p-type bi-layered structures can be explained by a very small proportion of Si columns (2%). In addition, they used an oversimplified effective model, and these results suggested how sensitive the refractive index can be to the structural morphology, even for the p-type material, which was expected to be isotropic orientation (Smith and Collins 1992).

Having a refractive index strongly dependent on the morphology and keeping in mind the relation between the pores' direction and the crystallographic orientation Smith and Collins (1992) made possible the fabrication of a very peculiar optical component as shown in Figure 8.6b. Dichroic Bragg reflectors were reported in 2002 by Diener et al. (2001). They used the standard

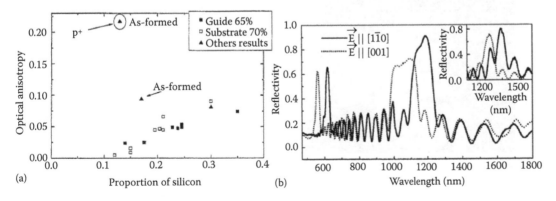

**FIGURE 8.6** Optical anisotropy in porous silicon: (a) In-plane optical anisotropy as measured from light propagation in meso- and nanoporous silicon planar waveguides. (Reprinted from *Thin Solid Films* 297, Mihalcescu et al., 245, Copyright 1997, with permission from Elsevier.) (b) Direct evidence of normal incidence anisotropy in polarization sensitive Bragg mirrors anodized from highly doped (110) silicon. (Reprinted with permission from Diener et al., *Appl. Phys. Lett.* 78, 3887, 2001. Copyright 2001. American Institute of Physics.)

current modulated anodization process but started from a (110)-oriented silicon wafer. As a result, a strong in-plane anisotropy was introduced, leading to two clear distinct stop bands for the TE and TM polarizations at normal incidence.

This property is rather remarkable and illustrates like the luminescent narrowing using all PSi optical microcavity (cf. 8.2.4), the potential of PSi for advanced photonics, with both engineered optical thicknesses as discussed later and engineered intrinsic properties. Optical anisotropy in PSi has been the subject of further studies and more details can be found in dedicated reviews like Golovan et al. (2007) where PSi is compared to porous gallium phosphite and alumina.

We briefly discussed the passive optical properties, but the same morphological anisotropy was also considered in the early stages to explain light polarization effects in the emission (Kovalev et al. 1995).

We will not discuss light-induced properties such as coherent and incoherent nonlinear optical properties and would rather refer to a recently published dedicated review paper (Golovan and Timoshenko 2013). One of the main features of birefrigent mesoporous silicon is the possibility of phase matching for second and even third harmonics generation.

## 8.2 THIN FILM OPTICAL PROPERTIES

### 8.2.1 COLOR AND OPTICAL THICKNESS

The first optical property of microporous silicon observable with naked eyes is its color. Indeed, when anodizing at low current densities below the electropolishing critical current density, Uhlir (1956) and Turner (1958) referred to a homogenous shiny dark or brown surface different from the metallic aspect of bulk silicon after polishing. Figure 8.7 shows examples of single and multi-layered PSi structures, evidencing the interferometric definition of the colors achievable using PSi optical thickness engineering as described later.

For periodic multilayered structures, the color is given by the so-called optical diffraction peak corresponding to the wavelength where the maximum of light is reflected:

(8.4) $$2nd \cos \theta = \lambda_{max}$$

with $n$ the average refractive index and $d$ the period ($d_1 + d_2$). The formula is the same as Bragg's law except that in optics, $\theta$ is the incident angle with respect to the normal to the sample. A multi-layered structure can be seen as a stack of optical interfaces. Each interface will reflect light and if all reflected waves are in phase, their respective amplitudes will sum up to give rise to almost

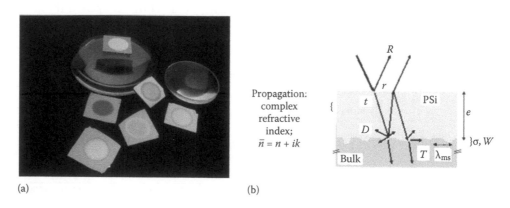

(a)                                                            (b)

**FIGURE 8.7**    Colorful porous silicon: (a) Examples of thin films and multilayered structures. (Courtesy of Peter Reece.) (b) Schematic view of the light interaction with a PSi thin film. While microporous thin films can be described as homogenous material of well-defined thickness and complex effective refractive index, light scattering at the PSi/bulk interface as well as large-scale layer thickness fluctuations have to be included in the modeling of the thin film reflectivity. (Data extracted from Lerondel et al., *Appl. Phys. Lett.* 71(2), 196, 1997b.)

100% reflectivity. The optimum phase condition is obtained if both optical thicknesses are equal to a quarter of the incident wavelength. For a typical refractive index modulation in mesoporous silicon ($n_1 = 2$ and $n_2 = 1.5$), 14% of the incident amplitude is reflected at each interface leading to a maximum reflectivity of 99.6% for only 10 periods.

It is amazing noting here that almost 100% reflectivity is obtained thanks to interferences (different from metallic reflection) and therefore require transparent materials.

To explain the color of a single layer of PSi on bulk silicon, the situation is in fact more complicated, yet the optical system is simpler as shown in Figure 8.7b. It consists of two interfaces separated by a distance of the order of the wavelength. If the material were transparent, one would expect a maximum in reflection when the front and back reflected waves are in phase, that is, for each wavelength $\lambda_k$ so that

$$(8.5) \qquad\qquad 2n_{PSi}d \cos\theta = k\lambda_k$$

Note the similitude with the Bragg relation, while replacing the average refractive index $n$ with the PSi thin film index $n_{PSi}$. It is also worth noting that the maximum of the fringe is fixed by the reflectivity of the substrate: 33% in the red part of the visible spectrum. Because PSi, as it was already discussed, sees its absorption increase as the wavelength gets closer to UV, the fringe maximum will decrease until it reaches a value equal to the normal incident Fresnel coefficient $(n_{PSi} - 1)^2/(n_{PSi} + 1)^2$, that is, around 5% for a typical nanoporous silicon film of 70% porosity ($n_{PSi} = 1.6$). The reflectivity will further increase down to the UV but the eye is less sensitive there. Such a spectrum is shown in Figure 8.8 together with a visible scale bar. This spectrum will be further discussed, especially the effect of light scattering at the porous silicon/bulk interface, which can be neglected as far as the color analysis is concerned. Because of the reflectivity decrease toward the UV, the color of the sample should be mainly determined by the red part of the spectrum. However, because the eye is more sensitive in the green, the final color of thick samples will lie somewhere between dark yellow or brown. If the sample thickness decreases the number of peaks and the absorption will decrease. Therefore, the sample will have an increasingly better-defined color from red to blue for the thinnest ones. As an example, a typical reflectivity spectrum of a 200-nm thick nanoporous film of 57% of porosity will exhibit three peaks—370, 440, and 800 nm. It will therefore appear blue as shown in Figure 8.7.

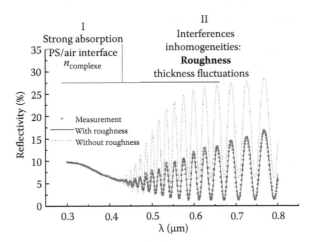

**FIGURE 8.8** Typical reflectivity spectrum of a nanoporous silicon thin film of 71% porosity and 4 µm thickness (dots). The spectrum can be divided into two zones—a metallic-type reflection at small wavelengths and Fabry-Perot cavity type reflection (fringes) at larger wavelengths. To account for the reflectivity contrast and amplitude in the fringes area, a contribution of both large-scale and small-scale layer thickness fluctuations called, respectively, waviness and roughness has to be taken into account. The effect of light scattering is illustrated by comparing spectra obtained for samples with (solid line) and without (small dots) interface roughness. A roughness amplitude (rms) of 40 nm as measured by profilometry was used. A contribution due to large-scale layer thickness fluctuations (waviness) of 0.5% of the initial film thickness was also introduced. While waviness induces incoherent effects, interface roughness induces light scattering. (Data extracted from Lerondel et al. 1997a.)

This color tuning property of thin PSi films strongly suggests that they can be used as sensors. By drying the sample after anodization, one could easily observe the color change. The idea of using a PSi thin film as an optical sensor was indeed demonstrated in 1997 (Lin et al. 1997).

### 8.2.2 SPECULAR REFLECTION AND DIRECT TRANSMISSION

Reflection and light scattering have already been discussed in detail (Lerondel et al. 1997). If it was clear from the early reflectance measurements of Pickering in 1984 that PSi can be seen as a homogeneous thin film with an effective refractive index, it took awhile to get a clear figure of what is really measured in reflection when looking at PSi thin films. A typical reflection spectrum for p-type nanoporous silicon sample is shown in Figure 8.8. The spectrum can be divided into two parts due to a rapid increase of the absorption as the function of the wave number. In the visible, part I, one observes interferences. By analyzing the position and the contrast of these interferences (fringes), both the refractive index and the absorption can be retrieved. However, it was shown here that the absorption was not the only source of optical losses unless one would accept that the absorption in PSi is higher than in bulk silicon. Indeed, the reflection spectrum is well accounted for by using a homogenous thin film with a given complex refractive index and a given thickness (here 4 µm), taking into account two kinds of layer thickness fluctuations depending on their lateral size. Large-scale thickness fluctuations induce incoherent effect (intensity averaging) whereas small-scale ones (wavelength scale) are responsible for light scattering (field averaging).

While waviness will simply lead to a decrease of the interference contrast, light scattering due to the interface roughness leads to a decrease of the average signal as shown by the spectra in Figure 8.8. To account for the losses, the Davies-Bennett relation has been introduced in the reflectivity calculations, taking into account the attenuation of the reflected amplitude. As already mentioned, the roughness effect has to be taken into account using a coherent contribution at the micron scale because at this scale any source of light will be coherent. To quantify the effect for a 5-µm thick sample of 70% of porosity, this is more than 50% of the light at 633 nm that is scattered at the PSi/silicon interface. The formalism was later extended to account for the attenuation in the transmitted amplitude (Lerondel et al. 2000), which is needed to account for the reflectivity of multilayered structures and especially rugate filters (Foss 2005). It is worth noting that the latter formalism developed to calculate the Fresnel coefficient of a rough interface in the case of PSi thin films has finally been of a great interest for other systems such as radar waves (Rodriguez-Alvarez et al. 2009), LEDs (Matioli and Weisbuch 2011), and solar cells (Law et al. 2008).

### 8.2.3 LIGHT SCATTERING

With pore and silicon skeleton sizes much smaller than the wavelength of light (even in the UV), light scattering is not expected to be very efficient in microporous silicon. Indeed, to account for the reflectivity of nanoporous and mesoporous silicon films as shown before, the material was considered homogeneous and no contribution of volume scattering was needed. However, the latter point needed to be confirmed by measuring on purpose the light scattering efficiency from the films. This was done in 1996 (Lerondel et al. 1996). The study was also motivated by a simple observation made by naked eyes. After dissolution of the nanoporous silicon film, it was noticed that the back interface (PSi/silicon interface) had a milky aspect depending on the porous layer thickness (the thicker the film, the more milky would be the back interface). Figure 8.9 summarizes the results obtained by Lerondel et al. (1997b).

Dealing with a nanostructured porous thin film, there are three possibilities for light scattering to occur: air/PSi interface, volume or inner surface of the material, and finally the PSi/bulk Si interface.

Data obtained on thick PSi films, either nano- or mesoporous, exactly follow the Fresnel Law, strongly suggesting that there is no scattering at the air/PSi interface nor in volume. Consequently, the antireflective effect observed on porous silicon by Chen et al. (1994) is simply due to the very low refractive index of the material.

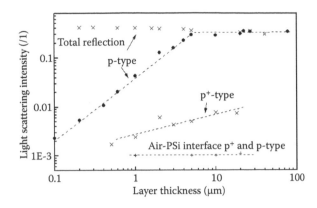

**FIGURE 8.9**    Light scattering in microporous silicon. Measurements performed using an illumination wavelength of 457 nm. (From Lerondel et al., *Properties of Porous Silicon*, INSPEC, London, UK, p. 241. Copyright 1997: The Institution of Electrical Engineers. With permission.)

These observations have been later confirmed for mesoprous silicon by a modeling using the vector radiative theory considering highly doped porous silicon as a uniaxial effective medium consisting of two components—hollow cylindrical silicon shells and free cylindrical pores (Abouelsaood 2002). In the case of p-type nanoporous silicon, the nanometric size of the crystallite leads to an absorption cross-section of 0.12% and a negligible (not measurable) scattering cross-section (Le Cunff 2015).

On the other hand, in the case of p-type nanoporous silicon, one observes that the scattering efficiency at the PSi/bulk Si interface increases linearly with the layer thickness before saturating at a level corresponding to the Fresnel reflection coefficient. Light is redistributed but the amount of light being back scattered is limited by the amount of reflected light.

Light scattering was found to originate from interface roughness as probed by profilometry (Lerondel et al. 1997). The intensive study of the interface has shown that this roughness is isotropic in the layer plane. The roughness originated from fluctuations in the electrochemical dissolution front and is therefore very sensitive to the current density and the electrolyte viscosity. The latter property was found to be very efficient to reduce these fluctuations by anodizing at low temperature (Setzu et al. 1998). Interface fluctuations are very sensitive to the doping level of the initial substrate. While for nanoporous silicon one has to deal with roughness (layer thickness fluctuations with a periodicity in the order of the micron), for mesoporous silicon, one has to deal with waviness, that is, layer thickness fluctuations, with a periodicity in the order of 200 µm, which turns out to be anisotropic-forming concentric rings. Amazingly, the root mean square amplitudes of both fluctuations are comparable. The difference in the periodicity justifies that two distinct coherent and incoherent contributions have to be taken into account to account for the fringes contrast in the reflectivity (cf. Figure 8.8).

### 8.2.4 ENGINEERED OPTICAL THICKNESSES

The ability to tune the refractive index as a function of the current density (porosity), together with the possibility of getting thin films thanks to the existence of an anodization front, makes microporous silicon a unique material for optical thickness engineering. Indeed, starting from single layer structures to bi-layers and then mutlilayered structures, numerous optical components have been demonstrated. Early structures include passive and active waveguides (Araki et al. 1996 and Loni et al. 1996, respectively), Bragg mirrors (Berger et al. 1994; Vincent 1994), luminescent microcavities (Pellegrini et al. 1995), Rugate filters (Berger et al. 1997), and lateral superlattices (Lerondel et al. 1997c). Interestingly, one can obtain structures either directly integrated on silicon or free-standing. Since the 90s, there has been a lot of interest in refractive index modulation in mesoporous silicon and its numerous related applications especially in sensing (cf. *Porous Silicon: Biomedical and Sensor Applications*). More advanced structures have been demonstrated and will be discussed further (cf. also Section 8.2.5 on waveguiding properties and Section 8.2.6 on omnidirectionality).

For in-plane or surface photon confinement, electrochemical anodization was also combined with standard reactive ion etching after PSi formation as recently demonstrated for planar mesoporous photonic crystals supporting slow bloch surface waves (Jamois et al. 2010). Alternatively, it is worth mentioning that the porous nature of the thin film makes direct photostructuring by means of photo-(electro)chemistry also possible, which in turn can lead to volume photostructuring depending on the light penetration depth in the material (Lerondel et al. 1996b).

Figures 8.10 and 8.11 show a few examples of structures with engineered optical thicknesses emphasizing, respectively, early works and state-of-the-art structures (cf. the legend for references).

Actually, one should better use the expression "beyond state-of-the-art structures" when referring to multilayered structures with more than 100 stacked layers (Ghulinyan et al. 2003). The current modulation technique (cf. insert in Figure 8.11a) is the only technique allowing us to realize such structures. Not only are optical superlattices, such as coupled microcavities, unique samples, they also have made possible in the year 2000 the observation of numerous physical phenomena in analogy to the electron behavior in a crystal. These effects include photon Bloch oscillations (cf. Figure 8.11d), Zener tunneling of light waves, slow modes, and Anderson localization of light (for a review, see Ghulinyan and Pavesi 2008).

The versatility of PSi in making optical elements is no longer a question, and the same goes for the rapidity and simplicity of the electrochemical anodization as a fabrication process. Nevertheless, questions may still be raised 20 years after the first demonstration of multilayered porous silicon and one is about the quality of the structures. We will limit ourselves here to a few features and numbers.

Generally speaking, besides the refractive index modulation and optical absorption, the quality of refractive index modulated structures is limited by optical thickness fluctuations, which can arise either in the vertical direction (anodization direction) or laterally. In PSi, both are actually relevant.

As far as vertical optical thickness variations are concerned, the introduction of etch stops has been shown to be very efficient to allow for the HF regeneration at the dissolution front (Billat et al. 1997), and thus compensate for the increasing HF scarcity leading to an increase of the porosity as a function of the dissolution time. Indeed, Bruyant et al. (2003) reported for a periodicity of 354 nm, a dispersion in the layer thickness as small as ±10 nm as evaluated from the Fourier

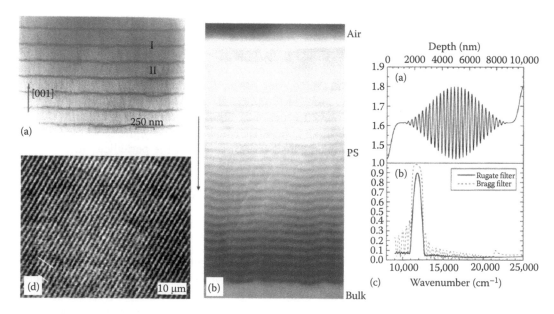

**FIGURE 8.10** Among the first examples of optical thickness engineering in porous silicon: (a) "Type II" superlattice obtained from highly doped silicon wafers. (Reprinted with permission from Berger et al., *J. Phys. D: Appl. Phys.* 27, 1333, 1994. Copyright 1994. American Institute of Physics.) (b) Highly luminescent superlattice obtained from low-doped p-type silicon wafers. (From Lerondel et al., *Mat. Res. Soc. Symp. Proc.* 452, 711, 1996b. Copyright 1997: Materials Research Society. With permission.) (c) Refractive index profile and reflectivity of a rugate filter made from highly doped silicon. (Reprinted from *Thin Solid Films* 297(1), Berger et al., 237, Copyright 1997, with permission from Elsevier.) (d) Lateral superlattices. (From Lerondel et al., *Mat. Res. Soc. Symp. Proc.* 452, 631, 1996a. Copyright 1997: Materials Research Society. With permission.)

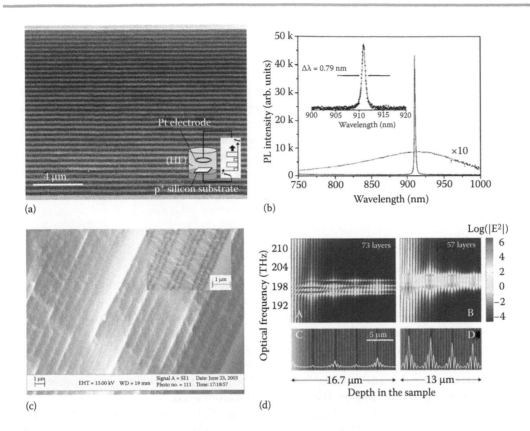

**FIGURE 8.11**  Current state-of-the-art optical thickness engineering in PSi: (a) chirped omnidirectional mirror. (Reprinted with permission from Bruyant et al., *Appl. Phys. Lett.* 82(19), 3227, 2003. Copyright 2003. American Institute of Physics.) (b) Subnanometer line width luminescent microcavity. (Reprinted with permission from Reece et al., *Appl. Phys. Lett.* 81(26), 4895, 2002. Copyright 2003. American Institute of Physics.) (c) Coupled superlattices for the observation of optical Bloch modes. (Reprinted with permission from Agarwal et al., *Phys. Rev. Lett.* 92, 097401, 2004. Copyright 2004 by the American Physical Society.) (d) Coupled superlattices for the observation of slow modes. (Reprinted with permission from Ghulinyan et al., *Appl. Phys. Lett.* 88, 241103, 2006. Copyright 2006. American Institute of Physics.)

transform of the SEM image. Another impressive example is the report of free-standing coupled microcavities corresponding to more than 109 stacked layers (Ghulinyan et al. 2003) where the drift compensation was confirmed by measuring both transmission and reflection (resonances in transmission are expected to be blueshifted if there is no compensation).

Let us now address the lateral optical thickness variations in the case of multilayered structures.

As discussed previously in Section 8.2.2, for a single layer, the lateral thickness fluctuations were found for nanoporous silicon to be less than 1.5% (0.5% for large-scale fluctuation waviness and 1% for roughness). For mesoporous silicon, values are even smaller as the roughness becomes negligible.

Interestingly, the amplitude of these lateral layer thickness fluctuations can be further decreased by anodizing at low temperatures, as first proposed for p-type nanoporous silicon to reduce the interface roughness (Setzu et al. 1998). The process was later extended to mesoporous silicon to reduce the interface waviness (Reece et al. 2002). In both cases, drastic improvements have been observed. In the case of nanoporous silicon mulitlayers, a maximum reflectivity of 99.5% was measured for DBR using the cavity ring down spectroscopy. For microcavities, a Q factor of 150 was found (Setzu et al. 2000). Surprisingly, the value at that time was as high as for mesoporous microcavities, until it was discovered that the limiting parameter for the Q factor in mesoporous microcavities was the interface waviness (Lerondel et al. 2003). Subnanometer line width in the NIR, leading to a Q factor of more than 1500, was then reported (Reece et al. 2002), which is comparable to structures obtained by molecular beam epitaxy or molecular chemical vapor deposition techniques. Interestingly, there is still some space for improvement. Taking into account the ultimate source of optical losses in mesoporous silicon structures, namely the free carrier absorption in the lower porosity layer, which is about 30 cm⁻¹ as reported for samples of similar porosity (Chan et al. 1996), a sub-nanometer line width of 0.2 nm has been observed leading to a Q factor of more than 7000 (Lerondel et al. 2003).

Besides quality, the structure stability is also an issue due to the oxidation of PSi in air as a function of time. For nanoporous multilayered structures, a blueshift of over 10% of the central wavelength can be observed with a stabilization time constant of 45 days typically. As for mesoporous structures, the aging effect is less pronounced in proportion, as expected from the reduced specific surface, and becomes visible after a year typically. To overcome this effect, a surface chemical modification via hydrosilylation has been proposed and was found to be very efficient even after 36 months for thick structures such as coupled microcavities (Gelloz et al. 2008; Ghulinyan et al. 2008).

### 8.2.5 WAVEGUIDING OPTICAL PROPERTIES

Among the possibilities offered by optical thickness engineering in PSi, waveguiding is clearly of special interest. In terms of refractive index modulation and putting microporous silicon in the context of integrated optics, one potentially has access on the same platform to both glass technology such as ion exchange with very small $\Delta n$ (<0.01) and silicon on insulator (SOI) technology with very large $\Delta n$ (>2). In addition, due to the material being porous, not only the tail of the waveguided mode will be sensitive to the environment, but so will the mode be. In other words, not only the cladding index is modified but also is the guide index. This property is definitely of great interest for optical sensing (cf. the two examples given next).

After the first demonstration of light waveguiding in PSi in 1996 (Loni et al. 1996), there has been a rather limited number of reports on PSi waveguides until the year 2000, although different PSi-based approaches have been proposed: epitaxial growth onto PSi, ion implantation, multilayered porosification, and, finally, oxidation (see for one of the first reviews on the topic, Arrand et al. 1998). The development of waveguiding structures faces two main issues: the fabrication requiring micro- and nanofabrication and the characterization, which is made difficult because of the evanescent nature of the guided modes. Figure 8.12a illustrates a typical M-line measurement configuration for planar waveguides where the light is injected via a prism. When a mode is excited, energy is transferred into the waveguide leading to a minimum (dark line) in reflectivity. Figure 8.12b shows the angular dependence response of a nanoporous planar waveguide consisting of a guiding and a cladding layer (Mihalcescu et al. 1997). Three distinct angular ranges can be observed corresponding, respectively, to the fringes region, guiding region, and total reflection.

From the spectrum analysis, optical parameters such as the cladding and guide refractive indexes $n_c$ and $n_g$ and the guide thickness, $W$, can be retrieved. Note that the refractive indexes can

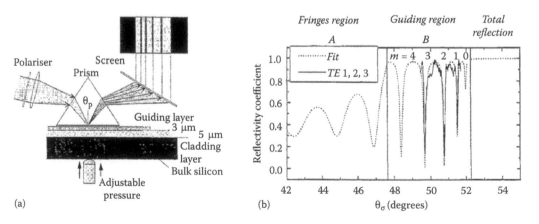

**FIGURE 8.12** Light propagation in PSi planar waveguide: (a) M-line technique and (b) typical reflectivity angular dependence. (Reprinted from *Thin Solid Films* 297, Mihalcescu et al., 245, Copyright 1997, with permission from Elsevier.)

be different between the TE and TM polarizations (cf. Section 8.2.3 on the optical anisotropy). To confine light waves in dielectric films, two conditions need to be fully met: (1) total internal reflection at both interfaces (air/guide and guide/cladding) and (2) a propagation constant ($\beta = kn \cdot \sin\theta$) obeying a discrete phase condition. These conditions can be summarized as follows:

$$(8.6) \qquad kn_g > \beta > kn_c \quad \text{and} \quad W\sqrt{(kn_g)^2 - \beta^2} - \Psi_{gc} - \Psi_{ga} = m\pi, \quad m = 0,1,2,3\ldots$$

with $k = 2\pi/\lambda$, $\lambda$ being the wavelength in vacuum and $\Psi_{gc}$ and $\Psi_{ga}$ half the reflection phases at the guide-cladding and guide-air interfaces, respectively (both are polarization dependent). These express the constructive interference condition (the same as in Section 8.3.1 but for total internal reflection). Measuring for each polarization at least three modes allows the retrieval of $n_c$, $n_g$, and $W$. The results have been presented in Figure 8.6a. Typical values for a TM polarization and as-formed low-doped nanoporous silicon waveguides are 1.5965 for $n_c$ and 1.504 for $n_g$. These values correspond to porosities of 65% and 70%, respectively. For almost fully oxidized samples (10% of remaining silicon as estimated from the EMA approximation), effective refractive indexes can be reduced down to 1.3456 for $n_c$ and 1.271 for $n_g$.

Another important parameter for waveguiding is the propagation loss. From the line width analysis, the losses can also be retrieved. In this case, the losses were estimated to be 100 dB/cm, which is higher than what was reported at that time, typically 10 dB/cm, but in the NIR (Loni et al. 1996). Loni et al. (1996) also pointed out the importance of light scattering at the cladding-core interface. It was confirmed later on by Pirasteh et al. (2007). Currently, the minimum losses ever reported are 0.5 dB/cm for oxidized mesoporous waveguides (Pirasteh et al. 2007). As pointed out by the authors and previously discussed, low temperature anodization could be used to further reduce the waviness amplitude, which was found here to be the limiting factor in the NIR range for planar mesoporous microcavities (cf. Section 8.3.4).

While the M-line technique is very powerful, it does not allow for a direct observation of the light propagation in the waveguide. This kind of observation has been made possible thanks to the development in the past 20 years of the so-called scanning near-field microscopies, and more specifically the scanning optical near-field microscopy (SNOM). SNOM and especially the phase sensitive type of SNOM have been found to be very powerful for modal analysis in the case of integrated optical waveguides. Not only does it allow light propagation to be observed, but it also allows a quantitative modal analysis (Stefanon et al. 2005).

Figure 8.13 shows the first and only thus far reported direct observation of light propagation in a mesoporous silicon waveguide as observed through SNOM (Bruyant et al. 2005). Typical field (A) and phase (B) mappings allow light propagation to be observed, where one can clearly evidence modes beating (interferences) occurring because of the excitation of several modes. The Fourier transform of the phase image (cf. the top graph) reveals that three modes are excited. Refractive indexes of these modes retrieved from the peak position are summarized in the inserted table. In the case of an oxidized mesoporous waveguide, effective refractive indexes as low as 1.22 were observed. The lowest value of 1.18 most probably corresponds to a substrate mode.

As an example of a PSi waveguiding platform, an integrated nanoporous Mach-Zehnder interferometric waveguide sensor was reported (Kim and Murphy 2013). Nanoporous silicon single mode integrated waveguides were successfully fabricated by electrochemical etching and direct-write laser oxidation as previously introduced (Rossi et al. 2001). Using isopropanol as a test analyte, a sensitivity of 13,000 rad/RIU-cm was reported. As another example of a PSi waveguiding platform, one may note the recent report of a ring resonator fabricated by RIE etching of a PSi planar waveguide, with a bulk detection sensitivity of 380 nm/RIU (Rodriguez et al. 2015).

## 8.2.6 OMNIDIRECTIONALITY

We will end this review with another unique property of PSi as a photonic material, namely omnidirectionality. Omnidirectionality is the ability to reflect light no matter the angle of incidence and for both TE and TM polarizations. For a long time, complete rejection of light has been thought to be only achievable using 3D photonic crystals with a full bandgap, meaning a compact

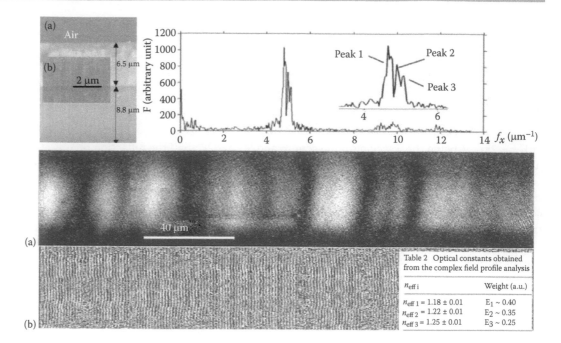

| Table 2 Optical constants obtained from the complex field profile analysis | |
| --- | --- |
| $n_{\mathrm{eff}\,i}$ | Weight (a.u.) |
| $n_{\mathrm{eff}\,1} = 1.18 \pm 0.01$ | $E_1 \sim 0.40$ |
| $n_{\mathrm{eff}\,2} = 1.22 \pm 0.01$ | $E_2 \sim 0.35$ |
| $n_{\mathrm{eff}\,3} = 1.25 \pm 0.01$ | $E_3 \sim 0.25$ |

**FIGURE 8.13** Light propagation in PSi planar waveguide as probed by near-field optical microscopy: (a) SEM view of the planar bilayer. (a and b) Amplitude and phase images obtained using a phase-sensitive scanning near-field microscope. Top graph Fourier transform of the complex field evidencing three peaks corresponding to three modes of different propagation constant and respective amplitude; cf. the inserted table. (Bruyant et al.: *Phys. Stat. Sol. (a)*. 2003. 202(8). 1417. Copyright Wiley-VCH Verlag GmbH & Co. KGaA. Reproduced with permission.)

structure and high refractive index difference. This condition is necessary if light is emitted by the material itself (high refractive index materials), but if light is impinging on the structure while coming from a low refractive index material like air, a 1D multilayered structure is enough to observe omnidirectionality (Fink et al. 1998; Yablonovitch 1998). However, omnidirectionality is only possible if the brewster angle remains outside the light cone and the bandgap is sufficiently large to avoid an overlap of the transmission bands.

Figure 8.14 shows the photonic band structure calculated using a refractive index modulation typical for mesoporous multilayered materials with $n_{\mathrm{L}} = 1.51$ and $n_{\mathrm{H}} = 2.56$, corresponding to a porosity modulation of 76% and 33%, respectively. As evidenced by the black rectangle, an omnidirectional bandgap should be observed. It was indeed demonstrated in 2003 (Bruyant et al. 2003). Figure 8.15a shows an SEM view of the first reported PSi omnidirectional mirror: a 50-layer periodic structure with a periodicity of 350 nm. As shown by the reflectivity spectra

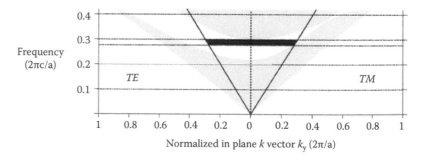

**FIGURE 8.14** Photonic band structure calculated for a typical refractive index modulation in mesoporous silicon, that is, low and high refractive indexes of 1.51 and 2.56, respectively. $k_y$ represents the in-plane component of the incident light wave vector. As indicated by the arrow, the Brewster angle lies outside the light cone for an ambient medium like air. The light cone in air is evidenced by the solid straight lines.

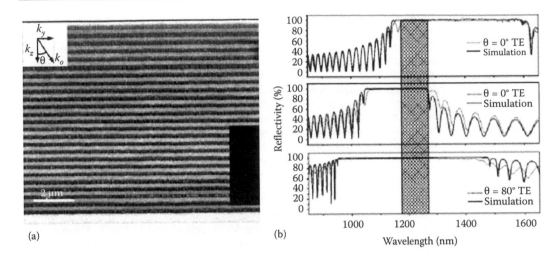

**FIGURE 8.15**    Mesoporous silicon omnidirectional mirror: (a) SEM image of a 50-layer structure with a periodicity of 350 nm and (b) optical response (reflectivity) as a function of the incident angle and the polarization evidencing the omnidirectional bandgap (zebra zone). (Reprinted with permission from Bruyant et al., *Appl. Phys. Lett.* 82(19), 3227, 2003. Copyright 2003. American Institute of Physics.)

obtained at 0° and 80° for the TM and TE polarizations, a reflexion of nearly 100% is achieved for all the wavelengths in the zebra zone no matter the angle of incidence or the polarization (cf. Figure 8.15b).

Prior to this work, omnidirectionality had only been observed using heterostructures based on two different materials for the high and low refractive index layers. Materials include $SiO_2$/$TiO_2$, $TiSu/SiO_2$, and $ZnSe/Na_3AlF_6$ (see references 3 to 8 in Bruyant et al. 2003). Unfortunately for semiconductor materials like AlGaAs, the minimum achievable refractive index is too high and the $\Delta n$ is not large enough to avoid the overlap of the transmission bands. Optimum values are, respectively, $n_L \sim 1.49 n_0$ and $n_H > 2.25 n_0$, $n_0$ being the ambient medium refractive index, usually air (Southwell 1999).

Besides the ability of making monolithic structures, here "all-silicon" structures, the porosity modulation technique allows for the fine-tuning of the optical thickness, that is, refractive index, and physical thickness. As an example of the fine-tuning, a second structure with a low refractive index, non-optimized but presenting a linear increase of the periodicity, has also been realized (chirped structure cf. Figure 8.11a). Compared to periodic structures, the spectral range corresponding to the full rejection band has been multiplied by 3, that is, 350 nm, which is one of the largest omnidirectional bandgaps ever reported in the visible-NIR spectral range.

# REFERENCES

Abouelsaood, A.A. (2002). Modeling light scattering from mesoporous silicon. *J. Appl. Phys.* 91, 2753.

Agarwal, V., del Rio, J.A., Malpuech, G. et al. (2004). Photon bloch oscillations in porous silicon optical superlattices. *Phys. Rev. Lett.* 92, 097401.

Araki, M., Koyama, H., and Koshida, N. (1996). Fabrication and fundamental properties of an edge emitting device with step-index porous silicon waveguide. *Appl. Phys. Lett.* 58(21), 2999.

Arrand, H.F., Benson, T.M., Sewell, P. et al. (1998). The application of porous silicon to optical waveguiding technology. *IEEE J. Selected Topics Quant. Electron.* 4(6), 975–982.

Aspnes, D.E., Theeten, J.B., and Hottier, F. (1979). Investigation of effective-medium models of microscopic surface roughness by spectroscopic ellipsometry. *Phys. Rev. B* 20, 3292.

Berger, M.G., Dieker, C., Thonissen, M. et al. (1994). Porosity superlattices: A new class of Si heterostructures. *J. Phys. D: Appl. Phys.* 27, 1333.

Berger, M.G., Frohnhoff, S., Theiss, W., Rossow, U., and Münder, H. (1995). Porous Si: From single porous layers to porosity superlattices. In: Vial J.C. and Derrien J. (eds.), *Porous Silicon Science and Technology*. Springer, Berlin, Heidelberg, pp. 345–355.

Berger, M.G., Arens-Fischer, R., Thönissen, M. et al. (1997). Dielectric filters made of PS: Advanced performance by oxidation and new layer structures. *Thin Solid Films* 297(1), 237–240.

Bergman, D.J. (1978). The dieletric constant of a composite material—A problem in classical physics. *Phys. Rep.* 43(9), 377–407.

Billat, S., Thönissen, M., Arens-Fischer, R., Berger, M.G., Krüger, M., and Lüth, H. (1997). Influence of etch stops on the microstructure of porous silicon layers. *Thin Solid Films* 297(1), 22–25.

Bruyant, A., Lerondel, G., Reece, P.J., and Gal, M. (2003). All-silicon omnidirectional mirrors based on one-dimensional photonic crystals. *Appl. Phys. Lett.* 82(19), 3227.

Bruyant, A., Stefanon, I., Lerondel, G. et al. (2005). Light propagation in a porous silicon waveguide: An optical modes analysis in near-field. *Phys. Stat. Sol. (a)* 202(8), 1417–1421.

Canham, L.T. (1993). Progress toward crystalline-silicon-based light emitting diodes. *MRS Bull.* 18(7), 22–28.

Canham, L.T., Cullis, A.G., Pickering, C., Dosser, O.D., Cox, T.I., and Lynch, T.P. (1994). Luminescent anodized silicon aerocrystal networks prepared by supercritical drying. *Nature* 368, 133–135.

Chen, L.Y., Hou, X.H., Huang, D.M. et al. (1994). Optical study of photon-trapped porous silicon layer. *Jpn. J. Appl. Phys.* 33, 1937–1943.

Cody, G.D., Tiedje, T., Abeles, B., Brooks, B., and Goldstein, Y. (1981). Disorder and the optical-absorption edge of hydrogenated amorphous silicon. *Phys. Rev. Lett.* 47(20), 1480.

De Filippo, F., De Lisio, C., Maddalena, P., Solimeno, S., Lerondel, G., and Altucci, C. (2002). Application of VUV laser harmonic radiation to the measurement of porous silicon dielectric function. *Opt. Lasers Eng.* 37(5), 611–620.

Diener, J., Künzner, N., Kovalev, D. et al. (2001). Dichroic Bragg reflectors based on birefringent porous silicon. *Appl. Phys. Lett.* 78, 3887.

Fink, Y., Winn, J.N., Fan, S. et al. (1998). A dielectric omnidirectional reflector. *Science* 282, 1679–1682.

Foss, S.E. (2005). Graded optical filters in porous silicon for use in MOEMS applications. PhD thesis, University of Oslo.

Gelloz, B., Murata, K., Ohta, T. et al. (2008). Stabilization of porous silicon free-standing coupled optical microcavities by surface chemical modification. *ECS Trans.* 16(3), 211–219.

Ghulinyan, M. and Pavesi, L. (2008). Porous multilayers as a dielectric host for photons manipulation. *ECS Trans.* 16(3), 307–321.

Ghulinyan, M., Oton, C.J., Gaburro, Z., Bettotti, P., and Pavesi, L. (2003). Porous silicon free-standing coupled microcavities. *Appl. Phys. Lett.* 82, 1550.

Ghulinyan, M., Gelloz, B., Ohta, T., Pavesi, L., Lockwood, D.J., and Koshida, N. (2008). Stabilized porous silicon optical superlattices with controlled surface passivation. *Appl. Phys. Lett.* 93, 061113.

Golovan, L.A. and Timoshenko, V.Y. (2013). Nonlinear-optical properties of porous silicon nanostructures. *J. Nanoelectron. Optoelectron.* 8, 223–239.

Golovan, L.A., Kashkarov, P.K., and Timoshenko, V.Y. (2007). From birefringence in porous semiconductors and dielectrics: A review. *Crystallography Rep.* 52(4), 672–685.

Hybertsen, M.S. (1994). Absorption and emission of light in nanoscale silicon structures. *Phys. Rev. Lett.* 72(10), 1514.

Jamois, C., Li, C., Orobtchouk, R., and Benyattou, T. (2010). Slow Bloch surface wave devices on porous silicon for sensing applications. *Photonics Nanostruct. Fundam. Appl.* 8, 72–77.

Kim, K. and Murphy, T.E. (2013). Porous silicon integrated Mach-Zehnder interferometer waveguide for biological and chemical sensing. *Opt. Express* 21(17), 19488–19497.

Koshida, N. (1993). Visible luminescent properties of porous silicon. *Nonlinear Opt. Principles Mater. Phenom. Dev.* 4(2) 143.

Koshida, N., Koyama, H., Suda, Y. et al. (1993). Optical characterisation of porous silicon by synchrotron radiation reflectance spectra analyses. *Appl. Phys. Lett.* 63(20), 2774.

Kovalev, D., Ben Chorin, M., Diener, J. et al. (1995). Porous Si anisotropy from photoluminescence polarization. *Appl. Phys. Lett.* 67, 1585.

Kovalev, D., Polisski, G., Ben Chorin, M., Diener, J., and Koch, F. (1996). Temperature dependance of the absorption coefficient of porous silicon. *J. Appl. Phys.* 80(10), 5978.

Kux, A. and Ben Chorin, M. (1995). Photoluminescence excitation spectroscopy of porous silicon. *Mater. Res. Soc. Symp. Proc.* 358, 447.

Law, M., Beard, M.C., Choi, S., Luther, J.M., Hanna, M.C., and Arthur, J.N. (2008). Determining the internal quantum efficiency of PbSe nanocrystal solar cells with the aid of an optical model. *Nano Lett.* 8(11), 3904–3910.

Le Cunff, L. (2015). Unpublished results.

Lerondel, G. and Romestain, R. (1997a). Reflection and light scattering in porous silicon. In: Canham L. (ed.), *Properties of Porous Silicon*. INSPEC, London, UK, pp. 241–246.

Lerondel, G. and Romestain, R. (1999). Fresnel coefficients of a rough interface. *Appl. Phys. Lett.* 74, 2740.

Lerondel, G., Thönissen, M., Setzu, S., Romestain, R., and Vial, J.C. (1996a). Holographic grating in porous silicon. *Mater. Res. Soc. Symp. Proc.* 452, 631.

Lerondel, G., Ferrand, P., and Romestain, R. (1996b). Holographic grating in porous silicon. *Mater. Res. Soc. Symp. Proc.* 452, 711.

Lerondel, G., Romestain, R., Madéore, F., and Muller, F. (1996c). Light scattering from porous silicon. *Thin Solid Films* 276(1), 80–83.

Lerondel, G., Romestain, R., and Barret, S. (1997b). Roughness of the porous silicon dissolution interface *J. Appl. Phys.* 81, 6171–6178.

Lerondel, G., Thönissen, M., Romestain, R., and Vial, J.C. (1997c). Porous silicon lateral superlattices. *Appl. Phys. Lett.* 71(2), 196.

Lerondel, G., Madeore, F., Romestain, R., and Muller, F. (2000). Direct determination of the absorption of porous silicon by photocurrent measurement at low temperature. *Thin Solid Films* 366, 216–224.

Lerondel, G., Reece, P., Bruyant, A., and Gal, M. (2003). Strong light confinement in microporous photonic silicon structures. *Mater. Res. Soc. Symp. Proc.* 797, W1.7.

Lin, V.S.-Y., Motesharei, K., Dancil, K.-P.S., Sailor, M.J., and Reza Ghadiri, M. 1997. A porous silicon-based optical interferometric biosensor. *Science* 278(5339), 840–843.

Loni, A., Canham, L.T., Berger, M.G. et al. (1996). Porous silicon multilayer optical waveguides. *Thin Solid Films* 276, 143–146.

Matioli, E. and Weisbuch, C. (2011). Direct measurement of internal quantum efficiency in light emitting diodes under electrical injection. *J. Appl. Phys.* 109, 073114.

Mazzoleni, C. and Pavesi, L. (1995). Application to optical components of dieletric porous multilayers. *Appl. Phys. Lett.* 67(20), 2983.

Mihalcescu, I., Lerondel, G., and Romestain, R. (1997). Porous silicon anisotropy investigated by guided light. *Thin Solid Films* 297, 245–249.

Mulloni, V. and Pavesi, L. (2000). Porous silicon microcavities as optical chemical sensors. *Appl. Phys. Lett.* 76, 2523.

Ookubo, N. (1993). Depth dependent porous silicon photoluminescence. *J. Appl. Phys.* 74(10) 6375.

Pellegrini, V., Tredicucci, A., Mazzoleni, C., and Pavesi, L. (1995). Enhanced optical properties in porous silicon microcavities. *Phys. Rev. B* 52, R14328.

Pickering, C., Beale, M.J., Robbins, D.J., Pearson, P.J., and Greef, R. (1984). Optical studies of the structure of porous silicon films formed in p-type degenerate and non-degenerate silicon. *J. Phys. C: Solid State Phys.* 17, 6535.

Pirasteh, P., Charrier, J., Dumeige, Y., Haesaert, S., and Joubert, P. (2007). Optical loss study of porous silicon and oxidized porous silicon planar waveguides. *J. Appl. Phys.* 101, 083110.

Reece, P.J., Lerondel, G., Zheng, W.H., and Gal, M. (2002). Optical microcavities with subnanometer linewidths based on porous silicon. *Appl. Phys. Lett.* 81(26), 4895.

Rodriguez, G.A., Hu, S., and Weiss, S.M. (2015). Porous silicon ring resonator for compact, high sensitivity biosensing applications. *Opt. Express* 23(6), 7111–7119.

Rodriguez-Alvarez, N., Rodriguez-Alvarez, N., Bosch-Lluis, X. et al. (2009). Soil moisture retrieval using GNSS-R techniques: Experimental results over a bare soil field. *IEEE Trans. Geosci. Remote Sens.* 47(1), 3616.

Rosenbauer, M., Finkbeiner, S., Bustarret, E., Weber, J., and Stutzmann, M. (1995). Resonantly excited photoluminescence spectra of porous silicon. *Phys. Rev. B* 51(16), 10539.

Rossi, A.M., Amato, G., Camarchia, V., Boarino, L., and Borini, S. (2001). High-quality porous-silicon buried waveguides. *Appl. Phys. Lett.* 78, 3003.

Sagnes, I., Halimaoui, A., Vincent, G., and Badoz, P.A. (1993). Optical absorption evidence of quantum size effect in porous silicon. *Appl. Phys. Lett.* 62(10), 1155.

Setzu, S., Lerondel, G., and Romestain, R. (1998). Temperature effect on the roughness of the formation interface of p-type porous silicon. *J. Appl. Phys.* 81, 6171–6178.

Smith, R.L. and Collins, S.D. (1992). Porous silicon formation mechanisms. *J. Appl. Phys.* 71(8), R1.

Southwell, W.H. (1999). Omnidirectional mirror design with quarter-wave dielectric stacks. *Appl. Opt.* 38, 5464–5467.

Stefanon, I., Blaize, S., Bruyant, A. et al. (2005). Heterodyne detection of guided waves using a scattering-type scanning near-field optical microscope. *Opt. Express* 13(14), 5553–5564.

Theiss, W. (1997). Optical properties of porous silicon. *Surf. Sci. Rep.* 29, 91–192.

Theiss, W., Henkel S., and Arntzen M. (1995). Connecting microscopic and macroscopic properties of porous media: Choosing appropriate effective medium concepts. *Thin Solid Films* 255, 177–180.

Thönissen, M., Berger, M.G., Theiß, W., Hilbrich, S., Krüger, M., Arens-Fischer, R., Billat, S., Lerondel, G., and Lüth, H. (1996). Depth Gradients in Porous Silicon: How to Measure Them and How to Avoid Them. *MRS Proceedings*, 452, 431 doi:10.1557/PROC-452-431.

Thönissen, M., Berger, M.G., Billat, S. et al. (1997). Analysis of the depth homogeneity of p-PS by reflectance measurements. *Thin Solid Films* 297, 92–96.

Turner, D.R. (1958). Electropolishing silicon in hydrofluoric acid solutions. *J. Electrochem. Soc.* 105, 402–408.

Uhlir, A. (1956). Electrolytic shaping of germanium and silicon. *Bell Syst. Tech. J.* 5, 333–347.

Vincent, G. (1994). Optical properties of porous silcion superlattices. *Appl. Phys. Lett.* 64(18), 2367.

Von Behren, J., Fauchet, P.M., Chimowitz, E.H., and Lira, C.T. (1997). Optical properties of free standing ultrahigh porosity silicon films prepared by supercritical drying. *Mater. Res. Soc. Symp. Proc.* 452, 565.

Wang, L., Wilson, M.T., Goorsky, M.S., and Haegel, N.M. (1992). Photoluminescence excitation spectroscopy (PLE) of porous silicon. *Mater. Res. Soc. Symp. Proc.* 229, 73.

Wolkin, M.V., Jorne, J., Fauchet, P.M., Allan, G., and Delerue, C. (1999). Luminescence in porous silicon quantum dots: The role of oxygen. *Phys. Rev Lett.* 82(1), 198.

Yablonovitch, Y. (1998). Engineered omnidirectional external-reflectivity spectra from one-dimensional layered interference filters. *Opt. Lett.* 3, 1648–1649.

Yang, Z., Zhu, D., Zhao, M., and Cao, M. (2004). The study of a nano-porous optical film with the finite difference time domain method. *J. Optics A: Pure Appl. Opt.* 6(6), 564–568.

# Thermal Properties of Porous Silicon

**9**

Pascal J. Newby

CONTENTS

## 9.1 INTRODUCTION

Bulk crystalline silicon is a good thermal conductor, with a thermal conductivity of 156 W/(m·K) (Glassbrenner and Slack 1964), which is comparable to that of copper (400 W/(m·K)). However, porous silicon has a low thermal conductivity, which can reach values up to 100–1000 times lower than that of crystalline silicon, depending on its morphology.

The low thermal conductivity of porous Si makes it an ideal material for thermal insulation in micro electro-mechanical systems (MEMS) (Lysenko et al. 2002), and indeed, a wide variety of devices have been fabricated using porous silicon in this way. This application of porous Si is covered in detail in Chapter 8 in Porous Silicon as a Material for Thermal Insulation in MEMS. Porous silicon has also been studied as a potential thermoelectric material (De Boor et al. 2012), which is discussed in Chapter 17 in *Porous Silicon: Opto- and Microelectronic Applications.*

The thermal properties of porous silicon must be known in order to design devices where it is used as thermal insulation and to tailor its morphology and geometry to the required performance. In addition, porous Si is used in RF devices for its low electrical conductivity (see Chapter 5, *Porous Silicon: Opto- and Microelectronic Applications*), and in this case its thermal properties should also be known to avoid any undesirable self-heating effects (Siegert et al. 2012).

In this chapter we will review the work that has been published regarding characterization of the main thermal properties of porous silicon, namely thermal conductivity, thermal diffusivity, and heat capacity. We will see how these properties are affected by the morphological properties of porous Si, as well as how its thermal properties can be modified by subsequent treatments. We will then analyze the causes of the differences in thermal properties between bulk crystalline and porous silicon, and finally discuss the work that has been carried out in order to model the thermal properties of porous silicon.

As we will see, a variety of methods has been used to characterize the thermal properties of porous silicon; however, we will not discuss the details of these methods, as this topic is covered in Chapter 5 of this book. This chapter focuses on porous silicon fabricated by electrochemical etching. However, we should note that the thermal properties of other forms of porous silicon have received much attention recently. These forms include so-called "holey silicon," which is formed by fabricating periodic holes in crystalline silicon, generally by photo- or e-beam-lithography, followed by plasma etching (Tang et al. 2010; Yu et al. 2010; Hopkins et al. 2011; Marconnet et al. 2012; Ma et al. 2014), and sintered nanocrystalline silicon (Wang et al. 2011).

## 9.2 MEASUREMENT OF THE THERMAL PROPERTIES OF POROUS SILICON

### 9.2.1 THERMAL PROPERTIES

The three main thermal properties that will be studied in this chapter are thermal conductivity, heat capacity, and thermal diffusivity. Thermal conductivity, λ, given in W/(m·K), describes how well a material conducts heat. Specific heat, $C_p$, given in J/(g·K), is the amount of heat required to increase the temperature of 1 g of a material by 1 degree. The third property, thermal diffusivity (α), depends on the two other properties:

(9.1)
$$\alpha = \frac{\lambda}{\rho C_p}$$

where ρ is the density of the material. $\rho C_p$ can be defined as volumetric heat capacity.

### 9.2.2 OVERVIEW OF THERMAL CONDUCTIVITY MEASUREMENTS

Numerous articles have published results of measurements of the thermal conductivity of porous silicon. Table 9.1 summarizes the measurement technique used and the important properties of

**TABLE 9.1    List of Articles Published with Results of Measurements of the Thermal Conductivity of Porous Silicon**

| References | Measurement Technique | Si Wafer Properties (Doping Type, Resistivity, Crystalline Orientation) | Post-Anodization Treatment | Morphology + Crystallite Size | Porosity + Layer Thickness |
|---|---|---|---|---|---|
| Lang et al. 1994; Drost et al. 1995 | Microsensors | p-type, 38–52 Ω.cm, (100) | As-anodized and $O_2$, 1 h, 300°C | Nano | 40% 10 μm |
| | | n-type, 1–2 Ω.cm, (100) | As-anodized and $O_2$, 1 h, 300°C | Nano + macro | 53% 10 μm |
| | | p-type, 10–18 mΩ.cm, (100) | As-anodized and $O_2$, 1 h, 300°C | Meso | 45% 10 μm |
| Gesele et al. 1997 | 3 omega | p-type, 0.2 Ω.cm, (100) | As-anodized | Nano, 1.7–4.5 nm | 64–89% 31–46 μm |
| | | p-type, 0.01 Ω.cm, (100) | As-anodized | Meso, 9 nm | 64% 21 μm |
| Benedetto et al. 1997a,b | Photoacoustic | p-type, 0.01 Ω.cm, (111) | As-anodized | Meso | 50, 60% 10, 23 μm |
| | | n-type, 10 Ω.cm, (100) | As-anodized | Nano + macro complex layer | 40% 75, 175 μm |
| Obraztsov et al. 1997 | Photoacoustic | p-type, 10 Ω.cm, (100) | Air for several weeks | 4–5 nm | 80% 15 μm |
| Calderón et al. 1997 | Photoacoustic | n-type, 1–5 Ω.cm, (100) | As-anodized | Macro + nano complex layer | |
| Bernini et al. 1999a, 2003; Lettieri et al. 2005 | Photoacoustic | n-type, 1 Ω.cm, (100) | As-anodized | Meso + macro complex layer | 40–72% 40–250 μm |
| Bernini et al. 1999b | Thermal wave interferometry | n-type, 1 Ω.cm | As-anodized | Meso + macro complex layer | 42, 57% 50–250 μm |
| Boarino et al. 1999 | Photoacoustic | p-type, 4–6 Ω.cm, (100) | As-anodized | Nano | 53–85% 2–48 μm |
| Lysenko et al. 1999b | Raman | p-type, 10–20 mΩ.cm, (100) | As-anodized and 1 h $O_2$, different temperatures | 8.3–7 nm | 38–74% 100 μm |
| Maccagnani et al. 1999; Amato et al. 2000 | Photoacoustic | p-type 0.02 Ω.cm, (100) | As-anodized and nitridation, 2 min, 1000, 1100°C | Meso | 55, 75% 26 μm |
| Lysenko and Volz 2000 | Scanning thermal microscopy | p-type 1–10 Ω.cm | | Nano | |
| | | p-type 10 mΩ.cm | | Meso | |
| Périchon et al. 2000 | Raman | p-type, 0.01–0.02 Ω.cm, (100) | As-anodized, and $O_2$, 1 h, different temperatures | Meso | 50, 70 5–100 μm |
| Bernini et al. 2001 | Optical pump-probe | p-type, 0.7–1.3 Ω.cm | As-anodized | Nano | 61–73% 19–38 μm |
| Shen and Toyoda 2002, 2003 | Photo-acoustic | p-type 5–15 Ω.cm, (100) | As-anodized | Macro- and meso-complex layer | 23–61% 27–30% |
| Bernini et al. 2005 | Thermal lens | p-type 1 Ω.cm | As-anodized and 3 weeks in ambient | Nano, 4.9 nm | 50% 18 μm |
| Lettieri et al. 2005 | Optical pump-probe | p-type, 0.2 Ω.cm, (100) | As-anodized | Nano | 61–79% 40–250 μm |

*(Continued)*

**TABLE 9.1 (CONTINUED)** **List of Articles Published with Results of Measurements of the Thermal Conductivity of Porous Silicon**

| References | Measurement Technique | Si Wafer Properties (Doping Type, Resistivity, Crystalline Orientation) | Post-Anodization Treatment | Morphology + Crystallite Size | Porosity + Layer Thickness |
|---|---|---|---|---|---|
| Kihara et al. 2005 | 3 omega | p type 10–30 Ω.cm, (100) | As-anodized, and oxidized by $O_2$ plasma | Nano, 3 nm | 55% 22 μm |
| Wolf and Brendel 2006 | Lock-in thermography | p-type, 0.01 Ω.cm, (100) | Sintered, 30 min, 1000°C, $H_2$ | ~100 nm | 28–66% 3–27 μm |
| Gomès et al. 2007 | Raman, scanning thermal microscopy | p-type, 0.01 Ω.cm, (100) | As-anodized | Meso, 10–20 nm | 30–80% 100 nm – 8 μm |
| Fang et al. 2008 | Raman | p-type, 10–20 mΩ.cm, (100) | As-anodized and oxidized (1h, 300–600°C) | Meso, 5–40 nm | 60–79% 48–51 μm |
| De Boor et al. 2011 | 3 omega | n-type, 15–50 mΩ.cm, (111) | | Meso, 7–44 nm | 17–66% 100 μm |
| | | n-type, 100 mΩ.cm, (100) | Oxidized and HF-etched | Macro, 114 nm | 42% 100 μm |
| Amin-Chalhoub et al. 2011 | Pulsed-photothermal | n-type, 12–15 mΩ.cm, (100) | As-anodized | Meso, 5–10 nm | 15, 35% 10 μm |
| Siegert et al. 2012 | DC + FEM simulation, Raman | n-type, 10 mΩ.cm, (100) | As-anodized | Meso | 20% 100 μm |
| Behzad et al. 2012b | Photoacoustic method | p-type, 1–10 Ω.cm, (100) | As-anodized | | 62–80% 12–69 μm |
| Behzad et al. 2012a | Photoacoustic method | n-type, 1–10 Ω.cm, (100) | As-anodized | Macro | 47–94% 2–102 μm |
| Newby et al. 2013 | Raman, scanning thermal microscopy | p-type, 0.01–0.02 Ω.cm, (100) | As-anodized, ion irradiation, partial oxidation | Meso, 7–14 nm | 41–75% 10 μm |
| Isaiev et al. 2014 | Photoacoustic | | | | |
| Valalaki and Nassiopoulou 2013, 2014 | DC + FEM simulation | p-type, 1–10 Ω.cm, (100) | As-anodized | Nano, 3 nm | 63% 40 μm |
| Lucklum et al. 2014 | 3 omega | p-type, 10–50 mΩ.cm, (100) | As-anodized | Meso, ~15 nm | 20–85% |
| Andrusenko et al. 2014 | Photoacoustic | p-type, 10–20 mΩ.cm, (100) | As-anodized | Meso | 60% 240 μm |
| Massoud et al. 2014 | Scanning thermal microscopy | p-type, 0.01–0.02 Ω.cm, (100) | As-anodized, ion irradiation | Meso, ~10 nm | 56% 10 μm |

the porous Si samples. It is impractical to also present the thermal conductivity results in this table; these will be presented next for groups of samples with similar characteristics.

Certain measurement techniques, such as the photoacoustic method, measure thermal diffusivity rather than thermal conductivity. However, it is trivial to calculate thermal conductivity, using Equation 9.1, and the vast majority of authors who measure thermal diffusivity also present their results as thermal conductivity values. In Table 9.1, we have made no distinction between thermal diffusivity and conductivity measurements.

### 9.2.3 THERMAL CONDUCTIVITY OF AS-ANODIZED POROUS SILICON

Figure 9.1 shows the results of thermal conductivity measurements for as-anodized porous silicon as a function of porosity. This graph only contains the results for samples where the morphology was clearly identified by the authors, and samples containing different morphologies within a single porous layer were also excluded.

Two trends can be observed in this figure: first, thermal conductivity decreases as porosity increases, and second, the thermal conductivity of nanoporous silicon is lower than that of mesoporous silicon. Two results go against these trends: the measurements presented by Amin-Chalhoub et al. (2011) and the measurements for mesoporous silicon in Drost et al. (1995). The latter is surprisingly high and is questionable (Chung and Kaviany 2000), the former stands out less.

Figure 9.2 plots experimental thermal conductivity values from the literature as a function of crystallite size. There is less data than in Figure 9.1, as this parameter is not always characterized, and indeed is more difficult to measure than porosity. This data shows that thermal conductivity increases with crystallite size, which also confirms the difference between nano- and

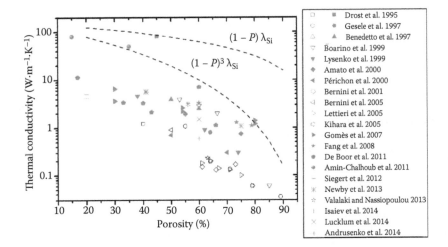

**FIGURE 9.1**  Experimental data from the literature for thermal conductivity as a function of porosity, for as-anodized nano- and meso-porous silicon. Full symbols correspond to mesoporous silicon, and hollow symbols to nanoporous Si. The dashed lines show two functions calculated using the thermal conductivity of bulk crystalline silicon, $\lambda_{Si}$, and porosity, $P$.

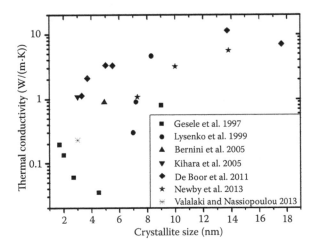

**FIGURE 9.2**  Experimental data from the literature for thermal conductivity as a function of crystallite size, for as-anodized nano- and meso-porous silicon.

mesoporous silicon observed in Figure 9.1. The physical reasons for this behavior will be discussed in Section 9.3.

### 9.2.3.1 EFFECT OF TEMPERATURE ON THE THERMAL CONDUCTIVITY OF POROUS SILICON

The vast majority of published data for the thermal properties of porous Si is at room temperature. However, some work has been published with thermal conductivity measured from 35 to 320 K (Gesele et al. 1997), 120–450 K (De Boor et al. 2011) and 4.2–350 K (Valalaki and Nassiopoulou 2013, 2014), and thermal diffusivity from 300 to 600 K (Bernini et al. 2005).

For porous silicon constituted of nanostructures smaller than 100 nm, these measurements all show a monotonic decrease with temperature below room temperature. Above room temperature, thermal conductivity remains constant with temperature. The results from De Boor et al. (2011) are shown in Figure 9.3a (note that these values are normalized by $(1 - P)$, where $P$ is porosity). At very low temperature, between 4.2 and 20 K, the thermal conductivity of porous Si reaches a plateau (Valalaki and Nassiopoulou 2014). This behavior shows a strong contrast with that of bulk crystalline silicon (see Figure 9.3b).

In bulk crystalline silicon, the thermal conductivity initially increases when temperature is lowered, as the mean free path of phonons increases. In porous silicon, the phonon mean free path is already limited by the size of its nanostructures, so it simply decreases due to the heat capacity decreasing. These mechanisms will be discussed further in Section 9.3

### T9.2.3.2 EFFECT OF THICKNESS ON THE THERMAL CONDUCTIVITY OF POROUS SILICON

Thermal conductivity measurements are usually carried out on films with thicknesses varying between the µm range and the 100 µm range. However, the effect of this parameter is generally not studied. Gesele et al. (1997) measured the thermal conductivity between 35 and 320 K of 71% porosity nanoporous with two different layer thicknesses (31 and 46 µm) and found no significant difference between the two thicknesses.

A comprehensive study of the effect of layer thickness on thermal conductivity was published by Gomès et al. (2007). The authors used scanning thermal microscopy to measure the thermal conductivity of porous Si films, with porosities of 30, 54, and 80%, and thicknesses between 38 nm and 7.2 µm. Thermal conductivity remains constant, for thicknesses greater than 1 µm. Below 1 µm, the measured thermal conductivity decreases. This is due to the thermal interface

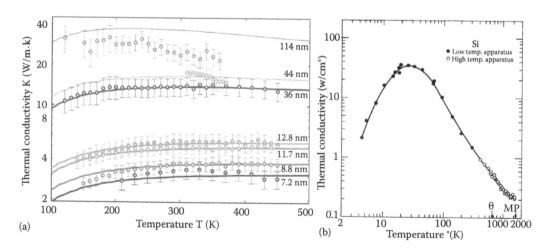

**FIGURE 9.3**   (a) Thermal conductivity versus temperature for different nanostructure size, measured by the 3-omega method. Note that these values are normalized by $(1 - P)$, where $P$ is the porosity. (b) Thermal conductivity as a function of temperature for bulk crystalline silicon. ([a] From De Boor J. et al., *Europhys. Lett.* 96(1), 16001, 2011. Copyright 2011: European Physical Society. With permission. [b] Reprinted with permission from Glassbrenner C.J. and Slack G.A., *Phys. Rev.* 134, A1058, 1964. Copyright 1964 by the American Physical Society.)

resistance between the porous layer and silicon substrate, whose effect becomes measurable for thinner films. This has been observed for other materials (Lee and Cahill 1997).

When studying the effect of thickness, one should bear in mind that porosity can also vary with thickness, if precautions to ensure uniform layers are not taken. The most common method to ensure uniform layers is to introduce etch stops during anodization (Thönissen et al. 1996; Billat et al. 1997).

### 9.2.4 EFFECT OF POST-ANODIZATION TREATMENTS ON THE THERMAL CONDUCTIVITY OF POROUS SILICON

#### 9.2.4.1 OXIDATION

Figure 9.4 shows the thermal conductivity of meso-porous silicon as a function of oxidation temperature (Périchon et al. 2000) and as a function of the fraction of porous Si which has been oxidized (Lysenko et al. 1999b), for different porosities. In both cases, the measurements were carried out using Raman spectroscopy. Both sets of measurements show that the thermal conductivity initially decreases as the porous Si layer is oxidized, before reaching a minimum value around 300°C, and increasing again. Based on these measurements, oxidation for 1 h at 300°C seems to be optimal for reducing thermal conductivity. This same treatment had already been shown to stabilize the structure of porous silicon, and prevent it from reorganizing itself when being annealed under an inert atmosphere, for temperatures up to 800°C (Herino et al. 1984).

An explanation for this behavior is that initially, oxidation leads to a reduction in the size of the silicon crystallites, which causes a reduction of thermal conductivity. This effect is stronger for samples with lower porosity, as the crystallites are bigger than for higher porosities. When oxidation continues, it causes swelling of the crystallites and therefore a reduction in porosity, so thermal conductivity increases again (Lysenko et al. 1999b; Périchon et al. 2000).

A few other articles have been published, which study the effect of oxidation, but with a small number of samples (Drost et al. 1995; Kihara et al. 2005; Fang et al. 2008).

#### 9.2.4.2 NITRIDATION

A process for partial nitridation of porous silicon was developed by Maccagnani et al. (1999), by annealing it in an $NH_3$ atmosphere at 1000–1100°C. Measurements show that thermal conductivity decreases following nitridation for a sample with 55% porosity, and increases for a sample with 75% porosity. This behavior is similar to that observed when oxidizing porous silicon and it seems likely that similar mechanisms are involved.

Nitridation has been proposed as a method of passivating porous Si against attack by alkaline solutions (Lai et al. 2011), and seems to cause less stress in samples than oxidation (Maccagnani et al. 1999).

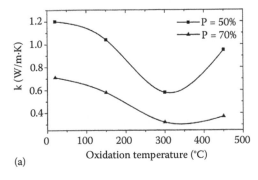
(a)

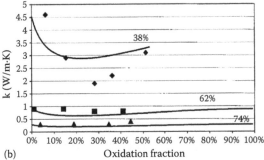
(b)

**FIGURE 9.4** (a) Thermal conductivity of meso-porous silicon as a function of oxidation temperature, for 50 and 70% porosity. (b) Thermal conductivity of meso-porous silicon as a function of oxidized fraction, for 38, 62, and 70% porosity. ([a] Reprinted from *Sens. Actuators A* 85, Périchon S. et al., 335, Copyright 2000, with permission from Elsevier. [b] Reprinted with permission from Lysenko V. et al., *J. Appl. Phys.* 86(12), 6841, 1999. Copyright 1999. American Institute of Physics.)

### 9.2.4.3 ANNEALING

Wolf and Brendel (2006) studied the effect of thermal annealing on mesoporous silicon, in an atmosphere of $H_2$ for 30 min at 1000°C. This study lacks thermal conductivity measurements of the unannealed samples, but the thermal conductivity values obtained after annealing are higher than those generally found in the literature for unannealed mesoporous silicon. Indeed, it is well known that annealing porous silicon leads to a coarsening of its structure (Herino et al. 1984), thus increasing the size of its structures, which explains the increase in thermal conductivity.

### 9.2.4.4 HEAVY ION IRRADIATION

Newby et al. (2013) showed that mesoporous silicon could be made amorphous through heavy ion irradiation, with uranium ions at 110 MeV. This treatment makes porous Si amorphous, but without modifying the porous meso-structure. Amorphization is partial or total, depending on the irradiation dose. Raman spectroscopy and scanning thermal microscopy measurements showed that this treatment led to a reduction in thermal conductivity by up to a factor of 3. By carrying out a partial oxidation (1 h 300°C) before irradiation, a fivefold reduction of thermal conductivity was achieved. These samples were also characterized by the photo-acoustic method (Isaiev et al. 2014), and were observed for irradiation with Xe ions at 29 and 91 MeV (Massoud et al. 2014).

## 9.2.5 HEAT CAPACITY MEASUREMENTS

The majority of authors simply consider that the specific heat $C_p$, in J/(Kg·K) of porous Si is simply equal to that of bulk silicon. This means that the volumetric heat capacity of porous Si can be related to that of bulk Si by its porosity, as follows:

$$\text{(9.2)} \qquad\qquad C_{\text{Porous Si}} = \rho_{\text{Porous Si}}\, C_p = (1 - P)\rho_{\text{Si}}\, C_p$$

where $\rho$ is density.

However, two authors have measured the heat capacity of porous silicon. Kihara et al. (2005) used the 3-omega method to measure the volumetric heat capacity of porous silicon. For a 55% porosity nanoporous sample, they measured $\rho C_p = 0.18$ MJ/(K·m³). From this we can calculate $C_p = 172$ J/(K·m³), which is around four times less than bulk silicon 700 J/(K·m³). The authors ascribe this to confinement effects and carrier depletion, caused by the very small crystallite size in their porous Si sample, which was 3 nm according to photoluminescence measurements.

The second result for the specific heat capacity of porous silicon was obtained by lock-in thermography (Wolf and Brendel 2006). This technique allows the measurement of both thermal conductivity and diffusivity, so heat capacity can be calculated from these two values (Wolf et al. 2004). The authors study mesoporous silicon samples, which have been annealed in hydrogen for 30 min at 1000°C, with initial porosities between 27 and 66%. They find that the specific heat capacity is close to the bulk value. The structures in annealed mesoporous silicon are considerably larger than in nanoporous silicon, so this is consistent with the observations of Kihara et al. (2005).

## 9.3 WHY DOES POROUS SILICON HAVE A LOW THERMAL CONDUCTIVITY?

Three factors are responsible for the thermal conductivity reduction in porous silicon. First, the material removed from the silicon substrate, when pores are etched, no longer participates in thermal transport. The effective thermal conductivity can be calculated as follows, where $\lambda_{\text{PSi}}$ and $\lambda_{\text{Si}}$ are the thermal conductivities of, respectively, porous Si and bulk Si:

$$\text{(9.3)} \qquad\qquad \lambda_{\text{PSi}} = (1 - P)\, \lambda_{\text{Si}}$$

However, this is not enough to explain the difference between the thermal conductivity of bulk silicon and porous silicon. Equation 9.3 is plotted in Figure 9.1, and we can see that the majority of measured thermal conductivities are 1 to 2 orders of magnitude lower than Equation 9.3.

The first model published describing the thermal conductivity of porous silicon was an analytical model, proposed by Gesele et al. (1997). This model can help explain qualitatively the different phenomena involved in the thermal conductivity reduction in porous silicon. According to this model, the thermal conductivity of porous Si can be described by

$$\lambda_{PSi} = (1 - P)\, g_0\, \lambda_{Si}$$
(9.4)

Compared to Equation 9.3, there is an additional factor, $g_0$, which is a *percolation factor*. This factor expresses the fact that heat cannot take a direct path between two points, but must instead follow a more tortuous path due to the porous morphology. This is the second factor that contributes to the low thermal conductivity of porous Si. The value of $g_0$ depends on the porous morphology. Gesele et al. (1997) found that $g_0 = (1 - P)^2$ was a good fit with their measurements. This value of $g_0$ is used to plot Equation 9.3 in Figure 9.1. We can see that the fit is closer to the experimental data, but still too high.

In silicon, thermal transport is dominated by phonon heat transport. Therefore, the thermal conductivity of bulk crystalline silicon is given by

$$\lambda_{Si} = \frac{1}{3}\rho C_p vl$$
(9.5)

where $v$ is the speed of sound and $l$ is the phonon mean free path. In silicon, the mean free path of phonons calculated with the Debye model is 41 nm, and 260 nm if calculated using a dispersion model (Chen 1998). Both these values are higher than crystallite size generally observed in nano- and mesoporous silicon. Therefore, the phonon mean free path in porous silicon is limited due to the nanoscale size of its structure, so the thermal conductivity of the silicon phase is also reduced. This is the third and final factor responsible for the low thermal conductivity of porous silicon.

Therefore, in summary, three factors are responsible for the reduced thermal conductivity of silicon: (1) the presence of pores, (2) percolation due to the crystallite interconnections, and finally (3) phonon scattering due to the nanometer-scale dimensions of the porous silicon crystallites.

## 9.4 MODELING OF POROUS SILICON THERMAL CONDUCTIVITY

Several different models have been developed to describe the thermal conductivity of porous silicon, using a variety of analytical and numerical approaches. The analytical model proposed by Gesele et al. (1997) was explained qualitatively in the previous paragraph; its complete expression is given below.

$$\lambda_{PSi} = \frac{1}{3}(1-P)^3 \rho C_p v d_{cr}$$
(9.6)

All the terms are as defined earlier, and $d_{cr}$ is the diameter of the crystallites forming the porous silicon structure, which has simply replaced the phonon mean free path.

Some refinements to this model have also been proposed (Lysenko et al. 1999a,b). The first modification was to better describe heat transport and phonon scattering effects in a single nanocrystallite, by introducing an effective phonon mean free path, $l_{eff}$, based on work on heat transport in thin films (Majumdar 1993):

$$l_{eff} = \frac{l_{Si}}{1 + \dfrac{4l_{Si}}{3r_{cr}}}$$
(9.7)

where $l_{Si}$ is the phonon mean free path in silicon and $r_{cr}$ is the crystallite radius. This is introduced into Gesele's model such as

$$\lambda_{PSi} = \frac{1}{3}(1-P)^3 \rho C_p v l_{eff}$$
(9.8)

A model using the same correction factor has been proposed, but without the percolation factor $g_0$ (De Boor et al. 2011). This model is used to fit the experimental data in Figure 9.3a. The model described in Equation 9.8 was extended to also describe partially oxidized porous silicon (Lysenko et al. 1999a,b). This model fit well with the measurements presented in Figure 9.4a and b, and successfully predicted the minimum thermal conductivity for an oxidized fraction around 20%.

As mentioned in Section 9.3, the percolation factor depends on the geometry of the material. Indeed, Lettieri et al. (2005) found that a percolation factor of $g_0 = (1 - P)^2$ was suitable for fitting the thermal conductivity of n-type porous silicon, but for their measurements on p-type porous silicon, the best fit was obtained with $g_0 = (1 - P)^{1.3}$.

Chantrenne and Lysenko (2005) represented mesoporous silicon as nanowires formed of interconnected cubic silicon nanoparticles, with the interconnection area smaller than the particle cross-section, and calculated phonon heat conduction using kinetic theory of gases. The model agrees well with experimental data, and shows that the interconnection area between crystallites is an important factor for controlling thermal conductivity. Phonon hydrodynamics have been used to develop an analytical model describing thermal conductivity and using porous silicon porosity and pore size as the main material parameters (Alvarez et al. 2010; Sellitto et al. 2012).

Numerical studies have also been carried out. Chung and Kaviany (2000) solved the Boltzmann transport equation, and found that the main factors causing thermal conductivity reduction were phonon pore scattering and pore randomness. This work is interesting as it is one of the rare works to explicitly study the difference between thermal conductivity in the cross-plane and in-plane directions. The authors' simulations show that for porosities between 10 and 30%, in-plane thermal conductivity is between 12 and 33 times lower than in the cross-plane direction. Randrianalisoa and Baillis (2008a) generated a 3D medium porosity mesoporous structure, and then used the Monte Carlo (Randrianalisoa and Baillis 2008b) method to simulate phonon transport and thermal conductivity, based on several material parameters. The same authors also used a hybrid model (Randrianalisoa and Baillis 2009), combining an analytical solution to the Boltzmann equation and the phonon tracking method. The main factors reducing thermal conductivity in their model are porosity and pore size, as well as pore arrangement and pore-wall roughness. Results of these last two models are shown in Figure 9.5, and compared to experimental data from the literature.

We have only mentioned in this section the models that have been specifically developed to describe the thermal conductivity in porous silicon formed by electrochemical etching.

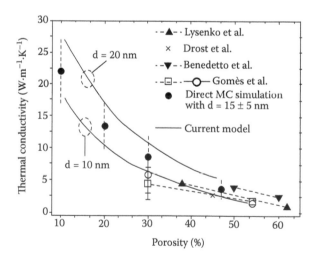

**FIGURE 9.5** Thermal conductivity as a function of porosity, simulated using a combined analytical and phonon-tracking approach. The simulation results are compared to experimental data from the literature (From Drost A. et al.., *Sensors Mater.* 7(2), 111–120, 1995; Benedetto G. et al., *Appl. Phys. A Mater. Sci. Process.* 64(2), 155–159, 1997; Lysenko V. et al., *J. Appl. Phys.* 86(12), 6841–6846, 1999; Gomès S. et al., *J. Phys. D. Appl. Phys.* 40(21), 6677–6683, 2007.) and Monte Carlo simulations previously published by the same authors (From Randrianalisoa J. and Baillis D., *J. Heat Transfer* 130(7), 072404, 2008.). (Randrianalisoa J. and Baillis D.: *Adv. Eng. Mater.* 2009. 11(10). 852. Copyright Wiley-VCH Verlag GmbH & Co. KGaA. Reproduced with permission.)

Nevertheless, a large body of literature has been published on the subject of thermal transport in porous media, often based on effective medium theory. However, these models may fail to account for the phonon scattering effects caused by the nanoscale structure of porous silicon (Randrianalisoa and Baillis 2008a).

We mentioned holey silicon at the beginning of this chapter. Many articles have been published with models for this material (Prasher 2006), including molecular dynamics simulations (He et al. 2011; Yang et al. 2014). These are often for regular structures, rather than the random structures that constitute nano- and mesoporous silicon formed by electrochemical etching, but the structure size is comparable, so some of these models may be applicable to porous silicon formed by anodization. Work on the modeling of thermal transport in silicon nanowires and nanoparticles may also be relevant, as mesoporous silicon can be described as an array of interconnected nanowires, albeit with a less regular structure, and nanoporous silicon is essentially a network of connected nanoparticles (Chantrenne and Lysenko 2005).

## 9.5 SUMMARY

We mentioned for the first time in the previous section the anisotropy of porous silicon thermal conductivity. Mesoporous silicon in particular has a strongly anisotropic structure, which can be described as being formed of quasi-columns that are laterally interconnected. It would be highly surprising if this does not have an effect on thermal conductivity. Intuitively, one can expect that thermal conductivity should be lower in the lateral, in-plane direction than in the cross-plane direction (i.e., along the columns in the mesoporous structure). Indeed, this is confirmed by the work of Chung and Kaviany (2000). However, to our knowledge, this has never been confirmed experimentally. All the thermal conductivity results given in Section 9.2 are effective thermal conductivity values, averaged over the in-plane and cross-plane values.

Apart from this, the thermal conductivity of porous silicon has been well characterized. A large number of articles have been published on the subject, using a variety of measurement techniques, and the different results generally agree well with each other. The main trends observed are that the thermal conductivity of porous Si is significantly lower than that of bulk silicon, and shows a strong dependence on porosity and crystallite size. We have also shown that different processes can be used to further tune its thermal conductivity.

Heat capacity, however, has received little attention. The majority of authors assume that the specific heat of porous silicon is equal to that of bulk silicon. Very little work has been published concerning measurement of this property, and to our knowledge, only one article shows a dependence of heat capacity on crystallite size, for nanoporous silicon. This could be an interesting area to explore.

Finally, the low thermal conductivity of porous silicon means it is an ideal material for thermal insulation in MEMS and other silicon-based devices that require thermal insulation. Indeed, porous silicon has successfully been integrated into several devices, which will be presented in Chapter 8 of the book Porous Silicon: Biomedical and Sensor Applications.

## REFERENCES

Alvarez, F.X., Jou, D., and Sellitto, A. (2010). Pore-size dependence of the thermal conductivity of porous silicon: A phonon hydrodynamic approach. *Appl. Phys. Lett.* **97**(3), 033103.

Amato, G., Angelucci, R., Benedetto, G. et al. (2000). Thermal characterisation of porous silicon membranes. *J. Porous Mater.* **7**(1), 183–186.

Amin-Chalhoub, E., Semmar, N., Coudron, L. et al. (2011). Thermal conductivity measurement of porous silicon by the pulsed-photothermal method. *J. Phys. D. Appl. Phys.* **44**(35), 355401.

Andrusenko, D., Isaiev, M., Tytarenko, A., Lysenko, V., and Burbelo, R. (2014). Size evaluation of the fine morphological features of porous nanostructures from the perturbation of heat transfer by a pore filling agent. *Microporous Mesoporous Mater.* **194**, 79–82.

Behzad, K., Yunus, W.M.M., Talib, Z.A., Zakaria, A., and Bahrami, A. (2012a). Effect of preparation parameters on physical, thermal and optical properties of n-type porous silicon. *Int. J. Electrochem. Sci.* **7**, 8266–8275.

Behzad, K., Yunus, W.M.M., Talib, Z.A., Zakaria, A., Bahrami, A., and Shahriari, E. (2012b). Effect of etching time on optical and thermal properties of p-type porous silicon prepared by electrical anodisation method. *Adv. Opt. Technol.* **2012**, 581743.

Benedetto, G., Boarino, L., Brunetto, N., Rossi, A., Spagnolo, R., and Amato, G. (1997a). Thermal properties of porous silicon layers. *Philos. Mag. Part B* **76**(3), 383–393.

Benedetto, G., Boarino, L., and Spagnolo, R. (1997b). Evaluation of thermal conductivity of porous silicon layers by a photoacoustic method. *Appl. Phys. A Mater. Sci. Process.* **64**(2), 155–159.

Bernini, U., Maddalena, P., Massera, E., and Ramaglia, A. (1999a). Photo-acoustic characterization of porous silicon samples. *J. Opt. A Pure Appl. Opt.* **1**, 210–213.

Bernini, U., Maddalena, P., Massera, E., and Ramaglia, A. (1999b). Thermal characterization of porous silicon via thermal wave interferometry. *Opt. Commun.* **168**(1–4), 305–314.

Bernini, U., Lettieri, S., Maddalena, P., Vitiello, R., and Di Francia, G. (2001). Evaluation of the thermal conductivity of porous silicon layers by an optical pump-probe method. *J. Phys. Condens. Matter* **13**, 1141–1150.

Bernini, U., Lettieri, S., Massera, E., and Rucco, P.A. (2003). Investigation of thermal transport in n-type porous silicon by photo-acoustic technique. *Opt. Lasers Eng.* **39**(2), 127–140.

Bernini, U., Bernini, R., Maddalena, P., Massera, E., and Rucco, P. (2005). Determination of thermal diffusivity of suspended porous silicon films by thermal lens technique. *Appl. Phys. A Mater. Sci. Process.* **81**(2), 399–404.

Billat, S., Thönissen, M., Arens-Fischer, R., Berger, M.G., Krüger, M., and Lüth, H. (1997). Influence of etch stops on the microstructure of porous silicon layers. *Thin Solid Films* **297**, 22–25.

Boarino, L., Monticone, E., Amato, G. et al. (1999). Design and fabrication of metal bolometers on high porosity silicon layers. *Microelectron. J.* **30**(11), 1149–1154.

Calderón, A., Alvarado-Gil, J.J., Gurevich, Y.G. et al. (1997). Photothermal characterization of electrochemical etching processed n-type porous silicon. *Phys. Rev. Lett.* **79**(25), 5022–5025.

Chantrenne, P. and Lysenko, V. (2005). Thermal conductivity of interconnected silicon nanoparticles: Application to porous silicon nanostructures. *Phys. Rev. B* **72**(3), 035318.

Chen, G. (1998). Thermal conductivity and ballistic-phonon transport in the cross-plane direction of superlattices. *Phys. Rev. B* **57**(23), 14958–14973.

Chung, J.D. and Kaviany, M. (2000). Effects of phonon pore scattering and pore randomness on effective conductivity of porous silicon. *Int. J. Heat Mass Transf.* **43**(4), 521–538.

De Boor, J., Kim, D.S., Ao, X. et al. (2011). Temperature and structure size dependence of the thermal conductivity of porous silicon. *Europhys. Lett.* **96**(1), 16001.

De Boor, J., Kim, D.S., Ao, X. et al. (2012). Thermoelectric properties of porous silicon. *Appl. Phys. A* **107**(4), 789–794.

Drost, A., Steiner, P., Moser, H., and Lang, W. (1995). Thermal conductivity of porous silicon. *Sensors Mater.* **7**(2), 111–120.

Fang, Z., Hu, M., Zhang, W., Zhang, X., and Yang, H. (2008). Thermal conductivity and nanoindentation hardness of as-prepared and oxidized porous silicon layers. *J. Mater. Sci. Mater. Electron.* **19**(11), 1128–1134.

Gesele, G., Linsmeier, J., Drach, V., Fricke, J., and Arens-Fischer, R. (1997). Temperature-dependent thermal conductivity of porous silicon. *J. Phys. D. Appl. Phys.* **30**(21), 2911–2916.

Glassbrenner, C.J. and Slack, G.A. (1964). Thermal conductivity of silicon and germanium from 3 K to the melting point. *Phys. Rev.* **134**(4A), A1058–A1069.

Gomès, S., David, L., Lysenko, V., Descamps, A., Nychyporuk, T., and Raynaud, M. (2007). Application of scanning thermal microscopy for thermal conductivity measurements on meso-porous silicon thin films. *J. Phys. D. Appl. Phys.* **40**(21), 6677–6683.

He, Y., Donadio, D., Lee, J.-H., Grossman, J.C., and Galli, G. (2011). Thermal transport in nanoporous silicon: Interplay between disorder at mesoscopic and atomic scales. *ACS Nano* **5**(3), 1839–1844.

Herino, R., Perio, A., Barla, K., and Bomchil, G. (1984). Microstructure of porous silicon and its evolution with temperature. *Mater. Lett.* **2**(6), 519–523.

Hopkins, P.E., Reinke, C.M., Su, M.F. et al. (2011). Reduction in the thermal conductivity of single crystalline silicon by phononic crystal patterning. *Nano Lett.* **11**, 107–112.

Isaiev, M., Newby, P.J., Canut, B. et al. (2014). Thermal conductivity of partially amorphous porous silicon by photoacoustic technique. *Mater. Lett.* **128**, 71–74.

Kihara, T., Harada, T., and Koshida, N. (2005). Precise thermal characterization of confined nanocrystalline silicon by 3ω method. *Jpn. J. Appl. Phys.* **44**(6A), 4084–4087.

Lai, M., Parish, G., Liu, Y., Dell, J.M., and Keating, A.J. (2011). Development of an alkaline-compatible porous-silicon photolithographic process. *J. Microelectromech. Syst.* **20**(2), 418–423.

Lang, W., Drost, A., Steiner, P., and Sandmaier, H. (1994). The thermal conductivity of porous silicon. *MRS Proc.* **358**, 561–566.

Lee, S.-M.-M. and Cahill, D.G. (1997). Heat transport in thin dielectric films. *J. Appl. Phys.* **81**(6), 2590–2595.

Lettieri, S., Bernini, U., Massera, E., and Maddalena, P. (2005). Optical investigations on thermal conductivity in n- and p-type porous silicon. *Phys. Stat. Sol. (c)* **2**(9), 3414–3418.

Lucklum, F., Schwaiger, A., and Jakoby, B. (2014). Highly insulating, fully porous silicon substrates for high temperature micro-hotplates. *Sens. Actuators A* **213**, 35–42.

Lysenko, V. and Volz, S. (2000). Porous silicon thermal conductivity by scanning probe microscopy. *Phys. Stat. Sol. A* **182**(2), R6–R7.

Lysenko, V., Boarino, L., Bertola, M. et al. (1999a). Theoretical and experimental study of heat conduction in as-prepared and oxidized meso-porous silicon. *Microelectron. J.* **30**(11), 1141–1147.

Lysenko, V., Perichon, S., Remaki, B., Barbier, D., and Champagnon, B. (1999b). Thermal conductivity of thick meso-porous silicon layers by micro-Raman scattering. *J. Appl. Phys.* **86**(12), 6841–6846.

Lysenko, V., Perichon, S., Remaki, B., and Barbier, D. (2002). Thermal isolation in microsystems with porous silicon. *Sens. Actuators A* **99**(1–2), 13–24.

Ma, J., Sadhu, J.S., Ganta, D., Tian, H., and Sinha, S. (2014). Thermal transport in 2- and 3-dimensional periodic 'holey' nanostructures. *AIP Adv.* **4**, 124502.

Maccagnani, P., Angelucci, R., Pozzi, P. et al. (1999). Thick porous silicon thermo-insulating membranes. *Sensors Mater.* **11**(3), 131–147.

Majumdar, A. (1993). Microscale heat conduction in dielectric thin films. *J. Heat Transfer* **115**, 7–16.

Marconnet, A.M., Kodama, T., Asheghi, M., and Goodson, K.E. (2012). Phonon conduction in periodically porous silicon nanobridges. *Nanoscale Microscale Thermophys. Eng.* **16**(4), 199–219.

Massoud, M., Canut, B., Newby, P., Frechette, L., Chapuis, P.O., and Bluet, J.M. (2014). Swift heavy ion irradiation reduces porous silicon thermal conductivity. *Nucl. Instrum. Methods Phys. Res. Sect. B* **341**, 27–31.

Newby, P.J., Canut, B., Bluet, J.-M. et al. (2013). Amorphization and reduction of thermal conductivity in porous silicon by irradiation with swift heavy ions. *J. Appl. Phys.* **114**(1), 014903.

Obraztsov, A.N., Timoshenko, V.Y., Okushi, H., and Watanabe, H. (1997). Photoacoustic spectroscopy of porous silicon. *Semiconductors* **31**(5), 534–536.

Périchon, S., Lysenko, V., Roussel, P. et al. (2000). Technology and micro-Raman characterization of thick meso-porous silicon layers for thermal effect microsystems. *Sens. Actuators A* **85**(1–3), 335–339.

Prasher, R. (2006). Transverse thermal conductivity of porous materials from aligned nano- and microcylindrical pores. *J. Appl. Phys.* **100**(3), 064302.

Randrianalisoa, J. and Baillis, D. (2008a). Monte Carlo simulation of cross-plane thermal conductivity of nanostructured porous silicon films. *J. Appl. Phys.* **103**(5), 053502.

Randrianalisoa, J. and Baillis, D. (2008b). Monte Carlo simulation of steady-state microscale phonon heat transport. *J. Heat Transfer* **130**(7), 072404.

Randrianalisoa, J. and Baillis, D. (2009). Combined analytical and phonon-tracking approaches to model thermal conductivity of etched and annealed nanoporous silicon. *Adv. Eng. Mater.* **11**(10), 852–861.

Sellitto, A., Jou, D., and Cimmelli, V.A. (2012). A phenomenological study of pore-size dependent thermal conductivity of porous silicon. *Acta Appl. Math.* **122**, 435–445.

Shen, Q. and Toyoda, T. (2002). Characterization of thermal properties of porous silicon film/silicon using photoacoustic technique. *J. Therm. Anal. Calorim.* **69**(3), 1067–1073.

Shen, Q. and Toyoda, T. (2003). Dependence of thermal conductivity of porous silicon on porosity characterized by photoacoustic technique. *Rev. Sci. Instrum.* **74**(1), 601.

Siegert, L., Capelle, M., Roqueta, F., Lysenko, V., and Gautier, G. (2012). Evaluation of mesoporous silicon thermal conductivity by electrothermal finite element simulation. *Nanoscale Res. Lett.* **7**(1), 427.

Tang, J., Wang, H.-T., Lee, D.H. et al. (2010). Holey silicon as an efficient thermoelectric material. *Nano Lett.* **10**(10), 4279–4283.

Thönissen, M., Billat, S., Krüger, M. et al. (1996). Depth inhomogeneity of porous silicon layers. *J. Appl. Phys.* **80**(5), 2990–2993.

Valalaki, K. and Nassiopoulou, A.G. (2013). Low thermal conductivity porous Si at cryogenic temperatures for cooling applications. *J. Phys. D* **46**(29), 295101.

Valalaki, K. and Nassiopoulou, A.G. (2014). Thermal conductivity of highly porous Si in the temperature range 4.2 to 20 K. *Nanoscale Res. Lett.* **9**(1), 318.

Wang, Z., Alaniz, J.E., Jang, W., Garay, J.E., and Dames, C. (2011). Thermal conductivity of nanocrystalline silicon: Importance of grain size and frequency-dependent mean free paths. *Nano Lett.* **11**(6), 2206–2213.

Wolf, A. and Brendel, R. (2006). Thermal conductivity of sintered porous silicon films. *Thin Solid Films* **513**(1–2), 385–390.

Wolf, A., Pohl, P., and Brendel, R. (2004). Thermophysical analysis of thin films by lock-in thermography. *J. Appl. Phys.* **96**(11), 6306–6312.

Yang, L., Yang, N., and Li, B. (2014). Extreme low thermal conductivity in nanoscale 3D Si phononic crystal with spherical pores. *Nano Lett.* **14**, 1734–1738.

Yu, J.-K., Mitrovic, S., Tham, D., Varghese, J., and Heath, J.R. (2010). Reduction of thermal conductivity in phononic nanomesh structures. *Nat. Nanotechnol.* **5**(10), 718–721.

# Alternative Methods of Silicon Porosification and Properties of These PSi Layers

## 10

Ghenadii Korotcenkov and Vladimir Brinzari

CONTENTS

Electrochemical etching was the first method that showed the possibility of porous silicon formation. However, later experiments have shown that porous silicon can also be formed by using other approaches.

## 10.1 CHEMICAL STAIN ETCHING

Chemical stain etching was the first alternative method used for fabrication of porous Si without applying an electric field. Turner (1960) and Archer (1960) were the first to study stain etching of Si in HF-HNO$_3$ solution, which allowed forming porous Si (Beckmann 1965; Canham 1991; Sarathy et al. 1992; Shih et al. 1992a; Lin et al. 1993; Steckl et al. 1993). During the last decades, stain etching solutions with the ratio HF/HNO$_3$ changing from 1:3 to 600:1 were tested (Aoyagi et al. 1993; Xu and Steckl 1995). Along with its simplicity, the stain etching method is interesting because it does not require special equipment and gives the possibility to prepare a very thin (<100 nm) PSi layer. The stain etching can be performed in ambient light at room temperature with no intentional heating. Moreover, PSi produced by stain etching in HF:HNO$_3$-based solutions exhibited similar PL to that prepared by anodization in HF-based electrolytes (see Figure 10.1). The PL spectrum, taken from PSi under UV 365 nm excitation, usually has a broad emission band, peaked at around ~635 nm.

In addition, according to Xu and Steckl (1995), the stain-etching process possesses some additional advantages over anodization, which include submicron PSi pattern formation and fabrication of poly-PSi thin films on glass. The capability of fabricating luminescing PSi patterns embedded in conventional Si is very important for monolithic integration of optically active Si components onto an Si substrate (Dimova-Malinovska 1999). In addition, for application in flat panel display devices, the stain-etching technique might well be the only practical method to produce luminescent thin poly-Si films on quartz and glass because stain etching is performed without the electrode and electrolytic bath required by anodization. Vazsonyi et al. (2001) believe that the stain etching method can also be advantageous in porous silicon layer formation on the surface of bulk micro-machined membranes, which can be used for sensor applications. The production of thin PSi layers for photovoltaic applications is also a promising area for stain etching (Dimova-Malinovska 1999; Yerokhov et al. 2002). Research has shown that the incorporation of a PSi layer in monocrystalline-based solar cells gives the possibility to significantly increase the cells' efficiency (Bilyalov et al. 2000; Unal et al. 2001; Guerrero-Lemus et al. 2002).

Turner (1960) suggested that stain etching is actually electrochemical in its action; that is, there are anodic and cathodic sites on the surface of the semiconductor with local cell currents flowing between them. This assertion is supported by structural studies, such as those of Beale et al. (1986) and Schoisswohl et al. (1995), in which it was reported that the structure of anodically

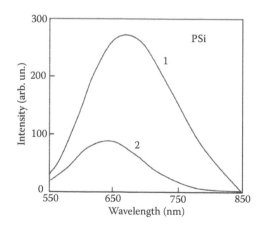

**FIGURE 10.1** Photoluminescent emission spectra of (1) conventionally anodized and (2) chemically etched porous silicon at 400-nm excitation. (Reprinted with permission from Kelly M.T. et al., *Appl. Phys. Lett.* 64, 1693, 1994. Copyright 1994. American Institute of Physics.)

and stain-etched PSi is similar and, therefore, their formation mechanisms must share much in common. Si goes into solution at the anodic sites while the oxidant is reduced at the cathodic areas. If the etching process is nonpreferential and material is removed uniformly, any given area on the surface continually alternates between being anodic and cathodic. When one spot is anodic much more than it is cathodic, an etch pit will form. Conversely, hillocks are formed on areas that are cathodic more than they are anodic. From such anisotropy, pores develop.

Turner (1960) proposed the following reactions for etchants composed of $HNO_3$, HF, and $H_2O$:

Anode:

(10.1)    $$Si + 2H_2O + mh^+ \rightarrow SiO_2 + 4H^+ + (4 - m)e^-$$

(10.2)    $$SiO_2 + 6HF \rightarrow H_2SiF_6 + 2H_2O$$

Cathode:

(10.3)    $$HNO_3 + 3H^+ \rightarrow NO + 2H_2O + 3h^+$$

Overall:

(10.4)    $$3Si + 4HNO_3 + 18HF \rightarrow 3H_2SiF_6 + 4NO + 8H_2O + 3(4 - m)h^+ + 3(4 - m)e^-$$

Therefore, chemical etching of Si can be considered a localized electrochemical process in which the reaction is started chemically. Microscopically, anode and cathode sites are formed on the etched surface with local cell currents flowing between them during the etching. Therefore, the chemical etching mechanism should incorporate a source of excess holes and electrons in order to describe the charge transfer between electrodes (Guerrero-Lemus et al. 2002). The anode reaction is proposed to consist mainly of the dissolution of Si, while the cathode reaction is a complicated reduction of $HNO_3$, which causes holes to be injected into the Si (Shih et al. 1992b). In this manner, the oxidant, or more precisely its electrochemical potential, takes on the role of the voltage in electrochemical etching. However, it is necessary to note that the above-mentioned reactions are composite reactions rather than elementary steps (Kolasinski 2005). The anode and cathode sites are not generally fixed during the process. A constant switching between the two electrode sites results in an almost uniform material removal or polishing. The attainment of PSi requires a preferential etching at the localized anode sites (Shih et al. 1992b). Shih et al. (1992b) supposed that the holes from the cathodic reaction might move to another site or recombine with free electrons from the solution before they could react with Si at the original site. In this case, the original site remains a cathode for a longer time, and the corresponding anode located at the other site remains on the surface in order to keep the overall reaction neutral.

It is necessary to admit that the PSi formation mechanism during stain etching is less steerable by external process variables (solution concentration and temperature, etching time) than the electrochemical one. With a given set of conditions, the chemical etching is self-regulating so that the simultaneous control of the PSi layer parameters (thickness and porosity) is difficult to achieve (Menna et al. 1995). The concentration of oxidizing species in the chemical process seems to play the same role that the anodic current density plays in the electrochemical etching process.

Archer (1960) first elucidated the effect of reactant concentration on the rate of colored stain film growth. Later it was found that this film had a porous structure. Archer concluded that the rate of colored stain film growth was proportional to the concentration of nitric acid (or, alternatively, sodium nitrite) employed and essentially independent of substrate resistivity, with the exception of $p$-type Si in the range from 0.001 to 0.10 $\Omega$·cm. Concentrations of nitric acid ranging from 0.10 M to 0.001 M in concentrated HF were examined, and the rate of growth was determined to fit a parabolic law:

(10.5)    $$\frac{dL}{dt} = \frac{3.73 \times 10^8 [HNO_3]^{1.48}}{L}$$

where $L$ is thickness in angstrom, $t$ is in sec, concentration in M/l. The parabolic kinetics of the staining reaction suggests that the rate limiting step is the diffusion of a reactant through the growing film, but the 3/2 power dependence on the initial quantity of nitric acid or of sodium nitrite used to make the solutions has no obvious interpretation (Archer 1960). Later, Di Francia and Citarella (1995) reexamined this issue. The study essentially confirmed the parabolic rate law determined by Archer, albeit with a slightly smaller rate constant.

As it was shown by Vazsonyi et al. (2001), during stain etching the mass of the dissolved silicon is a linear function of time in $p$- and $n$-type samples with both high and low doping concentration (see Figure 10.2). At that, the mass of dissolved silicon strongly depends on the composition of the etching solution: $HNO_3$ concentration has high impact and the additive substance $NaNO_2$ has a lower effect. It is important that the top surface of PSi layer be continuously removed during etching.

Several dependence characteristics for silicon stain etching in $HF:HNO_3:H_2O$ solution are shown in Figure 10.3. Starostina et al. (2002), who carried out this research, established that porous-layer thickness depends on the proportion in which the three ingredients are mixed and is determined by the ratio of rates for the formation and erosion of porous silicon. At that, the porous-layer thickness was found to be largest for the etchant with the highest proportion of hydrofluoric acid, namely, for $HF:HNO_3:H_2O=1200:1:0$, that is, for etchants without water addition. With such an etchant, pore formation had a low degree of selectivity. The maximum thickness was independent of the type of electric conduction and the value of resistivity for the silicon.

Usually, PSi prepared using stain etching was very smooth and uniform and showed a shiny, mirror-like dark-blue interference color (Fathauer et al. 1992; George et al. 1992; Shih et al. 1992b). There are also vertical pores with diameters of 10–40 nm, existing through the entire thickness of the PSi layer (Xu and Steckl 1995). Cross-sectional transmission electron microscopy (TEM) micrographs have shown that the PSi layers are typically characterized by a graded transition from bulk to a shallow high porosity region, with single crystalline Si filaments, embedded in a low-density amorphous matrix (Schirone et al. 1999). Schirone et al. (1999) established that very different morphologies can be obtained, changing $HNO_3$ content and etching time. It was found that an increasing of $HNO_3$ content leads to the decrease of microcrystallities and an increase of pore size. As a result, the PSi prepared in the solution with the highest nitric acid concentration shows a sharp transition between the bulk and a highly wide porosity region, while when a lower concentration of $HNO_3$ are used in the solution, a rather smooth transition is observed.

Regular pores propagating perpendicular to the wafer surface, which shows some similarity to the columnar structure of electrochemically formed PSi, was also observed by Vazsonyi et al. (2001) for $p^+$ PSi layers. In the case of low-doped $p$-type and low- and highly doped $n$-type silicon, the morphology of the final structure was characterized by a random pore propagation direction. Vazsonyi et al. (2001) have found that as a rule all types of stain-etched PS layers exhibit a porosity gradient with decreasing

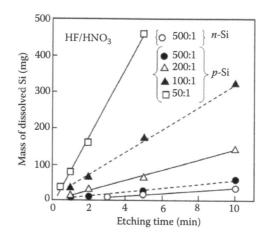

**FIGURE 10.2**   The linear function of the mass of dissolved silicon during stain etching of $p$- and $n^+$-type wafers. Stain etching of silicon was performed in solutions containing different $HF/HNO_3$ ratios (0.1 g/l $NaNO_2$ and neutral tenzide were added to the solutions). (Reprinted from *Thin Solid Films* 388, Vazsonyi E. et al., 295, Copyright 2001, with permission from Elsevier.)

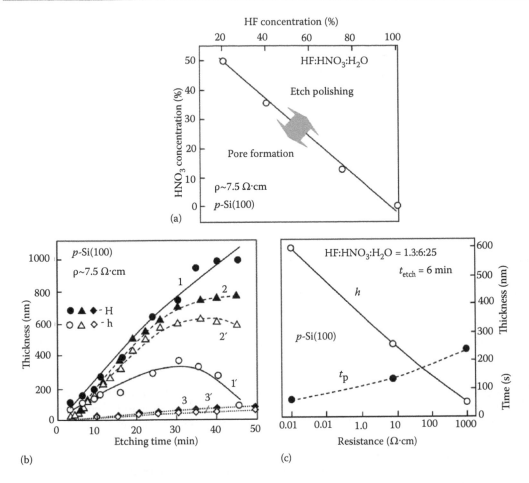

**FIGURE 10.3**    (a) Borderline between etch polishing and pore formation for *p*-Si(100) with ρ = 7.5 Ω·cm for HF:HNO$_3$:H$_2$O solution; (b) the etching depth H and the porous-layer thickness h versus etching time in (1, 1')–HF:HNO$_3$:H$_2$O = 4:1:5, (2, 2')–HF:HNO$_3$:H$_2$O = 1200:1:0, and (3, 3')–HF:HNO$_3$:H$_2$O = 1:6:9 solutions; and (c) the time to pore formation $t_p$ and the porous-layer thickness *h* versus silicon resistivity for 6-min growth in HF:HNO$_3$:H$_2$O = 1.3:6:25 solution. (Data extracted from Starostina E.A. et al., *Russ. Microelectron.* 31(2), 88, 2002.)

porosity in depth. The existence of a porosity gradient in the layers is conditioned by the partial dissolution of the pore walls and the top surface during formation. Transmission electron diffraction patterns (George et al. 1992) and valence band spectra obtained by X-ray photoelectron spectroscopy (XPS) (Vasquez et al. 1992) of the stained PSi were found to be similar to amorphous silicon. However, the electron diffraction patterns measured by Vazsonyi et al. (2001) have shown that *p*⁺-PSi skeleton had the crystalline structure. The chemical composition of the as-prepared samples was mainly characterized by Si-H$_x$ bonds. The increase of HNO$_3$ concentration caused partial oxygen incorporation into the PSi pore walls. In addition, Vazsonyi et al. (2001) observed that PSi contained an upper part with a high porosity and a bottom part with much lower porosity. At the porous/crystalline interface, the skeleton size of the porous structure was larger. They have also found that the PSi layers formed on *n*(100) and *n*⁺(111) wafers had similar porosity and thickness values for the sublayers.

Di Francia and Citarella (1995) studied the porosity of PSi layers prepared by stain etching and found that the porosity strongly depends on the etching time and can achieve 80% (see Figure 10.4). However, for too long etching the porosity measurements were not reproducible. For sufficiently long times, some part of the porous film could be etched off and reporosized during the etching itself, originating the large uncertainties observed.

It was also observed that there is an "incubation" time for the etching that depends on the composition of the etching solution (Sarathy et al. 1992), the type of doping, and the resistivity of the Si substrates (Steckl et al. 1993). The incubation time, $t_i$, was defined (Steckl et al. 1993) as a time delay that occurs between the insertion of the Si into the etching solution and the onset of

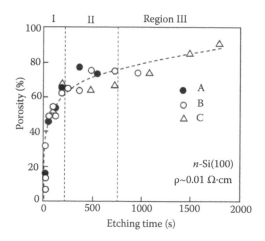

**FIGURE 10.4** Sample porosity as a function of the etching duration in HF/HNOs-based solutions (*n*-Si(100), 0.01 Ω·cm). The diagram has been divided into three regions: region I where a sharp increase of the porosity with the etching time is observed, region II where porosity keeps almost constant, and region III where a monotonic increase in the porosity can be still observed for specific series. A, B, and C correspond to different $HNO_3$ concentrations in HF; 0.024 (A), 0.012 (B), and 0.0048 (C) moles/l. (Reprinted with permission from Di Francia G. and Citarella A., *J. Appl. Phys.* 77, 3549, 1995. Copyright 1995. American Institute of Physics.)

the PSi production. Incubation times of 60–800 s have been reported (Sarathy et al. 1992; Steckl et al. 1993). For example, Xu and Steckl (1995) found that during Si etching in $HF:HNO_3:H_2O=1:3:5$ solution, from 4 min, necessary for PSi (~200 nm) forming, 3 min was incubation time. It was established that incubation time increases with resistivity for *p*-type Si and decreases for *n*-type Si. For *n*-type $t_i$ is higher (Steckl et al. 1993).

Some attempts were made to decrease incubation time using the introduction of an additional additive such as sodium-nitrite with a concentration between 0.1 and 0.6 g/l (Canham 1991; Steckl et al. 1993). In particular, Vazsonyi et al. (2001) observed that the incubation time of the process decreased when $n^-$, $n^+$-, and *p*-type Si substrates were etched in such solutions. Moreover, during etching of $p^+$-type substrate no incubation time was obtained.

The limitation of PSi layer thickness during stain etching is another important feature of this process (Eukel et al. 1990; Kalem and Rosenbauer 1995). During stain etching, the thickness of PSi layers increases until it reaches a saturation level. For example, from the experiments carried out under different conditions, Kalem and Rosenbauer (1995) have found that it was not possible to grow more than 0.3 µm porous layer with stain etching. However, the origin of this phenomenon has not been elucidated yet (Vazsonyi et al. 2001). One possible explanation was that the diffusion of the ionic species was limited by the increasing thickness of the PSi layer according to Fick's law. Consequently, the etching rate dropped to a very low level (Winton et al. 1997). With increasing doping level of the substrate, a higher thickness for PSi layer was obtained, and this phenomenon was related to as the existence of more chemically active defect sites that can be attacked by etching (Kelly et al. 1994). For example, Vazsonyi et al. (2001) established that the increase of the boron concentration up to $10^{18}$ atom/cm³ leads to an unlimited thickness of the formed porous layers, for instance after 5 min of etching time in $HF/HNO_3$ (500:1 with 0.1 g/l $NaNO_2$) solution it is in the range of 6–8 µm.

The reproducibility of the final thickness of PSi layers always has some uncertainty as well. According to Vazsonyi et al. (2001), this is due to two reasons: during etching, the evolving bubbles have a masking effect, and the PSi layer formation stops where the bubbles stick temporarily. As it is known, the reduction of $HNO_3$(aq) is extremely complex and leads to the formation of various gaseous nitrogen oxides (Kolasinski and Gogola 2011). The complexity and disadvantages of the nitric acid system have led to the conclusion that stain etching is irreproducible and capable of only producing inhomogeneous thin porous silicon layers. For this effect's reduction, some researchers introduce into etching solution a surface-active substance to ensure that the evolving bubbles do not stick to the silicon surface (Canham 1991). The mixing of the solution is also disadvantageous because at the local anode, the interaction between $HNO_3$ and silicon becomes inhibited.

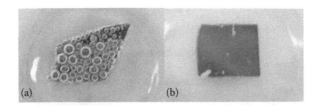

**FIGURE 10.5** Stain etching in (a) HF/HNO$_3$/H$_2$O, and (b) HF/Fe$^{3+}$/H$_2$O. (From Nahidi M. and Kolasinski K.W., *J. Electrochem. Soc.* 153, C19, 2006. Copyright 2006: The Electrochemical Society. With permission.)

For resolving the above-mentioned problem and the problem of low luminescent efficiency of PSi formed by stain etching, Kolasinski and co-workers (Nahidi and Kolasinski 2006; Dudley and Kolasinski 2009; Kolasinski and Gogola 2011) tested several new classes of strain etchants, prepared from mixtures of HF, NH$_4$HF$_2$, HNO$_3$, FeCl$_3$, NaMnO$_4$, H$_2$O, and HCl. They found that no bubbles were formed when FeCl$_3$ and NaMnO$_4$ were used as oxidants (see Figure 10.5). As it is known, stain etching of Si is an electroless form of PSi formation that involves an acidic aqueous mixture of fluoride and an oxidant. The role of the oxidant is to inject holes into the Si valence band (Kolasinski 2003). This result demonstrates that gas evolution is not inherent to PSi formation. In the absence of bubbles, more structurally and optically homogeneous films were produced. Moreover, compared to a "standard" HF:HNO$_3$:H$_2$O=2:1:2 by volume solution, Dudley and Kolasinski (2009) have shown that the substitution of NH$_4$HF$_2$ for HF and/or Fe$^{3+}$ or MnO$_4^-$ for HNO$_3$ promoted the formation of PSi films with significantly more intense PL. However, it was established that these solutions, especially those formed from NH$_4$HF$_2$ + HCl + FeCl$_3$ and HF + FeCl$_3$ exhibited a substantial induction period before the initiation of PSi formation. The induction period leads to great variability in the results of etching when etching times are short. However, in later studies, Kolasinski (2010) showed that the optimization of new formulations of stain etchants that utilize Fe$^{3+}$, VO$_2^+$, and Ce$^{4+}$ as the oxidant allows getting around many of the problems associated with common nitrate/nitrite based stain etchants. The new formulations needed no "activation," exhibited short if any induction time, produced homogeneous films, and were quite reproducible. Film thicknesses of 10–20 μm were easily obtained in as short as 60 min of etching. Subsequently, Loni et al. (2011) showed that this approach could be used for porosification of metallurgical-grade silicon powder (10 m$^2$/g surface area). Using stain etching in FeCl$_3$:HF solution at reduced temperature (up to –25°C), mesoporous powders with surface areas up to 480 m$^2$/g and very high chemical reactivity were prepared. The reduced temperature, and its subsequent control, favored pore nucleation and propagation while minimizing bulk chemical etching.

Kolasinski and Barclay (2013) showed also that further optimization of the etching process can be achieved by combining stain etching with metal-assisted etching. This conjunction promotes the increase of the etch rate and changes the structure of the film in a manner that depends on the chemical identity and morphology of the metal. However, it was found that the film etched using Pd were generally not photoluminescent. Etching using other metals (Ag, Pt) gave photoluminescent porous silicon films.

## 10.2 CHEMICAL VAPOR ETCHING

An interesting, but not well-understood variation on stain etching is that of chemical vapor etching (VE) (Saadoun et al. 2002, 2003; Ben Jaballah et al. 2004, 2005; Ben Rabha et al. 2005; Kolasinski 2005). In chemical VE, a solution similar to stain etching is made up from concentrated HF plus concentrated HNO$_3$. However, instead of dipping the Si substrate in the solution, the substrate is held above the solution. For this application, a Teflon cell usually is used in order to expose the Si surface to solution vapors. An example of an experimental setup used for VE is shown in Figure 10.6. As a rule, it includes a hermetic reactor placed in a special hood having a filter exhaust fan system. At a given temperature, the volume of the free acid container part (filled with air, comprised between the acid solution surface and the substrate) is saturated with acid vapors. Experiments have shown that for VE it is possible to use *p*-type or *n*-type substrates of low or high resistivity.

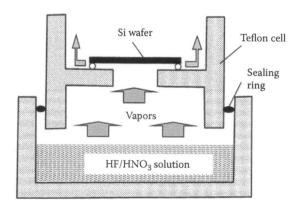

**FIGURE 10.6** Set-up scheme of the vapor-etching technique.

According to Boughaba and Wang (2006), the development of a gas (vapor)-based method for the fabrication of porous silicon layers presents two significant advantages. From the perspective of process integration, using conventional integrated circuits technology, the wet etching methods are not compatible with the widespread use of gas cluster tools. Furthermore, in a manufacturing environment, wet etching techniques generate large quantities of hazardous waste that require costly environmental management and disposal programs.

It is necessary to note that the mechanism of VE is not well understood. Infrared spectroscopy reveals a combination of hydrogen termination and oxidation of the surface (Saadoun et al. 2003; Ben Jaballah et al. 2004; Ben Rabha et al. 2005). Similar to stain etching, VE has an incubation period. Saadoun et al. (2002) observed that for $p$-type Si, the formation of the PSi layer begins at least 2 or 3 min after exposing the substrate to $HNO_3$/HF vapors.

Saadoun et al. (2002) have shown that in order to obtain homogeneous PSi layers, it is necessary to control the etch kinetics, and hence, all the determining experimental factors such as the $HNO_3$/HF volume ratio, the temperature of the acid mixture, and the exposure time of the Si substrate to the acid vapors. At that, last three parameters should vary in a limited range of values in order to prevent the damage of the forming PSi layers. As the temperature increases, the amount of the acid vapors may exceed the limit corresponding to saturation, and thus it forms a fog leading to the appearance of small drops on the container walls and particularly on the Si substrate. The size of the drops depends on the roughness of the Si substrate surface and on the saturation degree of the acid vapors. In the case where the Si substrate surface is rather rough (e.g., textured), the acid drops may coalesce leading to the formation of a continuous liquid film. The latter could destroy the PSi surface or prevent the progress of the VE process. At that, it is necessary to take into account that the capillary condensation in the pores may be occurring before droplets are observed on the external surface.

In order to resolve the problem of inhomogeneity and edge effects, Saadoun et al. (2003) proposed to enlarge the container and reduce the area of the Si substrate. Slight raising of the temperature of the Si substrate is other useful advice of Saadoun et al. (2003), which helps to prevent undesirable acid vapor condensation that may limit the etching progress and hence the formation of rather thick and controllable thickness.

Data from different articles indicate that the thickness of PSi layers may vary from 200 nm to 1–2 μm depending on temperature and the etching time. Usually a temperature equaled to 30–50°C is used. According to estimations (Kalem and Yavuzcetin 2000; de Vasconcelos et al. 2005), a growth rate is about 100 nm per hour.

Another variant of VE technology was proposed by Boughaba and Wang (2006). For etching, they used a mixture of oxygen ($O_2$) and nitrogen dioxide ($NO_2$) gases, combined with HF and water vapors.

The etching mechanism of Si during the proposed method of Si porosification is still under investigation, but the possible underlying processes that were taken into account in the selection of these gases could be represented by a combination of the following chemical reactions.

Formation of nitric acid:

(10.6)
$$4NO_2 + 2H_2O + O_2 \rightarrow 4HNO_3$$

(10.7)                    $$2NO_2 + H_2O \rightarrow HNO_2 + HNO_3$$

(10.8)                    $$3HNO_2 \rightarrow HNO_3 + 2NO + H_2O$$

(10.9)                    $$2NO + O_2 \rightarrow 2NO_2$$

Oxidation of silicon:

(10.10)                   $$4HNO_3 + Si \rightarrow SiO_2 + 2H_2O + 4NO_2$$

(10.11)                   $$Si + O_2 \rightarrow SiO_2$$

Etching of silicon dioxide:

(10.12)                   $$SiO_2 + 6HF \rightarrow H_2SiF_6 + 2H_2O$$

Schematic of the gas etching setup is shown in Figure 10.7.

A gas distribution plate was incorporated for improvement of the uniformity of the gas flow. The chamber, tray, and distribution plate were made of chemically inert Teflon. Pure oxygen was flown through a scrubber containing HF (47–51%), then merged with a flow of diluted nitrogen dioxide (2% in air) before entering the etching chamber. The outlet of the chamber was connected to a scrubber containing a 2 N sodium hydroxide (NaOH) solution made from electrolytic pellets. The role of the NaOH solution was to neutralize the HF. The HF scrubber could be kept at room temperature or heated up to 70°C. The $NO_2$ cylinder was heated at its base to a temperature of 40°C to avoid accumulation of nitrogen dioxide at its bottom and to enhance the mixing of $NO_2$ and air. Similarly, the stainless steel tubing connecting the $NO_2$ cylinder to the chamber was heated to a temperature of 30°C to avoid condensation of $NO_2$ on the tubing wall. Following the etching process, the samples were rinsed using ethyl alcohol; the substrates were dipped in ethyl alcohol for 5 min, then removed and left to dry in a high-purity (99.95%) nitrogen environment.

From our point of view, the approach mentioned above is more promising in comparison with etching in $HF/HNO_3$ vapors because it gives better control of the etching process. However, it needs further optimization.

Research carried out by Ben Rabha et al. (2005) has shown that the VE technique for Si porosification can be very attractive for fabrication of antireflective coatings in solar cells. The use of conventional PSi formation techniques, like anodization or "stain etching" for this purpose presents some difficulties compared to the VE method (Saadoun et al. 2002). While using

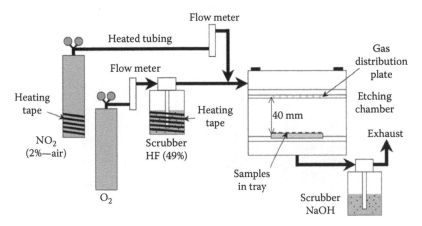

**FIGURE 10.7**    Schematic of the gas etching setup. (Reprinted from *Thin Solid Films* 497, Boughaba S. and Wang K., 83, Copyright 2006, with permission from Elsevier.)

conventional techniques to elaborate PSi on the frontal surface of Si solar cells, the wafer must be immersed in acid solutions. This may cause serious problems, like the destruction of the metallic front grid. Thus, supplementary and delicate operations like contact masking and time etching minimization should be taken into account. In the VE method, Si wafer is not immersed into acid solutions, only a vapor phase etching is applied. The technique enables us to grow porous semiconductors from the gas phase etchants instead of chemical solution. These results suggest that it can be effectively used to process selectively on semiconductor surfaces (Kalem and Yavuzcetin 2000). Besides, it was shown that this very simple and low-cost method is easily adapted to cover large surfaces (Ben Rabha et al. 2005).

### 10.2.1 STRUCTURE CHARACTERIZATION

Preliminary results indicate that the morphology of the VE-based PSi layers are different from what has been reported for electrochemically or stain etching formed PSi. SEM investigations (see Figure 10.8) showed that PSi prepared from $HNO_3/HF$ VE had a mesoporous structure formed by interconnected seeds (Saadoun et al. 2002, 2003).

In contrast to electrochemical etching, morphological investigations made on VE-based PSi show that there are no significant differences between $p$- and $p^+$-type PSi structures (Saadoun et al. 2002). As it was established $p$- and $p^+$-type PSi layers prepared using the conventional electrochemical etching (EE) technique exhibit some differences regarding their structures. SEM images of prepared porous layers usually are homogeneous with quite rough PSi/Si interface.

SEM images of PSi layers prepared during etching in the flow of $O_2 + NO_2$ gases and HF vapors are shown in Figure 10.9.

It is seen that the etching resulted in the formation of a porous layer, which consisted of islands (Figure 10.9b through e) embedded in a matrix of different morphology (Figure 10.9f). A pyramidal texture was observed inside the islands, while a smoother surface was obtained in the matrix. For 30 min etching, the thickness was found to have an average value of ~1.2 μm inside the islands and ~300 nm in the matrix. Boughaba and Wang (2006) have found that such an island pattern is representative of all samples etched at various conditions. They believe that the islands are attributed to the condensation of reactant-laden drops onto the surface during the etching process, because at the end of the process, upon opening of the chamber, the surface was always found to be uniformly covered by droplets. The size of the islands is likely to be directly related to the size of the drops, which depends on the surface tension and liquid drop coalescence driven by surface tension. So, two concurrent etching modes could be considered to describe the gas etching process: a liquid-based one, occurring within the areas covered by condensed drops, and a vapor phase-based one, taking place between the drops. The first process could be attributed to anisotropic etching of silicon, while the second one to an isotropic etching. The density and size of the islands was found to be affected by the $O_2:NO_2$ flow rate ratio for a given $O_2$ flow rate. It was found that the island size increased when decreasing the $O_2:NO_2$ ratio, while the island density decreased concurrently.

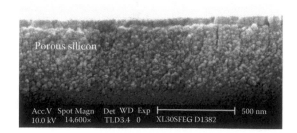

**FIGURE 10.8**    SEM cross-section view of a PSi layer prepared by VE $p$-type (100) oriented Si wafer. The $HNO_3/HF$ volume ratio is 1/8, the exposure time is 15 min, and the temperature of the acid solution is 30°C. (Reprinted from *Appl. Surf. Sci.* 210, Saadoun M. et al., 240, Copyright 2003, with permission from Elsevier.)

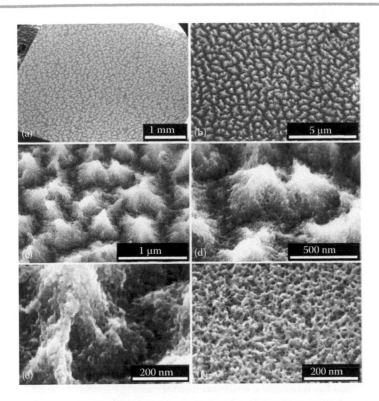

**FIGURE 10.9**  Scanning electron micrographs of a silicon sample etched using $O_2$ and $NO_2$ flow rates of 100 and 10 ml/min, respectively. (a) Low magnification of sample; (b–e) increasing magnifications of the surface inside an island; (f) surface between islands. (Reprinted from *Sens. Actuators B* 120, Tessier D.C. et al., 220, Copyright 2006, with permission from Elsevier.)

## 10.2.2  PL MEASUREMENTS

De Vasconcelos et al. (2005) have found that the overall appearance (color and intensity) of the light emitted by the PSi layers is not significantly affected by the type and resistivity of the exposed substrate. Both *p*- and *n*-type of substrates with resistance from 0.020 Ω·cm to 1000 Ω·cm for *p*-type and from 0.020 Ω·cm to 100 Ω·cm for *n*-type were tested and gave similar results. The roughness of the exposed Si surface is also not important.

Usually, the PL peak emission of VE-based PSi is located in the range from 750 to 630 nm (see Figure 10.10) (Kanemitsu et al. 1994; Kalem and Rosenbauer 1995; Saadoun et al. 2002). The position of the PL maximum depends on the etching parameters. It is important that the samples obtained with the stain-etching technique from the same solution exhibit a PL peak at around 630 nm for a series of samples grown under different conditions. The 750 nm emission has also been reported from thermally oxidized porous Si samples with oxyhydrides formed at the surface of nanocrystallites (Kanemitsu et al. 1994).

Mentioned behavior of PL spectra of VE-based PSi is being connected with peculiarity of chemical composition of PSi layer formed during the process of VE. Research has shown that depending on etching parameters, either a PSi layer or a layer composed primarily of $(NH_4)_2SiF_6$ (white powders [WP]) with a thin PSi transition layer is being formed. According to Saadoun et al. (2003), PSi was found to be a major phase above a $HNO_3/HF$ volume ratio of 1/9, while the WP phase becomes major at $HNO_3/HF$ volume ratio ranging between 1/6 and 1/2. Similar results were obtained by Aouida et al. (2005). In addition, Saadoun et al. (2003) have found that the increase of etching time increases the contents of WP. At an early stage of formation, this powder is formed by very small particles; visible only by SEM. Depending on preparation parameters, the $(NH_4)_2SiF_6$ could contain Si nanocrystallites having different size distribution (Aouida et al. 2005).

Both PSi and WP layers are photoluminescent. Koker et al. (2002) have demonstrated that the luminescence associated with a hexafluorosilicate/PSi interface is blue shifted compared to the PL from the pure PSi layer. A similar trend was found by Saadoun et al. (2003).

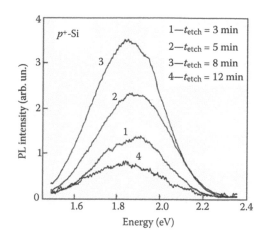

**FIGURE 10.10**   Spectral evolution of the PL of VE-based PSi layers formed on $p^+$-type Si substrate with the exposure to HNO$_3$/HF vapors: The HNO$_3$/HF volume ratio and the temperature of the acid mixture are fixed at 1/4 and 40°C, respectively. The PL spectra were recorded at 300 K under vacuum (10$^{-6}$ torr). The excitation source is the 514.5 nm line of an Ar+ laser. (Reprinted from *Thin Solid Films* 405, Saadoun M. et al., 29, Copyright 2002, with permission from Elsevier.)

According to Aouida et al. (2005), a photoluminescence (PL) band around 1.93 eV is conditioned by the emission from dot-like Si particles with sizes not exceeding 5 nm. The shoulder at 2.09 eV was attributed to an excitonic emission from the energy levels of the SiO$_x$ surrounding the smallest Si nanocrystallites. This PL band becomes more significant after oxidation in air. Aouida et al. (2005) assumed also that the PL band emission of the (NH$_4$)$_2$SiF$_6$ powder presents two peaks. The first one was attributed to Si nanocrystallites emitting at 1.98 eV. The second peak could be associated with the smallest nanocrystallites (≤1.5 nm). For these crystallites, excitons are trapped on the SiO$_x$ energy levels, leading to a maximum PL band emission around 2.1 eV. This PL band seems to have the same origin than the small shoulder observed in the PL emission of PSi.

As ammonium hexafluorosilicate is water-soluble and PSi is soluble in alkaline solutions, CVE can be used to form grooves in Si, which is of interest for the use of PSi in solar cell technology (Khedher et al. 2004; Ben Rabha et al. 2005; Ben Jaballah et al. 2005; Hajji et al. 2005).

PL spectra of PSi layers prepared during etching in the flow of O$_2$ + NO$_2$ gases and HF vapors are shown in Figure 10.11.

It is seen that PL peak emission in contrast to etching in HF/HNO$_3$ vapors is located in the range from 650 to 670 nm. The average peak wavelength was determined to be 658 nm. At that, Boughaba and Wang (2006) have found that there was no dependence of the average peak

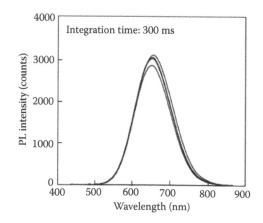

**FIGURE 10.11**   Photoluminescence spectra measured on four samples etched using O$_2$ and NO$_2$ flow rates of 200 and 10 ml/min, respectively. HF scrubber kept at room temperature. Integration time during PL measurements: 300 ms. The excitation source with the 470-nm line was used. (Reprinted from *Thin Solid Films* 497, Boughaba S. and Wang K., 83, Copyright 2006, with permission from Elsevier.)

wavelength on the processing conditions, in conformity with the fact that comparable porosity was obtained for all conditions. Furthermore, no dependence of the PL peak intensity on the gas flow rate was observed.

## 10.3 LASER-INDUCED ETCHING

The method of photochemical or laser-induced etching (LIE) was first reported by Noguchi and Suemune (1993) and Cheah and Choy (1994). However, it is necessary to note that, in fact, LIE of silicon is not new; it was reported by Chuang (1981) and Houle (1989). The main feature of this fabrication was that the silicon wafer was etched without any electrodes for anodization. The source of charge carriers for the chemical reactions only came from photogeneration. Excess holes are generated when an $n$-type Si wafer is irradiated with laser light in HF acid. These laser-generated holes diffuse toward the surface due to an electric field produced by band bending at the Si/electrolyte interface and enhance the spontaneous etching of silicon (Koker and Kolasinski 2000, 2001). As it is known, the etching of $n$-type Si is initiated by a holes' capture at the surface (Smith and Collins 1992). According to Lehmann and Gosele (1991), the capture of a hole at the surface leads to nucleophilic attack on Si–H bonds by fluoride ions and forming of Si–F bonds. Due to polarization induced by Si–F groups, the electron density of Si–Si backbonds is lowered and these weakened bonds will now be attacked by HF or $HF_2^-$. The surface reaction is assumed to produce $SiF_4$ or $SiHF_3$. Since these are not stable in water and Si atoms are removed in the form of $H_2SiF_6$, the final product and overall reaction is commonly written as

$$(10.13) \qquad Si + 6HF + 2H^+ = SiF_6^{2-} + 2H_2 + 4H^+.$$

After removal of Si atoms, the atomic-sized dip remains at the surface, which changes the electric field distribution in such a way that the holes transport to the bottom of pores rather than to pore walls. This process reconstructs the surface to form nanocrystallites. Without these charge carriers, the etching rate will be considerably slowed down (Choy and Cheah 1995).

The typical experimental set-up used for laser-induced etching is shown in Figure 10.12. The mixture of HF and $H_2O_2$ as an oxidant (HF:$H_2O_2$ = 6:1) usually is used for Si etching. Laser irradiation of Si substrate takes place through etching solution. The photochemical etching time may be varied from several minutes to several hours. In shown set-up, the He-Ne laser is used. However, it is necessary to say that besides He-Ne laser other lasers may be applied as well. For example, Cheah and Choy (1994) have fabricated Si nanocrystallites using different laser photon energies from 1.59 to 2.71 eV.

The influence of laser power density on the etching rate of the Si wafer is shown in Figure 10.13. It is seen that the etching rate increases by increasing the etching laser power density. Naturally,

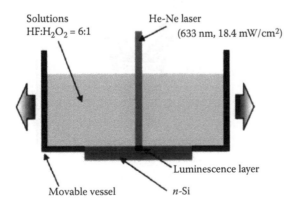

**FIGURE 10.12** Photo-chemical etching method. (Reprinted from *Thin Solid Films* 359, Yamamoto N. and Takai H., 184, Copyright 2000, with permission from Elsevier.)

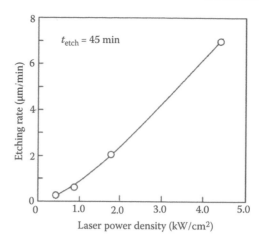

**FIGURE 10.13**    Etching rate for the porous silicon samples as a function of the laser power density. Laser etching was performed by using an argon ion laser beam (photon energy of 2.41 eV) focused onto the Si wafer. (Reprinted from *Micron.* 39, Kumar R. et al., 287, Copyright 2008, with permission from Elsevier.)

when the laser power density is increased, the number of the photo-generated electron hole pairs is also increased. Therefore, etching rate increases with the number of holes created per second.

It was established that both the Si etching and the PSi layer forming take place only in the region of laser irradiation. At that, the profile of etching layer corresponds to a Gaussian intensity profile of the laser beam (Kolasinski et al. 2000). Figure 10.14 displays a typical film grown after exposure to an HeNe laser at normal incidence.

Koker and Kolasinski (2000) established that during photoelectrical etching there are three distinct etching processes—two photoelectrochemical and one chemical. These processes are distinct in that they achieve three distinguishable outcomes: porous silicon formation (anisotropic photoelectrochemical etching), porous silicon removal (isotropic photoelectrochemical etching), and porous silicon removal in the dark (isotropic chemical etching).

Profilometry studies carried out by Kolasinski et al. (2000) revealed that both the upper and lower interfaces of the film descend deeper into the film as etching time proceeds. However, the PSi formation rate exceeds the rate of photochemical removal, and therefore, the thickness of PSi layer increases with time. In this manner, films up to 10 μm thick have been grown (see Figure 10.15). We need to note that the feature mentioned above (of PSi layer formed by laser-induced etching) may be considered a disadvantage of the discussed method of Si porosification.

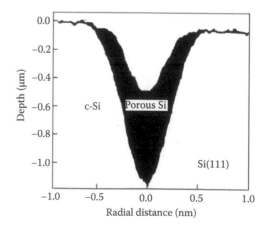

**FIGURE 10.14**    A cross-sectional profile of a PSi film grown photochemically by irradiation of Si(111) in 48% HF with a 15-mW HeNe laser (633 nm, 3 mW). (Reprinted from *Mater. Sci. Eng. B* 69–70, Kolasinski K.W. et al., 157, Copyright 2000, with permission from Elsevier.)

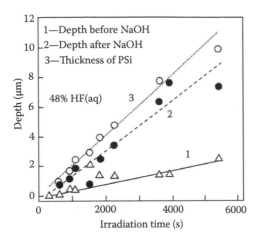

**FIGURE 10.15** The dependence of the maximum interface depth as determined by profilometry on etching time in 48% HF(aq) for irradiation with a 0.55-W/cm² HeNe laser. The thickness of the PSi is obtained by subtracting the maximum depth of the upper interface from the maximum depth of the lower interface. (From Koker L. and Kolasinski K.W., *Phys. Chem. Chem. Phys.* 2, 277, 2000. Reproduced by permission of the Royal Society of Chemistry.)

### 10.3.1 PL CHARACTERIZATION

Typical PL spectra of PSi prepared by laser-induced etching by Mavi et al. (2006) are shown in Figure 10.16.

It is seen that the PL spectra of the laser-etched samples exhibit broad visible luminescence at room temperature. PL spectra have two peaks at 1.91 and 2.05 eV arising from two predominant size distributions of nanocrystallites along the depth. The same PL spectra were observed by Kumar et al. (2008). It is necessary to note that the peak position and the wide full width at half maximum (FWHM) from these layers are considered to be similar to those from porous silicon or stain-etched layers. As is known, a visible luminescence layer formed on the Si wafer by the stain-etching method in HF solution with $HNO_3$ as oxidant shows strong yellow or red luminescence (Shih et al. 1992b; Schoisswohl et al. 1995).

It was established that PL peak intensity increases with increasing the etching time and then decreases (Choy and Cheah 1995). Yamamoto and Takai (2000) have shown that the peak wavelength becomes shorter with increasing etching time, and then saturates (see Figure 10.17). The layer etched for 5 min shows red luminescence with a peak at 695 nm, while the etched layer

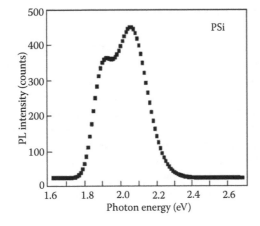

**FIGURE 10.16** PL spectra of laser-etched Si samples recorded with excitation photon energy of 2.71 eV: The PL spectra of laser-etched Si were recorded using the 457.9-nm (2.71 eV) line of an argon ion laser with 50 mW power and integration time of 1 s. (Mavi H.S. et al.: *Phys. Stat. Sol. (a)*. 2006. 203(10). 2444. Copyright Wiley-VCH Verlag GmbH & Co. KGaA. Reproduced with permission.)

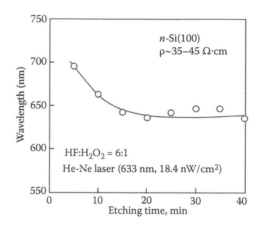

**FIGURE 10.17** The etching time dependence of PL peak wavelength from etched layer. Photoluminescence from the PSi layers at RT was measured using He-Cd laser ($\lambda = 325$ nm) for excitation. (Reprinted from *Thin Solid Films* 359, Yamamoto N. and Takai H., 184, Copyright 2000, with permission from Elsevier.)

for 30 min shows the strong yellow luminescence with a peak at 637 nm. In the photochemical etching method, therefore, it is clear that the peak wavelength between 695 and 637 nm can be controlled by the etching time.

Kolasinski et al. (2000) also observed the shift of PL maximum in the blue range for PSi layers grown photochemically by irradiation with a 15-mW HeNe laser. However, this shift was smaller and observed in the range from 570 to 530 nm. In addition, Kolasinski et al. (2000) have demonstrated that the PL spectrum obtained depends not only on the duration of irradiation but also on the wavelength of the incident light ($\lambda_{fab}$) used to fabricate the film (for $\lambda_{fab} > 473$ nm). Both 365- and 473-nm light resulted in the same PL spectrum (within spot to spot variations) with $\lambda_{peak}$: 550 nm. For longer wavelengths, on the other hand, $\lambda_{peak}$ was found to shift redder as $\lambda_{fab}$ was made progressively redder. Specifically, $\lambda_{fab} = 633$, 685, and 730 nm resulted in $\lambda_{peak}$: 630, 650, and 670 nm, respectively. This value of $\lambda_{peak}$, however, is only an approximation because of photochemistry induced by the excitation laser. Even though the in situ PL spectra exhibit a dependence on $\lambda_{fab}$, we observed that all films exhibit more or less the same PL spectrum once they are exposed to air. In air, $\lambda_{peak} = 670$–700 nm independent of $\lambda_{fab}$ but dependent somewhat on the length of exposure to air. Kolasinski et al. (2000) assumed that the changes in the spectrum must be the result of structural modifications to the film caused by etching.

## 10.3.2 STRUCTURE CHARACTERIZATION

Analysis of the PSi layer morphology has shown that the structure of the porous layer during laser-induced etching can be controlled by the excitation laser power density, etching time, concentration of etchant, and wavelength of laser source used (Wellner et al. 2000; Rasheed et al. 2001; Mavi et al. 2006). In particular, Kumar et al. (2008) have shown that depending on the laser power density, different etching stages naming cracks, pores, and pillar-like structures can be formed on the Si wafer. To get the porous structures, the laser power density should be neither very low nor very high. Cracks are formed on the wafer at very low laser power density and pillar-like structures are formed at very high laser power density. The pillar-type structures can be understood as follows. As the laser power density increases, the pore walls are getting thinner and thinner and these pore walls collapse at very high laser power density due to very high etching rates.

If Si wafers are etched at a faster rate, wider and deeper pores can be formed. It means that the pore width and pore depth can be controlled by controlling the etching laser power density and thus the etching rate. At very high laser power density, a hole can be drilled on the Si wafer. Based on observed low and high magnification SEM and AFM images, Kumar et al. (2008) concluded that macro- and micro-surface morphology reconstruction take place simultaneously during laser etching of Si wafers because of increased etching rate. Macrosurface morphology reconstruction takes place at the wafer surface and microsurface morphology reconstruction occurs inside the pore walls.

Cheah and Choy (1994) established that laser photon energy also influences PSi layer morphology. They have fabricated PSi layers using different laser photon energies from 1.59 to 2.71 eV. Long and regular column-like structures of porous Si were produced using a 2.61-eV laser line with fine Si wires on top of these columns with diameters of 200–300 nm. Noguchi and Suemune (1993), who have used an He–Ne laser (630 nm), observed nanoparticles with diameters in the range 5–20 nm.

However, one should admit that the discussed laser-induced method of Si porosification has small prospects to be introduced in mass production. It comes from disadvantages, such as layer inhomogeneity over the area, the presence of optical radiation source, optical system, and system of either scanning or sample's displacement. Irreproducibility of PSi layer parameters is another disadvantage of this method (Koker and Kolasinski 2000). According to Koker and Kolasinski (2000), hydrocarbon contamination is one contributor to irreproducibility.

## 10.4 METAL-ASSISTED ETCHING

As it was indicated above, stain etching, used for porous silicon forming, is a slow enough process. However, research made by Dimova-Malinovska et al. (1997) demonstrated that stain etching in the presence on the Si surface of an evaporated Al layer (200 nm) can be considerably accelerated. For more details, one could refer to the recent paper of Kolasinski (2005), Chartier et al. (2008), Huang et al. (2011), and the references therein. However, it was found that produced PSi was ~10 times weaker in luminescence than anodically etched PSi of similar thickness. This approach gained further development in research of Kelly and co-workers (Ashruf et al. 1999; Xia et al. 2000), Bohn and co-workers (Li and Bohn 2000; Harada et al. 2001; Chattopadhyay et al. 2002; Chattopadhyay and Bohn 2004), and many others (Mitsugi and Nagai 2004; Hadjersi et al. 2005a). It was shown that noble metals (Pd, Pt, Au, Ag) also promote the stain etching in the solution, containing HF and an oxidant ($H_2O_2$). This process is a controlled one, applicable for forming both microporous and nanoporous Si with intense PL and required structure. A thin metallic film, generally deposited directly on a silicon surface prior to immersion in solution, facilitates the etching in HF and $H_2O_2$, and of the metals investigated, Pt yields the fastest etch rates and produces PSi with the most intense PL. At that, gas evolution from the metal-coated area was clearly observed, especially for Pt and Au/Pd. For these metals, no metal dissolution was observed, in contrast to the behavior of Ag and Al. It has been demonstrated that a dissolution and/or redeposition process occurred for Ag particles, takes place due to the relatively low electrochemical potential of Ag. Pt and Au particles were stable in the etchant even for concentrations of $H_2O_2$ as high as 8.1 M and they maintained their initial shapes during the etching.

Noble metals, Ag, Au, Pt, and Pd, can be deposited on the Si substrate via various methods, which include thermal evaporation, sputtering, electron beam evaporation, electroless deposition, focused-ion-beam (FIB)-assisted deposition, or spin-coating of particles via other methods (Huang et al. 2011). We need to note that physical deposition in vacuum (e.g., thermal evaporation, sputtering, and e-beam evaporation) is favorable for preparing patterned structures of PSi by metal-assisted chemical etching because the morphology of the resulting noble metal film can more easily be controlled in these methods. However, electroless deposition is a simpler method for the deposition of noble metals on the surface of Si, and therefore this method is usually utilized to deposit noble metals if there is no strict demand on the morphology of the resulting etched structures.

A possible mechanism of porous growth using Ag-assisted etching in 10 mM $Fe(NO_3)_3$ + 4.6 M HF solution (Peng and Zhu 2004) is shown in Figure 10.18. Deposited silver particles catalyze the reduction of ferric ions and function as a local cathode, where electrons to be consumed are supplied by the oxidation reaction of silicon. Bare silicon just beneath the silver particles serves as an effective local anode. HF dissolves the formed oxides and finally leaves silicon nanostructures. The behavior is independent of the type of dopant.

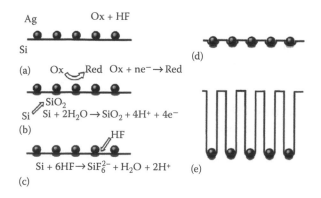

**FIGURE 10.18** Selective growth of pores. (a) Deposition of metal particles on silicon. (b) Reduction reaction is catalyzed on the particles. Counter reactions accompanying silicon oxidation proceed beneath the particles. (c) Formed oxides are dissolved by HF. (d) Metal particles sink into the silicon substrate. (e) The continuation bores deep holes into silicon leaving nanowires structure. (Reprinted from *Curr. Opin. Solid State Mater. Sci.* 10, Ogata Y.H. et al., 163, Copyright 2006, with permission from Elsevier.)

It was found that every metal layer used for Si etching has a different normal potential and, therefore, this results in a different current in combination with silicon. It is the same effect as in different current densities utilized by anodization (Splinter et al. 2000). This metallization is performed by various techniques such as sputtering, thermal evaporation, electrochemical deposition, or electroless deposition in HF solutions.

For explanation of observed effects during metal-assisted stain-etching, Li and Bohn (2000) proposed a reaction scheme involving local coupling of redox reactions with the metal.

Cathode:

$$(10.14) \qquad H_2O_2 + 2H^+ \rightarrow 2H_2O + 2h^+$$

$$(10.15) \qquad 2H^+ + 2e^- \rightarrow 2H_2 \uparrow$$

Reaction:

$$(10.16) \qquad Si + 4h^+ + 4HF \rightarrow SiF_4 + 4H^+$$

$$(10.17) \qquad SiF_4 + 2HF \rightarrow H_2SiF_6$$

Overall:

$$(10.18) \qquad Si + H_2O_2 + 6HF \rightarrow 2H_2O + H_2SiF_6 + H_2 \uparrow$$

Si-HF etching may occur as a localized electrochemical process, with the nanometer-sized metal acting as a local cathode. In other words, when oxidants ($H_2O_2$) are reduced on the surfaces of noble metal catalysts, positive holes ($h^+$) are generated. After the removal of electrons from metal particles, the potential of the metal shifts toward a positive value to a level enabling the injection of $h^+$ into the silicon substrate. The holes then diffuse away from the metal particle explaining why etching is confined to the near-particle area. Finally, anodic oxidation and the dissolution of silicon take place in the chemical etchant containing HF. The observations of much higher etch rates for Pt and Pd than for Au suggest a catalytic role of noble metals in etching.

A number of studies have appeared in the past years involving electroless metal-particle-assisted etching. Depending on the type of metal deposited and Si doping type and doping level, PSi films with different morphologies and light-emitting properties were produced on both *p*- and *n*-type Si. As it is seen from the results presented in Table 10.1, besides $H_2O_2$ nitric acid

**TABLE 10.1    Parameters of Si Porosification Using Metal-Assisted Etching**

| Metal | Si | Solution | Pores | References |
|---|---|---|---|---|
| Au, Pt, Au/Pt (3–20 nm) | Si(100) 0.005–10 $\Omega \cdot$cm | $HF/H_2O_2/CH_3OH = 1:2:1$ (or $CH_3CH_2OH$) solution ($t \sim 2$–20 s) | Nano | Li and Bohn 2000 |
| Au (~100 nm) Spin coating | $p$-Si(100) ~0.01 $\Omega \cdot$cm | 4.8 M HF + 0.4 M $H_2O_2$ ($t$ – 1–60 min) | Nano | Scheeler et al. 2012 |
| Pt | $n$-Si(100) 10–15 $\Omega \cdot$cm | 5% HF + 1% $H_2O_2$ | Macro | Xia et al. 2000 |
| Al/Au or Pd Chemical deposition | $n$-Si(100) 2–6 $\Omega \cdot$cm | (50% $HBF_4$):$HNO_3$ = 600:1 ($t$ – 2–8 min) | Nano | Unal et al. 2001 |
| Ag | $p$-Si | 4.6 M HF + 0.02 M $AgNO_3$ | Macro | Peng et al. 2005 |
| Ag (20 nm) Vacuum evaporation | $n$-Si(100) 1.6 $\Omega \cdot$cm | 0.05 M HF + 22.5 M ($K_2Cr_2O_7$, or $KMnO_4$ or ($Na_2S_2O_8$) ($t$ – 1–80 min) | Macro | Douani et al. 2008 |
| Ag (20 nm) Vacuum evaporation | $p$-Si(100) 100 $\Omega \cdot$cm | 22.5 M HF–0.05 M $K_2Cr_2O_7$–$H_2O$ ($t$ – 30 s–20 min) | | Hadjersi and Gabouze 2007 |
| Pd or Ag (200 nm) Vacuum evaporation | $p$-Si(111) 1–2 $\Omega \cdot$cm $p$-Si(100) 100 $\Omega \cdot$cm | 22.5 M HF:0.05 M $KMnO_4$:$H_2O$ or 22.5 M HF:0.05 M $Na_2S_2O_8$:$H_2O$ | Nano | Hadjersi et al. 2004 |
| Pd (10–80 nm) Immersing | Si(111) 0.5–2 $\Omega \cdot$cm | $HF$:$H_2O_2$:$H_2O$ = 25:100:20 ($t$ = 1–3 min) | Nano | Lipinski et al. 2009 |
| Bi (20 nm) Vacuum evaporation | $p$-Si(100) 100 $\Omega \cdot$cm | 22.5 M HF–0.03 M $Co(NO_3)2$–$H_2O$ ($t$ – 15–75 min) | | Megouda et al. 2009 |
| Bi (20 nm) Vacuum evaporation | $p$-Si(100) 100 $\Omega \cdot$cm | 22.5 M HF – 0.03 M $Co(NO_3)2$–$H_2O$ ($t$ – 5–75 min) | Macro | Megouda et al. 2009 |
| Cu Immersing | $p$-Si 200 $\Omega \cdot$cm | $HF$:$H_2O_2$:$H_2O$ = 5:1.25:30 ($t \sim 120$ min) | P~80% | Zheng et al. 2014 |

($HNO_3$), $K_2Cr_2O_7$, $Fe(NO_3)_3$, $KBrO_3$, $KMnO_4$, and sodium persulphate ($Na_2S_2O_8$) can be used as oxidizing agents (Douani et al. 2008; Huang et al. 2011). Dissolved $O_2$ can also play the role of oxidant but leads to etching at a very low rate (Yae et al. 2003).

Peng and co-workers (2005) have used Ag-assisted etching in $HF/AgNO_3$ solutions to form films composed of aligned Si nanowires. Etching for 20 min at 50°C created a film approximately 10 μm thick. After etching, Ag particles remain in the film. The films exhibit very low reflectivity, which makes them attractive for solar cell applications (Peng et al. 2005). Detailed study of $p$-Si stain etching with Ag assistance (20 nm) using 22.5 M HF/0.05 M $K_2Cr_2O_7$/$H_2O$ etching solution was carried out by Hadjersi and Gabouze (2007).

$HF/H_2O_2$ mixtures and Ag, Pt, Pd, or Cu particles deposited by electroless plating have been used by Tsujino and Matsumura (2005a,b) to etch cylindrical and helical pores in c-Si. A microporous layer with visible PL had thickness up to 3 μm for Pt assistance and only 300 nm when Ag was used. Sometimes for Pt, cylindrical or helical pores are found below the microporous region. The helical macropores are also being observed sometimes for Ag. Switching from cylindrical to helical pores is accomplished by changing the solution concentrations, and the walls of the macropores are lined with microporous silicon. Ag particles are found at the bottoms of these macropores, with a diameter matching that of the pore. If the etching time is extended to 10 h, pores as deep as 500 μm and ~50 nm in diameter are found.

Cruz et al. (2005) have studied $HF/H_2O_2/CH_3CH_2OH$ etching with Au or Pt particles and found that the etch depth and film morphology respond to doping level but not doping type. The metal films (1 nm < $h$ < 8 nm) were deposited by vacuum sputtering. Au is found to form a more columnar structure at a higher rate as opposed to a spongy structure for Pt (Cruz et al. 2005). They did observe the formation of some straight macropores but always in the presence of interconnecting lateral pores.

Thus, studies testify that pore morphology of PSi formed using metal-assisted chemical etching strongly depends on the type of noble metal used. If we generalize as before, we can conclude that the use of Ag and Au particles usually is accompanied by the formation of straight pores (Cruz et al. 2005), while the behavior of Pt particles is somewhat complex. In dependence of conditions of Pt-assisted etching straight pores, helical pores or curvy pores without a uniform etching direction can be formed in silicon substrates (Cruz et al. 2005; Huang et al. 2011). Moreover, the pores in or wires on substrates etched in the presence of Pt were usually surrounded by a porous layer (Tsujino and Matsumura 2005a), while no observable porous layer was found around the pores or wires etched from Au-coated (Lee et al. 2008) or Ag-coated (Tsujino and Matsumura 2005b) substrates under otherwise identical conditions. It was also established that compared to Ag-assisted etching, smaller pore sizes and depths are usually created using Au as the metal catalyst layer (Tsao and Chang (2011). The specific type of noble metal also influences the etching rate (Li and Bohn 2000; Cruz et al. 2005; Yae et al. 2012). Results related to Si etching in HF:$H_2$O solution saturated with oxygen are shown in Figure 10.19. However, we have to note that the difference in the etching rate and morphologies of the etched structures has not yet been well explained in literature (Huang et al. 2011). The difference in the catalytic activity of the noble metal for the $H_2O_2$ reduction might be a possible reason, although there is no literature directly comparing the catalytic activities of Pt, Au, and Ag particles on Si substrates for the $H_2O_2$ reduction.

It was established that the use of thinner films (3 nm < $h$ < 20 nm) significantly improves the parameters of PSi, decreases pore's size, and improves the uniformity of porous layer. AFM images have shown that thin metal coatings on Si with indicated above thickness usually appear as nanometer-size (~10 nm) islands, with thermal annealing producing larger islands. Li and Bohn (2000) stated that produced PSi pore sizes have no direct correspondence with the size or spacing of the deposited metal islands. However, it looks like this statement has essential limitations. SEM images' analysis testifies that just size of clusters and their density determine quantity and size of micropores observed in the PSi layer. The increase of noble metal islands usually increased the size of macropores. Tsao and Chang (2011) also studied metal-assisted etching of $p$-Si(100) (0.01~0.02 Ω·cm) in HF:$H_2O_2$:EtOH=1:1:1 and found that the thickness of the catalyst metal layer has obvious effects on the silicon nanostructure pore size, depth, and surface morphology. They established that as the Au or Ag layer thickness increased from 3 nm to 10 nm, the metal surface coverage percentages also increased from 35% to 80% and 39% to 85% for Au and Ag, respectively. In addition, it was found that at low metal deposition thicknesses with low metal surface coverage rates, porous silicon nanostructure surfaces were created, while when the metal layer thickness increased with higher metal surface coverage rate, the surface morphology of silicon nanostructures transformed from porous to wire or filament nanostructures. Typical SEM images of fabricated PSi using metal-assisted stain etching are shown in Figure 10.20.

It is seen that the surface of PSi presents a high density of macropores that penetrate into the bulk of the silicon. At that, the inner of the pore clearly appears nanoporous. In addition, the walls are seen to be parallel. Moreover, it has been shown that the pore formation develops with the etching time and etching temperature. Increased temperature (4°C vs. room temperature)

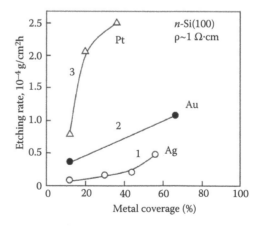

**FIGURE 10.19** Etching rate of $n$-Si(100) wafers (1 Ω·cm) during metal-assisted etching ($T$ = 298 K, $t$ = 24 h) as a function of metal coverage. 7.3 M HF:$H_2$O solution was saturated with oxygen. The following marks represent the deposited metals: 1: Ag, 2: Au, 3: Pt. (With kind permission from Springer Science+Business Media: *Nanoscale Res. Lett.* 7, 2012, 352, Yae S. et al.)

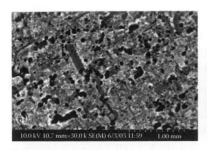

**FIGURE 10.20**    Plan (a) and cross-sectional (b) view of SEM images of Ag-assisted chemically etched silicon (100) with a resistivity of 100 Ω·cm in 22.5 HF-0.05 M $K_2Cr_2O_7$-$H_2O$ solution for 20 min. (Hadjersi T. and Gabouze N.: *Phys. Status Solidi (c)*. 2007. 4(6). 2155. Copyright Wiley-VCH Verlag GmbH & Co. KGaA. Reproduced with permission. )

increases the etch rate. Etch depth usually is proportional to etch time. Both macropore density and pore diameter increase with etching time as well. In addition, the surface appears rough and nonuniformly etched. Probably, after an optimal etching time, the macro-dissolution of the silicon takes place. Then nanoporous and macroporous silicon can coexist. It is important that the porous layer remains crystalline after etching.

It was found that the morphology of PSi layer strongly depends on the oxidant used. For example, Douani et al. (2008) have found that the Ag-assisted etching in $HF/Na_2S_2O_8$ solution gives macropores with a diameter ~3.5 μm; after etching in $HF/K_2Cr_2O_7$ solution, pores had diameter ~0.4 μm, and after etching in $HF/KMnO_4$ solution diameter of macropores was about 0.15–0.2 μm. Douani et al. (2008), while analyzing the morphology of porous films obtained by using metal-assisted stain etching, have made a conclusion that the etching depends on the rate at which the oxidant can generate holes under the metal particles. Indeed, if the oxidant generated a large number of holes, some of them will be consumed for oxidizing the silicon under the metal particles and the excess holes will diffuse laterally to oxidize the silicon in the near-particle area. The oxidized silicon then reacts with HF to form a water-soluble complex. In this case, the etching occurs at the same time under the metal particles and in the near-particle area, but with an etching rate higher in the former region. If a rate of hole generation is low, the etching takes place only under the metal particles.

The quality and uniformity of the metal layer also influences film morphology. For example, Li and Bohn (2000) studying Au-assisted stain etching in $HF/H_2O_2$ solution have found that large (~30 nm) interconnected pores propagating anisotropically perpendicular to the surface are observed on the Au-coated areas, while in areas between the Au on the same wafer, a much more compact structure with random arrays of small pores (~3 nm spaced by about 3 nm) can be seen (see Figure 10.21). The same situation was observed for other metals. At that, for all doping concentrations examined ($p^+$, $p^-$, and $n^+$), metal-coated areas always exhibit larger pores with columnar structure, while the off-metal areas always display smaller pores (3–5 nm) and randomly oriented structures. In each case, the exact pore sizes and connectivity vary with doping

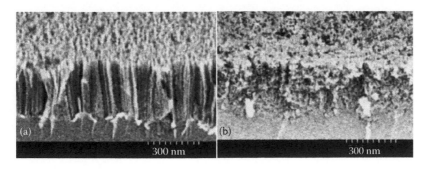

**FIGURE 10.21**    SEM images of Au-coated Si(100) after etching in $HF/H_2O_2$ for 30 s. (a) Au-coated area on $p^+$ Si, (b) off the Au-coated area on $p^+$ Si. (Reprinted with permission from Li X. and Bohn P.W., *Appl. Phys. Lett.* 77, 2572, 2000. Copyright 2000. American Institute of Physics.)

levels, producing large differences in luminescence properties. This means that by patterning the noble metal deposit, the resulting porous film can also be patterned (Chattopadhyay and Bohn 2004). An example of such patterned PSi layer is shown in Figure 10.22.

Regarding other parameters of silicon metal-assisted etching, there is information that the length of Si pores and nanowires fabricated by metal-assisted chemical etching in $HF/AgNO_3$ solution or $HF/H_2O_2$ solution increased approximately linearly with the etching time. At that, the etching rate increased with increasing etching temperature (Cheng et al. 2008). It was also established that etching of $p$- and $n$-type substrates of Si(100) can be conducted successfully both in the dark and with illumination. It was found that for the same etching times, the difference between etching depths in the dark and with room light illumination was less than 5% for both $p$- and $n$-type substrates, while the etching depth during illumination with a 20-W bulb was about 1.5 times the etching depth in the dark or with room light illumination, clearly demonstrating the influence the presence of illumination has on the etching rate.

It has been speculated that the metal-assisted etching is isotropic and the noble metal always catalyzes the etching along the vertical direction relative to the substrate surface. However, experiments carried out with (100) and (111) silicon substrates have shown that in reality the etching can be along the vertical direction, non-[100] directions, or the etching direction in Si(111) can be switch from the vertical direction to one of the <100> directions (Peng et al. 2006, 2008; Chen et al. 2008). Different conclusions concerning the relationship between etching rate and doping type or doping level of the Si substrate have been reported. Li et al. (2000) found that under identical conditions, Au-covered regions on a $p^+$-Si substrate (0.01–0.03 $\Omega$·cm) and Au-covered regions on a $p^-$-Si substrate (1–10 $\Omega$·cm) showed only small variations in pore size and etching depth, while Cruz et al. (2005) reported that the etching depth in Au-covered regions of a $p^-$ (10 $\Omega$·cm) Si substrate was 1.5 times larger than that of a $p^+$-Si substrate (0.01 $\Omega$·cm) under identical conditions. The reason for different etching rates for substrates with different doping levels remains unclear thus far. Concerning the doping type, Zhang et al. (2008) found that a $p$-type (7–13 $\Omega$·cm) substrate was etched more slowly than an $n$-type (7–13 $\Omega$·cm) substrate. This relationship was valid for both (100) and (111) substrates. More detailed discussions related to analysis of metal-assisted chemical etching and inter-correlation between PSi morphology and technological parameters of metal-assisted strain etching can be found in published papers (Li and Bohn 2000; Kolasinski 2005; Douani et al. 2008; Asoh et al. 2009; Huang et al. 2011).

It is interesting that PL of PSi prepared using metal-assisted etching has specific features in comparison with PSi prepared with using anodic etching. Hadjersi and Gabouze (2007) have shown that the evolution of PL intensity with etching time can be divided into two phases. In the first phase, the emission is in the blue region. In this region, two peaks were obtained, centered at 435 nm (2.85 eV) and 550 nm (2.25 eV). On other hand, in the second phase, the emission is in the red region with a peak at about 630 nm. In addition, it has been observed that the PL peak intensity increased with increasing etching time. The previous result is very interesting because it is not easy to obtain blue-light-emitting-layer directly by the anodization method. It is important that below certain etching time no photoluminescence be observed. After too long a time period of etching, the intensity of PL is being decreased as well. For processes discussed by Hadjersi and Gabouze (2007) mentioned above, times were equaled ~10 min and ~60 min, correspondingly.

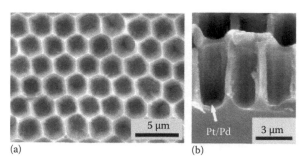

**FIGURE 10.22**   SEM images of Pt–Pd-coated silicon after chemical etching in $HF/H_2O_2$ for 2 min: (a) top view; (b) cross view. (Reprinted from *Electrochim. Acta* 54, Asoh H. et al., 5142, Copyright 2009, with permission from Elsevier.)

It is necessary to note that mentioned effect is typical for all metals and solutions used at metal-assisted stain etching (Megouda et al. 2009).

### 10.4.1 ADVANTAGES AND DISADVANTAGES OF METAL-ASSISTED STAIN ETCHING

Discussed analysis testifies that metal-assisted stain etching is really a perspective method of porous silicon forming. First, metal-assisted chemical etching is a simple and low-cost method for fabricating various Si nanostructures with the ability to control various parameters (e.g., cross-sectional shape, diameter, length, orientation, doping type, and doping level) (Huang et al. 2011). Ashruf et al. (1999) and Splinter et al. (2000) believe that the main advantage of this porous formation technique is that a special sample holder to contact the Si is not required. The process of anodization is easy to handle; it has no etching cell and needs an external current source. The etch rate may be controlled by the metal/Si area ratio and the concentration of oxidizing agent in solution. Almost all procedures can be accomplished in a chemical lab without expensive equipment. Second, metal-assisted chemical etching enables control of the orientation of Si nanostructures (e.g., nanowire, pore) relative to the substrate. Thus, pores ranging from mesopores to macropores could be obtained. This makes the technique suitable for batch fabrication of porous silicon devices. Moreover, using metal-assisted stain etching, porous silicon layers have been formed on highly resistive silicon, which are difficult to produce by the electrochemical etching (Hadjersi and Gabouze 2007). This feature is especially important for SOI technology (Backes et al. 2013). Furthermore, compared to stain-etched layers, galvanically formed porous layers have better uniformity and much bigger thickness (Li and Bohn 2000). Therefore, metal-assisted chemical etching has become increasingly important in the last decade (Huang et al. 2011; Scheeler et al. 2012). In the past, this method was successfully used in photonic, photovoltaic, or diffusion membrane applications (Splinter et al. 2000; Yae et al. 2003, 2005; Cruz et al. 2005; Hadjersi et al. 2005b; Peng et al. 2005; Tsujino and Matsumura 1996, 2005a,b). It is important that highly oriented arrays of Si nanowires (SiNW) can also be prepared using this technology (Peng et al. 2005, 2006, 2008; Geyer et al. 2012). This direct approach allows the rapid fabrication of high-quality, well-aligned SiNW arrays with large-area homogeneity and tunable depths. An example of such SiNWs is shown in Figure 10.23. One can find models explaining growth of SiNWs during metal assisted etching in the literature (Abouda-Lachiheb et al. 2012; Geyer et al. 2012; Smith et al. 2013).

However, one should admit that the present method is not well investigated yet. Therefore, until now, a well-established mechanism that describes the metal-assisted electroless etching process does not exist. There are considerable limitations on controllability and uniformity of porous layers' parameters along area and thickness as well. While conducting metal-assisted stain etching, it is necessary to control not only etching conditions and silicon parameters, but

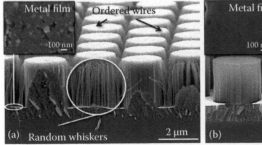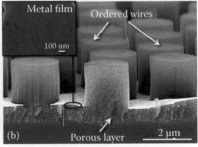

**FIGURE 10.23**   SEM images of Si wires etched in 5.65 M/L HF:0.10 M/L H$_2$O$_2$ solution using (a) a 15-nm and (b) a 55-nm thick Ag film. The etching using thin Ag films with random pores (see inset: top view of 15-nm thick metal film) leads to the formation of disordered nanowires (marked white). With increasing thickness of the metal, closed films (see inset: top view of 55-nm metal film) are obtained, and random nanowires can be avoided. In the two images, the metal film was not removed. (Reprinted with permission from Geyer N. et al., *J. Phys. Chem. C* 116, 13446, 2012. Copyright 2012. American Chemical Society.)

also characteristics of metal film (thickness, size of clusters) deposited on the surface of Si. It reduces even greater reproducibility of porous Si parameters. Furthermore, stain etching cannot be used to prepare stratified structures such as double layers or multilayered photonic crystals.

## 10.5  SPARK PROCESSING

In addition to the previously discussed methods of chemical etching, spark processing was also shown to generate a silicon-based substance that strongly photoluminesces in several visible bands at room temperature (Hummel and Chang 1992; Hummel et al. 1995a; John et al. 1996, 1997). The production of spark-processed Si (*sp*-Si) is quite simple. It does not utilize aqueous solutions, nor does it use anodic etching. It is some kind of erosion, taking place during spark processing. For this purpose, high frequency (several kHz), high voltage (several thousand volts, usually 10–15 kV), and low average current (typically 5–10 mA) electric pulses are applied for a certain length of time between a silicon substrate and a counter electrode (Sigel 1973/1974). For example, Hummel and Ludwig (1996) used unipolar pulses with of 0.02-μs duration and a frequency of 16.7 kHz. This counter-electrode may be a pointed piece of an Si wafer or a metal tip such as a tungsten wire (Hummel and Ludwig 1996). Wafer (cathode) and counter-electrode (anode) characteristically are separated by a 1–2 mm gap.

The affected area usually is between 1 and 5 mm in diameter and between hundreds of nanometers (nm) and tens of micrometers (μm) in depth, depending on the processing duration (typically 1–4 h). The resulting product is a grayish layer (see Figure 10.24). At that, the surface is very rough and the surface morphology consists of "cauliflower" shaped clusters with large holes regularly penetrating the Si surface (Weis et al. 2002).

Pulse frequency and applied voltage are only two of the many processing parameters that influence the spark-processing time and the PL intensity. Other processing parameters include either the environment in which the sparking is conducted or the wafer temperature. Results showing the influence of sparking processing on PL spectra are presented in Figure 10.25.

It was established that there is no need to conduct spark-processing necessarily in air. For example, nitrogen, $CO_2$, helium, argon, hydrogen, oxygen, or organic gas atmospheres have been used. Some of these processing atmospheres suppress the photoluminescence considerably (Ar: He), whereas others yield essentially similar PL intensities compared to ordinary air. At that, the largest energy of photons and the highest light intensity has been observed for Si, which has been spark-processed when a stream of dried, compressed air was directed toward the front surface of the wafer (Hummel and Ludwig 1996). For indicated conditions, a PL maximum near 3 eV (410 nm) is observed when *sp*-Si is illuminated by a continuous wave (CW) HeCd laser, having a wavelength of 325 nm. If, however, the spark processing is conducted in stagnant air, a blue-green PL is detected whose maximum wavelength centers around 525 nm (2.36 eV). It was also found that the wafer temperature during spark processing influences the green PL band. Specifically,

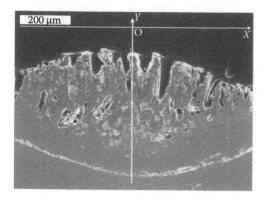

**FIGURE 10.24**    Cross-sectional SEM micrograph of *sp*-Si at a magnification of 120×. Samples were cut from a single *p*-type Si:B wafer, which contained (3–9) × 1018 cm⁻³. *Y* is the axis of symmetry, *X* is a radial axis. (Reprinted from *Mater. Sci. Eng. B* 107, Polihronov J.G. et al., 124, Copyright 2004, with permission from Elsevier.)

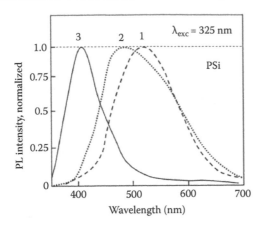

**FIGURE 10.25**  Normalized photoluminescence spectra of spark-processed silicon wafers. The following preparation conditions were utilized: (1) Spark processing at elevated temperatures (green luminescing sample); (2) spark-processing at somewhat lower temperatures compared to (1) (see text); (3) compressed, dry air was blown on the Si wafer during spark-processing (blue-luminescing *sp*-Si). (Reprinted from *Solid State Commun.* 93(3), Hummel R.E. et al., 237, Copyright 1995, with permission from Elsevier.)

a heated substrate yields an increase in peak intensity. For Si, which has been spark-processed while a stream of nitrogen gas was directed toward the front of the substrate, PL peak has been found to be 560 nm (compared to 410 nm for compressed air) (Hummel et al. 1995c). Thus, one can see that spark processing (sp) really creates a substance that allows Si to exhibit strong room temperature photoluminescence (PL) and cathodoluminescence in the violet to red spectral range. At that, it was found that crystalline Si, amorphous Si, and porous Si have a weaker PL intensity and poorer stability than *sp*-Si (Kwanghoon and Hummel 2008). More detailed comparison of porous Si with nanocrystalline Si, prepared using spark processing, can be found in a review prepared by Ludwig (1996). However, in reality, Si fabricated by this method is not porous, in the frame of ideas accepted for standard porous Si. Analysis of *sp*-Si TEM images and PL testify that spark processing yields essentially spherical, non-oriented silicon nanocrystals or Si clusters, which are imbedded in a "thick" matrix of Si oxide or oxinitride (Kovalev et al. 1994). According to Polihronov et al. (2004), the *sp*-Si sample has a characteristic cylindrical symmetry due to the uniform surface resistance of the Si substrate and to the random nature of spark processing. However, *sp*-Si is not isotropic, uniform, and random, exhibiting radial and axial anisotropy of porosity. Hummel et al. (1993) have assumed that the high-energy electric spark causes localized redeposition of silicon leading eventually to the nanometer-sized crystallites. Hummel et al. (1993) believe that the $SiO_2$-surrounded Si nanocrystallites are responsible for the observed visible room temperature photoluminescence. In other words, the pores in the silicon matrix are probably only of secondary importance for the light emission.

Besides, the morphology of *sp*-Si has the following features: *sp*-Si contains a large portion of internally embedded, closed pores (see Figure 10.29 later in the chapter). The size of pores can be changed in a very wide range from nm to μm. The porosity is not constant along the layer depth and the area. Moreover, the porosity of *sp*-Si is not easy to estimate. The traditional methods for porosity measurements cannot be applied due to the nature of *sp*-Si growth and morphology (Polihronov et al. 2004). Of course, such properties of *sp*-Si limit possibilities of its application, in spite of good luminescent characteristics.

## 10.6  REACTIVE ION (PLASMA) ETCHING

Reactive ion (plasma) etching (RIE) can be also used for porous Si forming. RIE is the most common dry etching method, which combines the effects of chemically active gaseous radicals and physical ion bombardment (Rangelow 1996, 2001; Elwenspoek and Jansen 1998; Franssila and Sainiemi 2008). Due to the combinations of reactive neutrals and an ion bombardment, the etch rate may be 10 times greater than that obtained by considering these contributions separately. Other dry etching methods include spontaneous chemical etching (e.g., $XeF_2$ etching) and ion beam etching. The application of RIE for PSi fabrication is conditioned by better anisotropy of the etching process. As it is known, spontaneous chemical etching without ion bombardment is a process of isotropic etching of silicon (Ibbotson et al. 1984). Deep reactive ion etching (DRIE) is considered an extension of RIE, but DRIE enables the fabrication of deeper and narrower structures with a higher etch rate than conventional RIE. DRIE reactors are also equipped with two power sources, an inductive coupled plasma (ICP) source for high-density plasma generation, and a capacitive coupled plasma (CCP) source for controlling the ion energies. However, unlike traditional RIE system, DRIE employs sidewall passivation to enhance process anisotropy. The passivation layer improves the directionality of the etching. The layer is removed from horizontal surfaces by sputtering, but sidewalls remain protected. DRIE techniques are typically utilized to create sidewalls that are as vertical as possible. DRIE can be used to produce tilted sidewalls as well. By controlling the amount of passivation during the process, both positively and negatively tapered sidewalls are attainable. If passivation is not used, DRIE is capable of producing completely isotropic etch profiles, which is especially beneficial for the release of freestanding structures (Sainiemi 2009).

### 10.6.1 FEATURES OF REACTIVE ION ETCHING

Detailed analysis of RIE can be found in published review articles and books (Winters 1978; Oehrlein 1990; Jansen et al. 1996, 2009; Elwenspoek and Jansen 1998; Kiihamaki and Franssila 1999; Cardinaud et al. 2000; Rangelow 2001; Franssila and Sainiemi 2008; Sainiemi 2009). Analysis of processes taking place during RIE can be found there as well. According to Li et al. (1994) these processes include: (1) mass transportation of reactive species from plasma to the substrate surface; (2) adsorption of the reactive species to the substrate surface; (3) chemical reaction of the species with the substrate material; (4) desorption of the reaction product from the substrate surface; and (5) removal of the reaction product from the system. Some typical characteristics of RIE process are in Table 10.2.

For RIE, $SF_6$ (Arens-Fischer et al. 2000), $Cl_2$ (Fischer and Chou 1993), or gas mixtures such as $SF_6/O_2$, $SF_6/C_4F_8$, $SF_6/CHF_3$ (Tserepi et al. 2003), and $Cl_2/BCl_3/H_2$ (French et al. 1997) are usually being used. The use of $SF_6/O_2$ plasma at various flow rates makes it possible to vary the etching directionality between isotropic and anisotropic ones, while $Cl_2$ offers lower etching rates but provides much better profile control (Yoo 2010). During $SF_6/O_2$ plasma etching, F* radicals provide the chemical etching of silicon materials by formation of volatile $SiF_4$. $O_2$ plasma produces O* radicals for sidewall passivation with $Si_xO_yF_z$, which helps to control the etching profiles. A low amount of oxygen in $CF_4$ or $SF_6$ plasma also increases the etch rate of silicon because the

**TABLE 10.2    Technological Parameters of Si Reactive Ion Etching**

| Plasma | Protection Layer | The Parameter of the RIE Process | Ref. |
|---|---|---|---|
| $SF_6/O_2$ | Photoresist, Al, $SiO_2$, $Al_2O_3$ | Plasmalab System 100 reactor (Oxford Instruments), ICP, CCP-13.56 MHz | Sainiemi 2009 |
| $SF_6$ | Ti (100 nm) | Pressure: 20 µbar; power: 70 W; voltage: 240 V and gas flow: 20 ml/min | Arens-Fischer et al. 2000 |
| $Cl_2$ | Cr (50 nm) | $Cl_2$ and $SiCl_4$ flow rates, 76.6 and 13.3 seem, respectively, a power density of 0.32 W/cm$^2$, and a pressure of 40 mTorr | Fischer and Chou 1993 |
| $SF_6$; $SF_6/C_4F_8$ (1:1); $SF_6/CHF_3$ | | High density plasma (HDP) reactor (micromachining etching tool of alcatel) and in a reactive ion etching (RIE) reactor (NE330 of Nextral) | Tserepi et al. 2003 |
| $Cl_2/BCl_3/H_2$ | | Plasma Therm PK-1250 parallel-plate RIE Power 200–400 W; Pressure 20–40 mtorr | French et al. 1997 |
| $Cl_2(Br_2)/BCl_3$ $SF_6/$ $O_2/CHF_3$ (Ar) | Cr, Ni, Al, or $SiO_2$ | RF power density: 0.05 W/cm$^2$; ion current density: 0.35 mA/cm$^2$ | Rangelow and Hudek 1995 |

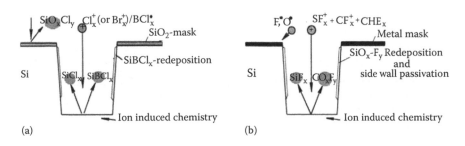

**FIGURE 10.26**   The chemistry during the etching of Si in (a) $Cl_2$ (or $Br_2$)/$BCl_3$ plasma, and (b) $SF_6$/$O_2$/$CHF_3$ plasma. (Reprinted from *Microel. Eng.* 27, Rangelow I.W. and Hudek P., 471, Copyright 1995, with permission from Elsevier.)

recombination of fluorine radicals with $CF^{3+}$ or $SF^{5+}$ ions is reduced, which increases the amount of free fluorine (Mogab et al. 1978). Too high oxygen concentration in plasma results in a thick passivation layer, which leads to a reduced etch rate and formation of silicon nanograss (or black silicon, silicon nanoturf, and columnar microstructures) (Jansen et al. 1995).

Formation of a non-volatile passivation layer on the substrate surface during the RIE process has been reported, especially when exploiting $CHF_3$ (or $CF_4$/$H_2$) plasma or plasma composed of molecules that contain fluorine (e.g., $SF_6$ or $CF_4$) and oxygen (Jansen et al. 1996). In the case of $CHF_3$ plasma, the passivation layer is composed of carbon and fluorine. The role of hydrogen is to catalyze the formation of polymeric precursors such as CF. Hydrogen also reduces the density of free fluorine radicals by forming HF. The quality of the passivation layer strongly depends on the process temperature as well. At cryogenic temperatures, the passivation layer is more stable and less oxygen is required for its formation (Jansen et al. 2009). Examples of RIE processes' chemistry are shown in Figure 10.26.

## 10.6.2  DEEP REACTIVE ION ETCHING

There are two main deep reactive ion etching (DRIE) processes. The most common one is the Bosch process (Lärmer and Schilp 1996), which is also known as "switched process" or "time multiplexed process." The other one is known as a "cryogenic process" due to its low process temperature (Tachi et al. 1988; de Boer et al. 2002). The Bosch process utilizes a separate passivation step followed by an etching step, while in the cryogenic process the passivation occurs simultaneously with the etching.

The Bosch process is the most widely used DRIE technique. It provides maximal etching rate, which attains 50 μm/min. The processing of masked silicon wafer starts with a short etching step that utilizes $SF_6$ plasma. After this etching step, a thin fluorocarbon film is deposited on the wafer. The fluorocarbon film passivates the surface and prevents etching. Octofluoro cyclobutane ($C_4F_8$) is commonly used in the passivation step. It generates ($CF_2$) n radicals and results in a Teflon-like soft polymer film. At the beginning of the next short etching step, the fluorocarbon film is removed from horizontal surfaces. The $SF_6$ etching step is not anisotropic, but the polymer still etches preferentially from the horizontal surfaces due to directional ion bombardment, while the vertical sidewalls remain protected. The repetition of etching and passivation cycles results in almost vertical sidewalls (Kiihamäki 2005).

The drawback of the process is the scalloping of the sidewalls due to the alternating etching and passivation steps (Andersson et al. 2000). The sidewall roughness can be reduced by shortening the duration of the etching and passivation steps (Sainiemi 2009) or by postprocessing: thermal oxidation followed by oxide etching (Matthews and Judy 2006) or annealing in a hydrogen atmosphere at high temperature (Lee and Wu 2006) reduces the size of the scallops. In the Bosch process, it is also important to have an adequate ratio between ions and radicals. A relative ion concentration that is too high degrades the sidewall profiles. The reactor issues are discussed in a more detailed manner by Walker (2001).

The cryogenic DRIE process does not have separate etching and passivation steps, as they both occur simultaneously. The etching is performed in $SF_6$/$O_2$ plasma. At cryogenic temperatures

($T < -100°C$), a passivating $SiO_xF_y$ layer forms on the top of the silicon surface (Dussart et al. 2004; Mellhaoui et al. 2005), which again is sputtered away from horizontal surfaces by directional ion bombardment. The substrate temperature plays a key role in cryogenic processes. Therefore, the possibility of controlling the substrate temperature accurately at very low temperatures is crucial. When the temperature is fixed, the thickness of the passivation layer is mainly determined by the $O_2$ flow rate. Too low oxygen flow results in the failing of the passivation layer and isotropic etching profiles, whereas too high oxygen content in the plasma leads to over-passivation, a reduction to the silicon etch rate, and the creation of black silicon (de Boer et al. 2002; Suni et al. 2008; Sainiemi 2009).

Changing the $SF_6/O_2$ ratio is the most convenient way to optimize passivation layer thickness and, ultimately, the sidewall angles. The etch rate of silicon is mainly dependent on $SF_6$ flow rate and the power of the ICP source. Higher $SF_6$ flow and ICP power increase the quantity of free fluorine radicals that result in the higher etch rate of silicon. The etch rate of the masking material is mainly dependent on the ion energies that are determined by CCP source. The ions have to have sufficient energy to remove the passivation layer from horizontal surfaces, but when a certain threshold is reached, an increase in CCP power only increases the etch rate of masking material and undercutting (Sainiemi 2009).

The main advantage of the cryogenic process over the Bosch process is smooth sidewalls. The sidewall quality of structures etched using cryogenic DRIE is superior to the Bosch process. Cryogenic processes typically have higher selectivity than Bosch processes because ions that are at a low energy are already enough to sputter the thin passivation layer (de Boer et al. 2002). However, the etch rate during cryogenic processes is almost 10 times smaller than during Bosch process.

We need to note that the RIE process is complicated enough. In order to obtain high aspect ratio, 2D- and 3D-structures in silicon with reactive ion etching, several process conditions should be controllable and properly chosen. They are (Rangelow 1996, 2001): (a) control the ratio of ion flux to radical flux and their density and energy; (b) choice of plasma chemistry; (c) mechanisms for forming sidewall passivants (sidewall passivation engineering) and good knowledge about the surface kinetics; and (d) control of the substrate temperature in a wide range ($-100 - +150°C$).

### 10.6.3 MASKING DURING REACTIVE ION ETCHING

The mask pattern is an important element of RIE technology. The etching processes applied to etch required material should not affect the mask. Pattern transfer requires that a substrate material should be preferentially etched with respect to the masking layer. This parameter is called the etch selectivity. It means that the mask material should be stable under the etching conditions. Thin photoresist (1 µm) is a good mask material for the RIE of silicon, if only shallow features (<100 µm) are required and the etching is performed at room temperature (Jansen et al. 1996). Thicker photoresist masks (>1.5 µm) make it possible to create deeper structures, but the use of thick photoresist is typically not desirable due to line width limitations and possible cracking problems (Walker 2001). Photoresists are also unable to tolerate harsh wet etching conditions such as heated potassium hydroxide solutions. The photoresist masks also have a quite limited temperature range because high and low temperatures are known to harm the resist (Walter 1997; Walker 2001; Sainiemi 2009).

If photoresist cannot be used during the silicon-etching step, a hard mask is needed. Usually, hard masks are utilized when deep structures are required. Typically, the selectivity of hard mask materials is at least one order of magnitude higher than the selectivity of photoresists. The temperature range permitted by hard masks is also much greater than in the case of photoresists. Hard mask materials do not suffer from cracking because their coefficients of thermal expansion (CTE) are better matched to the CTE of silicon (Sainiemi 2009). Hard mask materials also have better mechanical properties than photoresists. The obvious drawback of all hard masks is the increased amount of complexity in the process because extra deposition and etching steps are required. The isotopic etching of hard mask material also results in poor dimensional control (Rakhshandehroo and Pang 1996). Deposition of a hard mask may also require the inclusion of high-temperature steps in the process.

The most common hard mask material is silicon dioxide. The popularity of silicon dioxide is based on its well-explored material properties and designed growth, deposition, and etching techniques. The etch rate of $SiO_2$ during DRIE of silicon is very low, which makes the creation of deep structures possible. According to French et al. (1997), the ratio of etch rates $Si/SiO_2$ could change from 370 for etching in $SF_6/O_2$ up to 20 for etching in $Cl_2/BCl_3/H_2$. The thermal growth of $SiO_2$ requires temperatures around 1000°C, which limits its use. Plasma-enhanced chemical vapor deposition (PECVD) can be done at considerably lower temperatures (ca. 300°C). In plasma etching, the etch rate of PECVD oxide is comparable with thermal oxide (Bühler et al. 1997). After the lithography, the $SiO_2$ layer is etched and the photoresist is removed. The patterned $SiO_2$ layer now acts as an etch mask. The thickness chosen for this dielectric mask is a compromise and usually is around 100–200 nm. The thicker the mask, the longer it withstands plasma and the deeper one can etch the semiconductor. The thinner the mask, the thinner resist one can use to pattern it and thus achieve higher resolution (Krauss and De La Rue 1999). The temperature range allowed by the $SiO_2$ mask is much wider in comparison with standard photoresists. The stability of $SiO_2$ during the wet etching of silicon is also reasonable. Silicon dioxide has not been reported to inflict surface roughening, changes in etch rate, or pronounced undercutting like some of the metal masks.

Other common hard mask materials used during DRIE include metals (Tian et al. 2000). Many metal masks such as aluminum and nickel offer easy deposition at room temperature by sputtering or evaporation. Aluminum is commonly utilized as an etch mask because of its wide availability, reasonable price, and the fact that it can easily be etched anisotropically in chlorine-based plasmas (Fedynyshyn et al. 1987a; Mansano et al. 1996; Boufnichel et al. 2005). It was established that metal masks are even more selective than $SiO_2$, but some metals have been reported to affect etch rate (Fedynyshyn et al. 1987a,b), undercutting (Mansano et al. 1996; Boufnichel et al. 2005), and the surface quality of the etched features (Fedynyshyn et al. 1987a; Fleischman et al. 1998). Fedynyshyn et al. (1987a,b) noted that the etch rate of silicon increased in fluorine-containing plasmas when aluminum was used as an etch mask. According to Sainiemi (2009), aluminum catalyzes the generation of free fluorine radicals and thus increases the etch rate. It is also known that using aluminum masks can be accompanied by formation of micro- or nanograss on the etched silicon surfaces due to sputtering and redeposition of the aluminum on the etch field (Fedynyshyn et al. 1987a; Fleischman et al. 1998).

Metal oxides, such as aluminum and titanium oxides (Dekker et al. 2006; Sainiemi 2009), and silicon nitride can be used. Amorphous alumina ($Al_2O_3$) combines the good properties of aluminum and silicon dioxide. This material has extreme selectivity (66,000:1), a fully conformal deposition profile, a deposition temperature of 85°C, and it does not inflict micromasking (Dekker et al. 2006; Chekurov et al. 2007; Grigoras et al. 2007; Sainiemi 2009). The pattern transfer to the $Al_2O_3$ layer can be done accurately, even when isotropic wet etchants such as phosphoric acid or hydrofluoric acid are used, because an alumina layer that is just a few nanometers thick is enough for through-wafer etching. According to Chekurov et al. (2007), $Al_2O_3$ mask only a few angstroms thick ($d \sim 0.3$–0.9 nm) is required for etching a thickness, $h$, which is defined by the processed structure (typically 1 μm $< h <$ 20 μm). Dekker et al. (2006) has shown that for $TiO_2$, the etch rate using high-frequency operation is roughly 10 times higher than that of $Al_2O_3$. In addition, $TiO_2$ etch rates were not as reliable as the $Al_2O_3$ etch rates (more scatter). In addition, $TiO_2$ is less resistant and appears to suffer more from chemical attack. Silicon nitride is an excellent masking material in KOH etching, but it is consumed quite rapidly during DRIE of silicon. Nevertheless, silicon nitride is also sometimes utilized as an etch mask for DRIE due to its low built-in stresses. Therefore, nitride is well suited for membrane applications (Leivo and Pekola 1998; Zhang et al. 2000).

### 10.6.4 FIELD OF APPLICATION

It is necessary to note that the field of RIE application for Si porosification is limited. Because of existing limitations of modern photolithography, usually with the help of RIE just macroporous Si can be formed. Photonic crystals are a perspective trend of such macroporous Si's application. As it is known, the lattice of photonic crystals is characterized by periodicity between 200 and

700 nm, with sub-100 nm control of feature size desirable (Krauss and De La Rue 1999). Various 2D structures fabricated in the frame of micromachining technology can also be considered as porous materials (Zijlstra et al. 1999). Etch depth of such structures could be in the range from 20 to 300 µm, while aspect ratio may exceed 30 (Clerc et al. 1998).

For nanoporous silicon's forming, one should use special technologies. For example, nanoporous silicon can be fabricated using nanoimprint lithography combined with cryogenic deep reactive ion etching (Shingubara et al. 2010). Block copolymer and nanosphere masks combined with DRIE can also be used for nanoporous silicon production (Lazzari and Lopez-Quintela 2003; Lu et al. 2005). The fabrication strategy for providing macroporous polysilicon using nanosphere masks is illustrated in Figure 10.27.

The generation of microstructures in order to improve optical properties of monocrystalline silicon used in solar cells is another possible field of RIE application (El Amrani et al. 2008). As it is known, the aim of silicon texturization is to achieve the lowest light reflectance at the surface and the highest light absorption particularly in the infrared region by generating randomly distributed microstructures on the surface, such as micropyramids, spikes, or pores, which trap the incident light.

One can use RIE technology for preparing initial pits for deep electrochemical etching during macroporous silicon microstructures fabrication as well (see Figure 10.28). RIE offers the opportunity to fabricate complex pattern shapes, for instance curved or round structures, whereas KOH pits are limited to rectangles. Additionally, when carefully selecting the etching conditions and masking layer, RIE can offer much smaller and deeper initial pits compared to alkaline etching (Grigoras et al. 2001).

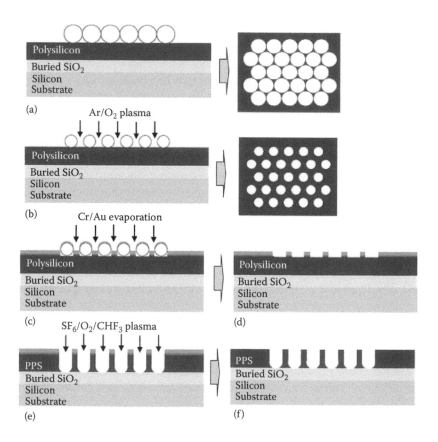

**FIGURE 10.27** Schematic diagram of the fabrication process of macroporous polysilicon using RIE: (a) Deposition of nanospheres template; (b) size reduction of nanospheres template using RIE with $Ar/O_2$ plasma; (c) thermal evaporation of Cr/Au metal mask; (d) removal of nanospheres template; (e) formation of macroporous polysilicon using RIE with $SF_6/O_2/CHF_3$ plasma; and (f) wet chemical etching to remove the Cr/Au metal layer. (Reprinted from *Nucl. Instrum. Meth. Phys. Res. B* 269, Perez-Bergquist A.G. et al., 561, Copyright 2011, with permission from Elsevier.)

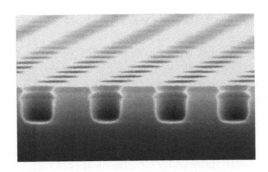

**FIGURE 10.28**    Initial pits (1.2 μm deep) made by RIE through a photoresist (top layer, 250 nm) and amorphous silicon (200 nm) layers. (From Grigoras K. et al., *J. Micromech. Microeng.* 11, 371. 2001. Copyright 2001: IOP. With permission.)

## 10.7 ETCH-FREE FORMATION OF POROUS SILICON

There are several attempts to form pores in silicon without etching. Most of them are based on the use of implantation. In particular, it was established that pores can be fabricated in silicon by implantation of a gaseous ion species, H or He, creating bubbles, followed by an annealing stage to out-diffuse the gas (Raineri et al. 2000; Williams et al. 2001a). Romanov and Smirnov (1978) and Lulli et al. (1993) reported that implantation of non-gaseous ion species also can be applied for this purpose. Small voids in Si were obtained after ion implantation with $P^+$ and $As^+$ at room temperature. However, the voids were sparsely distributed and did not resemble the extended porous networks (Romanov and Smirnov 1978). In addition, the voids only appeared following annealing at temperatures above 400°C, which is above the evaporation point of phosphorus, drawing into question whether the reported voids were in fact phosphorus bubbles. Zhang et al. (1999) were also able to fabricate oblong surface pores in Si following high fluence $Co^+$ irradiation, but the structure could more aptly be described as columnar rather than porous. Williams et al. (2001b) have shown that implantation with $Si^+$ ions and following annealing at 850°C was able to produce some distinct voids in Si as well. Recently, Perez-Bergquist et al. (2011) repeated experiments with self-ion implantation and found that the irradiated silicon samples, which remained crystalline under high temperature ion irradiation, exhibited an increased porous fraction with increasing sample temperature at a given fluence, up to the maximum tested temperature of 650°C. However, it was established that extremely high ion fluences of at least $2 \times 10^{18}$ ions/cm$^2$ were necessary to produce significant void growth. Typical SEM images of PSi formed at indicated conditions are shown in Figure 10.29.

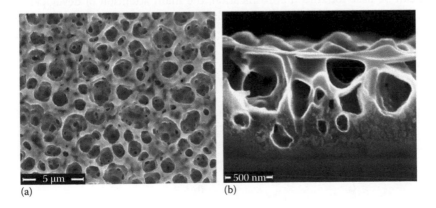

(a)    (b)

**FIGURE 10.29**    (a) Plan view SEM image showing significant exposed surface porosity of Si implanted with 300 keV $Si^+$ (~$4 \times 10^{18}$ ions/cm$^2$) at a surface temperature of ~650°C. (b) XSEM image showing the presence of large, interconnecting voids. (Reprinted from *Nucl. Instrum. Meth. Phys. Res. B* 269, Perez-Bergquist A.G. et al., 561, Copyright 2011, with permission from Elsevier.)

## 10.8 SUMMARIZING GENERAL CONSIDERATION

As it was shown in previous sections, there are many different methods that could be used for Si porosification. Stain etching, in which the silicon sample is simply immersed in an HF-based solution, is the easiest way of producing PSi (Coffer 1997). It has been reported that the physical structure of those layers was similar to the one fabricated by the anodization method (Shih et al. 1992a). The resulting pores are in the range of 1 nm up to the micrometer. Stain etching is an electroless process, and therefore, it has several advantages in comparison with electrochemical etching. It yields, however, layers of low photoluminescence efficiency, deficient homogeneity, and poor reproducibility (Cullis et al. 1997). HF spray and VE methods are also electroless processes. They have been developed to produce porous silicon layers as emitters in solar cells, since the use of EE was found to yield low conversion efficiencies (Saadoun et al. 1999, 2002). VE was also investigated to address the difficulty of isolating metal contacts of devices from the electrolyte solution in anodic etching (Kalem and Yavuzcetin 2000). Spark erosion was tested for preparing porous Si as well (Hummel and Chang 1992; Riiter et al. 1994). Noble metal assisted etching of Si to form PSi is also a simple process that does not require the attachment of any electrodes and can be performed on objects of arbitrary shape and size (Jones et al. 1995; Cullis et al. 1997; Kolasinski 2005).

A combination of two techniques has also been used for Si porosification (Kang and Jorne 1998; Juhasz and Linnros 2002). For example, Splinter et al. (2001) proposed a PSi formation technique that combines the layer anodization and electroless stain etching (see Figure 10.30). A current generated by a galvanic element of silicon and a precious metal deposited on the backside of a silicon wafer in a hydrofluoric acid (HF)/hydrogen peroxide ($H_2O_2$)/ethanol electrolyte was utilized to generate porous silicon. In this case, the silicon operates as anode and the metal as cathode for current generation. This current is similar to the external current needed for anodization. Splinter et al. (2001) believe that this technique has advantages because (1) similar to stain etching, no complicated equipment like an etching cell or an external current source is needed, and (2) high etch rates and thick porous silicon layers are achievable similar to anodization.

As it follows from previous discussions, all of the above-mentioned methods allow forming porous silicon with its unique parameters, but none of them have reached a stage of maturity, similar to anodization. Besides, as it was shown earlier, many methods have essential limitations on reproducibility, on attainment of required porosity and thickness of formed layer. For example, reactive ion etching (RIE), as we mentioned before, has limitations in preparing nanoporous Si. Besides, RIE becomes quite complicated in the case of deep structures (tens to hundreds of micrometers) because of the aspect ratio dependent on etching rate and various pattern density effects (Kiihamaki and Franssila 1999; Cardinaud et al. 2000). Etch-free methods of PSi formation also are very complicated and expensive. Moreover, these methods have limited ability for controlled formation of the pores with the required parameters. One can find disadvantages of other methods in previous sections. Therefore, in spite of the presence of such a variety of methods, which can be used for silicon porosification, the main attention of developers is attracted to anodic oxidation, which is the most controllable process and provides reproducible parameters. For example, only several scientific teams use spark processing or ion implantation, while

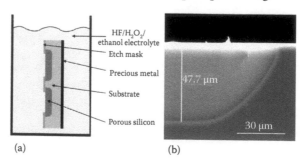

(a)                          (b)

**FIGURE 10.30** (a) Schematic view of the porous silicon fabrication technology proposed by Splinter et al. (2001); (b) SEM image of a porous layer fabricated with 10-μm gold layer on the wafer backside. This mesoporous structure has a porosity of approximately 60%. (Reprinted from *Mater. Sci. Eng. C* 15, 109, Splinter A. et al., Copyright 2001, with permission from Elsevier.)

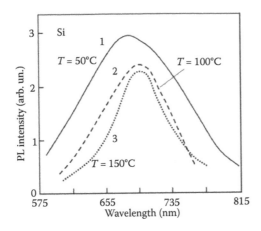

**FIGURE 10.31** PL spectra of nanocrystalline Si deposited by PeCVD at substrate temperature *T*, of 50°C (1), 100°C (2), and 150°C (3), respectively. (Reprinted with permission from Liu X.-N. et al., *Appl. Phys. Lett.* 64, 220, 1994. Copyright 1994. American Institute of Physics.)

electrochemical etching is used in hundreds of labs. We have to note that the standard method of electrochemical etching has appreciably more resources for fabrication of high quality PSi layers in comparison with the above-mentioned methods, such as stain etching, metal-assisted stain etching, laser-induced etching, and spark processing. Electrochemical etching is simple, inexpensive, and gives designers a sufficiently free hand for fabrication of PSi layers with required structure.

Although, other fabrication methods have been explored for fabrication of Si with visible PL:glass melt (Risbud et al. 1993), and plasma deposition (Liu et al. 1994). Si quantum structures can also be synthesized using a variety of techniques, which include rf co-sputtering (Fujii et al. 1990; Mishra and Jain 2002), plasma decomposition of compounds (Banerjee et al. 2003), high-vacuum electron beam evaporation (Wan et al. 2003), and the most widely used Stranski–Krastanov mode of molecular beam epitaxy (MBE) (Peng et al. 1998; Hirai and Itoh 2004). However, we need to note that silicon, deposited or synthesized while using these methods, is not a porous one in traditional understanding, in spite of the presence of luminescence in visible areas of the spectrum (see Figure 10.31). As research has shown, deposited layers mainly represent nanocrystallites of silicon embedded in amorphous Si, $SiO_2$, or glass matrix. For example, Liu et al. (1994) have observed visible PL from nanocrystallites embedded in PECVD a-Si:H films. Usually the average size of such Si nanocrystallites was about 2–4 nm. Taking into account all of the above-mentioned, those methods were not analyzed in the present review.

## ACKNOWLEDGMENTS

This work was supported by the Ministry of Science, ICT and Future Planning (MSIP) of the Republic of Korea, and partly by the Moldova Government under grant 15.817.02.29F and ASM-STCU project #5937.

## REFERENCES

Abouda-Lachiheb, M., Nafie, N., and Bouaicha, M. (2012). The dual role of silver during silicon etching in HF solution. *Nanoscale Res. Lett.* 7, 455.

Andersson, H., van der Wijngaart, W., Enoksson, P., and Stemme, G. (2000). Micromachined flow-through filter-chamber for chemical reactions on beads. *Sens. Actuators B* 67, 203–208.

Aouida, S., Saadoun, M., Ben Saad, K., and Bessais, B. (2005). Structural and luminescence properties of vapour-etched porous silicon and related compounds. *Phys. Stat. Sol. (c)* 2(9), 3409–3413.

Aoyagi, H., Motohashi, A., Kinoshita, A., and Aono, T. (1993). A comparative study of visible photoluminescence from anodized and from chemically stained silicon wafers. *Jpn. J. Appl. Phys.* 32, L1–L4.

Archer, R.J. (1960). Stain films on silicon. *J. Phys. Chem. Solids* 14, 104–110.

Arens-Fischer, R., Kruger, M., Thonissen, M. et al. (2000). Formation of porous silicon filter structures with different properties on small areas. *J. Porous Mater.* 7, 223–225.

Ashruf, C.M.A., French, P.J., Bressers, P.M.M.C., and Kelly, J.J. (1999). Galvanic porous silicon formation without external contacts. *Sens Actuators A* 74, 118–122.

Asoh, H., Arai, F., and Ono, S. (2009). Effect of noble metal catalyst species on the morphology of macroporous silicon formed by metal-assisted chemical etching. *Electrochim. Acta* 54, 5142–5148.

Backes, A., Walkoun, P., Patocka, F., and Schmid, U. (2013). Generation of a porous device layer on SOI substrate, In: *Proceedings of Transducers 2013*, June 16–20, 2013, Barcelona, Spain, T3P.033, pp. 1048–1050.

Banerjee, S., Salem, M.A., and Oda, S. (2003). Conducting-tip atomic force microscopy for injection and probing of localized charges in silicon nanocrystals. *Appl. Phys. Lett.* 83, 3788.

Beale, M.I.J., JBenjamin, J.D., Uren, M.U., Chew, N.G., and Cullis, A.G. (1986). The formation of porous silicon by chemical stain etches. *J. Cryst. Growth* 75, 408–414.

Beckmann, K.H. (1965). Investigation of the chemical properties of stain films on silicon by means of infrared spectroscopy. *Surf. Sci.* 3, 314–332.

Ben Jaballah, A., Saadoun, M., Hajji, A., Ezzaouia, H., and Bessais, B. (2004). Silicon dissolution regimes from chemical vapour etching: From porous structures to silicon grooving. *Appl. Surf. Sci.* 238, 199–203.

Ben Jaballah, A., Hassen, M., Hajji, M., Saadoun, M., Bessais, B., and Ezzaouia, H. (2005). Chemical vapour etching of silicon and porous silicon: Silicon solar cells and micromachining applications. *Phys. Stat. Sol. (a)* 202, 1606–1610.

Ben Rabha, M., Saadoun, M., Boujmil, M.F., Bessais, B., Ezzaouia, H., and Bennaceur, R. (2005). Application of the chemical vapor-etching in polycrystalline silicon solar cells. *Appl. Surf. Sci.* 252, 488–493.

Bilyalov, R.R., Ludemann, R., Wettling, W. et al. (2000). Multicrystalline silicon solar cells with porous silicon emitter. *Sol. Energy Mater. Sol. Cells* 60, 391–420.

Boufnichel, M., Lefaucheux, P., Aachboun, S., Dussart, R., and Ranson, P. (2005). Origin, control and elimination of undercut in silicon deep plasma etching in the cryogenic process. *Microelectron. Eng.* 77, 327–336.

Boughaba, S. and Wang, K. (2006). Fabrication of porous silicon using a gas etching method. *Thin Solid Films* 497, 83–89.

Bühler, J., Steiner, F.-P., and Baltes, H. (1997). Silicon dioxide sacrificial layer etching in surface micromachining. *J. Micromech. Microeng.* 7, R1–R13.

Canham, L.T. (1991). Silicon quantum wire array fabrication by electrochemical and chemical dissolution of wafers. *Appl. Phys. Lett.* 57, 1046–1048.

Cardinaud, C., Peignon, M.C., and Tessier, P.Y. (2000). Plasma etching: Principles, mechanisms, application to micro- and nano-technologies. *Appl. Surf. Sci.* 164, 72–83.

Chartier, C., Bastide, S., and Levy-Clemen, C. (2008). Metal-assisted chemical etching of silicon in HF–$H_2O_2$. *Electrochim. Acta* 53, 5509–5511.

Chattopadhyay, S. and Bohn, P.W. (2004). Direct-write patterning of microstructured porous silicon arrays by focused-ion-beam Pt deposition and metal-assisted electroless etching. *J. Appl. Phys.* 96, 6888–6894.

Chattopadhyay, S., Li, X., and Bohn, P.W. (2002). In-plane control of morphology and tunable photoluminescence in porous silicon produced by metal-assisted electroless chemical etching. *J. Appl. Phys.* 91, 6134–6140.

Chau, C.F. and Melvin, T. (2008). The fabrication of macroporous polysilicon by nanosphere lithography. *J. Micromech. Microeng.* 18, 064012 (1–9).

Cheah, K.W. and Choy, C.H. (1994). Wavelength dependence in photosynthesis of porous silicon dot. *Solid State Commun.* 91, 795–797.

Chekurov, N., Koskenvuori, M., Airaksinen, V.-M., and Tittonen, I. (2007). Atomic layer deposition enhanced rapid dry fabrication of micromechanical devices with cryogenic deep reactive ion etching. *J. Micromech. Microeng.* 17, 1731–1736.

Chen, C.Y., Wu, C.S., Chou, C.J., and Yen, T.J. (2008). Morphological control of single-crystalline silicon nanowire arrays near room temperature. *Adv. Mater.* 20, 3811–3815.

Cheng, S.L., Chung, C.H., and Lee, H.C. (2008). A study of the synthesis, characterization, and kinetics of vertical silicon nanowire arrays on (001)Si substrates. *J. Electrochem. Soc.* 155, D711–D714.

Choy, C.H. and Cheah, K.W. (1995). Laser-induced etching of silicon. *Appl. Phys. A,* 61, 45–50.

Chuang, T.J. (1981). Infrared laser radiation effects on $XeF_2$ interaction with silicon. *J. Chem. Phys.* 74, 1461.

Clerc, P.-A., Dellmann L., Gretillat F. et al. (1998). Advanced deep reactive ion etching: A versatile tool for microelectromechanical systems. *J. Micromech. Microeng.* 8, 272–278.

Coffer, J.L. (1997). Porous silicon formation by stain etching. In: Canham L.T. (ed.) *Properties of Porous Silicon.* INSPEC, London, p. 23–28.

Cruz, S., Honig-d'Orville, A., and Muller, J. (2005). Fabrication and optimization of porous silicon substrates for diffusion membrane applications. *J. Electrochem. Soc.* 152, C418–C424.

Cullis, A.G., Canham, L.T., and Calcott, P.D.G. (1997). The structural and luminescence properties of porous silicon. *J. Appl. Phys.* 82, 909–965.

de Boer, J.M., Gardeniers, J.G.E., Jansen, H.V. et al. (2002). Guidelines for etching silicon MEMS structures using fluorine high-density plasmas at cryogenic temperatures. *J. Microelectromech. Syst.* 11, 385–401.

Dekker, J., Kolari, K., and Puurunen, R.L. (2006). Inductively coupled plasma etching of amorphous $Al_2O_3$ and $TiO_2$ mask layers grown by atomic layer deposition. *J. Vac. Sci. Technol. B* 24, 2350–2355.

De Vasconcelos, E.A., da Silva, E.F., dos Santos, B.E.C.A. Jr., de Azevedo W.M., and Freire J.A.K. (2005). A new method for luminescent porous silicon formation: Reaction-induced vapor-phase stain etch. *Phys. Stat. Sol. (a)* 202(8), 1539–1542.

Di Francia, G. and Citarella, A. (1995). Kinetic of the growth of chemically etched porous silicon. *J. Appl. Phys.* 77, 3549.

Dimova-Malinovska, D. (1999). Application of stain-etched porous silicon in light emitting diodes and solar cells. *J. Lumin.* 80, 207–211.

Dimova-Malinovska, D., Sendova-Vassileva, M., Tzenov, N., and Kamenova, M. (1997). Preparation of thin porous silicon layers by stain etching. *Thin Solid Films* 297, 9–12.

Douani, R., Si-Larbi, K., Hadjersi, T., Megouda, N., and Manseri, A. (2008). Silver-assisted electroless etching mechanism of silicon. *Phys. Stat. Sol. (a)* 205(2), 225–230.

Dudley, M.E. and Kolasinski, K.W. (2009). Structure and photoluminescence studies of porous silicon formed in ferric ion containing stain etchants. *Phys. Stat. Sol. (a)* 206(6), 1240–1244.

Dussart, R., Boufnichel, M., Marcos, G. et al. (2004). Passivation mechanisms in cryogenic $SF_6/O_2$ etching process. *J. Micromech. Microeng.* 14, 190–196.

El Amrani, A., Tadjine, R., and Moussa, F.Y. (2008). Microstructures formation by fluorocarbon barrel plasma etching. *Intern. J. Plasma Sci. Eng.* 2008, 371812.

Elwenspoek, M. and Jansen, H. (1998). *Silicon Micromachining*. Cambridge University Press, Cambridge.

Eukel, C.J.M., Branebjerg, J., Elwenspoek, M., and van De Pol, F.C.M. (1990). A new technology for micromachining of silicon: Dopant selective HF anodic etching for the realization of low-doped monocrystalline silicon structures. *IEEE Electron. Dev. Lett.* 11, 588–590.

Fathauer, R.W., George, T., Ksendkov, A., and Vasquez, R.P. (1992). Visible luminescence from silicon wafers subjected to stain etches. *Appl. Phys. Lett.* 60, 965.

Fedynyshyn, T.H., Grynkewich, G.W., Hook, T.B., Liu, M.-D., and Ma, T.-P. (1987a). The effect of aluminium vs. photoresist masking on the etching rates of silicon and silicon dioxide in $CF_4/O_2$ plasmas. *J. Electrochem. Soc.* 134, 206–209.

Fedynyshyn, T.H., Grynkewich, G.W., and Ma, T.-P. (1987b). Mask dependent etch rates II. The effect of aluminium vs. photoresist masking on the etching rates of silicon and silicon dioxide in fluorine containing plasmas. *J. Electrochem. Soc.* 134, 2580–2585.

Fischer, P.B. and Chou, S.Y. (1993). Sub-50 nm high aspect-ratio silicon pillars, ridges, and trenches fabricated using ultrahigh resolution electron beam lithography and reactive ion etching. *Appl. Phys. Lett.* 62(12), 1414.

Fleischman, A.J., Zorman, C.A., and Mehregany, M. (1998). Etching of 3C-SiC using $CHF_3/O_2$ and $CHF_3/O_2/$ He plasmas at 1.75 Torr. *J. Vac. Sci. Technol. B* 16, 536–543.

Franssila, S. and Sainiemi, L. (2008). Reactive ion etching. In: Li D. (ed.) *Encyclopedia of Micro and Nanofluidics*. Springer, New York, pp. 1772–1781.

French, P.J., Gennissen, P.T.J., and Sarro, P.M. (1997). New silicon micromachining techniques for microsystems. *Sens. Actuators A* 62, 652–662.

Fujii, M., Hayashi, S., and Yamamoto, K. (1990). Raman scattering from quantum dots of Ge embedded in $SiO_2$ thin films. *Appl. Phys. Lett.* 57, 2692.

George, T., Anderson, M.S., Pike, W.T. et al. (1992). Microstructural investigations of light-emitting porous Si layers. *Appl. Phys. Lett.* 60, 2359.

Geyer, N., Fuhrmann, B., Huang, Z., de Boor, J., Leipner, H.S., and Werner, P. (2012). Model for the mass transport during metal-assisted chemical etching with contiguous metal films as catalysts. *J. Phys. Chem. C* 116, 13446–13451.

Grigoras, K., Niskanen, A.J., and Franssila, S. (2001). Plasma etched initial pits for electrochemically etched macroporous silicon structures. *J. Micromech. Microeng.* 11, 371–375.

Grigoras, K., Sainiemi, L., Tiilikainen, J., Säynätjoki, A., Airaksinen, V.-M., and Franssila, S. (2007). Application of ultra-thin aluminum oxide etch mask made by atomic layer deposition technique. *J. Phys.: Conf. Series.* 61, 369–373.

Guerrero-Lemus, R., Hernandez-Rodriguez, C., Ben-Hander, F., and Martinez-Duart, J.M. (2002). Anodic and optical characterisation of stain-etched porous silicon antireflection coatings. *Sol. Energy Mater. Sol. Cells* 72, 495–501.

Hadjersi, T. and Gabouze, N. (2007). Luminescence from porous layers produced by Ag-assisted electroless etching. *Phys. Stat. Sol. (c)* 4(6), 2155–2159.

Hadjersi, T., Gabouze, N., Kooij, E.S. et al. (2004). Metal-assisted chemical etching in $HF:Na_2S_2O_8$ OR $HF:KMnO_4$ produces porous silicon. *Thin Solid Films* 459, 271–275.

Hadjersi, T., Gabouze, N., Ababou, A., Boumaour, M., Chergui, W., and Cheraga, H. (2005a). Metal-assisted chemical etching of multicrystalline silicon in $HF/Na_2S_2O_8$ produces porous silicon. *Mater. Sci. Forum.* 139, 480–481.

Hadjersi, T., Gabouze, N., Yamamoto, N., Benazzouz, C., and Cheraga, H. (2005b). Blue luminescence from porous layers produced by metal-assisted chemical etching on low-doped silicon. *Vacuum* 80, 366–370.

Hajji, M., Ben Jaballah, A., Hassen, M., Khedher, N., Rahmouni, H., and Bessais, B. (2005). Silicon gettering: Some novel strategies for performance improvements of silicon solar cells. *J. Mater. Sci.* 40, 1419–1422.

Harada, Y., Li, X., Bohn, P.W., and Nuzzo, R.G. (2001). Catalytic amplification of the soft lithographic patterning of Si. Nonelectrochemical orthogonal fabrication of photoluminescent porous Si pixel arrays. *J. Am. Chem. Soc.* 123, 8709–8717.

Hirai, A. and Itoh, L.M. (2004). Site selective growth of Ge quantum dots on AFM-patterned Si substrates. *Physica E* 23, 248–252.

Houle, F.A. (1989). Photochemical etching of silicon: The influence of photogenerated charge carriers. *Phys. Rev. B* 39, 10120–10132.

Huang, Z., Geyer, N., Werner, P., de Boor, J., and Gösele, U. (2011). Metal-assisted chemical etching of silicon: A review. *Adv. Mater.* 23, 285–308.

Hummel, R.E. and Chang, S.-S. (1992). Novel technique for preparing porous silicon. *Appl. Phys. Lett.*, 61, 1965–1967.

Hummel, R.E. and Ludwig, M.H. (1996). Spark-processing—A novel technique to prepare light-emitting nanocrystalline silicon. *J. Lumin.* 68, 69–76.

Hummel, R.E., Morrone, A., Ludwig, M., and Chang, S.-S. (1993). On the origin of photoluminescence in spark-eroded (porous) silicon. *Appl. Phys. Lett.* 63, 2771–2773.

Hummel, R.E., Ludwig, M., Chang, S.-S., and LaTorre, G. (1995a). Comparison of anodically etched porous silicon with spark-processed silicon. *Thin Solid Films* 255, 219–223.

Hummel, R.E., Ludwig, M.H., and Chang, S.-S. (1995b). Strong, blue, room-temperature photoluminescence of spark-processed silicon. *Solid State Commun.* 93(3), 237–241.

Hummel, R.E., Ludwig, M.H., Hack, J., and Chang, S.-S. (1995c). Does the blue/violet photoluminescence of spark-processed silicon originate from hydroxyl groups? *Solid State Commun.* 96(9), 683–687.

Ibbotson, D.E., Mucha, J.A., and Flamm, D.L. (1984). Plasmaless dry etching of silicon with fluorine-containing compounds. *J. App. Phys.* 56, 2939.

Jansen, H., de Boer, M., Legtenberg, R., and Elwenspoek, M. (1995). The black silicon method: A universal method for determining the parameter setting of a fluorine-based reactive ion etcher in deep silicon trench etching with profile control. *J. Micromech. Microeng.* 5, 115–120.

Jansen, H., Gardeniers, H., de Boer, M., Elwenspoek, M., and Fluitman, J. (1996). A survey on the active ion etching of silicon in microtechnology. *J. Micromech. Microeng.* 6, 14–28.

Jansen, H.V., de Boer, M.J., Unnikrishnan, S., Louwerse, M.C., and Elwenspoek, M.C. (2009). Black silicon method: A review on high speed and selective plasma etching of silicon with profile control: An in-depth comparison between Bosch and cryostat DRIE processes as roadmap to next generation equipment. *J. Micromech. Microeng.* 19, 033001 (1–41).

St. John, J.V., Coffer, J.L., Rho, Y., and Pinizzotto, R.F. (1996). Formation of rare-earth oxide doped silicon by spark processing. *Appl. Phys. Lett.* 68, 3416–3418.

St. John, J.V., Coffer, J.L., Rho, Y. et al. (1997). Erbium doped SiO2 layers formed on the surface of silicon by spark processing. *Chem. Mater.* 9, 3176–3180.

Jones, L.A., Taylor, G.M., Wei, F.X., and Thomas, D.F. (1995). Chemical etching of silicon: Smooth, rough and glowing surfaces. *Prog. Surf. Sci.* 50, 283–293.

Juhasz, R. and Linnros, J. (2002). Silicon nanofabrication by electron beam lithography and laser-assisted electrochemical size-reduction. *Microelectron. Eng.* 61/62, 563–568.

Kalem, S. and Rosenbauer, M. (1995). Optical and structural investigation of stain-etched silicon. *Appl. Phys. Lett.* 67, 2551.

Kalem, S. and Yavuzcetin, O. (2000). Possibility of fabricating light-emitting porous silicon from gas phase etchants. *Opt. Express* 6(1), 7–11.

Kanemitsu, Y., Futagi, T., Matsumoto, T., and Mimura, H. (1994). Origin of the blue and red photoluminescence from oxidized porous silicon. *Phys. Rev. B* 49, 14732–14735.

Kang, Y. and Jorne, J. (1998). Photoelectrochemical dissolution of n-type silicon. *Electrochim. Acta* 43, 2389–2398.

Kelly, M.T., Chun, J.K.M., and Bocarsly, A.B. (1994). High efficiency chemical etchant for the formation of luminescent porous silicon. *Appl. Phys. Lett.* 64, 1693.

Khedher, N., Ben Jaballah, A., Hassen, M., Hajji, M., Ezzaouia, H., and Bessais, B. (2004). Gettering by heat thermal processing: Application in crystalline silicon solar cells. *Mater. Sci. Semicond. Process.* 7, 439–442.

Kiihamäki, J. (2005). *Fabrication of SOI Micromechanical Devices.* Otamedia Oy, Espoo.

Kiihamaki, J. and Franssila, S. (1999). Pattern shape effects and artefacts in deep silicon etching. *J. Vac. Sci. Technol. A* 17, 2280–2285.

Koker, L. and Kolasinski, K.W. (2000). Photoelectrochemical etching of Si and porous Si in aqueous HF. *Phys. Chem. Chem. Phys.* 2, 277–281.

Koker, L. and Kolasinski, K.W. (2001). Laser-assisted formation of porous Si in diverse fluoride solutions: Reaction kinetics and mechanistic implications. *J. Phys. Chem. B* 105, 3864–3871.

Koker, L., Wellner, A., Sherratt, P.A.J., Neuendorf, R., and Kolasinski, K.W. (2002). Laser-assisted formation of porous silicon in diverse fluoride solutions: Hexafluorosilicate deposition. *J. Phys. Chem. B* 106, 4424–4431.

Kolasinski, K.W. (2003). The mechanism of Si etching in fluoride solutions. *Phys. Chem. Chem. Phys.* 5, 1270–1278.

Kolasinski, K.W. (2005). Silicon nanostructures from electroless electrochemical etching. *Curr. Opin. Solid State Mater. Sci.* 9(1–2), 73–83.

Kolasinski, K.W. (2010). Charge transfer and nanostructure formation during electroless etching of silicon. *J. Phys. Chem. C* 114, 22098–22105.

Kolasinski, K.W. and Gogola, J.W. (2011). Rational design of etchants for electroless porous silicon formation. *ECS Trans.* 33(16), 23–28.

Kolasinski, K.W. and Barclay, W.B. (2013). Stain etching of silicon with and without the aid of metal catalysts. *ECS Trans.* 50(37), 25–30.

Kolasinski, K.W., Barnard, J.C., Koker, L., Ganguly, S., and Palmer, R.E. (2000). In situ photoluminescence studies of photochemically grown porous silicon. *Mater. Sci. Eng. B* 69–70, 157–160.

Kovalev, D.I., Yoroshetzkii, I.D., Mushik, T., Petrova-Koch, V., and Koch, F. (1994). Fast and slow visible luminescence bands of oxidized porous Si. *Appl. Phys. Lett.* 64, 214–216.

Krauss, T.F. and De La Rue, R.M. (1999). Photonic crystals in the optical regime—Past, present and future. *Prog. Quant. Electron.* 23, 51–96.

Kumar, R., Mavi, H.S., and Shukla, A.K. (2008). Macro and microsurface morphology reconstructions during laser-induced etching of silicon. *Micron.* 39, 287–293.

Kwanghoon, K. and Hummel, E.R. (2008). Infrared luminescence from spark-processed silicon. *J. Phys. Chem. Solids* 69, 199–205.

Lärmer, F. and Schilp, A. (1996). Method for anisotropically etching silicon. US-Patent 5,501,893.

Lazzari, M. and Lopez-Quintela, M.A. (2003). Block copolymers as a tool for nanomaterial fabrication. *Adv. Mater.* 19(19), 1583–1594.

Lee, M.C.M. and Wu, M.C. (2006). Thermal annealing in hydrogen for 3-D profile transformation on silicon-on-insulator and sidewall roughness reduction. *J. Microelectromech. Syst.* 15, 338–343.

Lee, C.L., Tsujino, K., Kanda, Y., Ikeda, S., and Matsumura, M. (2008). Pore formation in silicon by wet etching using micrometre-sized metal particles as catalysts. *J. Mater. Chem.* 18, 1015–1020.

Lehmann, V. and Gosele, U. (1991). Porous silicon formation: A quantum wire effect. *Appl. Phys. Lett.* 58(8), 856–858.

Leivo, M.M. and Pekola, J.P. (1998). Thermal characteristics of silicon nitride membranes at sub-Kelvin temperatures. *App. Phys. Lett.* 72, 1305.

Li, X. and Bohn, P.W. (2000). Metal-assisted chemical etching in $HF/H_2O_2$ produces porous silicon. *Appl. Phys. Lett.* 77, 2572–2574.

Li, Y.X., Wolffenbuttel, M.R., French, P.J., Laros, M., Sarro, P.M., and Wolffenbuttel, R.F. (1994). Reactive ion applications. *Sens. Actuators A* 41–42, 317–323.

Lin, T., Sixta, M.E., Cox, J.N., and Delaney, M.E. (1993). Optical studies of porous silicon. *Mat. Res. Soc. Symp. Proc.* 298, 379.

Lipinski, M., Cichoszewski, J., Socha, R.P., and Piotrowski, T. (2009). Porous silicon formation by metal-assisted chemical etching. *Acta Phys. Polonica A* 116, S117–S119.

Liu, X.-N., Wu, X.-W., Bao, X.-M., and He, Y.-L. (1994). Photoluminescence from nanocrystallites embedded in hydrogenated amorphous silicon films prepared by plasma enhanced chemical vapor deposition. *Appl. Phys. Lett.* 64, 220.

Loni, A., D. Barwick, D., L. Batchelor, L. et al. (2011). Extremely high surface area metallurgical-grade porous silicon powder prepared by metal-assisted etching. *Electrochem. Solid-State Lett.* 14(5), K25–K27.

Lu, Y., Aguilar, C.A., and Chen, S. (2005). Shaping biodegradable polymers as nanostructures: Fabrication and applications. *Drug Discov. Today: Technol.* 2(1), 97–102.

Ludwig, M.H. (1996). Optical properties of silicon-based materials: A comparison of porous and spark-processed silicon. *Crit. Rev. Sol. St. Mater. Sci.* 21(4), 265–351.

Lulli, G., Merli, P.G., Migliori, A., Brusatin, G., and Drigo, A.V. (1993). Dynamics of void formation during implantation of Si under self-annealing conditions and their influence on dopant distribution. *Nucl. Instrum. Methods Phys. Res. B* 80–81, 559–563.

Mansano, R.D., Verdonck, P., and Maciel, H.S. (1996). Deep trench etching in silicon with fluorine containing plasmas. *Appl. Surf. Sci.* 100/101, 583–586.

Matthews, B. and Judy, J.W. (2006). Design and fabrication of a micromachined planar patch-clamp substrate with integrated microfluidics for single-cell measurements. *J. Microelectromech. Syst.* 15, 214–222.

Mavi, H.S., Prusty, S., Kumar, M., Kumar, R., Shukla, A.K., and Rath, S. (2006). Formation of Si and Ge quantum structures by laser-induced etching. *Phys. Stat. Sol. (a)* 203(10), 2444–2450.

Megouda, N., Hadjersi, T., Elkechai, O., Douani, R., and Guerbous, L. (2009). Bi-assisted chemical etching of silicon in HF/Co(NO$_3$)$_2$ solution. *J. Lumin.* 129, 221–225.

Mellhaoui, X., Dussart, R., Tillocher, T. et al. (2005). SiO$_x$F$_y$ passivation layer in silicon cryoetching. *J. Appl. Phys.* 98, 104901-1–10.

Menna, P., Di Francia, G., and La Ferrara, V. (1995). Porous silicon in solar cells: A review and a description of its application as an AR coating. *Sol. Energy Mater. Sol. Cells* 37, 13–24.

Mishra, P. and Jain, K.P. (2002). Raman, photoluminescence and optical absorption studies on nanocrystalline silicon. *Mater. Sci. Eng. B* 95, 202–213.

Mitsugi, N. and Nagai, K. (2004). Pit formation induced by copper contamination on silicon surface immersed in dilute hydrofluoric acid solution. *J. Electrochem. Soc.* 151, G302–G306.

Mogab, C.J., Adams, A.C., and Flamm, D.L. (1978). Plasma etching of Si and SiO$_2$—The effect of oxygen additions to CF$_4$ plasmas. *J. Appl. Phys.* 49, 3796.

Nahidi, M. and Kolasinski, K.W. (2006). Effects of stain etchant composition on the photoluminescence and morphology of porous silicon. *J. Electrochem. Soc.* 153, C19–C26.

Noguchi, N. and Suemune, I. (1993). Luminescent porous silicon synthesized by visible light irradiation. *Appl. Phys. Lett.* 62, 1429.

Oehrlein, G.S. (1990). Reactive etching. In: Rossnagel S.M. (ed.) *Handbook of Plasma Processing Technology: Reactive Ion Etching.* Noyes, Park Ridge, NJ, pp. 196–232.

Ogata, Y.H., Kobayashi, K., and Motoyama, M. (2006). Electrochemical metal deposition on silicon. *Curr. Opin. Solid State Mater. Sci.* 10, 163–172.

Peng, K. and Zhu, J. (2004). Morphological selection of electroless metal deposits on silicon in aqueous fluoride solution. *Electrochim. Acta* 49, 2563–2568.

Peng, C.S., Huang, Q., Cheng, W.Q. et al. (1998). Optical properties of Ge self-organized quantum dots in Si. *Phys. Rev. B* 57, 8805–8808.

Peng, K., Xu, Y., Wu, Y., Yan, Y., Lee, S.T., and Zhu, J. (2005). Aligned single crystal Si nanowire arrays for photovoltaic applications. *Small* 1, 1062–1067.

Peng, K.Q., Fang, H., Hu, J.J. et al. (2006). Metal-particle-induced, highly localized site-specific etching of Si and formation of single-crystalline Si nanowires in aqueous fluoride solution. *Chem. Eur. J.* 12, 7942–7947.

Peng, K., Lu, A., Zhang, R., and Lee, S.T. (2008). Motility of metal nanoparticles in silicon and induced anisotropic silicon etching. *Adv. Funct. Mater.* 18, 3026–3035.

Perez-Bergquist, A.G., Naab, F.U., Zhang, Y., and Wang, L. (2011). Etch-free formation of porous silicon by high-energy ion irradiation. *Nucl. Instrum. Meth. Phys. Res. B* 269, 561–565.

Polihronov, J.G., Dubroca, T., Manuel, M., and Hummel, R.E. (2004). Porosity and density of spark-processed silicon. *Mater. Sci. Eng. B* 107, 124–133.

Raineri, V., Saggio, M., and Rimini, E. (2000). Voids in silicon by He implantation: From basic to applications. *J. Mater. Res.* 15, 1449–1477.

Rakhshandehroo, M.R. and Pang, S.W. (1996). Fabrication of Si field emitters by dry etching and mask erosion. *J. Vac. Sci. Technol. B* 14, 612–616.

Rangelow, I.W. (1996). *Deep Etching of Silicon.* Oficyna Wydawnicza Politekchniki, Wroclav.

Rangelow, I.W. (2001). Dry etching-based silicon micro-machining for MEMS. *Vacuum* 62, 279–291.

Rangelow, I.W. and Hudek, P. (1995). MEMS fabrication by lithography and reactive ion etching (LIRIE). *Microel. Eng.* 27, 471–474.

Rasheed, B.G., Mavi, H.S., Shukla, A.K., Abbi, S.C., and Jain, K.P. (2001). Surface reconstruction of silicon and polysilicon by Nd:YAG laser etching: SEM, Raman and PL studies. *Mater. Sci. Eng. B* 79, 71–77.

Risbud, S.H., Liu, L.-C., and Shackelford, J.F. (1993). Synthesis and luminescence of silicon remnants formed by truncated glassmelt-particle reaction. *Appl. Phys. Lett.* 63, 1648.

Riiter, D., Kunz, T., and Bauhofer, W. (1994). Blue light emission from silicon surfaces prepared by spark-erosion and related techniques. *Appl. Phys. Lett.* 64, 3006.

Romanov, S.I. and Smirnov, L.S. (1978). Voids in ion-implanted silicon. *Radiat. Eff.* 37, 121–126.

Saadoun, M., Ezzaouia, H., Bessais, B., Boujmil, M.F., and Bennaceur, R. (1999). Formation of porous silicon for large-area silicon solar cells: A new method. *Sol. Energy Mater. Solar Cells* 59, 377–385.

Saadoun, M., Mliki, N., Kaabi, H. et al. (2002). Vapour-etching-based porous silicon: A new approach. *Thin Solid Films* 405, 29–34.

Saadoun, M., Bessais, B., Mliki, N., Ferid, M., Ezzaouia, H., and Bennaceur, R. (2003). Formation of luminescent (NH$_4$)$_2$SiF$_6$ phase from vapour etching-based porous silicon. *Appl. Surf. Sci.* 210, 240–248.

Sainiemi, L. (2009). Cryogenic Deep Reactive Ion Etching of Silicon Micro and Nanostructures. PhD Thesis, Helsinki University of Technology.

Sarathy, J., Shin, S., Jung, K. et al. (1992). Demonstration of photoluminescence in nonanodized silicon. *Appl. Phys. Lett.* 60, 1532.

Scheeler, S.P., Ullrich, S., Kudera, S. and Pacholski, C. (2012). Fabrication of porous silicon by metal-assisted etching using highly ordered gold nanoparticle arrays. *Nanoscale Res. Lett.* 7, 450.

Schirone, L., Sotgiu, G., and Montecchi, M. (1999). On the morphology of stain-etched porous silicon films. *J. Lumin.* 80, 163–167.

Schoisswohl, M., Cantin, J.L., von Bardeleben, H.J., and Amato, G. (1995). Electron paramagnetic resonance study of luminescent stain etched porous silicon. *Appl. Phys. Lett.* 66, 3660.

Shih, S., Jung, K.H., Hsieh, T.Y. et al. (1992a). Photoluminescence and structure of chemically etched Si. *Mat. Res. Soc. Symp. Proc.* 256, 27.

Shih, S., Jung, K.H., Hsieh, T.Y., Sarathy, J., Campbell, J.C., and Kwong, D.L. (1992b). Photoluminescence and formation mechanism of chemically etched silicon. *Appl. Phys. Lett.* 60, 1863.

Shingubara, S., Maruo, S., Yamashita, T., Nakao, M., and Shimizu, T. (2010). Reduction of pitch of nanohole array by self-organizing anodic oxidation after nanoimprinting. *Microelectron. Eng.* 87, 1451–1454.

Sigel, G.H. Jr. (1973/1974). Ultraviolet spectra of silicate glasses: A review of some experimental evidence. *J. Non-Cryst. Solids* 13, 372–398.

Smith, R.L. and Collins, S.D. (1992). Porous silicon formation mechanisms. *J. Appl. Phys.* 71(8), R1–R22

Smith, Z.R., Smith, R.L., and Collins, S.D. (2013). Mechanism of nanowire formation in metal assisted chemical etching. *Electrochem. Acta* 92, 139–147.

Splinter, A., Stürmann, J., and Benecke, W. (2000). New porous silicon formation technology using internal current generation with galvanic elements. In: CD Proceeding of the 13th European Conference on Solid-State Transducers, EUROSENSORS XIIV, August 27–30, 2000, Copenhagen, Denmark, Abstract T2P03, pp. 423–426.

Splinter, A., Sturmann, J., and Benecke, W. (2001). Novel porous silicon formation technology using internal current generation. *Mater. Sci. Eng. C* 15, 109–112.

Starostina, E.A., Starkov, V.V., and Vyatkin, A.F. (2002). Porous-silicon formation in HF–HNO$_3$–H$_2$O etchants. *Russ. Microelectron.* 31(2), 88–96.

Steckl, A.J., Xu, J., Mogul, H.C., and Morgen, S. (1993). Doping-induced selective area photoluminescence in porous silicon. *Appl. Phys. Lett.* 62, 1982.

Suni, N., Haapala, M., Mäkinen, A. et al. (2008). Rapid and simple method for selective surface patterning with electric discharge. *Angew. Chem. Int. Ed.* 47, 7442–7445.

Tachi, S., Kazunori, K., Okudaira, S. (1988). Low-temperature reactive ion etching and microwave plasma etching of silicon. *Appl. Phys. Lett.* 52, 616.

Tessier, D.C., Boughaba, S., Arbour, M., Roos, P., and Pan, G. (2006). Improved surface sensing of DNA on gas-etched porous silicon. *Sens. Actuators B* 120, 220–230.

Tian, W.-C., Weigold, J.W., and Pang, S.W. (2000). Comparison of Cl$_2$ and F-based dry etching for high aspect ratio Si microstructures etched with an inductively coupled plasma source. *J. Vac. Sci. Technol. B* 18, 1890–1896.

Tsao, C.-W. and Chang, C.-P. (2011). Effects of silicon nanostructure morphology at different metal catalyst layer thicknesses in metal assisted etching. In: Proceedings of the 2011 6th IEEE International Conference on Nano/Micro Engineered and Molecular Systems, February 20–23, 2011, Kaohsiung, Taiwan, pp. 1220–1223.

Tserepi, A., Tsamis, C., Gogolides, E., and Nassiopoulou, A.G. (2003). Dry etching of porous silicon in high density plasmas. *Phys. Stat. Sol. (a)* 197(1), 163–167.

Tsujino, K. and Matsumura, M. (1996). Texturization of multicrystalline silicon wafers for solar cells by chemical treatment using metallic catalyst. *Sol. Energy Mater. Sol. Cell* 90, 100–110.

Tsujino, K. and Matsumura, M. (2005a). Helical nanoholes bored in silicon by wet chemical etching using platinum nanoparticles as catalyst. *Electrochem. Solid State Lett.* 8, C193–C195.

Tsujino, K. and Matsumura, M. (2005b). Boring deep cylindrical nanoholes in silicon using silver nanoparticles as a catalyst. *Adv. Mater.* 17, 1045–1047.

Turner, D.R. (1960). On the mechanism of chemically etching germanium and silicon. *J. Electrochem. Soc.* 107, 810–816.

Unal, B., Parbukov, A.N., and Bayliss, S.C. (2001). Photovoltaic properties of a novel stain etched porous silicon and its application in photosensitive devices. *Opt. Mater.* 17, 79–82.

Vasquez, R.P., Fathauer, R.W., George, T., Ksendzov, A., and Lin, T.L. (1992). Electronic structure of light-emitting porous Si. *Appl. Phys. Lett.* 60, 1004.

Vazsonyi, E., Szilagyi, E., Petrik, P. et al. (2001). Porous silicon formation by stain etching. *Thin Solid Films* 388, 295–302.

Walker, M.J. (2001). Comparison of Bosch and cryogenic processes for patterning high aspect ratio features in silicon. *Proc. SPIE*, 4407, 89–99.

Walter, L. (1997). Photoresist damage in reactive ion etching. *J. Electrochem. Soc.* 144, 2150–2154.

Wan, Q., Wang, T.H., Liu, W.L., and Lin, C.L. (2003). Ultra-high-density Ge quantum dots on insulator prepared by high-vacuum electron-beam evaporation. *J. Cryst. Growth* 249, 23–27.

Weis, R., Chen, Y., and Coffer, J.L. (2002). The use of spark ablation to produce calcium phosphate films on silicon. *Electrochem. Solid State Lett.* 5, C22–C24.

Wellner, A., Koker, L., Kolasinski, K.E., Aindow, M., and Palmer, R.E. (2000). The effect of etchant composition on film structure during laser-assisted porous Si growth. *Phys. Stat. Sol. (a)* 182, 87–91.

Williams, J.S., Ridgway, M.C., Conway, M.J. et al. (2001a). Interaction of defects and metals with nanocavities in silicon. *Nucl. Instrum. Methods Phys. Res. B* 178, 33–43.

Williams, J.S., Conway, M.J., Williams, B.C., and Wong-Leung, J. (2001b). Direct observation of voids in the vacancy excess region of ion bombarded silicon. *Appl. Phys. Lett.* 78, 2867.

Winters, H.F. (1978). The role of chemisorption in plasma etching. *J. Appl. Phys,* 49, 5165.

Winton, M.J., Russel, S.D., and Gronsky, R. (1997). Observation of competing etches in chemically etched porous silicon. *J. Appl. Phys.* 82, 436.

Xia, X.H., Ashruf, C.M.A., French, P.J., and Kelly, J.J. (2000). Galvanic cell formation in silicon/metal contacts: The effect on silicon surface morphology. *Chem. Mater.* 12, 1671–1678.

Xu, J. and Steckl, A.J. (1995). Stain-etched porous silicon visible light emitting diodes, A comparative study of visible photoluminescence from anodized and from chemically stained silicon wafers. *J. Vac. Sci. Technol. B* 13(3), 1221–1224.

Yae, S., Kawamoto, Y., Tanaka, H., Fukumuro, N., and Matsuda, H. (2003). Formation of porous silicon by metal particle enhanced chemical etching in HF solution and its application for efficient solar cells. *Electrochem. Commun.* 5, 632–636.

Yae, S., Tanaka, H., Kobayashi, T., Fukumuro, N., and Matsuda, H. (2005). Porous silicon formation by HF chemical etching for antireflection of solar cells. *Phys. Stat. Sol. (c)* 2(9), 3476–3480.

Yae, S., Morii, Y., Fukumuro, N., and Matsuda, H. (2012). Catalytic activity of noble metals for metal-assisted chemical etching of silicon. *Nanoscale Res. Lett.* 7, 352.

Yamamoto, N. and Takai, H. (2000). Visible luminescence from photo-chemically etched silicon. *Thin Solid Films* 359, 184–187.

Yerokhov, V.Y., Hezel, R., Lipinski, M. et al. (2002). Cost-effective methods of texturing for silicon solar cells. Solar Energy Mater. *Solar Cells* 72, 291–298.

Yoo, J. (2010). Reactive ion etching (RIE) technique for application in crystalline silicon solar cells. *Solar Energy* 84(4), 730–734.

Zhang, Y., Winzell, T., Zhang, T. et al. (1999). High-fluence Co implantation in Si, $SiO_2$/Si and $Si_3N_4$/Si: Part III: heavy-fluence Co bombardment induced surface topography development. *Nucl. Instrum. Methods Phys. Res. B* 159, 158–165.

Zhang, T.-Y., Su, Y.-J., Qian, C.-F., Zhao, M.-H., and Chen, L.-Q. (2000). Microbridge testing of silicon nitride thin films deposited on silicon wafers. *Acta Mater.* 48, 2843–2857.

Zhang, M.L., Peng, K.Q., Fan, X. et al. (2008). Preparation of large-area uniform silicon nanowires arrays through metal-assisted chemical etching. *J. Phys. Chem. C* 112, 4444–4450.

Zheng, H., Han, M., Zheng, P., Zheng, L., Qin, H., and Deng, L. (2014). Porous silicon templates prepared by Cu-assisted chemical etching. *Mater. Lett.* 118, 146–149.

Zijlstra, T., van der Drift, E., de Dood, M.J.A., Snoeks, E., and Polman, A. (1999). Fabrication of two-dimensional photonic crystal waveguides for 1.5 μm in silicon by deep anisotropic dry etching. *J. Vac. Sci. Technol. B* 17(6), 2734–2739.

# The Mechanism of Metal-Assisted Etching of Silicon

Kurt W. Kolasinski

## 11

CONTENTS

## 11.1 CATALYZING STAIN ETCHING WITH METALS

Silicon is the central material of the digital age. After all, we call it Silicon Valley. Yet many integrated circuit technologists will tell you that it is not the properties of Si but those of its oxide, or more accurately, the properties of the $SiO_2/Si$ interface that ushered in the Silicon Age. In much the same manner, while Si is essential to the characteristics, properties, and performance of porous silicon (PSi) and Si nanowires (SiNW) in applications, it is the nature of the H-terminated Si interface in HF that makes Si so versatile. Understanding the role of this interface is essential to understanding etching and the formation of porous silicon (Gerischer et al. 1993; Kolasinski 2003, 2009).

If the etching of Si in HF were controlled by thermodynamics, Si would simply dissolve upon being immersed in HF(aq). Instead, etching is kinetically controlled because of the lack of reactivity of the H/Si interface (Kolasinski 2010). This chemical passivation means that etching of Si in HF requires activation with an applied bias (Uhlir 1956; Lehmann 2002), a strong oxidant (Fuller and Ditzenberger 1957; Archer 1960; Turner 1960), or photons (Noguchi and Suemune 1993; Zhang et al. 1993). The self-limiting nature of Si etching can then be understood by the intersection of the surface chemistry of H/Si in HF with quantum confinement (Canham 1990; Lehmann and Gösele 1991) and heterogeneous charge transfer, which can be interpreted in terms of Marcus theory (Kolasinski 2010; Kolasinski et al. 2012).

Another way to phrase this is that while what is forbidden by thermodynamics is forbidden, that which is allowed by thermodynamics is only a limit that, while it cannot be exceeded, need not necessarily be approached. In this respect, equilibrium electrochemistry is of little use in understanding the etching of silicon in HF. The standard cell potential of the reaction

$$(11.1) \qquad\qquad 4H^+ + Si + 6F^- \rightarrow SiF_6^{2-} + 2H_2$$

is $E° = 1.24$ V (Haynes 2010). This corresponds to $\Delta_r G_m° = -478.6$ kJ/mol. Even with this tremendous thermodynamic driving force, the etching of Si is kinetically constrained. The rate-determining step is the injection of holes into the Si valence band (Koker and Kolasinski 2001). Therefore, we must look to an understanding of the *rate* of electron transfer from the Si valence band to understand Si etching.

The low chemical reactivity in acidic solutions and the slow kinetics of charge transfer at the H/Si interface prime this system for another type of activation: catalysis of charge transfer and the concomitant etching in the presence of a metal. Tzenov and co-workers (Dimova-Malinovska et al. 1997) were the first to intentionally introduce a metal layer onto the Si substrate before it was stain etched in an aqueous mixture of $HNO_3$ + HF. They found that an Al layer 150–200 nm thick facilitated immediate initiation of stain etching, removing the variable induction period that was normally observed with stain etching using $HNO_3$. They observed that the Al layer was completely removed during the formation of a photoluminescent por-Si layer.

Kelly and co-workers (Ashruf et al. 1999, 2000; Xia et al. 2000; Kelly et al. 2001) developed the combination of a metal layer with HF etching in the presence of an oxidant into a more controlled technique by shifting to metals that were not removed from the surface such as Ag, Au, or Pt. Building off of work intended to explain the formation of a galvanic etch stop technique involving etching in alkaline solutions (Ashruf et al. 1998) as well as reports of por-Si formation during illumination of Pt coated Si (Zhang et al. 1993), they investigated what they called galvanic etching by either connecting the Si wafer to an external metal electrode or by directly depositing the metal on the Si wafer. Galvanic etching has been comprehensively reviewed elsewhere (Kolasinski 2014c). By patterning the metal layer with standard lithographic techniques, patterned layers of por-Si can be etched. Galvanic etching can be expected to occur anytime metal/Si interfaces with sufficiently large surface areas are exposed to HF solutions. This can be critically important during the formation of microelectromechanical systems (MEMS) (Miller et al. 2005, 2007, 2008; Pierron et al. 2005; Becker et al. 2010a,b, 2011), in particular because dissolved oxygen can act as the oxidant.

Contemporaneously, Bohn and co-workers (Li and Bohn 2000; Harada et al. 2001; Chattopadhyay et al. 2002; Chattopadhyay and Bohn 2004) also developed metal-assisted etching (MAE) first using sputtered layers of Au, Au/Pd, and Pt, and then progressing to patterned layers and galvanic deposition from solution. Micrographs from scanning electron microscopy (SEM) displayed in their first report show evidence for the formation of SiNW but their emphasis was on the formation of photoluminescent microporous Si.

Peng and co-workers (Peng et al. 2002, 2003; Peng and Zhu 2003) were the first to notice the importance of metal nanoparticle deposition and its relevance to SiNW formation. This group also recognized that ordered arrays of nanowires could be formed using nanosphere lithography to pattern the metal catalyst (Peng et al. 2007). The discovery of the versatility of MAE to produce not only por-Si but also SiNW, especially in ordered arrays, has attracted increased attention to the field (Huang et al. 2008, 2009, 2011; Peng et al. 2008; Zhang et al. 2008; Li 2012; Scheeler et al. 2012).

Silicon nanowires exhibit interesting electrical and optical properties (Holmes et al. 2000; Ponomareva et al. 2005; Schmitt et al. 2012) and are of considerable interest for applications in thermoelectrics (Hochbaum et al. 2008), sensing (He et al. 2011; Peng et al. 2014), photovoltaics and photoelectrochemistry (Peng et al. 2009; Boettcher et al. 2010; Wang et al. 2011; Hu et al. 2014), surface functionalization (Kolasinski 2014b) (including formation of superhydrophobic surfaces) (Xiu et al. 2007), catalysis (Rykaczewski et al. 2011), nanoelectronics (Cui et al. 2003; Peng et al. 2004; Hochbaum et al. 2009; Schmidt et al. 2009, 2010) and energetic materials (Becker et al. 2011). SiNW are of particular interest in energy storage devices such as batteries and supercapacitors (Hochbaum and Yang 2010; Mai et al. 2014).

Ogata and co-workers (Kawamura et al. 2008; Chourou et al. 2010) investigated the influence of metal nanoparticles on anodic etching of Si. Studies were performed with and without addition of $H_2O_2$ to HF after the loading of a $p$-type Si(100) substrate with Ag, Pd, or Pt nanoparticles. For Ag, a microporous layer surrounds the pore originated from the intrusion of an Ag particle. Of the metals deposited on Si under anodic conditions, only Ag nanoparticles promote the formation of straight pores in HF solution without $H_2O_2$. Anodization of Pt or Pd deposited on Si forms a microporous layer similar to the result obtained with the bare surface. Importantly, they concluded that preferential pore formation at the position of the metal particle depended on the type of metal in contact with Si. They noted that because of work function differences, the electronic nature of the metal/Si interface changes. We will return to the influence of the metal and Si work functions on the electronic structure of the metal/Si interface next.

Note again that naïve application of thermodynamics leads to a completely misleading conclusion concerning MAE. $H_2O_2$ is an interesting oxidant to use for MAE because its ability to induce etching of Si in HF solutions is quite low (Gondek et al. 2014), but its reduction is catalyzed by metals. Consider the following standard reduction potentials (Haynes 2010) for $H_2O_2$, the most commonly used oxidant, and metals

$$(11.2) \qquad H_2O_2 + 2H^+ + 2e^- \rightarrow 2H_2O \quad E° = 1.776 \text{ V}$$

$$(11.3) \qquad Ag^+ + e^- \rightarrow Ag \quad E° = 0.7996 \text{ V}$$

$$(11.4) \qquad Au^{3+} + 3e^- \rightarrow Au \quad E° = 1.498 \text{ V}$$

$$(11.5) \qquad Pd^{2+} + 2e^- \rightarrow Pd \quad E° = 0.951 \text{ V}$$

$$(11.6) \qquad Pt^{2+} + 2e^- \rightarrow Pt \quad E° = 1.18 \text{ V}$$

Since the reduction potential of $H_2O_2$ is substantially more positive than any of the metals, equilibrium electrochemistry would conclude that MAE of Si *cannot occur* because none of the metals most commonly used in MAE are stable in $H_2O_2$ + HF. They should all simply dissolve into solution. Just as for stain etching, if we are to understand MAE, we must understand the kinetics of charge transfer and not look to thermodynamic arguments. We will first describe a few fundamental aspects of heterogeneous charge transfer both at the solution/metal and the metal/Si interfaces. We will then describe the results of etching in a number of model systems in terms of the observed chemistry as well as the structures produced.

## 11.2  KINETICS OF HETEROGENEOUS CHARGE TRANSFER

Marcus (1965, 1990, 1991) and Gerischer (1960, 1961, 1997) concurrently laid out the foundations of a theory of interfacial electron transfer. Lewis and co-workers revisited these foundations to make significant advances in our understanding of interfacial electron transfer (Lewis 1991, 1998, 2005; Pomykal et al. 1996; Fajardo and Lewis 1997; Royea et al. 1997, 1998; Gao et al. 2000; Gstrein et al. 2002; Hamann et al. 2005). Developments in the field and open questions have been elegantly reviewed by Fletcher (2010).

An understanding of heterogeneous charge transfer is greatly facilitated by recognition of the following fundamental idea. Electron transfer does not occur between states that are degenerate in electronic energy. Instead, electron transfer occurs between donor and acceptor states – $|D\rangle$ and $|A\rangle$, respectively—, of equal Gibbs energy after rearrangement of the solvation shells about the acceptor and donor. The distinction is important because only then can the dynamics be properly described. Electron transfer occurs after the attainment of a transition state in which the electron to be transferred is degenerate either in the acceptor or in the donor species.

There are two system-dependent parameters of paramount importance for determining the charge transfer probability per collision. The first is the reorganization energy $\lambda$, defined in Figure 11.1 as the Gibbs energy required to take the acceptor from its equilibrium solvation shell geometry to the geometry of the equilibrium solvation shell of the donor. The second is the coupling matrix element $H_{DA} = \langle D|H|A \rangle$ between the donor and acceptor levels. The character of the orbitals involved in charge transfer determines the magnitude of $H_{DA}$ as well as the related tunneling range parameter $\beta$ since they are related exponentially by

(11.7)
$$H_{DA} = V_0 \exp[-\beta(r - r_m)/2].$$

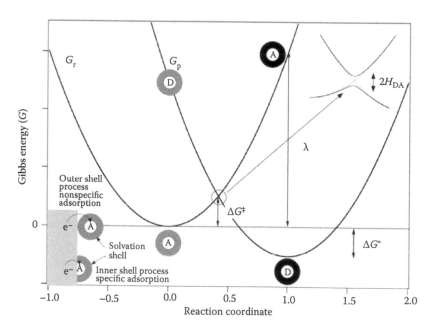

**FIGURE 11.1**    The left insert schematically represents the inner and outer sphere processes at an electrode surface. A description of the potential energy curves involved in charge transfer between an acceptor A and donor D is shown in the middle. Reorganization of the equilibrium solvation shell of A (black) to the equilibrium solvation shell of the donor (gray) would come at the expense of the reorganization energy $\lambda$. The right insert shows that electron transfer occurs along an avoided crossing. The splitting near the avoided crossing is directly proportional to twice the coupling matrix element $H_{DA}$. Also defined are the Gibbs energy of activation $\Delta G^{\ddagger}$ and Gibbs energy of reaction $\Delta G°$. (Reprinted with permission from Kolasinski K.W. et al., *J. Phys. Chem. C* 116, 21472, 2012. Copyright 2012. American Chemical Society.)

The magnitude of $H_{DA}$ determines the separation at the avoided crossing as shown in Figure 11.1. In the weak coupling limit, the Gibbs energy curves maintain their diabatic nature and the curves intersect. As the coupling grows stronger, the crossing becomes avoided. The curves split into an excited state and an adiabatic ground state that smoothly connects the reactants through the transition state to the products.

Orbital overlap is required for electron transfer, which leads to a strong distance dependence characterized by $\beta \approx 1$ Å (Marcus and Sutin 1985; Barbara et al. 1996). Since the diameter of a water molecule is 2.76 Å and the radius of a typical hexaaquo transition metal ion is ~3–4 Å, then >90% of electron transfer occurs by tunneling from an occupied band in the electrode (Si or metal on Si) into the acceptor level of the oxidant as it collides with the surface and moves from a distance of about 6 Å away from the surface (the distance of the solvated ion core with an intervening water molecule) to 3 Å (the minimum distance of the ion core from the surface).

The discussion of electron transfer rates can be made quantitative by formulating the rate constant for heterogeneous electron transfer $k_{et}$ quantum mechanically within the framework of transition state theory as done by Marcus (Marcus and Sutin 1985; Marcus 1990), Gerischer (1997) and Lewis (Lewis 1991, 1998; Royea et al. 1997). The rate constant is

(11.8)
$$k_{et} = \frac{2\pi}{\hbar} |H_{DA}|^2 (4\pi\lambda k_B T)^{-1/2} \exp(-\Delta G^{\ddagger}/k_B T)$$

where $\hbar$ is the reduced Planck constant, $k_B$ is the Boltzmann constant, $T$ is the absolute temperature, and $\Delta G^{\ddagger}$ is the Gibbs activation energy for electron transfer.

There are two limiting cases for electron transfer with an electrode occurring either with specific or nonspecific adsorption (Kolasinski 2012). Specific adsorption is analogous to inner shell homogeneous electron transfer in which the solvation shell of the acceptor is lost at least partially before electron transfer. Nonspecific adsorption is analogous to outer shell electron transfer in which the solvation shell remains intact. These are illustrated in Figure 11.1.

Electron transfer occurs by tunneling from an occupied state at or below $E_F$ into the not-fully-occupied acceptor level $|A\rangle$. In the case of a metal, we expect the electron to come from a state close to $E_F$. From a conventionally doped semiconductor, the electron originates from the top of the valence band at energy $E_V$.

There are several differences between electron transfer at metal and semiconductor electrodes that must be made clear. In general, the rate of electron transfer can be written

(11.9)
$$r_{et} = c_{red} c_{ox} k_{et}$$

where $c_{red}$ is the concentration of the reductant (the donor D), and $c_{ox}$ is the concentration of the oxidant (the acceptor A). Because the density of electron states at and below $E_F$ is so large and constant, electron transfer kinetics at a metal electrode is pseudo first-order. Thus, the rate equation becomes

(11.10)
$$r_{metal} = c_{ox} k_{metal}.$$

At a semiconductor electrode, the electron density is comparatively low and variable. Thus, the rate equation is second-order

(11.11)
$$r_{sc} = n_s c_{ox} k_{sc}.$$

Here $n_s$ is the density of occupied states at the band edge (the valence band since we are interested in hole injection into the valence band for catalytic etching). This means that the units of the rate constant for hole injection into a metal $k_{metal}$ (m/s) differ from those for hole injection into a semiconductor $k_{sc}$ (m⁴/s) (Lewis 1991; Royea et al. 1997).

The next major difference between electron transfer kinetics at metal electrodes and semiconductors is that *the voltage drop near a metal electrode occurs completely in the solution* because the electrons in the metal are so highly polarizable. Changing the voltage on the metal electrode changes the current because the relative position of the Fermi level with respect to the donor/acceptor level in the solution varies. Since this energy difference influences the activation energy for charge transfer, the current changes exponentially. On the other hand, *the voltage drop at a semiconductor electrode occurs completely in the space charge region.* Changing the voltage does not change the relative positions of the band edges at the surface compared to the donor/acceptor level in solution. Instead, the current varies exponentially with applied voltage because the surface density of states changes with bias. In a similar fashion, the polarizability of the electrons in the metal also changes the nature of the bands at the metal/semiconductor interface. We will deal with this effect next.

Lewis (1991) has examined the implications of these differences for the ferrocenium/ferrocene ($Fe^{+/0}$) redox system. The difference in dielectric constant between a metal electrode such as Pt and a semiconductor such as Si is quite large. Consequently, the image charge effects in the two systems are much different (Marcus 1990; Smith and Koval 1990) with efficient screening at a metal surface. This reduces $\lambda_0$, the inner shell contribution to $\lambda$, at a Pt electrode to about half the value found for a homogeneous electron transfer, but leaves $\lambda_0$ little changed at an Si electrode. Thus, $\lambda_0 = 0.5$ eV for a Pt electrode versus 1.0 eV for an Si electrode. Using $n_{s0} = N_c \exp[(-0.9 \text{ eV})/k_B T]$, with $N_c = 10^{25}$ m$^{-3}$ for the effective density of states in the Si conduction band, $n_{s0}$ is $10^{10}$ m$^{-3}$, which yields $k_{sc}^\circ = k_{et} n_{s0} = 10^{-15} - 10^{-14}$ ms$^{-1}$, and exchange current densities of $J_0 = 10^{-7}$ A m$^{-2}$ at $[Fe^+] = 1.0$ M. In contrast, $k_{metal}^\circ = 4$ ms$^{-1}$ for $Fe^{+/0}$ at Pt, which would lead to an exchange current density of $J_0 = 4 \times 10^8$ A/m$^2$ at $[Fe^+] = 1.0$ M. The latter value, of course, cannot be measured experimentally as the current would first become diffusion limited. The extraordinary difference is related primarily to the reduction in the electron density in the semiconductor electrode relative to that at the metal, not to a change in mechanism. This example highlights the expected increase in rate of electron transfer in the presence of a metal on a semiconductor surface.

Within a transition state theory framework (Marcus 1965; Lewis 1991, 1998), the rate constant of electron transfer can be written in terms of a preexponential factor and Gibbs energy of activation as usual

(11.12)
$$k_{et} = A \exp(-\Delta G^\ddagger/k_B T).$$

The units of $A$ are different for a metal (s$^{-1}$) and for a semiconductor (m$^4$/s) as discussed above. Due to the differences in band structure, the expression for the activation energy is also different. For a metal electrode held at a potential $E$

(11.13)
$$\Delta G^\ddagger = (E - E_{ox} + \lambda)^2/4\lambda.$$

For a semiconductor

(11.14)
$$\Delta G^\ddagger = (E_V - E_{ox} + \lambda)^2/4\lambda.$$

$E_{ox}$ is the Nernst potential of the oxidant, $\lambda$ is the reorganization energy, and $E_F$ or $E_V$ is referenced to a common vacuum level along with $E_{ox}$. Note that the Nernst potential of the oxidant is dependent on the solution composition according to

(11.15)
$$E_{ox} = E^\circ - \frac{RT}{nF} \ln Q,$$

where $R$ is the gas constant, $n$ is the valence of the redox process, $F$ is the Faraday constant, and $Q$ is the reaction quotient. This means that the rate of electron transfer to the oxidant depends on the solution composition not only because of the concentration appearing in the rate equation, but also because the position of the acceptor level relative to the band from which the electron originates depends on the composition of the electrolyte. As long as the acceptor lies below the energy of the top of the band, there is no change in the electron transfer rate. However, should the acceptor shift above the energy of the band, the rate will drop precipitously. It should be noted that rarely is the extent of reaction, the volume of etchant, the total number of moles oxidant consumed, or the potential for change in the oxidant concentration mentioned in the experimental description of MAE. This factor has the potential to lead to major issues in reproducibility and control of MAE.

As long as the acceptor level $|A\rangle$ approaches the donor level $|D\rangle$ from above, that is, from a higher energy, electron transfer occurs either at the Fermi level for a metal or else the appropriate band edge for a semiconductor after the acceptor level has fluctuated to an appropriate energy to facilitate degenerate electron transfer (Gerischer 1960, 1961; Marcus 1965). In this case, all holes are injected into the electrode in a narrow energy range at $E_F$, the valence band maximum or conduction band minimum regardless of the ground state energy of the acceptor level on the oxidant. If the acceptor level lies below the energy of an occupied band, there are always states that are degenerate with respect to electronic energy available for electron transfer from the electrode. Nonetheless, the solvation shell still needs to reorganize; therefore, there is still an activation energy that must be overcome before electron transfer can occur.

## 11.3 STRUCTURES FORMED BY CATALYTIC ETCHING

The structure of the material produced by MAE depends on the conditions under which etching is performed. Variables include the type of metal and oxidant as well as the concentration of both the oxidant and HF. The most commonly chosen oxidant is $H_2O_2$. Chartier et al. (2008) have shown that the ratio of electropolishing to por-Si formation can be changed by changing the ratio of $H_2O_2$ to HF concentration. Nitrates (Peng et al. 2002) can also be used as well as $V_2O_5$ (Kolasinski and Barclay 2013), $S_2O_8^{2-}$, $Cr_2O_7^{2-}$, and $MnO_4^-$ (Hadjersi 2007; Douani et al. 2008). However, precipitation of hexafluorosilicates is a recurrent problem if counter ions other than $H^+$ are used (Koker et al. 2002)

An extremely important parameter is the structure of the metal catalyst (Milazzo et al. 2012). Galvanic etching (Kolasinski 2014c) utilizes a planar metal film and leads to the formation of more or less planar porous silicon films or electropolishing in the vicinity of the metal film depending on the composition of the etchant and ratio of metal surface area to the surface area of the Si exposed to solution. Incorporation of etch stop techniques, for example fabricating $p/n$ junctions by doping and/or epilayer growth, allows for the formation of membranes and free-standing beams (Kelly et al. 2001; Sun et al. 2012). Becker et al. (2010a, 2011) have optimized galvanic etching to produce films up to ~150 μm thick with specific surface area as high as 900 m²/g and pore diameters ~3 nm.

Isolated metal nanoparticles etch primarily by burrowing into the Si substrate (Tsujino and Matsumura 2005a,b; Lee et al. 2008). As shown in Figure 11.2 (Scheeler et al. 2012), the burrowing initially is random but localized to the region beneath the nanoparticle. Eventually, the etch

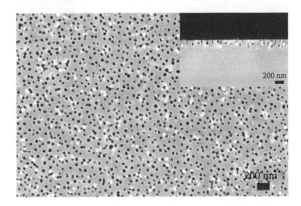

**FIGURE 11.2**   Metal-assisted etching performed by etching with 4.8 M HF(aq) + 0.4 M H₂O₂(aq) in the presence of 55 nm Au nanoparticles for 1 min. (With kind permission from Springer Science+Business Media: *Nanoscale Res. Lett.* 7, 2012, 450, Scheeler S.P. et al.)

tracks of isolated nanoparticles develop into a network of more or less parallel etch track pores, the size of which is controlled by the diameter of the metal nanoparticle.

Porous powders can also be made in a similar manner. Nielsen et al. (2007) deposited Pt from solution on Si powder grains, which were dispersed in an aqueous mixture of HF and $H_2PtCl_6$ for 15 min. This procedure produced dispersions of photoluminescent Si nanoparticles in the size range of 3–6 nm after sonication of the grains in iso-propanol. Nakamura et al. produced both Ag/por-Si (Nakamura et al. 2010) and Pt/por-Si (Nakamura et al. 2011) composite powders from metallurgical grade Si powder. The powder was added to the 18% HF, followed by $AgNO_3$. The powders were photoluminescent.

As shown in Figure 11.3 (Chiappini et al. 2010), the structure of the etched material depends sensitively on the structure of the metal but also the solution composition and the resistivity of the Si substrate. A connected array of metal (or metal dendrites) leads to the formation of Si nanowires (SiNW). The connected array can be formed by dendritic growth, as occurs for Ag deposition from $AgNO_3$ (Peng et al. 2002, 2003). Alternatively, lithography can be used to make ordered arrays of holes within an otherwise flat thin film of metal, which then leads to ordered arrays of SiNW (Harada et al. 2001; Chattopadhyay and Bohn 2004; Huang et al. 2008, 2009, 2011).

The diversity of structures obtainable with MAE is shown in Figure 11.3. These structures include a layer composed solely of solid Si nanowires, Figure 11.3a; a layer of porous Si nanowires, Figure 11.3b; a layer of rigid porous Si nanowires on top of a por-Si layer, Figure 11.3c; thin spaghetti-like porous Si nanowires on top of a por-Si layer, Figure 11.3d; a layer of por-Si alone, Figure 11.3e; and a rough etched layer resulting from electropolishing, Figure 11.3f.

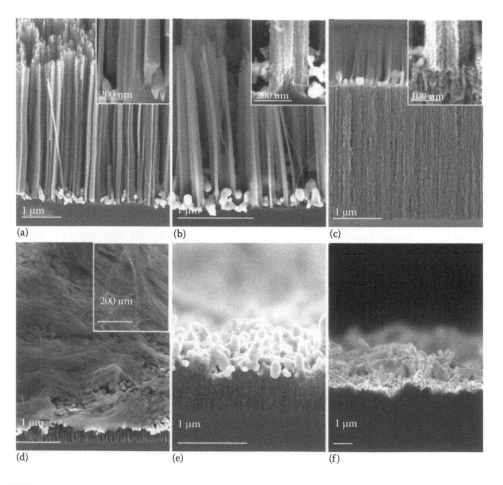

**FIGURE 11.3**  The range of structures achievable with metal-assisted etching. (a) Solid Si nanowires. (b) Porous Si nanowires. (c) Rigid porous Si nanowires on top of a por-Si layer. (d) Thin spaghetti-like porous Si nanowires on top of por-Si. (e) Porous Si. (f) Electropolishing. (Chiappini et al.: *Adv. Func. Mater.* 2010. 20. 2231. Copyright Wiley-VCH Verlag GmbH & Co. KGaA. Reproduced with permission.)

## 11.4 CATALYTIC ETCH CHEMISTRY AND MECHANISMS

It is commonly proposed (Chattopadhyay et al. 2002; Peng et al. 2008; Huang et al. 2011; Li 2012) that electron transfer from Si to the oxidant is facilitated directly by the metal nanoparticle. That is, a hole is injected into the metal by electron transfer from the metal to the oxidant. The hole then diffuses to the metal/Si interface and enters the Si valence band. Once the hole is present in the Si valence band, it induces electrochemistry along one or more of the three electrochemical pathways mentioned above. Near the metal electropolishing occurs. Holes that diffuse more deeply into the Si lead to por-Si formation by either divalent or tetravalent etching. According to this picture, there is but one unbroken chain of charge transfer from oxidant to metal to Si that leads to both local and nonlocal etching.

It has been pointed out (Kolasinski 2014a) that there are problems with this mechanism of charge transfer induced catalytic chemistry. Holes relax to the top of a band. The Fermi level of the metal lies above the Si valence band maximum. Electron transfer into the metal often occurs close to the Fermi energy. Even if it does not, relaxation of hot holes is an ultrafast process that occurs on the femtosecond time scale. Hot holes travel no more than a few nanometers before they relax to the top of the valence band. Therefore, few if any hot holes will arrive at the metal/Si interface. Furthermore, the transfer of electrons and holes across the metal/Si interface is influenced by band bending.

Ideal Schottky barrier heights ($E_{b,n}^{ideal}$ and $E_{b,p}^{ideal}$ for $n$-type and $p$-type Si, respectively) and band positions in the absence of formation of surface states and reconstruction at the interface can be calculated according to the Schottky-Mott relationships (Sze 1981; Kolasinski 2014b; Tung 2014). Calculations of band bending at the metal/Si interface reveals that it is energetically unfavorable for holes to diffuse away from the metal/Si interface and into the bulk Si (Kolasinski 2014a). The results of such calculations are shown in Figure 11.4 for Ag, Au, and Pt. They show that diffusion of holes deep into Si after they have been injected at the solution/metal interface by the oxidant is unlikely to be directly responsible for nonlocal etching.

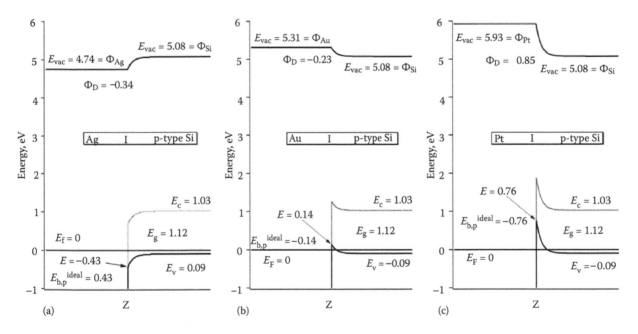

**FIGURE 11.4** Band bending for selected metal/Si interfaces calculated according to Schottky-Mott analysis. The Si is taken to be $p$-type with an acceptor density of $1 \times 10^{15}$ cm$^{-3}$, which corresponds to a resistivity of 14 Ω·cm. See Kolasinski (2014) for details. Three different examples are given to illustrate the effect of increasing metal work function. (a) Ag, $\Phi = 4.74$ V. (b) Au, $\Phi = 5.31$ V. (c) Pt, $\Phi = 5.93$ V. $E_{vac}$ = the vacuum energy. $\Phi_M$ = metal work function. $\Phi_{Si}$ = Si work function. $E_g$ = Si band gap. $E_F$ = Fermi energy. $E_c$ = Si conduction band energy. $E_V$ = Si valence band energy. $\Phi_D$ = maximum band bending. The value $E$ indicates the energy of the Si valence band directly at the metal/Si interface. $E_{b,p}^{ideal}$ is the Schottky barrier height from Schottky-Mott relationships. (With kind permission from Springer Science+Business Media: *Nanoscale Res. Lett.* 9, 2014, 432, Kolasinski K.W.)

Inspection of Figure 11.4a reveals that Ag corresponds to the textbook description of band bending at the metal/Si interface: Bands bend upward in *n*-type Si and downward in *p*-type Si. This relationship also holds for the two most common metals used as interconnects in semiconductor circuits, namely, Al and Cu. However, this behavior is a reflection of the work functions of these metals compared to the work functions of *n*-type and *p*-type Si. It is not a universal behavior. For higher work function metals, such as Au and Pt shown in Figure 11.4b and c, respectively, the bands bend upward for both *p*-type and *n*-type material. Upward band bending is also observed for Pd/Si.

Several groups, for instance (Peng et al. 2003; Qu et al. 2009; Chiappini et al. 2010; Geyer et al. 2012, 2013), have noted that $Ag^+$ may take part directly in the charge transfer that leads to Si etching. There are strong indications for $Ag^+$ dissolution with subsequent redeposition. Figure 11.4a demonstrates that because of the low work function of Ag, a hole injected into Ag is more stable in the metal nanoparticle than it is in Si. Therefore, the injected hole is available to form $Ag^+$. If redeposition were to occur away from the metal nanoparticle that released the ion, this pathway would lead to nonlocal etching. Metal oxidation and redeposition has been proposed to account for all (Qu et al. 2009) or at least a portion of the charge transfer that leads to etching (Chiappini et al. 2010; Geyer et al. 2012, 2013). Particularly strong evidence for the role played by redeposition was found in the observation (Chiappini et al. 2010) that formation of PSi occurred on wafers without deposited Ag when these wafers were placed in the same solution as the Ag-deposited wafers either during the MAE or after completion of an etch of a Ag-deposited wafer. However, no porosification occurs on blank wafers in $H_2O_2 + HF$ solution if that solution has not first been exposed to a wafer with deposited Ag.

Dissolution and redeposition, however, cannot be the only cause of nonlocal etching. While band bending is favorable for dissolution of Ag, Figure 11.4b and c demonstrate that it is not for Au and Pt (also for Pd, not shown). The high work functions of these metals mean that an injected hole is immediately transported to the metal/Si interface where it is more stable. Indeed, there is no evidence for redeposition of Au, Pd, and Pt (Li and Bohn 2000) when $H_2O_2$ is used as the oxidant. The oxidant $VO_2^+$ has a reduction potential of only 0.991 V. Therefore, it cannot dissolve Au and it can only lead to very low solubility of Pt when it is used as the oxidant. Nonetheless, $VO_2^+$ can be used to induce both local and nonlocal etching with Ag, Au, and Pt (Kolasinski and Barclay 2013; Kolasinski et al. 2015). In addition, both Ag and Au exhibit the same stoichiometry with approximately 2 moles of V(V) reduced and 1 mole of $H_2$ evolved for every mole of Si etched. It appears that either (1) dissolution and redeposition do not play a role or (2) dissolution and redeposition do not alter the mechanism of the etching processes and are not essential for local and nonlocal etching.

Another means of producing nonlocal etching must then be invoked for metals other than Ag. It has been proposed (Kolasinski 2014a) that holes build up on the metal nanoparticle. The accumulated charge polarizes the Si to affect electrochemistry just as does a bias applied from a power supply. Near the metal nanoparticle where the polarization is highest, etching is pushed into the electropolishing regime. Far from the nanoparticle, as the polarization drops off, etching proceeds via either the divalent or tetravalent etch pathways and PSi is formed.

Experiments (Kolasinski et al. 2015) examining the rate and stoichiometry of metal-assisted catalytic etching have confirmed that both divalent and tetravalent etching pathways can contribute to etching. Etching in the presence of metal nanoparticles is accelerated to the point of being diffusion limited as expected from the consideration presented in Section 11.2. Differences in the metal chosen do not lead to significant difference in the etch rate but they do lead to subtle differences in the balance between the different etch pathways as well as the structure of the material formed. These differences may well be related to differences in the band structure imposed by the specific metal chosen.

## 11.5 CONCLUSIONS

Metal assisted etching (MAE) is a versatile catalytic etch process that can lead to a wide range of porous and nonporous nano-, meso-, and macrostructures. These structures are of interest in a wide range of applications. The elucidation of the mechanisms of their production continues to present fundamental challenges. The presence of local and nonlocal hole injection has to be inferred from the structures observed. The balance of these two processes, their identification and control continue to be areas of active research. It is hoped that a better understanding of the charge transfer processes outlined here with specific application to the oxidant/Si, oxidant/metal, and metal/Si interfaces will lead to better reproducibility and control over MAE.

## DEDICATION

This chapter is dedicated to Yukio Ogata, who was a leader in porous silicon science, a gracious host, and someone I am proud to have called a colleague.

## ACKNOWLEDGMENT

Claudia Pacholski and Ciro Chiappini are thanked for providing original copies of figures.

## REFERENCES

Archer, R.J. (1960). Stain films on silicon. *J. Phys. Chem. Solids* **14**, 104–110.

Ashruf, C.M.A., French, P.J., Bressers, P.M.M.C., Sarro, P.M., and Kelly, J.J. (1998). A new contactless electrochemical etch-stop based on a gold/silicon/TMAH galvanic cell. *Sens. Actuators A* **66**, 284–291.

Ashruf, C.M.A., French, P.J., Bressers, P.M.M.C., and Kelly, J.J. (1999). Galvanic porous silicon formation without external contacts. *Sens. Actuators A* **74**, 118–122.

Ashruf, C.M.A., French, P.J., Sarro, P.M., Kazinczi, R., Xia, X.H., and Kelly, J.J. (2000). Galvanic etching for sensor fabrication. *J. Micromech. Microeng.* **10**, 505–515.

Barbara, P.F., Meyer, T.J., and Ratner, M.A. (1996). Contemporary issues in electron transfer research. *J. Phys. Chem.* **100**, 13148–13168.

Becker, C.R., Currano, L.J., Churaman, W.A., and Stoldt, C.R. (2010a). Thermal analysis of the exothermic reaction between galvanic porous silicon and sodium perchlorate. *ACS Appl. Mater. Interfaces* **2**, 2998–3003.

Becker, C.R., Miller, D.C., and Stoldt, C.R. (2010b). Galvanically coupled gold/silicon-on-insulator microstructures in hydrofluoric acid electrolytes: Finite element simulation and morphological analysis of electrochemical corrosion. *J. Micromech. Microeng.* **20**, 085017.

Becker, C.R., Apperson, S., Morris, C.J. et al. (2011). Galvanic porous silicon composites for high-velocity nanoenergetics. *Nano Lett.* **11**, 803–807.

Boettcher, S.W., Spurgeon, J.M., Putnam, M.C. et al. (2010). Energy-conversion properties of vapor-liquid-solid-grown silicon wire-array photocathodes. *Science* **327**, 185–187.

Canham, L.T. (1990). Silicon quantum wire array fabrication by electrochemical and chemical dissolution of wafers. *Appl. Phys. Lett.* **57**, 1046–1048.

Chartier, C., Bastide, S., and Levy-Clement, C. (2008). Metal-assisted chemical etching of silicon in $HF-H_2O_2$. *Electrochim. Acta* **53**, 5509–5516.

Chattopadhyay, S. and Bohn, P.W. (2004). Direct-write patterning of microstructured porous silicon arrays by focused-ion-beam Pt deposition and metal-assisted electroless etching. *J. Appl. Phys.* **96**, 6888–6894.

Chattopadhyay, S., Li X., and Bohn, P.W. (2002). In-plane control of morphology and tunable photoluminescence in porous silicon produced by metal-assisted electroless chemical etching. *J. Appl. Phys.* **91**, 6134–6140.

Chiappini, C., Liu, X.W., Fakhoury, J.R., and Ferrari, M. (2010). Biodegradable porous silicon barcode nanowires with defined geometry. *Adv. Funct. Mater.* **20**, 2231–2239.

Chourou, M.L., Fukami, K., Sakka, T., Virtanen, S., and Ogata, Y.H. (2010). Metal-assisted etching of p-type silicon under anodic polarization in HF solution with and without $H_2O_2$. *Electrochim. Acta* **55**, 903–912.

Cui, Y., Zhong, Z.H., Wang, D.L., Wang, W.U., and Lieber, C.M. (2003). High performance silicon nanowire field effect transistors. *Nano Lett.* **3**, 149–152.

Dimova-Malinovska, D., Sendova-Vassileva, M., Tzenov, N., and Kamenova, M. (1997). Preparation of thin porous silicon layers by stain etching. *Thin Solid Films* **297**, 9–12.

Douani, R., Si-Larbi, K., Hadjersi, T., Megouda, N., and Manseri, A. (2008). Silver-assisted electroless etching mechanism of silicon. *Phys. Stat. Sol. (a)* **205**, 225–230.

Fajardo, A.M. and Lewis, N.S. (1997). Free-energy dependence of electron-transfer rate constants at Si/liquid interfaces. *J. Phys. Chem. B* **101**, 11136–11151.

Fletcher, S. (2010). The theory of electron transfer. *J Solid State Electrochem.* **14**, 705–739.

Fuller, C.S. and Ditzenberger, J.A. (1957). Diffusion of donor and acceptor elements in silicon. *J. Appl. Phys.* **27**, 544–553.

Gao, Y.Q., Georgievskii, Y., and Marcus, R.A. (2000). On the theory of electron transfer reactions at semiconductor electrode/liquid interfaces. *J. Chem. Phys.* **112**, 3358–3369.

Gerischer, H. (1960). Über den ablauf von redoxreacktionen an metallen und an halbleitern I. Allgemeines zum elektronenübergang zwischen einem festkörper und einem redoxelektrolyten. *Z. Phys. Chem. N. F.* **26**, 233–247.

Gerischer, H. (1961). Über den ablauf von redoxreacktionen an metallen und an halbleitern III. Halbleiter elektroden. *Z. Phys. Chem. N. F.* **27**, 40–79.

Gerischer, H. (1997). Principles of electrochemistry. In: Gellings P. and Bouwmeester H. (eds.) *The CRC Handbook of Solid State Electrochemistry*. CRC Press, Boca Raton, FL.

Gerischer, H., Allongue, P., and Costa Kieling, V. (1993). The mechanism of the anodic oxidation of silicon in acidic fluoride solutions revisited. *Ber. Bunsen-Ges. Phys. Chem.* **97**, 753–756.

Geyer, N., Fuhrmann, B., Huang, Z.P., De Boor, J., Leipner, H.S., and Werner, P. (2012). Model for the mass transport during metal-assisted chemical etching with contiguous metal films as catalysts. *J. Phys. Chem. C* **116**, 13446–13451.

Geyer, N., Fuhrmann, B., Leipner, H.S., and Werner, P. (2013). Ag-mediated charge transport during metal-assisted chemical etching of silicon nanowires. *ACS Appl. Mater. Interfaces* **5**, 4302–4308.

Gondek, C., Lippold, M., Röver, I., Bohmhammel, K., and Kroke, E. (2014). Etching silicon with $HF-H_2O_2$-based mixtures: Reactivity studies and surface investigations. *J. Phys. Chem. C* **118**, 2044–2051.

Gstrein, F., Michalak, D.J., Royea, W.J., and Lewis, N.S. (2002). Effects of interfacial energetics on the effective surface recombination velocity of Si/liquid contacts. *J. Phys. Chem. B* **106**, 2950–2961.

Hadjersi, T. (2007). Oxidizing agent concentration effect on metal-assisted electroless etching mechanism in HF-oxidizing agent-$H_2O$ solutions. *Appl. Surf. Sci.* **253**, 4156–4160.

Hamann, T.W., Gstrein F., Brunschwig, B.S., and Lewis, N.S. (2005). Measurement of the dependence of interfacial charge-transfer rate constants on the reorganization energy of redox species at n-ZnO/$H_2O$ interfaces. *J. Am. Chem. Soc.* **127**, 13949–13954.

Harada, Y., Li, X., Bohn, P.W., and Nuzzo, R.G. (2001). Catalytic amplification of the soft lithographic patterning of Si. Nonelectrochemical orthogonal fabrication of photoluminescent porous Si pixel arrays. *J. Am. Chem. Soc.* **123**, 8709–8717.

Haynes, W.M. (ed.). (2010). *CRC Handbook of Chemistry and Physics*. CRC Press, Boca Raton, FL.

He, Y., Zhong, Y.L., Peng, F. et al. (2011). Highly luminescent water-dispersible silicon nanowires for long-term immunofluorescent cellular imaging. *Angew. Chem., Int. Ed. Engl.* **50**, 3080–3083.

Hochbaum, A.I. and Yang, P.D. (2010). Semiconductor nanowires for energy conversion. *Chem. Rev.* **110**, 527–546.

Hochbaum, A.I., Chen, R.K., Delgado, R.D. et al. (2008). Enhanced thermoelectric performance of rough silicon nanowires. *Nature (London)* **451**, 163–165.

Hochbaum, A.I., Gargas, D., Hwang, Y.J., and Yang, P. (2009). Single crystalline mesoporous silicon nanowires. *Nano Lett.* **9**, 3550–3554.

Holmes, J.D., Johnston, K.P., Doty, R.C., and Korgel, B.A. (2000). Control of thickness and orientation of solution-grown silicon nanowires. *Science* **287**, 1471–1473.

Hu, S., Shaner, M.R., Beardslee, J.A., Lichterman, M., Brunschwig, B.S., and Lewis, N.S. (2014). Amorphous $TiO_2$ coatings stabilize Si, GaAs, and GaP photoanodes for efficient water oxidation. *Science* **344**, 1005–1009.

Huang, Z.P., Zhang, X.X., Reiche, M. et al. (2008). Extended arrays of vertically aligned sub-10 nm diameter [100] Si nanowires by metal-assisted chemical etching. *Nano Lett.* **8**, 3046–3051.

Huang, Z.P., Shimizu, T., Senz, S. et al. (2009). Ordered arrays of vertically aligned [110] silicon nanowires by suppressing the crystallographically preferred etching directions. *Nano Lett.* **9**, 2519–2525.

Huang, Z., Geyer, N., Werner, P., De Boor, J., and Gösele, U. (2011). Metal-assisted chemical etching of silicon: A review. *Adv. Mater.* **23**, 285–308.

Kawamura, Y.L., Fukami, K., Sakka, T., and Ogata, Y.H. (2008). Electrochemically driven intrusion of silver particles into silicon under polarization. *Electrochem. Commun.* **10**, 346–349.

Kelly, J.J., Xia, X.H., Ashruf, C.M.A., and French, P.J. (2001). Galvanic cell formation: A review of approaches to silicon etching for sensor fabrication. *IEEE Sensors J.* **1**, 127–142.

Koker, L. and Kolasinski, K.W. (2001). Laser-assisted formation of porous silicon in diverse fluoride solutions: Reactions kinetics and mechanistic implications. *J. Phys. Chem. B* **105**, 3864–3871.

Koker, L., Wellner, A., Sherratt, P.A.J., Neuendorf, R., and Kolasinski, K.W. (2002). Laser-assisted formation of porous silicon in diverse fluoride solutions: Hexafluorosilicate deposition. *J. Phys. Chem. B* **106**, 4424–4431.

Kolasinski, K.W. (2003). The mechanism of Si etching in fluoride solutions. *Phys. Chem. Chem. Phys.* **5**, 1270–1278.

Kolasinski, K.W. (2009). Etching of silicon in fluoride solutions. *Surf. Sci.* **603**, 1904–1911.

Kolasinski, K.W. (2010). Charge transfer and nanostructure formation during electroless etching of silicon. *J. Phys. Chem. C* **114**, 22098–22105.

Kolasinski, K.W. (2012). *Surface Science: Foundations of Catalysis and Nanoscience.* Wiley, Chichester.

Kolasinski, K.W. (2014a). The mechanism of galvanic/metal-assisted etching of silicon. *Nanoscale Res. Lett.* **9**, 432.

Kolasinski, K.W. (2014b). Photochemical and nonthermal chemical modification of porous silicon for biomedical applications. In: Santos H. (ed.). *Porous Silicon for Biomedical Applications.* Woodhead Publishing, London, pp. 52–80.

Kolasinski, K.W. (2014c). Porous silicon formation by galvanic etching. In: Canham L.T. (ed.). *Handbook of Porous Silicon.* 2nd ed. Springer Verlag, Berlin.

Kolasinski, K.W. and Barclay, W.B. (2013). Stain etching of silicon with and without the aid of metal catalysts. *ECS Trans.* **50**, 25–30.

Kolasinski, K.W., Gogola, J.W., and Barclay, W.B. (2012). A test of Marcus theory predictions for electroless etching of silicon. *J. Phys. Chem. C* **116**, 21472–21481.

Kolasinski, K.W., Barclay, W.B., Sun, Y., and Aindow, M. (2015). The stoichiometry of metal assisted etching of Si in $V_2O_5$ + HF and HOOH + HF solutions. *Electrochim. Acta* **158**, 219–228.

Lee, C.-L.L., Tsujino, K., Kanda, Y., Ikeda, S., and Matsumura, M. (2008). Pore formation in silicon by wet etching using micrometre-sized metal particles as catalysts. *J. Mater. Chem.* **18**, 1015–1020.

Lehmann, V. (2002). *Electrochemistry of Silicon: Instrumentation, Science, Materials and Applications.* Wiley-VCH, Weinheim.

Lehmann, V. and Gösele, U. (1991). Porous silicon formation: A quantum wire effect. *Appl. Phys. Lett.* **58**, 856–858.

Lewis, N.S. (1991). An analysis of charge-transfer rate constants for semiconductor liquid interfaces. *Annu. Rev. Phys. Chem.* **42**, 543–580.

Lewis, N.S. (1998). Progress in understanding electron-transfer reactions at semiconductor/liquid interfaces. *J. Phys. Chem. B* **102**, 4843–4855.

Lewis, N.S. (2005). Chemical control of charge transfer and recombination at semiconductor photoelectrode surfaces. *Inorg. Chem.* **44**, 6900–6911.

Li, X.L. (2012). Metal assisted chemical etching for high aspect ratio nanostructures: A review of characteristics and applications in photovoltaics. *Curr. Opin. Solid State Mater. Sci.* **16**, 71–81.

Li, X. and Bohn, P.W. (2000). Metal-assisted chemical etching in $HF/H_2O_2$ produces porous silicon. *Appl. Phys. Lett.* **77**, 2572–2574.

Mai, L., Tian, X., Xu, X., Chang L., and Xu L. (2014). Nanowire electrodes for electrochemical energy storage devices. *Chem. Rev.* **114**, 11828–11862.

Marcus, R.A. (1965). On the theory of electron-transfer reactions. VI. Unified treatment for homogeneous and electrode reactions. *J. Chem. Phys.* **43**, 679–699.

Marcus, R.A. (1990). Reorganization free energy for electron transfers at liquid-liquid and dielectric semiconductor-liquid interfaces. *J. Phys. Chem.* **94**, 1050–1055.

Marcus, R.A. (1991). Theory of electron-transfer rates across liquid/liquid interfaces. 2. Relationships and Application. *J. Phys. Chem.* **95**, 2010–2013.

Marcus, R.A. and Sutin, N. (1985). Electron transfers in chemistry and biology. *Biochim. Biophys. Acta* **811**, 265.

Milazzo, R.G., D'arrigo, G., Spinella, C., Grimaldi, M.G., and Rimini, E. (2012). Ag-assisted chemical etching of (100) and (111) *n*-type silicon substrates by varying the amount of deposited metal. *J. Electrochem. Soc.* **159**, D521–D525.

Miller, D.C., Gall, K., and Stoldt, C.R. (2005). Galvanic corrosion of miniaturized polysilicon structures morphological, electrical, and mechanical effects. *Electrochem. Solid State Lett.* **8**, G223–G226.

Miller, D.C., Hughes, W.L., Wang, Z.L., Gall, K., and Stoldt, C.R. (2007). Mechanical effects of galvanic corrosion on structural polysilicon. *J. Microelectromech. Sys.* **16**, 87–101.

Miller, D.C., Becker, C.R., and Stoldt, C.R. (2008). Relation between morphology, etch rate, surface wetting, and electrochemical characteristics for micromachined silicon subject to galvanic corrosion. *J. Electrochem. Soc.* **155**, F253–F265.

Nakamura, T., Hosoya, N., Tiwari, B.P., and Adachi, S. (2010). Properties of silver/porous-silicon nanocomposite powders prepared by metal assisted electroless chemical etching. *J. Appl. Phys.* **108**, 104315.

Nakamura, T., Tiwari, B.P., and Adachi, S. (2011). Direct synthesis and enhanced catalytic activities of platinum and porous-silicon composites by metal-assisted chemical etching. *Jpn. J. Appl. Phys.* **50**, 081301.

Nielsen, D., Abuhassan, L., Alchihabi, M., Al-Muhanna, A., Host J., and Nayfeh, M.H. (2007). Current-less anodization of intrinsic silicon powder grains: Formation of fluorescent Si nanoparticles. *J. Appl. Phys.* **101**, 114302.

Noguchi, N. and Suemune, I. (1993). Luminescent porous silicon synthesized by visible light irradiation. *Appl. Phys. Lett.* **62**, 1429–1431.

Peng, K.Q. and Zhu, J. (2003). Simultaneous gold deposition and formation of silicon nanowire arrays. *J. Electroanal. Chem.* **558**, 35–39.

Peng, K.-Q., Yan, Y.-J., Gao, S.-P., and Zhu, J. (2002). Synthesis of large-area silicon nanowire arrays via self-assembling nanoelectrochemistry. *Adv. Mater.* **14**, 1164–1167.

Peng, K., Yan, Y., Gao, S., and Zhu, J. (2003). Dendrite-assisted growth of silicon nanowires in electroless metal deposition. *Adv. Func. Mater.* **13**, 127–132.

Peng, K.Q., Huang, Z.P., and Zhu, J. (2004). Fabrication of large-area silicon nanowire p-n junction diode arrays. *Adv. Mater.* **16**, 73–76.

Peng, K.Q., Zhang, M.L., Lu, A.J. et al. (2007). Ordered silicon nanowire arrays via nanosphere lithography and metal-induced etching. *Appl. Phys. Lett.* **90**, 163123.

Peng, K.Q., Lu, A.J., Zhang, R.Q., and Lee, S.T. (2008). Motility of metal nanoparticles in silicon and induced anisotropic silicon etching. *Adv. Func. Mater.* **18**, 3026–3035.

Peng, K.Q., Wang, X., Wu, X.L., and Lee, S.T. (2009). Platinum nanoparticle decorated silicon nanowires for efficient solar energy conversion. *Nano Lett.* **9**, 3704–3709.

Peng, F., Su, Y.Y., Zhong, Y.L., Fan, C.H., Lee, S.T., and He, Y. (2014). Silicon nanomaterials platform for bio-imaging, biosensing, and cancer therapy. *Acc. Chem. Res.* **47**, 612–623.

Pierron, O.N., Macdonald, D.D., and Muhlstein, C.L. (2005). Galvanic effects in Si-based microelectromechanical systems: Thick oxide formation and its implications for fatigue reliability. *Appl. Phys. Lett.* **86**, 211919.

Pomykal, K.E., Fajardo, A.M., and Lewis, N.S. (1996). Theoretical and experimental upper bounds on interfacial charge-transfer rate constants between semiconducting solids and outer-sphere redox couples. *J. Phys. Chem.* **100**, 3652–3664.

Ponomareva, I., Menon, M., Srivastava, D., and Andriotis, A.N. (2005). Structure, stability, and quantum conductivity of small diameter silicon nanowires. *Phys. Rev. Lett.* **95**, 265502.

Qu, Y.Q., Liao, L., Li, Y.J., Zhang, H., Huang, Y., and Duan, X.F. (2009). Electrically conductive and optically active porous silicon nanowires. *Nano Lett.* **9**, 4539–4543.

Royea, W.J., Fajardo, A.M., and Lewis, N.S. (1997). Fermi golden rule approach to evaluating outer-sphere electron-transfer rate constants at semiconductor/liquid interfaces. *J. Phys. Chem. B* **101**, 11152–11159.

Royea, W.J., Fajardo, A.M., and Lewis, N.S. (1998). Fermi golden rule approach to evaluating outer-sphere electron-transfer rate constants at semiconductor/liquid interfaces (101, 11152, 1997). *J. Phys. Chem. B* **102**, 3653.

Rykaczewski, K., Hildreth, O.J., Wong, C.P., Fedorov, A.G., and Scott, J.H.J. (2011). Guided three-dimensional catalyst folding during metal-assisted chemical etching of silicon. *Nano Lett.* **11**, 2369–2374.

Scheeler, S.P., Ullrich, S., Kudera, S., and Pacholski, C. (2012). Fabrication of porous silicon by metal-assisted etching using highly ordered gold nanoparticle arrays. *Nanoscale Res. Lett.* **7**, 450.

Schmidt, V., Wittemann, J.V., Senz, S., and Gösele, U. (2009). Silicon nanowires: A review on aspects of their growth and their electrical properties. *Adv. Mater.* **21**, 2681–2702.

Schmidt, V., Wittemann, J.V., and Gösele, U. (2010). Growth, thermodynamics, and electrical properties of silicon nanowires. *Chem. Rev.* **110**, 361–388.

Schmitt, S.W., Schechtel, F., Amkreutz, D. et al. (2012). Nanowire arrays in multicrystalline silicon thin films on glass: A promising material for research and applications in nanotechnology. *Nano Lett.* **12**, 4050–4054.

Smith, B.B. and Koval, C.A. (1990). An investigation of the image potential at the semiconductor electrolyte interface employing nonlocal electrostatics. *J. Electroanal. Chem.* **277**, 43–72.

Sun, N.N., Chen, J.M., Jiang, C., Zhang, Y.J., and Shi, F. (2012). Enhanced wet-chemical etching to prepare patterned silicon mask with controlled depths by combining photolithography with galvanic reaction. *Ind. Eng. Chem. Res.* **51**, 793–799.

Sze, S.M. (1981). *Physics of Semiconductor Devices.* John Wiley & Sons, New York.

Tsujino, K. and Matsumura, M. (2005a). Boring deep cylindrical nanoholes in silicon using silver nanoparticles as a catalyst. *Adv. Mater.* **17**, 1045–1047.

Tsujino, K. and Matsumura, M. (2005b). Helical nanoholes bored in silicon by wet chemical etching using platinum nanoparticles as catalyst. *Electrochem. Solid State Lett.* **8**, C193–C195.

Tung, R.T. (2014). The physics and chemistry of the Schottky barrier height. *Appl. Phys. Rev.* **1**, 011304.

Turner, D.R. (1960). On the mechanism of chemically etching germanium and silicon. *J. Electrochem. Soc.* **107**, 810–816.

Uhlir, A. (1956). Electrolytic shaping of germanium and silicon. *Bell Syst. Tech. J.* **35**, 333–347.

Wang, X., Peng, K.Q., Pan, X.J. et al. (2011). High-performance silicon nanowire array photoelectrochemical solar cells through surface passivation and modification. *Angew. Chem., Int. Ed. Engl.* **50**, 9861–9865.

Xia, X.H., Ashruf, C.M.A., French, P.J., and Kelly, J.J. (2000). Galvanic cell formation in silicon/metal contacts: The effect on silicon surface morphology. *Chem. Mater.* **12**, 1671–1678.

Xiu, Y., Zhu, L., Hess, D.W., and Wong, C.P. (2007). Hierarchical silicon etched structures for controlled hydrophobicity/superhydrophobicity. *Nano Lett.* **7**, 3388–3393.

Zhang, Z., Lerner, M.M., Iii, A.T., and Keszler, D.A. (1993). Formation of a photoluminescent surface on n-Si by irradiation without an externally applied potential. *J. Electrochem. Soc.* **140**, L97.

Zhang, M.-L., Peng, K.-Q., Fan, X. et al. (2008). Preparation of large-area uniform silicon nanowires arrays through metal-assisted chemical etching. *J. Phys. Chem. C* **112**, 4444–4450.

# Porous Silicon Processing

### 12

**Ghenadii Korotcenkov and Beongki Cho**

## CONTENTS

Experiment showed that the formation of porous silicon (PSi) with the required properties, and then keeping them unchanged for a long time regardless of the surrounding environment is a difficult task, which is not always acceptable. It was found that the properties of PSi not only depend on the etching conditions, but also depend on how the etching process ends, how PSi is dried and stored, and what treatment of PSi is used during fabrication and operation of the devices. All of this suggests that the achievement of reproducible results when using PSi needs careful consideration of all stages of the technological process, including the formation of PSi, drying, modification, stabilization, and storage. The process of PSi formation has been analyzed in detail in previous chapters. This chapter reviews the processes that form the final stage of PSi preparation for use in real devices.

## 12.1 CHEMICAL CLEANING OF POROUS SILICON

The problem of PSi cleaning is crucial for PSi-based technology, because utilizing PSi layers in microelectronic devices requires their high purity. PSi should be cleaned to an extent that it could be processed using industrial equipment and pilot lines. Thus, knowledge of how various chemical treatments influence the composition and properties of PSi is important for promotion of PSi-based technology. A comprehensive investigation using Auger electron spectroscopy (AES), secondary-ion mass spectroscopy (SIMS), and neutron activation analysis (NAA) has conclusively shown that PSi layers do not produce extra contamination and can be used for device production (Yakovtseva et al. 2000). However, despite this for determining chemical solutions suitable for PSi chemical cleaning, the processing of PSi in standard solutions developed for Si cleaning as well as rinsing in de-ionized water has been examined (Yakovtseva et al. 2000). It was established that solutions based on $H_2O_2$-$NH_4OH$-$H_2O$ have been very unsuited for PSi treatment because they have been harmful toward PSi. Yakovtseva et al. (2000), based on the results of SIMS and NAA, have also shown that the cleaning of PSi has been best achieved by rinsing in de-ionized water for 20–25 min, boiling in the solution based on $H_2O_2$-$HCl$-$H_2O$ (1:1:3) for 10–15 min, and rinsing in de-ionized water for 20–25 min, in succession. Moreover, it has been determined that the rinsing in de-ionized water should be performed immediately after PSi formation. Yakovtseva et al. (2000) believe that once chemical cleaning has been performed, PSi should be oxidized as quickly as possible. Fauchet et al. (2000) also used the $H_2O_2$-$HCl$-$H_2O$ (1:1:15) solution for PSi cleaning and established that after such cleaning the adhesion of the top electrical contact was improved, which is important in the fabrication of any PSi-based devices.

Gondek et al. (2012) have found that the HF-$H_2O_2$-$H_2O$ system designed for etching crystalline silicon is also promising for PSi cleaning. It was established that for low contaminated PSi material, HF-$H_2O_2$-$H_2O$-mixtures effectively remove metal impurities, including Fe, Cr, and Ag. Gole et al. (2000) reported that the treatments with anhydrous concentrated hydrazine (~30 M) could be used for removing fluorine from the PSi surface via a reaction that converts the fluorine and hydrazine to nitrogen and HF.

## 12.2 DRYING AND CRACKING OF POROUS LAYERS

Once the PSi has been formed, the etched wafer has to be dried. However, it was found that the drying of PSi layers, especially those with high porosities, is crucial and the most delicate step

(Bellet 1997; Lammel and Renaud 2000; Lerondel et al. 2000). After the formation of a highly porous or thick PSi layer, when the electrolyte evaporates out of the pores, a cracking of the layer is systematically observed.

### 12.2.1  POROUS SILICON CRACKING

Numerous studies have shown that the tendency of porous silicon to deform and crack during drying is one of the important factors restraining its use (Lerondel et al. 2000). As evaporation proceeds, the cracks clearly propagate across the sample forming isolated regions or "platelets." As it is seen in Figure 12.1, the cracks have characteristic collapsed "mesa" structures, with a compressed top. The cracking is accompanied by changing all parameters of PSi, including luminescence, optical, and electrophysical properties. For example, macroscopic or microscopic cracking are characterized by a loss of specular reflection on the PSi surface (Lerondel et al. 2000).

As a rule, the probability of cracking increases with decreasing HF concentration, increasing the anodization time, that is, layer thickness, and current density (Mason et al. 2002; Li et al. 2014). It was established that below a critical thickness $d_c$ no cracking appears and the layer preserves its cohesion, as revealed by SEM or X-ray diffraction (Belmont et al. 1996). However, cracking appears for thicknesses larger than $d_c$. The correlation between the critical thickness and porosity of the PSi films is plotted in Figure 12.2. These dependences were determined for (001)PSi layers, when the drying liquid was water or pentane (Bellet 1997). The existence of a critical thickness is apparently a general feature for supported films when subsequent drying is necessary (Scherer 1992). The pertinent parameters upon which $d_c$ depends are the surface tension $g_{LV}$ of the drying liquid, the porosity $P$, the pore size $r$, and the mechanical properties of the PSi layer. In particular, results presented in this figure testify that it is not possible to obtain uniform thick PSi layers of high porosity when water is allowed to evaporate.

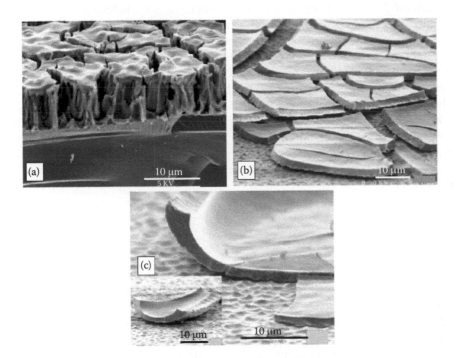

**FIGURE 12.1**    Morphology of porous $p$-Si(100) ($\rho = 5\ \Omega \cdot cm$) as a function of current density during etching in 15% HF dissolved in ethanol. The total charge ($Q = 6000\ mC/cm^2$) is kept constant for each of the three samples. Three different current densities are shown: low current density (5 mA/cm$^2$) shown in (a), moderate current density (10 mA/cm$^2$) shown in (b), and high current density (15 mA/cm$^2$) shown in (c). (Reprinted from *Thin Solid Films* 406, Mason M.D. et al., 151. Copyright 2002, with permission from Elsevier.)

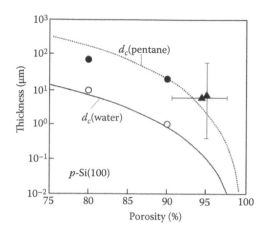

**FIGURE 12.2** Critical thickness $d_c$ of $p^+$-type PSi(100) layers versus the porosity. The lines correspond to the calculated critical thickness $d_c$. ○ corresponds to drying in water, ● corresponds to drying in pentane, and ▲ corresponds to supercritical drying. These results show that supercritical drying is more efficient than pentane drying. (Idea from Bellet D. In: *Properties of Porous Silicon*, Canham L. (ed). INSPEC, London, pp. 38, 1997; experimental data were extracted from Canham L.T. et al., *Nature*, 368, 133, 1994; Frohnhoff St. et al., *Thin Solid Films* 255, 115, 1995b; Belmont O. et al., *J. Appl. Phys.* 79, 7586, 1996; Von Behren J. et al., *Mat. Res. Soc. Symp. Proc.* 452, 565, 1997.)

It has been found that different pore structures, which are formed under influence of different etching conditions, are also characterized by different crack behaviors. For example, the results showed that the cracks do not occur when the porosity is relatively small (<55%), especially for the sample with columnar pores, while with the increase of porosity, the cracks become more serious and finally fall off. Moreover, Li et al. (2014) believe that the crack behaviors of PSi surface layer are a combined effect of the pore geometry and porosity; the pore geometry determines the crack types and the porosity determines the crack extent.

It was established that the cracking is caused by lateral stress during the evaporation process, and the increase of cracks can be attributed to increasing capillary stress with porosity. The relevant explanation model has been presented by Buratto and co-workers (Mason et al. 2002). During drying, when liquid evaporates from the highly porous network, surface tension overwhelms the porous network, causing it to collapse. As air replaces the liquid inside each pore, surface tension at the air–liquid interface strongly pulls the sides of the pores together and the porous network shrinks. As drying proceeds, the network continues to collapse, resulting in a decrease in porosity. The maximum pressure exerted on the pore walls is given by the Laplace equation:

(12.1)
$$\Delta p = -\frac{2\gamma}{r}$$

where $r$ is the radius of the pore, and $\gamma$ is the surface tension of the electrolyte (~22 mJ/m² for ethanol). Bellet and Canham (1998) and many other authors (Mason et al. 2002; Bouchaour et al. 2004) believe that local variation in porosity causes the increase of asymmetric capillary forces, which induce fracturing of the PSi network. Due to capillary tensions that occur on the branch surfaces during drying, the complete disintegration of the branches could happen. When the pore size is in the range of a few nanometers, capillary stresses, $g_{LV}$, can be as large as a few tens of megapascals (with $g_{LV}$ = 72 mJ/m² as for water) (Bellet and Canham 1998). Therefore, it is usually impossible to obtain a porosity of layers higher than 90–95% (Canham et al. 1994).

It was also established that the absence of cracking in PSi with little pore diameter and pore density is attributed to the massive existence of silicon skeleton that can effectively withstand capillary pressures. Instead, the PSi samples with bigger pore diameter and porosity are susceptible to pressures, so the surface layers are shrinking or peeling and finally falling off.

## 12.2.2 POROUS SILICON DRYING

As it was indicated before, cracking is not observed for PSi films with porosity lower than 55%. This means that atmospheric drying is an acceptable technique for the porous Si films with such porosity. However, the higher porosity (especially >70%) structures needed for optoelectronic applications often do not have the mechanical strength needed to survive solvent evaporation without undergoing degradation. As a result, to the present day, the PSi literature is strewn with reports of layers fabricated with too high a porosity or thickness, that when dried exhibit crazing, cracking, peeling, and shrinkage to various extents (Friedersdorf et al. 1992; Grivickas and Basmaji 1993; Zur Muhlen et al. 1996).

To avoid such cracking, several methods have been used to dry PSi layers other than by normal water evaporation. In particular, to reduce or eliminate the capillary stress methods such as pentane drying, supercritical drying, freeze drying, and slow evaporation rates were developed (Frohnhoff et al. 1995a; Fürjes et al. 2003). Pentane drying is the easiest to implement. Usually this process includes a rinse in methanol and then in pentane and finally drying in the presence of nitrogen gas (Pavesi and Mulloni 1999). Pentane has a very low surface tension and shows no chemical interaction with PSi in comparison with ethanol. Using pentane as a drying liquid reduces the capillary tension by a large amount, but since water and pentane are non-miscible liquids, either ethanol or methanol should be used as intermediary liquids. Using this drying technique, PSi layers with porosity values up to 90% and thicknesses up to 5 µm exhibit no cracking pattern after drying. The substitution of the water with either isopropanol or methanol also reduces the surface tension (Bisi et al. 2000; Lammel and Renaud 2000). However, this effect is not as strong as in the case of using pentane.

Another drying technique is so-called freeze-drying, in which the specimens are impregnated with water or iso-butanol followed by freezing to −50°C and sublimation in vacuum. This third method applies the slow evaporation of water or ethanol used for rinsing (Bellet 1997). Freeze-drying was successfully applied by Amato et al. (Amato and Brunetto 1996; Amato et al. 1996), at least on PSi derived from $p^+$-type substrates. PSi layers were produced by anodisation of (111) silicon wafers, and the porosity was about 80%. For such PSi samples, freeze-drying was shown to be a nondestructive process, while air-drying led to cracking. From this investigation (Amato and Brunetto 1996; Amato et al. 1996), freeze-drying appears to be simple to implement and is free of contaminating agents.

Research has shown that the forces promoting the destruction of porous materials could be avoided by supercritical drying (Canham et al. 1994). This technique was first used by Kistler (1931) in the 1930s and has been very useful for the processing of highly porous aerogels (Fricke 1986). Supercritical drying is because when the pressure is raised, the interface between the liquid and the gas phase becomes instable. However, when the pressure is larger than critical, the interface gas/liquid disappears, and a mixture of the two phases appears as a supercritical liquid (see phase diagram shown in Figure 12.3). In the supercritical state, there is no liquid/vapor interface and therefore capillary stresses are suppressed. High supercritical densities (comparable to organic solvents) are sufficient to provide good solvent capability, but low enough for high diffusivity. Liquid converts to a gas through pressure reduction at constant temperature. This means that the process of drying should be in an autoclave at increased pressure and temperature.

At present, supercritical drying is the most efficient method of PSi drying. As it is seen in Figure 12.2, supercritical drying is more efficient than pentane drying. In such technique, the HF solution is replaced by a suitable "liquid" under high pressure (Canham et al. 1994). Then the system is moved above the critical point by raising the pressure and the temperature. After that, gas is being simply removed by the supercritical liquid. In other words, the supercritical drying is the removal of pore liquid above its critical point to avoid the occurrence of liquid vapour interfaces. Pressure vessels are used to convert the pore liquid into a supercritical fluid and subsequently vent it as a gas (Cullis et al. 1997). Peculiarities of the above-mentioned process were discussed in detail by Bisi et al. (2000). Experiment has shown that supercritical carbon dioxide, due to gas-like transport properties and near liquid densities, is the most appropriate "liquid" for these purposes. Although carbon dioxide is a nonpolar molecule, its solvating capability in the supercritical can be controlled by varying pressure, temperature,

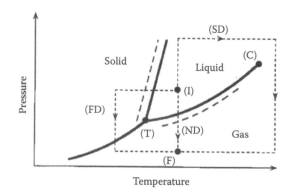

**FIGURE 12.3** Schematic representation of the phase diagram of a fluid: (T) and (C) are, respectively, the triple and critical points. To dry a porous material, one has to go from an initial point (I) in the liquid phase to the final point (F). Three different paths are possible: (ND) normal drying, (SD) supercritical drying, and (FD) freezedrying. For a fluid confined in pores of nanometric size, the solid-liquid and liquid-vapor lines are shifted (dotted lines) compared to the bulk fluid (continuous line). (Bellet D. and Canham L.: *Adv. Mater.* 1998. 10(6). 487. Copyright Wiley-VCH Verlag GmbH & Co. KGaA. Reproduced with permission.)

and cosolvents. Density of supercritical $CO_2$ can be tailored by changes in temperature and pressure; at low temperatures, relatively small changes in pressure result in significant changes in density. For $CO_2$, the critical pressure and temperature are, respectively, 7.36 MPa and 304.1 K (31°C). In addition, supercritical $CO_2$ is non-toxic and has a relatively low critical temperature and pressure. Moreover, high diffusivities and near liquid densities of supercritical $CO_2$, coupled with no surface tension permit rapid penetration and reaction with porous films (Bouchaour et al. 2004).

It was established that this technique allows producing layers with very high thickness and porosity values (up to 95%) and also improving optical flatness and homogeneity (Canham et al. 1994; Bellet 1997). Ethanol rinse liquid can be directly replaced with supercritical $CO_2$, whereas water rinse liquid must be replaced by surfactant-containing $CO_2$ or by surfactant-containing hexane prior to $CO_2$ exchange. Figure 12.4a shows how efficient supercritical drying can be. It is seen that after drying with supercritical $CO_2$, we do not observe any cracking in the PSi layer. However, supercritical drying is expensive and complicated to implement. Therefore, until now, in practice the most widespread method of final treatment remains washing in ethanol followed by air drying (Di Francia et al. 1998).

Finally, a fourth method is that of chemical modification of the surface prior to drying at ambient pressures, which was proposed by Linsmeier et al. (1997). TEOS (tetraethoxysilane) was used to mechanically strengthen the silicon network and after very lengthy treatment (about 120 h) prevented collapse. However, as one can see, the duration of such treatment is too long for practical application.

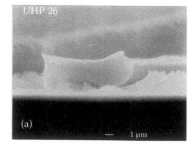
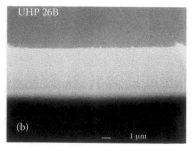

**FIGURE 12.4** Scanning electron microscopy observation of a 5-μm thick $p^+$-type porous silicon layer (porosity of 95%): (a) dried in ambient air, and (b) dried with supercritical $CO_2$. (Bellet D. and Canham L.: *Adv. Mater.* 1998. 10(6). 487. Copyright Wiley-VCH Verlag GmbH & Co. KGaA. Reproduced with permission.)

## 12.3  STORAGE OF POROUS SILICON

Numerous experiments with PSi have shown that the properties of highly porous silicon can be significantly influenced not only by fabrication and drying conditions, but even by the manner in which it is stored, prior to examination or use. As stated previously, PSi slowly reacts with the ambient air and consequently its chemical composition and its properties evolve continuously with storage time. The effects of "aging" on both the composition and structure of PSi are now well-documented (Sham et al. 1992; Xie et al. 1992; Zhu et al. 1992; Chelyadinsky et al. 1997; Dolino and Bellet 1997; Lee and Tu 2007; Sorokin et al. 2010). For example, as early as 1965 Beckmann (1965) had recorded the pronounced "ageing" (slow oxidation) of films that were stored in ambient air for prolonged periods. Changes in material properties such as electrical resistivity, $I$–$V$ and $C$–$V$ characteristics, strain, lattice parameters, optoelectronic properties, including refractive index, photoluminescence, and electroluminescence, accompany the aging process (see Table 12.1 and Figure 12.5).

According to Canham and co-workers (Canham et al. 1991; Cullis et al. 1997; Canham 1997b), the "aging" of porous silicon results from the reaction of the material with its environment: the surrounding ambient and sometimes the containment vessel, usually via oxidation of PSi internal surface. The speed of PSi oxidation depends on many factors. For example, at room temperature, native oxide growth is typically complete after about a year. However, Wolkin et al. (1999) have another opinion. They wrote that many researchers define "fresh samples" as samples that have been exposed to air from a few minutes up to half an hour. Results obtained by Wolkin et al. (1999) show that oxidation occurs in seconds, changing the sample's recombination mechanism and optical properties (see Figure 12.5d). In addition, they suggested that an oxide might not provide a good passivation in

## TABLE 12.1  Effects of Varying Storage Conditions

| Storage Conditions | Major Effect | References |
|---|---|---|
| Ambient air (15 min–15 month) | Contaminated native oxide growth | Canham et al. 1991 |
| Ambient air (0–40 days) | PL instability. Aging depends on the ratio of nanocrystalline and amorphous phase in PSi | Len'shin et al. 2011 |
| Ambient air (up to 6800 hours) | Structural changes. Distortion of the crystalline PSi skeleton | Sorokin et al. 2010 |
| Ambient air (20–500°C) | Surface oxidation. Low rectifying $I$–$V$ curves become strong rectifying, and $C$–$V$ curves become MIS-like | Ciurea et al. 1998 |
| Ambient air (20–500°C) | Strong relationship between doping density of the starting PSi and level of oxidation. The highly doped samples are more stable to oxidation | James et al. 2006 |
| $N_2$, $H_2$, forming gas, $O_2$ (min-h) | Widely varying PL stability | Tischler et al. 1992 |
| $O_2$, $N_2$, air and vacuum | Degradation of LED is connected mainly with oxidation and metal diffusion. The oxygen and metal in ionic state can diffuse quickly | Jumayev et al. 2002 |
| Air, vacuum | Changes in layer strain | Kim et al. 1991 |
| Dry $N_2$, then UHV | Avoids photostimulated oxidation | Munder et al. 1993 |
| Vacuum ($10^6$ torr) | Carbon and oxygen pickup | Loni et al. 1994 |
| Vacuum ($10^3$ torr) | Heavy hydrocarbon contamination | Teschke et al. 1994 |
| Keeping in anodizing bath (HF solution) | The formation of the passive layer at the bottom of the pores. Layers growing at $p$-Si and $n$-Si show different aging behavior | Parkhutik 2000 |
| HF, ethanol, freon, ether | Lowest carbon levels for HF storage | Earwaker et al. 1991 |
| Transport under propanol (<1 day) | Minimize oxidation by reducing air exposure | Lehmann et al. 1993 |
| Ethanol storage and removal in UHV ($<2 \times 10^9$ torr) | Minimize oxidation by completely avoiding air exposure | Suda et al. 1994 |
| Ethanol immersion (up to 60 h) | Reduction in Si grain size. Oxygen adsorption. Blue shift of PL. PL intensity change | Youssef 2001 |
| Cooled ethylene glycol | Green PL retained | Astrova et al. 1995 |
| Plastic and glass containment vessels | Blue PL due to plastic boxes outgassing | Loni et al. 1997 |

*Source:*  Data extracted from Canham L. (ed.), *Properties of Porous Silicon.* INSPEC, London, 1997, and other publications.

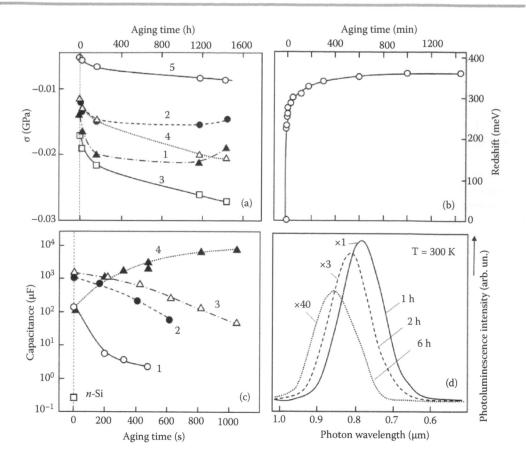

**FIGURE 12.5** Aging influence on the parameters of PSi: (a) Dependence of compressive stresses $\sigma\|$ in PSi on the air oxidation time for films formed using different parameters: (1) $J = 100$ mA/sm$^2$, $t = 1$ min; (2) 50–2; (3) 10–10; (4) 50–10; (5) 100–10. (From Sorokin L.M. et al., *J. Phys.: Conference Ser.* 209, 012059, 2010. Published by IOP as open access.) (b) The PL redshift for a blue-green sample as a function of air exposure time. The PSi samples were formed by electrochemical etching followed by photo-assisted stain etching of 6 Ω·cm *p*-type Si wafers at current densities of 8–50 mA/cm$^2$ using 10%–25% HF:ethanol solutions. Stain etching, accomplished under illumination with a 500-W halogen lamp. (Reprinted with permission from Wolkin M.V. et al., *Phys. Rev. Lett.* 82 (1), 197, 1999. Copyright 1999 by the American Physical Society.) (c) Kinetics of change of the capacitance parameters of the electrical equivalent circuit for *n*-PSi during aging: (1) sample grown at 1 mA/cm$^2$ during 300 s; (2) sample grown at 5 mA/cm$^2$ during 300 s; (3) sample grown at 5 mA/cm$^2$ during 600 s. (1–3) module $Q_1$, (4) module $Q_2$. (With kind permission from Springer Science+Business Media: *J. Porous Mater.* 7, 2000, 97, Parkhutik V.P.) (d) Room temperature PL of a freshly etched layer because of partial chemical dissolution in 40% aqueous HF for the times indicated from the time of anodical etching. (Reprinted with permission from Canham L.T. (1990) *Appl. Phys. Lett.* 57, 1046, 1990 Copyright 1990. American Institute of Physics.)

small Si crystallites, in contrast to bulk silicon. One should note that this observation might be very important for nanoelectronics applications. Len'shin et al. (2011) established that aging depends also on the ratio of nanocrystalline and amorphous phases in PSi. They have found that an increase in the content of amorphous phase and its defective oxides in PSi layer during aging lead to a significant decrease in the PL intensity. As it was found (James et al. 2006), the type of conductivity and the concentration of charge carries in silicon also influence the rate of PSi parameters aging.

Experiment has also shown that porous Si is susceptible to significant contamination by a variety of species present in the atmosphere in gaseous molecular form (Canham et al. 1991). Since there is considerable variability in the composition of ambient air, the composition and structure of the native oxide that develops can also be highly variable. As a result, the consequence of aging depends on both storage time and temperature, and is influenced by other factors such as pore morphology (macro-, micro-, or mesoporous) and storage conditions. In particular, properties of porous Si depend on the way in which the material is dried after fabrication, but they can also depend on the manner in which the material is stored. For example, most plastic wafer and sample containers contain complex molecules like phthalate esters and outgas over extended periods of time. Many such containers contain extremely efficient organic chromophores, which could gradually impregnate a highly porous material and actually give rise to luminescence (Canham et al. 1996).

To minimize the variability and extent of such storage effects, there are a number of options. According to Canham (1997b), for resolving the above-mentioned problem one could intentionally oxidize the material in controlled conditions (Cantin et al. 1996; Roussel et al. 1999; Mattei et al. 2000); isolate its internal surface by capping (Loni 1997); modify its surface (Chazalviel and Ozanam 1997; Salonen et al. 2000); or impregnate the pores (Herino 1997; Setzu et al. 1998). Alternatively, one could simply try to optimize storage time and conditions for the given application requirement.

One might assume that aging effects could be avoided by storing the PSi under vacuum. However, it is necessary to note that the effect of PSi aging is an unavoidable process. While the storing of PSi in vacuum does appreciably slow the aging process down, it does not eliminate it—the material still picks up C and O, mainly through a variety of residual species (hydrocarbons, water vapor) that can be present within a vacuum system. Experiments have shown that even with electrolyte drying via molecular sieves, vacuum distillation and outgassing, residual water at the ppm level was reported to result in oxide islands that after one week covered 60% of the surface (Ozanam and Chazalviel 1993).

This means that during our work with "bare" porous silicon, we have to carefully control the conditions of its storage. For example, according to Canham (1997b), storage in the dark is generally recommended to minimize photochemical effects. These can occur not only in air but also during storage in liquid media such as alcohols and HF. Ultra-dry inert gas or UHV storage is needed to minimize room temperature oxidation of freshly etched material. Ultrapure alcohol and highly concentrated HF are promising liquid storage media, at least for periods of hours to a few days (Lehmann et al. 1993; Mizuno et al. 1996). However, Salonen et al. (1997a) have found that the changes in PSi properties take place even during PSi keeping in methanol and ethanol. "Cold storage" also promotes better preservation of initial properties of PSi.

## 12.4 ETCHING OF POROUS SILICON

The etching used for shaping of necessary objects is standard technical procedure of any device production. However, one should note that the etching of PSi in contrast to bulk Si is not an easily controlled process. Moreover, some techniques such as wet chemical etching are not suitable for etching nanoscale, high aspect ratio Si structures due to undercutting of the mask and sloped sidewalls (Schwartz and Schaible 1979). As research has shown, reactive ion etching is the most suitable process for resolving this task. For example, Schwartz and Schaible (1979) have found that chlorine-based reactive ion etching (RIE) is well suited for etching nanoscale Si features with good control of undercutting and etch profiles.

Pore opening also requires PSi etching. This procedure is especially important at pores' filling (see next section). In most cases, it was shown that the deeper the penetration of material used in the porous layer, the higher the properties' improvement. Therefore, the pore opening process is a crucial point to ensure the complete filling of silicon mesoporous layer (pore diameter ranging from 20 to 50 nm). It was established that the etching in a 0.05 M NaOH solution fully opens the pores. The NaOH treatment seems to be a good process because it does not need much material or sophisticated equipment. However, the sodium hydroxide treatment strongly modifies the PSi chemistry and surface morphology. After this treatment, the oxidation of PSi is observed (Errien et al. 2007). Plasma etching is more effective in this application as well. Errien et al. (2007) have shown that the main advantage of the plasma etching process is that it allowed controlling precisely the process with a very good reproducibility compared to the NaOH treatment. The plasma treatment of PSi was performed by using the $SF_6$ gas under classical experimental conditions, namely, gas flow rate of 40 sccm (cm³/min at 273 K, 101.3 kPa), under a 10-mTorr pressure in the chamber, a polarization of 100 V, and a power set at 1500 W. Besides, it was established that the plasma treatment did not modify the chemistry of the PSi surface in nonreversible mode. The fluorination occurring during the plasma treatment is not stable in air. The surface morphology was not changed as well.

One of the important problems of PSi etching is use of photoresist. If photoresist is applied to a porous layer, it penetrates into the pores. In contrast to the easy removal of photoresist after the processing steps from the surface of bulk silicon with ethanol or acetone, you are not able to remove the photoresist completely out of the pores. In addition, the chemical reagents that are used to remove the exposed photoresist chemically etch and destroy the porous layer (Arens-Fischer et al. 2000). To resolve this problem, the metal protection layers such as Ti, Cr, or NiCr with thickness 40–100 nm usually are used. They cover PSi and ensure that the PSi does not come in contact with the photoresist.

Due to the sponge-like structure of PSi, it is possible that the etch process during RIE takes place not only at the surface of the porous layer but also in the pores. The physical component of the RIE-process (like the common sputter process) is anisotropic and the chemical component is isotropic. The chemical reactive gas can penetrate into the pores, which can lead to an increase of the porosity of the remaining layer and hence to the increasing etch rate with time. However, as a rule, the chemical etching is not significant in the RIE process (Arens-Fischer et al. 2000).

Tserepi et al. (2003) have shown that the etching properties of PSi are affected by PSi material properties (porosity, aging) and thermal treatment, as well as by the choice of plasma parameters. Specifically, PSi is etched much faster in a high-density plasma (HDP) reactor (micromachining etching tool of alcatel) (by almost an order of magnitude) than in an RIE reactor (NE330 of Nextral). The results of this comparison can be found in Figure 12.6.

One can see that the etching rate increases with increasing porosity. In detail, by changing the porosity from 57% to 74%, the etching rate increases by ~30%. It was also established that the aging of PSi in certain ambient might drastically decrease PSi etching rate. Thermal treatment affects PSi etching in a more drastic way than sample aging. Even mild thermal treatment is capable of diminishing the etching rate of PSi in $SF_6$ and $SF_6/C_4F_8$ plasmas. Thermally treated PSi is etched at rates comparable to $SiO_2$. It is evident from these results that the material properties of PSi influence considerably the etch rate in plasma. These results suggest that intentional modification of material properties of PSi may be used as a way to provide either selective removal of PSi with respect to c-Si or vice versa. For example, thanks to the difference between the etching rate of the silicon and the silicon dioxide, silicon dioxide layers are often used as etch stop layers (Clerc et al. 1998).

It should be noted that, despite the significant advantages of dry etching during operating with PSi, a wet chemical etching may also find application for resolving specific tasks. For example, Lehmann (2007) has shown that subsequent alkaline etching is well suited to produce certain macroporous geometries that cannot be produced by electrochemical etching in HF only. It is known that the variety of macropore cross-sections, which can be produced by electrochemical etching, is quite limited. It is therefore advantageous, if other cross-sections can be realized by subsequent chemical etching. Lehmann (2007) established that alkaline etching of macroporous silicon membranes could be used for the formation of square pore cross-sections and pore arrays of high porosity. Results of this approach to pore shaping are shown in Figure 12.7.

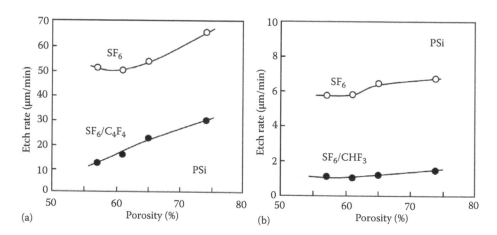

**FIGURE 12.6** Dependence of PSi etching rates from material porosity. Etching has been performed in an HDP (a) and an RIE (b) plasma reactor, with different gas compositions. (Tserepi A. et al.: *Phys. Stat. Sol.* (a) 2003. 197(1). 163. Copyright Wiley-VCH Verlag GmbH & Co. KGaA. Reproduced with permission.)

**FIGURE 12.7**   Macroporous silicon sample (12 µm pitch, 5.5 µm pore, 500 µm thick) after subsequent etching in (a) 50% KOH, 105°C for 10 s, and (b) 2% KOH + 10% propanol, high DOC, 14°C for 10 h. (Lehmann V.: *Phys. Stat. Sol.* (a) 2007. 204(5). 1318. Copyright Wiley-VCH Verlag GmbH & Co. KGaA. Reproduced with permission.)

It is known that all alkaline solutions show a well-known anisotropy of the etch rate for different crystal orientations. While the [111] plane is always the slowest etching plane, the etch rate relation between [100] and [110] planes depends on temperature and etchant composition. At high KOH concentrations (50%) and high temperatures (105°C), the [100] plane shows a lower etch rate and is therefore the remaining one (Figure 12.7a). This etch rate ratio is reversed for low alkaline concentrations (2% KOH + 10% propanol) and low temperatures (14°C). Now the [110] remain as pore-walls, as shown in Figure 12.7b. The addition of propanol is known to enhance the etch rate on [100] planes compared to [110] planes. A high dissolved oxygen concentration (DOC) suppresses formation of hillocks on [100] planes. According to Lehmann (2007), the results, shown in Figure 12.7, can be interpreted as a consequence of the different activation energies for the alkaline etching of [100] and [110] planes. The option to generate perfectly square pores, as shown in Figure 12.7b, is not only interesting for many micromechanical applications, it also enables generation of orthogonal pore arrays with a porosity in excess of 78%, which is not possible with circular pores.

Trifonov et al. (2007a) have shown that subsequent PSi etching with tetramethyl-ammonium hydroxide (TMAH) solution can also be used for the fabrication of macropores with square and eight-sided (octagonal) cross-section. The TMAH solution shows a well-known anisotropy of the etch rate for different crystal orientation (Sato et al. 1999). While the (111) plane always has the slowest etching rate, the relative etching rates between (110) and (100) planes depend on the temperature and etchant composition (Trifonov et al. 2007a,b). Trifonov et al. (2007a) established that if a macroporous silicon membrane is etched under the specified conditions (25 wt.% TMAH, 96°C, 25 sec), the obtained pore shape will be an octagon, with facets formed by (110) and (100) planes. At that, the (100) plane tends to be more developed, which means that its etching rate is slightly lower than that of a (110) plane. Nevertheless, conditions can be found for which the etching rates of both orientations are quite similar and, thus, pores with perfect octagonal shape can be obtained. In contrast, at low temperatures (5°C) the etching rate $v_{(110)}$ is considerably higher than $v_{(100)}$ and pore walls are formed mainly by the remaining (100) planes. Trifonov et al. (2007a) also found that the etch rate ratio can be reversed, that is, $v_{(110)}/v_{(100)} < 1$, at low TMAH concentrations and low temperatures. An alcohol addition, for instance propanol, is known to enhance the etch rate of (100) planes with respect to (110) ones.

## 12.5  OXIDATION OF POROUS SILICON

Freshly etched PSi is covered by hydrogen atoms that are covalently attached to the silicon surface. These hydrogen atoms provide a good electronic passivation layer that is crucial for efficient visible emission. However, after PSi exposition in air or any chemical treatment, the removal of some surface hydrogen atoms takes place. This process is accompanied by destabilizing surface electronic structure and changing electronic, electrophysical, and chemical properties of PSi. A procedure to restabilize the electronic structure is to replace the weakly bonded hydrogen atoms by another passivating species. Replacing the hydrogen by oxygen is one of the most common ways to passivate the electronic properties of PSi. It was established that oxidation does contribute to the stabilization properties of PSi, including optical, photoluminescence, and sensing characteristics (Shih et al. 1992; Amato et al. 1997; Canham 1997a; Astrova and Tolmachev 2000a; Gelloz et al. 2005; Aggarwal et al. 2014). For example, Aggarwal et al. (2014) have shown that PSi samples oxidized under $O_2$ atmosphere had excellent stable surface and no degradation observed even after one year. Many oxidation methods exist, such as aging in ambient (Salonen et al. 1997a), exposure to water vapor (Ogata et al. 1995; Salonen et al. 1997b; Ogata et al. 2000), anodic oxidation in a nonfluoride electrolyte (Yamana et al. 1990; Guerrero-Lemus et al. 1999; Mulloni and Pavesi 2000; Kochergin and Foell 2006; Sakly et al. 2006), chemical oxidation using $HNO_3$ (Leisenberger et al. 1997), $O_3$, $H_2O_2$ (Frotscher et al. 1996), vapor of pyridine (Mattei et al. 2000), water solution of $KNO_3$ and alcohol (Salonen et al. 1997a), and thermal oxidation in wet and dry $O_2$ fluxes (Unagami 1980; Maccagnani et al. 1998; Kim et al. 2001; Pap et al. 2004, 2005), including rapid thermal oxidation (Gelloz et al. 2005). A data-review regarding a wide range of oxidation treatments (e.g., thermal treatments, electrochemical and chemical oxidation, photochemical oxidation, etc.) has been reported (Canham 1997a; Riikonen et al. 2012). However, we need to note that electrochemical and especially thermal oxidations are the most commonly used methods. It was found that none of the chemical liquid-phase oxidation methods based on using $HNO_3$ and $H_2O_2$ solutions were able to oxidize the surface hydrides on PSi completely (Riikonen et al. 2012). A heat treatment of PSi at a high temperature in inert atmosphere causing a rearrangement of surface atoms and a reduction in the hydride concentration was necessary for the complete oxidation of the surface hydrides.

Mentioned studies have shown that oxidation of PSi did not modify the morphology of porous layers. It was only observed that the pore size decreased after oxidation (see Figure 12.8). However, the pore density was conserved. Therefore, the porosity after oxidation, of course, was lower than the porosity before oxidation. For example, after oxidation of PSi with porosity ~60%, the porosity was estimated to 33% (Pirasteh et al. 2006). Furthermore, volume expansion due to oxidation was isotropic and one could observe the increase in the thickness of the layer. For a single layer, when the thickness was equal to 5 μm before oxidation, it was nearly 5.8 μm thick after oxidation (Pirasteh et al. 2006). The refractive index of porous layers decreased after oxidation as well.

Based on the physical characterization of oxidized PSi, it was concluded that in addition to PSi stabilization, the oxidation of PSi layers is a good way for obtaining a lower optical loss of PSi waveguides by reducing both volume scattering and absorption in the near infrared wavelength range (Loni et al. 1996; Guerrero-Lemus et al. 1999; Charrier et al. 2000; Pirasteh et al. 2006). Besides that, the transformation of porous silicon into oxidized PSi also allows achieving light transmission in the visible wavelength range (Amato et al. 1997). As was shown, oxidized PSi could also be successfully used for the elaboration of gas sensors (Maccagnani et al. 1998; Ozdemir and Gole 2007), electrochemical sensors (Sakly et al. 2006), biosensors (Naveas et al. 2013), and device insulation (Park and Lee 2003). It was found that oxidation can produce biocompatible surfaces, which can be further functionalized by silanization (Riikonen et al. 2012).

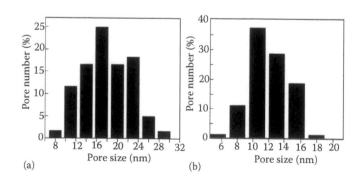

**FIGURE 12.8** Distribution of pore size of (a) PSi layer and (b) oxidized PSi layer. (From Reprinted from *Appl. Sur. Sci.* 253, Pirasteh P. et al., 1999, Copyright 2006, with permission from Elsevier.)

## 12.5.1 THERMAL OXIDATION

Air is the simplest oxidant. Air oxidation produces different types of surface species, depending on the temperature at which the reaction is performed and the humidity of air. The Si-Si bond is weaker than the Si-H bond, and mild oxidation tends to attack the Si-Si bonds preferentially. The rate of the reaction is highly dependent on the temperature. Experiment has shown that room temperature oxidation produces a fairly thin oxide layer within a few hours, which grows over the course of several months. Therefore, for forming thicker oxide layer or for complete oxidation of PSi the oxidation at elevated temperature usually is used. According to Unagami (1980), the rate of oxidation is related to the oxidation temperature by the equation

$$(12.2) \qquad \Delta W = A \ exp\left(-\frac{E_a}{kT}\right)$$

where $\Delta W$ is the amount of oxidation, $A$ is the coefficient related to the anodization conditions and the oxidation conditions, $k$ is the Boltzman coefficient, and $E_a$ corresponds to the activation energy of PSi oxidation. The value of $E_a$ changes slightly with resistivity of the silicon substrate. The value of $E_a$ for oxidation in dry $O_2$ is 0.10–0.13 eV for $\rho$ changed in the range of 0.02–7 $\Omega\cdot$cm. Moreover, Unagami (1980) found that this value of $E_a$ was independent of the anodic current density and PSi layer thickness. In addition, these values of $E_a$ were very small compared to the direct oxidation of bulk silicon. The activation energy of the surface oxidation reaction for bulk silicon is equaled ~2.0 eV. The presence of water vapor greatly accelerates the oxidation rate in comparison with the rate in dry air. For PSi oxidation in wet $O_2$ $E_a$ is ~0.65 eV (Unagami 1980). Partial desorption of hydrogen from porous silicon also promotes the oxidation (Ogata et al. 2000). This effect can be achieved by the thermal annealing of PSi in an inert atmosphere. Annealing causes surface hydrides to desorb from the surface, leading to dangling bonds (unpaired electrons) and a strained Si–Si structure on the surface (Ogata et al. 1998). Therefore, the annealed surface oxidizes rapidly under ambient conditions.

It was established that since porous Si has a very large surface-to-volume ratio, the oxidation cannot be neglected even under mild conditions. It was found that PSi is being oxidized already at 500–600°C (see Figure 12.9), and already oxidation at 800°C is sufficient to completely convert the PSi skeleton to silicon oxide, although the length of time needed to accomplish this transformation depends on the type of sample. For example, Pap et al. (2004) have shown already at $T =$ 800°C after 10 h oxidation the whole PSi structure with wall thickness ~30 nm was oxidized. The

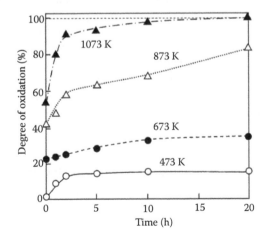

**FIGURE 12.9**   Oxidation extent of 70% porosity and 30-μm thick porous membranes versus time and temperature. (Reprinted with permission from Pap A.E. et al., *J. Phys. Chem. B* 108, 12744, 2004. Copyright 2004. American Chemical Society.)

PSi layer is porous, so oxidation occurs inside the PSi layer from the beginning of oxidation as well as at the surface of PSi. It was established that heating PSi samples in the presence of high pressure water vapor also gives a very stable oxide.

However, often the additional final annealing at $T$ = 1000–1100°C for oxide densification is being used (Maccagnani et al. 1998; Kim et al. 2001). It was found that after a short-term annealing at a high oxidation temperature above 1000°C in wet $O_2$, thick oxidized PSi (OPSi) films had the same properties as thermal $SiO_2$ of bulk silicon (Unagami 1980). Commonly used modes of thermal oxidation of PSi are presented in Table 12.2.

It is necessary to note that using a preliminary annealing at low temperatures sufficiently improves the parameters of oxidized PSi. It was established that it is not desirable to anneal as-prepared PSi at temperatures higher than 400°C because it could lead to a drastic restructuring of the porous layer (Yon et al. 1987). This is a consequence of the large surface area of the material: even at low temperature, surface migration of silicon atoms occurs and provokes pore coalescence (Bomchil et al. 1989). According to Yon et al. (1987), a preoxidation of PSi layers at low temperatures in the range of 300°C is sufficient for creating an oxide layer all along the pore walls, which prevents restructuring of the sample during further heating at higher temperatures up to 800–900°C (Yon et al. 1987). Therefore, it was established that the procedure: pre-oxidation at 300°C for structure stabilization; oxidation at 750–800°C for complete PSi transformation into silicon dioxide; and final densification at 1050–1090°C for obtaining a compact oxide equivalent to standard thermal silicon dioxide is the most optimal for reproducible forming of oxide layers from PSi (Yon et al. 1987; Maccagnani et al. 1998; Kim et al. 2001).

It was also found that the oxidation of PSi is being controlled by a surface reaction both at the surface of PSi and at the wall of pores in PSi layers. Both $Si_2O_3$ and $SiO_2$ are produced by low temperature oxidation below 400°C, and physically absorbed water exists in an OPSi film formed by wet $O_2$ oxidation at temperatures below 700°C. In addition, Mawhinney et al. (1997) have found that thermal oxidation at low temperatures (250–440°C) produces back-bond oxidized surfaces ($O_x$–Si–$H_{4-x}$, $x$ = [1, 3]). According to Salonen et al. (1997b), the surface hydrides oxidize to hydroxyl groups at temperatures above 440°C. It was found that oxidation at temperatures above 800°C causes complete oxidation of the whole PSi structure into Si–$O_2$. At a high oxidation temperature, Si is being increasingly transformed to $SiO_2$, with consequential loss of crystalline structure, resulting in a partly amorphous form. The etching rate, dielectric constant, $tan\ \delta$, and breakdown strength of the OPSi films are strongly dependent on the oxidation temperature, oxidation atmosphere, and substrate resistivity formed (Figure 12.10). OPSi films formed at a low temperature had a higher etching rate than thermal $SiO_2$ formed by the direct oxidation of bulk silicon. The etching rate decreases with raising oxidation temperature (see Figure 12.10).

As it was indicated before, using wet atmosphere contributes to a significant increase in the rate of oxidation. Mattei et al. (2000) established that PSi is also quickly oxidized in air in the presence of pyridine vapor. At that, this oxidation takes place even at room temperature. Figure 12.11a shows the oxidation degree of porous silicon as a function of the exposure time into the pyridine vapor for a sample of 70% porosity. It is seen that the logarithmic dependences of the oxidation degree, representing the fraction of Si converted into silicon oxide, on the exposure time to pyridine vapor was observed in these experiments. It was also found that the oxidation of PSi continues, although at lower rate, even when pyridine treatment is finished. This result is demonstrated in Figure 12.11b on the example of dielectric constant change.

**TABLE 12.2  Technological Parameters, Usually Used for Thermal Oxidation of PSi**

| 1 Step | 2 Step | 3 Step | References |
|---|---|---|---|
| Dry air, 800°c, 20 h | | | Pap et al. 2004 |
| Wet $O_2$; 1050°C; 3 h | | | Shen and Toyoda 2002 |
| Dry $O_2$, 1000°C, 60 s | | | Debarge et al. 1998 |
| Wet $O_2$; 300°C; 1 h | Wet $O_2$; 900°C, 1 h | | Pirasteh et al. 2006 |
| Dry $O_2$; 400°C; 30 min | Dry $O_2$; 1000°C; 1 h | Wet $O_2$, 1000°C; 30 min | Kim et al. 2001 |
| Dry $O_2$; 300°C | Wet $O_2$; 850°C | Wet $O_2$; 1100°C | Maccagnani et al. 1998 |

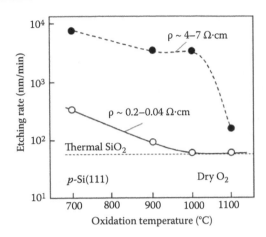

**FIGURE 12.10**  Etching rate of OPSi film for the buffered HF solution (50% HF: 40 wt.% NH$_4$F = 1:10). The thickness of the PSi layer was ~2 μm. The time of oxidation in dry O$_2$ was 1 h. (From Unagami T., *Jpn. J. Appl. Phys.* 19, 231, 1980. Copyright 1980: The Japan Society of Applied Physics. With permission.)

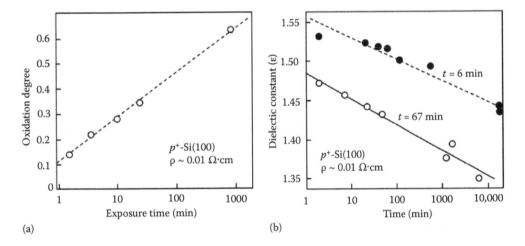

(a)                                                     (b)

**FIGURE 12.11**  (a) PSi (porosity 70%) sample oxidation degree as functions of the exposure time in air saturated by pyridine, (b) time dependence of the dielectric constant of PSi (porosity 75%) previously immersed in pyridine vapor as functions of the exposure time in air. Exposure time in pyridine before oxidation in air is denoted in the figure. The treatment in pyridine vapor was made by putting the samples and an opened glass vessel filled with pyridine into a closed box. (Mattei G. et al.: *Phys. Stat. Sol.* (a) 2000. 182. 139. Copyright Wiley-VCH Verlag GmbH & Co. KGaA. Reproduced with permission.)

Mattei et al. (2000) believe that the promoting role of pyridine molecules in quick oxidation of PSi is connected with establishment of an attractive interaction through its nitrogen lone pair with the hydrogen atoms bonded to the silicon and present in a large extent on the PSi inner surface. As it is known, pyridine is a basic heterocyclic organic compound with the chemical formula C$_5$H$_5$N. This interaction, proved by the slight change of the frequency of some pyridine absorption bands, should weaken the Si-H bond causing a fast interaction with the oxygen atoms. As a result, the silicon oxidation as well as the decrease of the density of the Si-H species on the PSi surface should be expected.

Research has shown that for obtaining dense SiO$_2$ from a PSi layer, the dimensions of the pores should balance the volume expansion during the thermal oxidation by a factor of 2.27, and by a pore wall thin enough to be fully oxidized. It was found that this condition could be achieved when the PSi porosity is in the range 55–90% (Yon et al. 1987; Bomchil et al. 1989; Shen and Toyoda 2002). A layer of lower porosity will never be fully oxidized, while an increased porosity would give a very fragile macro-PSi structure and consequently a very weak porous SiO$_2$.

Regarding disadvantages of thermal oxidation, Riikonen et al. (2012) believe that the main drawback of this process is a relatively high decrease in the pore volume as well as in the pore diameter. To overcome this problem, they proposed to use the combination of thermal oxidation and annealing in inert atmosphere ($N_2$, 700–1000°C). This treatment had the smallest effect on the pore volume, although the pore size significantly increased. However, Riikonen et al. (2012) established that the surface of such PSi was not inert and was unstable under ambient conditions. Riikonen et al. (2012) found that the treatment comprising a combination of thermal oxidation, annealing, and chemical oxidation in $H_2O_2$ solution is more efficient. This treatment had a very small effect on the pore size and caused a moderate decrease in the pore volume. Moreover, the surface did not show chemical reactions in their biomedical experiments and had a large number of –OH groups on the surface available for further modification. However, the authors admit that this oxidation method was the most time-consuming.

### 12.5.2 ELECTROCHEMICAL OXIDATION

For anodic oxidation of PSi, two types of electrolytes, based on 1 M $KNO_3$ (Cantin et al. 1996; Kleps et al. 1997) and 0.1 M $H_2SO_4$, aqueous solutions (Fan et al. 2003) usually were used. For example, Bisi et al. (2000) have proposed for PSi oxidation to use electrochemical oxidation in a nonfluoride electrolyte, for example, a 1:1 solution of 1 M $H_2SO_4$ with ethanol. Usually ethylic or methylic alcohol added in the anodic oxidation solution contributes to the better electrolyte penetration in the PSi pores. Typical current densities are in the range of 1–10 mA/cm². It was shown that 1M $Na_2SO_4$ aqueous solution ($J$ = 20 mA/cm²) (Guerrero-Lemus et al. 1999), as well as 1M HCl solutions ($J$ = 1–2 mA/cm²) (Kleps et al. 1997) also can be used for these purposes. In addition, Kleps et al. (1997) have found that for medium PSi porosity (P = 60%), the anodic oxidation must be effectuated without PSi sample drying. The monitoring of the potential during the oxidation helps to assess the time at which the sample is fully oxidized by looking at the step increase in the potential. The electrochemical oxidation occurs preferentially near the bottom of the pores. Larger currents yield better oxidation homogeneity. For further improvement of oxidation homogeneity, light assistance is required because anodic oxidation of silicon requires holes. PSi is depleted in holes and without illumination only the bottom layers and the crystalline silicon/PSi interface are oxidized. In this case, the oxidation stops when an insulating layer is produced at the interface, leaving most of the PSi unaffected (Bsiesy et al. 1991). With light illumination, holes are generated in PSi, and oxidation can take place in a more homogeneous way. The result is an oxide coating of a monolayer, which covers the PSi skeleton.

Bsiesy et al. (1991) believe electrochemical oxidation of PSi has the following advantages: (1) electrochemical oxidation of PSi can be achieved easily; (2) it is possible to oxidize either the lower part of the porous layer, or the whole depth, at a level that depends on the exchanged charge. This method therefore appears to be more attractive than thermal oxidation when incomplete oxidation is required. In particular, such requirement appears during silicon (or other material) epitaxy on porous silicon. These processes generally involve temperatures above 400°C and porous silicon must be stabilized by a preoxidation step in order to conserve its very thin microstructure. If this preoxidation is achieved by thermal oxidation, there is also oxide growth on top of the sample, which must be eliminated before subsequent epitaxy. Electrochemical oxidation, with an appropriate choice of experimental conditions, can lead to oxidation limited to the inner part of the porous layer.

## 12.6 PORE WIDENING

Currently, the pore widening of macroporous silicon membranes was performed either by multiple thermal oxidations and subsequent oxide stripping steps or by anisotropic etching in alkaline solutions as described earlier (Trifonov et al. 2007a,b). Usually the multiple oxidations were carried out in dry oxygen at temperatures varied in the range of 800–1100°C. Using indicated temperatures provides a reasonable rate of PSi oxidation. Experiment has shown that a short dipping of the sample in 5 wt.% HF solution is enough for removing the oxide layer formed after

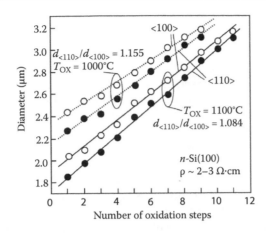

**FIGURE 12.12**    Pore diameter along <110> (squares) and <100> (circles) directions as a function of the number of oxidation/oxide-stripping steps. The pore size along <110> direction grows faster than that along the <100>. (Trifonov T. et al.: *Phys. Stat. Sol.* (a) 2007. 204(10). 3237. Copyright Wiley-VCH Verlag GmbH & Co. KGaA. Reproduced with permission.)

the oxidation step. Etching in $NH_4F$ solution also can be used for oxide removing. Figure 12.12 shows experimental data related to the pore diameter change as a function of the oxidation steps.

The advantage of this process for pore widening compared to an isotropic etching in, for example, a mixture of $HF/HNO_3$ is the precise thickness control: The amount of silicon removed by this procedure is determined by the thickness of the grown oxide layer. This allows for tuning the pore diameter after the etching. Although this process seems to be straightforward, some restrictions have to be kept in mind (Langner et al. 2011). First, it is necessary to take into account that the volume of silicon oxide is 2.25 times larger than that of the original silicon. Second, for thin oxide layers (short oxidation times), the oxide growth speed is very high. Hence, statistical fluctuations during the growth can have undesired effects such as large deviations in the oxide layer thickness. In the case of thick oxide layers, the dominating factor is the induced stress. Therefore, the possibility of damaging the macroporous structure has to be taken into account.

Data presented in Figure 12.12 demonstrate that the correction of the pore shape is also possible during this process because thermal oxidation is an anisotropic process as well. The oxidation rate along <110> crystallographic direction is greater than that of <100> (Kao et al. 1987). This means that in the subsequent oxidations/oxide-stripping steps, pore diameter will tend to increase faster along the <110> orientation than along the <100> one. The slope of the linear fits represents the mean diameter increase $d_{<110>,<100>}$ measured in µm/step. Therefore, the ratio $d_{<110>}/d_{<100>}$ can be considered a ratio between the oxidation rates along these two directions. As $d_{<110>}$ is always greater than $d_{<100>}$, pores become rounded after certain oxidation steps. Lower temperatures also give rise to enhanced anisotropy of the oxidation process and fewer oxidations are therefore necessary to reach circular pore shape. The described approach allows for pore shape and size to be finely tuned, which is of great importance for photonic crystal applications, where the geometry is a critical parameter (Trifonov et al. 2007a,b).

## 12.7  DOPING OF POROUS SILICON

Analysis of PSi doping should be divided into three parts: (1) doping of porous silicon to change its resistance, (2) doping of porous silicon with the purpose of its further use as an impurity source for diffusion in crystalline silicon, and (3) doping of the porous silicon by ions of rare earth elements, such as Er and Yb, for the appearance of the emission in the desired wavelength range.

It should be noted that the technology based on the use of PSi as an impurity source for diffusion in crystalline silicon is not directed on the change of PSi properties, and therefore this technology will be not considered in this chapter.

### 12.7.1 DOPING BY SHALLOW IMPURITIES

As is known, the pore size, porosity, and conditions of silicon anodic etching are strongly dependent on the doping level of the silicon substrate. Therefore, when choosing Si samples for porosification these factors primarily are taken into account. However, as was shown earlier, the width of the walls in porous silicon, formed by anodization, is commensurate with the double thickness of the space charge region. Therefore, at a certain porosity, PSi layers become high-resistance and resistance of PSi begins to depend weakly on the initial parameters of silicon substrate. The resistivity of PSi formed on silicon substrates has been found to be as high as $10^5$–$10^6$ $\Omega$·cm when measured in ambient air, even when the substrate is heavily doped as in $p^+$ and $n^+$ materials with resistivity $10^{-2}$–$10^{-3}$ $\Omega$·cm (Anderson et al. 1991; Mares et al. 1993; Deresmes et al. 1995).

Taking into account that pore formation does not affect the nature of the impurities behavior in PSi (Grosman and Ortega 1997), it becomes apparent that due to the PSi doping with shallow impurities, it is possible to control the width of the space charge region, and hence the resistance of PSi. An increase of the ionisation energy of the dopants is expected only in Si nanocrystallites where quantum confinement enlarges the energy gap. According to the calculations of Allan et al. (1995), this ionization energy can reach very large values ($\ll 1$ eV) for 3-nm large silicon nanocrystallites, which leads the authors to conclude that silicon nanocrystallites with sizes in the nanometer range are essentially intrinsic. However, such an explanation is not suitable for medium porosity $p^+$ or $n^+$ porous layers in which the silicon walls are too thick (~10 nm) to give quantum confinement effects.

However, despite the availability of such possibility, it should be noted that the doping of the PSi for influencing its electrical conductivity is practically not used. Only Nishimura et al. (1996), Sundaram et al. (1997), Astrova et al. (2000b), El-Bahar et al. (2000), and Kovalevskii et al. (2004) reported that doping by shallow impurities can be used for reduction of PSi layers resistance and influence on their photoluminescence properties. In particular, for these purposes Nishimura et al. (1996) used Sb-doped silicate glass (SbSG) as a source of Sb during thermal diffusion. SbSG was deposited onto a PSi layer by magnetron sputtering. El-Bahar et al. (2000) modified properties of PSi layer by thermal diffusion of phosphorous, using a phosphorus liquid source of phosphorous oxychloride ($POCl_3$) doping technique. They have found that five orders of magnitude decrease in the resistivity of the porous layer took place after the thermal doping ($T_{diff} = 850°C$), followed by an etching process to remove the oxide, which is formed during the diffusion and activation process (Table 12.3). A significant increase of the photoluminescence was also observed after the doping process. They have shown that the process can be easily controlled by starting and stopping the gas flow to the bubbler. Moreover, because the diffusion length of phosphorous atoms in silicon in the applied temperatures is much higher than the nanoporous crystalline size, all the volume of the porous silicon is expected to be diffused and doped uniformly. In addition, El-Bahar et al. (2000) established that this process does not introduce damage to the porous skeleton. Kovalevskii et al. (2004) have shown that the diffusion of arsenic or phosphorus from an RF plasma also is possible (see Figure 12.13). Moreover, it was found that plasma-assisted diffusion offers a way to significantly reduce the wafer temperature required for doping. Specifically, the plasma-assisted diffusion of arsenic or phosphorus at a substrate temperature within 400°C was as efficient as conventional diffusion at 800–850°C. In addition, doping concentration was more uniform over the surface if pores are arranged regularly (as with porous silicon produced by plasma etching).

**TABLE 12.3  Resistivity Measurement of PSi Layer before and after Phosphorus Diffusion from $POCl_3$-Source**

| Anodization | Diffusion | Conditions | $R, \Omega$ | $\rho, \Omega \cdot cm$ |
|---|---|---|---|---|
| $p$-Si(100) (5–7 $\Omega$· cm) | $T_{diff} = 850°C$ | As prepared PSi | $5\cdot10^6$ | $3\cdot10^8$ |
| HF:ethanol = 1:1 | | After doping | $6\cdot10^8$ | $4\cdot10^{10}$ |
| $J = 30$ mA/cm², $t = 10$ min, | | After oxide removal | 13 | $9\cdot10^2$ |
| Porosity ~65% | | | | |

*Source:*  Data extracted from El-Bahar et al. *IEEE Electron Dev. Lett.* 21(9), 436–438, 2000.

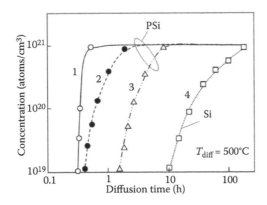

**FIGURE 12.13**   Arsenic concentrations at a depth of 100 µm PSi layer (porosity 25%) versus diffusion time for (1) plasma-assisted, (2) ampoule, and (3) open-tube diffusion. Plasma-assisted diffusion is carried out at a total pressure of 450 Pa and a gas flow rate of 600 dm³/min. AsH₃ was used as diffusion source. Curve 4 refers to open-tube diffusion into monocrystalline silicon to the same depth. (With kind permission from Springer Science+Business Media: *Russ. Microelectron.* 33(1), 2004, 13, Kovalevskii et al.)

Apparently, the lack of progress in this direction is connected with the fact that the process of doping is accompanied by changes in morphology and other properties of the PSi, which are unacceptable for its further use in the real device structures. In addition, an additional improved passivation process is required, following the formation of the contact.

### 12.7.2  DOPING BY RARE EARTH ELEMENTS

Rare earth elements like erbium (Er), ytterbium (Yb), and so on have generated great interest due to their unique electronic structure that provides the appearance of emission in the visible to infrared regions of the spectrum, that is, in the regions of the spectrum that are technologically important (Lopez 2001; Diaz-Herrera et al. 2009). For example, emission from the first excited state (1.54 µm) from erbium-doped materials has clearly set the telecommunication wavelength standard as 1.5 µm. This is an important wavelength because standard silica-based optical fibers have their maximum transparence at this wavelength. Moreover, emission from rare earth impurities is narrow and efficient and this emission is insensitive to the host material and surrounding environments (Shriver et al. 1990; Polman 1997). Due to the poor penetration of f-orbital electrons through the inner shells, there is a week coupling between the f-shell electrons and the electric field produced by the nucleus making them insensitive to the host matrix. Another reason for the relatively insensitive emission comes from the shielding of the 4f-shell electrons by the filled 5s- and 5p-shells. The shielding by the higher orbital electrons reduces the interactions felt by the f-shell electrons from the surrounding matrix and environment.

As it was indicated before, rare earth elements are usually incorporated as impurities in solids to take advantage of their electronic properties. Experiments have shown that PSi is an excellent host for these purposes (Kimura et al. 1994; Bondarenko et al. 1997). Its large surface area allows easy infiltration of the ions into the matrix and PSi readily oxidizes, thereby obtaining large concentrations of oxygen necessary for efficient Er emission (Michel et al. 1991). In addition, we have to take into account that silicon is the dominant material in the microelectronic industry. This means that a way to effectively integrate the emission from rare earth elements such as Er for optoelectronic applications is to use the silicon as the host.

At present, the doping of PSi with Er has been achieved by ion implantation (Namavar et al. 1995), spin-on technique with following diffusion (Dorofeev et al. 1995; Diaz-Herrera et al. 2009), and electro-migration (Kimura et al. 1994; Kazuchits et al. 1995; Lopez 2001). Cathodic electro-migration is the preferred process because it offers the advantages of deeper Er penetration, lower cost, and simplicity of processing (Lopez 2001). In particular, Mula et al. (2012) demonstrated that Er was present within the whole PSi layer even for layers with thickness ~30 µm. However, it should be noted that a linear decrease in the Er/Si signal ratio from the external surface toward the PSi/

crystalline Si interface was observed for this technology (Najar et al. 2009; Mula et al. 2012). Results obtained by Mula et al. (2012) are shown in Figure 12.14. For comparison, Er-doped crystalline silicon structures are usually prepared using techniques such as ion implantation, epitaxial growth, and chemical vapor deposition, which are expensive, time consuming, require specialized equipment, and are limited to producing very shallow doping profiles. This means that using PSi as the host and doping it with Er by cathodic electro-migration produces a deeper doping profile and is simpler and cost-effective, making the procedure very attractive and desirable for commercial applications.

Experiment has shown that in order to obtain efficient Er emission, it is important to have a large concentration of ions incorporated into the PSi matrix. Therefore, PSi layers should have maximum porosity and thickness (see Figure 12.15). In order to facilitate the dopant infiltration and increase the Er concentration, PSi surface and matrix should also be free of pore-blocking

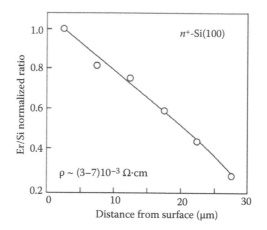

**FIGURE 12.14**  Er content as a function of the distance from the surface from SEM-EDS measurements in a 30-μm thick porous Si layer. PSi was formed using HF:H$_2$O:ethanol = 15:15:70 solution. The Er doping was also obtained electrochemically using a 0.1-M ethanolic solution of Er(NO$_3$)$_3$·5H$_2$O in a constant-current process. (Reprinted with permission from Mula G. et al., *J. Phys. Chem. C* 116, 11256, 2012. Copyright 2012. American Chemical Society.)

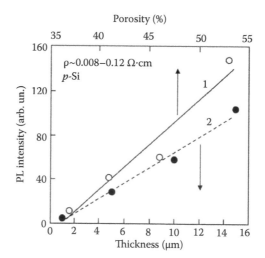

**FIGURE 12.15**  Preparation parameters affecting the 1.54 μm PL intensity of Er-doped PSi structures: 1 corresponds to different porosity samples with a constant thickness of 5 μm, and 2 corresponds to PSi samples of different thickness held at a constant porosity of ~46%. In order to activate the Er ions, the samples were oxidized at 950°C in a dilute oxygen environment and densified at 1100°C in a nitrogen environment. Luminescence was measured with a liquid nitrogen cooled Ge detector using the 514 nm Ar⁺ line as the excitation for PL, and a closed-cycle He cryostat. (Lopez H.A. and Fauchet P.M.: *Phys. Stat. Sol.* (a) 2000. 182. 413. Copyright Wiley-VCH Verlag GmbH & Co. KGaA. Reproduced with permission.)

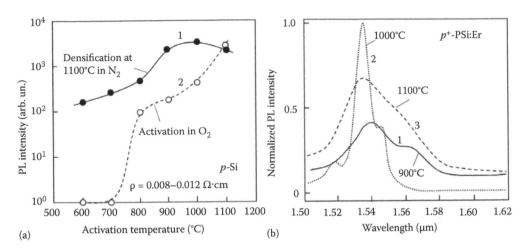

**FIGURE 12.16**   Preparation parameters affecting the 1.54 μm PL intensity of Er-doped. PSi structures: (a) PL intensities of PSi:Er samples after (2) activation and (1) densification, respectively. (b) Temperature annealing dependence of the PL intensity for an annealing duration of 1 h. ([a] Lopez H.A. and Fauchet P.M.: *Phys. Stat. Sol.* (a) 2000. 182. 413. Copyright Wiley-VCH Verlag GmbH & Co. KGaA. Reproduced with permission. [b] Reprinted from *Appl. Surf. Sci.* 256, Najar et al., 581, Copyright 2009, with permission from Elsevier.)

species (Lopez 2001). This can be achieved by performing a pore-cleaning step prior to electromigration of the Er ions. In particular, for these purposes Lopez (2001) used treatments in de-ionized water and in heated a hydrogen peroxide/hydrochloric acid/de-ionized water (1:3:15 by volume) solution. During this time, the solution was magnetically stirred to remove any bubbles on the surface of porous silicon. After cleaning, the samples were rinsed and placed in de-ionized water.

Usually for PSi doping by Er, using electromigration technique, the samples were rinsed additionally in ethanol and immersed in a saturated $ErCl_3 \cdot 6H_2O$/ethanol solution (Lopez 2001; Najar et al. 2009). However, other solutions also can be used. For example, Petrovich et al. (2000a,b) and Mula et al. (2012) used a 0.1-M ethanolic solution of $Er(NO_3)_3 \cdot 5H_2O$. The samples were then negatively biased relative to the platinum electrode in order to drive the erbium ions into the PSi matrix. The same experimental apparatus used to produce PSi can be applied for electromigration. The current density used should be chosen carefully so that the amount of Er ions are introduced at a rate that is not too slow or too fast. If the rate is too slow, then the resulting concentration of the infiltrated Er ions will be low. If the rate is too fast, then capping of the pores can occur, also resulting in a low Er concentration. For this process, Lopez (2001) used the current density varied in the range from 0.1 mA/cm² to 0.5 mA/cm² for times ranging from 15 to 30 min.

Activation of Er ions occurs during subsequent annealing at temperatures of 600–1100°C under controlled environments (argon, nitrogen, and oxygen) (see Figure 12.16a). For example, Bondarenko et al. (1997) and Najar et al. (2009) observed the increase of the photoluminescence (PL) intensity with diffusion temperature up to 1000°C and then a decrease of PL (Figure 12.16b). It has been shown that co-doping with oxygen, carbon, nitrogen, or fluorine results in a substantial increase in emission efficiency (Michel et al. 1991). An increase of PL takes place due to the formation of active Si–O–Er complexes (Najar et al. 2009). According to Najar et al. (2009), a decrease of Er PL, observed for long oxidation time and high diffusion temperature, takes place possibly due to segregation of these complexes. Bondarenko et al. (2003) have shown that electromigration technique can also be used for co-doping of PSi with Er and Fe. The Er PL spectrum of such PSi:(Er-Fe) samples after high temperature annealing had super fine structure of 20 sharp PL peaks with FWHM of about 0.4 meV.

## 12.8 ION IMPLANTATION

Ion implantation has been recognized as an effective and powerful tool to induce important modifications in the surface and near surface region in silicon and other materials. Therefore, at present, this technique is one of the basic technologies used in industrial semiconductor microelectronics for the formation of various types of silicon nano- and microdevices. The experiment showed that ion implantation has great potential for impact on the properties of PSi as well. The main advantage of ion implantation is apparently the possibility of controlled impurity introduction virtually at room temperature. A disadvantage is the production of defects in the course of ion bombardment, which implies the need for the postimplantation annealing. In addition, this technique requires expensive equipment and special masks.

It was established that the ion implantation could be used to solve many tasks related to PSi. They are the following:

1. Applying ion implantation, PSi can be formed without using any etching techniques. This method is described in Chapter 10.
2. Ion implantation can be used for PSi patterning. Ion doped layers can act as a mask during the electrochemical etching of silicon. For example, ion implantation of protons or helium ions, with energies of 250 keV to 2 MeV, causes localized increase in the resistivity of $p$-type Si wafers arising from the point defects created along the ion trajectories. Increased resistivity reduces the electrical hole current flowing through these regions during subsequent electrochemical anodization, slowing down the PSi formation This technique is described in Chapter 3. A silicon micromachining process based on high-energy ion beam irradiation and electrochemical anodization to form PSi has been used to fabricate various patterned PSi and silicon microstructures (see Figure 12.17) aimed for application in devices described in *Porous Silicon: Biomedical and Sensor Applications* and *Porous Silicon: Opto- and Microelectronic Applications*.
3. Ion implantation affects the conditions of PSi formation (Ambrazevicius et al. 1994; Pavesi et al. 1994; Ahmad and Naddaf 2011; Ow et al. 2011). For example, Pavesi et al. (1994) suggested that the porosity of silicon could be controlled by implanting Si ions into the initial wafer. Later, Manuaba et al. (2001) found that the MeV energy He implant accelerates the etching process, probably due to the bubbles or the remaining lattice damage. At a dose of $8 \times 10^{16}$ ions/cm² the He-containing PSi layer was formed with a significantly enhanced porosity due to the contribution of the large-sized bubbles. At

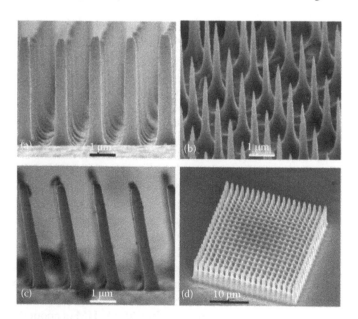

**FIGURE 12.17** SEM images of high-aspect-ratio pillars obtained by point irradiations with a 2-MeV proton dose of $5 \cdot 10^{16}$/cm² into a 3 Ω·cm Si wafer. The samples were then anodized for 15 min at $J = 40$ mA/cm². (a) Beam perpendicular to surface and not aligned with a major crystal axis. (b) Beam axially channeled along surface normal [001] axis. (c) Beam tilted 10° with respect to the perpendicular, not aligned with a major crystal axis. (d) Created under similar conditions as (a), but with smaller gaps between the point irradiations. (Reprinted with permission from Breese M.B.H. et al., *Phys. Rev. B 73*, 035428, 2006. Copyright 2006 by the American Physical Society.)

the highest dose of $32.5 \cdot 10^{16}$ ions/cm$^2$ flaking took place during the anodic etching. In contrast to He, N stopped the anodic etching at a depth of critical N concentration of ~0.9 at .%. For the lowest implantation dose, where the peak concentration was below this limit, the pores propagate through the implanted layer with an enhanced speed.

Studding MeV-Cu implantation into silicon, Ahmad and Naddaf (2011) have concluded that the size, shape, and density of the formed pores are also highly affected by the direction of beam implantation. In addition, the region of maximum defects which is located at 1.65 µm creates a barrier with higher resistivity, "preventing" or slowing down the formation of PSi in the vertical direction within the implanted regions. This agrees well with the results of Lehmann et al. (1991), who found that the implanted carbon at sufficiently high doses could be used as a very effective etch-stop.

4. Ion implantation can change the structure of already formed PSi. In particular, it was found that the pore structure could be effectively destroyed by ion implantation (Fried et al. 1996; Paszti et al. 1996; Manuaba et al. 1998). Depending on the implantation energy, ion type, and applied dose, the porosity of PSi gradually decreases and the PSi layer can be transformed back even into compact material with strong changes in structural, optical, and electrophysical properties. For example, after implantation, effects such as disorder and amorphization in the PSi layer (Roy et al. 1994), the decrease in Si nanocrystallite size (Prabakaran et al. 2004), difficulty in penetration of gas molecules (Manuaba et al. 1998), the decrease of the thickness of the layer (Paszti et al. 1996), and the increase in the refractive index (Fried et al. 1996) were observed. In particular, the results of the He and Ar gas retention measurements showed that noble gases can escape through the pores only at low fluences. As the implanted fluence increases, the pores gradually close and the gas retention rate reaches the full retention. This means that implantation can close the pores (Simon et al. 1999; Paszti et al. 1996, 2000; Manuaba et al. 1997, 1998; Paszti and Battistig 2000). Results obtained for columnar and spongy type PSi samples clearly indicated that the densification occurs most intensively in a narrow depth region around the penetration depth of the ions, that is, it is mainly caused by ion cascades (Simon et al. 1999). Therefore, Simon et al. (1999) believe that based on this phenomenon, production of deeply buried narrow compact layers in porous materials seem to be accessible. It was found that in the case of columnar type PSi layers, implantation can also tilt the pore walls (Paszti and Battistig 2000; Paszti et al. 2000).

5. Ion implantation may affect the composition of PSi. For example, ion beam synthesis was successfully applied to form silicide layers on PSi layers (Ramos et al. 2000). Naturally, ion implantation might be applied to dope the substrate before anodization, as well. Liao et al. (1995) successfully applied ion implantation to fabricate a structure of porous-SiC:porous-Si:Si-substrate, which emitted stable intense blue light by electro-luminescence. Zuk et al. (1999) established that oxygen implantation (225 keV, $10^{17}$ cm$^{-2}$, $T_{an} = 850°C$) was accompanied by the formation of oxide phase and the appearance of the strong blue band luminescence at 2.7 eV, peculiar for silica. Beloto et al. (2001) have found that implantation with nitrogen was accompanied by incorporation of nitrogen and formation of $SiO_2$ and $Si_3N_4$ compounds on the sample surfaces. Beloto et al. (2001) believe that exactly these changes are responsible for strong reduction of the reflectance of PSi structures in the ultraviolet region of the spectrum, and for tendency for the intensity of the peak of the ultraviolet excited photoluminescence to increase with the treatment time of plasma immersion ion implantation.

Stepanov et al. (2014) have shown that ion implantation can be used for preparing PSi layer functionalized by noble metals as well. In order to demonstrate this technique, they implanted in single crystalline silicon Ag$^+$-ions with an energy of 30 keV at a dose of $1.5 \times 10^{17}$ ion/cm$^2$. It was found that after this treatment silver nanoparticles with a diameter of 5–10 nm were observed on the surface of formed PSi layers.

6. Ion implantation may influence electrophysical properties of PSi layers. As it is known, PSi layers have high resistivity and the decrease of the resistance is necessary in order to obtain effective electroluminescent structures based on PSi. Peng et al. (1994) studied PSi implanted by silicon, boron, and phosphorus at an energy of 150, 75, and 170 keV, respectively. The implantation dose varied from $10^{12}$ to $10^{15}$ cm$^{-2}$. The resistivity of the reference PSi layer formed by electrochemical etching of $n$-Si(100) was about 1.5 MΩ·cm. Peng et al. (1994) have found that the resistivity of the boron implanted samples was

always much higher than the reference sample (e.g., 45.7 MΩ·cm for a dose of $10^{13}$ cm$^{-2}$) but it decreased when increasing the implantation dose. Phosphorous implanted samples, however, had a lower resistivity than the reference sample (e.g., 0.9 MΩ·cm for a dose of $10^{13}$ cm$^{-2}$) and the resistivity decreased with increasing implantation dose. Silicon implantation did not change the resistivity of the PSi layer.

7. Ion implantation can control the light-emitting properties of PSi. In particular, Peng et al. (1994) established that implantation of $n$-Si(100) with dopant ions such as phosphorus, boron, and silicon ions of different doses and energies did not significantly affect the light-emitting properties of PSi up to a dose of $10^{13}$–$10^{14}$ ions/cm$^2$ although it changed the resistivity. At higher doses and for implantation with silicon or other ions that do not provide doping, the photoluminescence (PL) was quenched. The same effect was observed for PSi doped by He, H, and Ne (Barbour et al. 1992; Jacobsohn et al. 2006) (see Figure 12.18).

Gokarna et al. (1999) have reported the increase in PL intensity of PSi with influence of 5 MeV phosphorous ions, whereas the PL intensity decreases with increasing fluence of 26 MeV phosphorous ions. Zuk et al. (1999) have observed the strong blue band at 2.7 eV, well known in silica, in the PL spectra following the oxygen implantation. 125 keV O$^+$ implantation of c-Si produced PL peak at 2.35 eV arising from a-Si nanozones created by O$^+$ single ion impact (Prabakaran et al. 2003). As it was shown (Wu et al. 1997), ion implantation can also be used for PSi doping by the Er ions. The incorporation of Er$^{3+}$ in PSi gives rise to a strong PL at 1.54 μm, which is quenched by less than a factor of two between 15 K and room temperature (Dorofeev et al. 1995).

The modification of the initial silicon surface (prior to the anodic treatment) by ion implantation also is an effective method for controlling the luminescent properties of PSi (Pavesi et al. 1994; Liao et al. 1995; Wu et al. 1996; Piryatinskii et al. 2000). For example, Piryatinskii et al. (2000) established that the samples implanted with B$^+$ ions exhibit a very sharp increase in the PL intensity at 450 nm after rapid thermal annealing (RTA). Demidov et al. (2008) also observed strong increase in PL intensity of PSi after joint doping of Si single crystals with shallow donors and acceptors, subsequent formation of PSi, and annealing. For these purposes, heavy doped $n$-Si(110) and $p$-Si(110) were irradiated by phosphorus and boron ions to a dose of $1$–$2 \times 10^{16}$ cm$^{-2}$. The intense blue PL from PSi prepared by anodizing silicon wafers implanted with carbon ions was also observed by Wu et al. (1996) and Liao et al. (1995). This effect was explained by the formation of β-SiC precipitates, leading to a decrease in the length of silicon fibers (Wu et al. 1996) and even by the emission from PSiC (Liao et al. 1995).

Wang et al. (2007) have shown that co-implantation also can be efficient. In particular, they observed enhanced PL emission from the PSi samples after co-implantation of Si$^+$ and C$^+$ ions

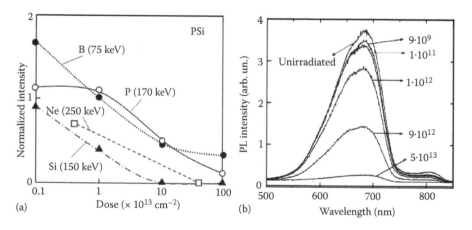

**FIGURE 12.18** (a) Changes of the PL intensity (average of several measurements) with implantation dose for different ions. The intensities are normalized to the PL intensity of unimplanted PSi. (Data extracted from Peng C. et al., *Appl. Phys. Lett.* 64(10), 1259–1261, 1994; and Barbour J.C. et al., *Nanotechnology* 3, 202–204, 1992.). (b) PL spectra of PSi irradiated He+ ions. Cumulative fluence is given in atoms/cm$^2$. (Reprinted from *Nucl. Instr. Meth. Phys. Res. B* 242, Jacobsohn L.G. et al., 164, Copyright 2006, with permission from Elsevier.)

in Si wafers. Moreover, due to co-implantation of Si⁺ and C⁺ ions much improved PL emission, compared with solo implantation of Si⁺ and C⁺ ions and un-implanted $n$-type porous silicon, was achieved. Wang et al. (2007) believe that the appearance of SiC (or amorphous $Si_{1-x}C_x$) nanostructures with localized recombination of optically excited holes and electrons in the $SiO_2$ matrix contributes to the enhanced PL emission.

## 12.9  STRESS REDUCTION IN POROUS SILICON

One of the potential issues associated with using PSi in Si VLSI technology is the stress ($\sigma$) generated in the fabrication process including anodization, drying, and storage. As it was shown earlier, this stress can result in film cracking and peeling (see Section 12.2.1). This means that the development of methods, which can help in reducing stress in PSi, represents an increased interest. However, it should be noted that most of the studies have concentrated on changes of the chemical state of internal surfaces and the related strain and stress evolution in PSi films during thermal annealing, without suggesting a solution to control the stress. The real deal how to control the stress in PSi films was made only by Kim et al. (2002). The approach proposed by Kim et al. (2002) consisted of multithermal annealing and oxidation cycles. In carried out experiments, it was found that the mechanism responsible for the stress in thick PSi films is related to the chemical structure changes at the internal surfaces of PSi films. It is well known that a PSi surface can be in a number of chemical states, each with the associated stress (Sugiyama and Nittono 1990; Ohring 1991; Unagami 1997). For example, as it is shown in Figure 12.19a the as-prepared PSi is under compression. According to Sugiyama and Nittono (1990) and Unagami (1997), a compressive stress is associated with an H-terminated state of PSi surface. Upon annealing in $N_2$ to 325°C, the film stress evolves into tension (see Figure 12.19a). Moreover, the magnitude of the tensile stress increases with the annealing time at 325°C and saturates at around 20 MPa after 20 min. Cooling from 325°C to 80°C resulted in less than 1 MPa of change in $\sigma$, indicating the unimportance of thermal expansion at the level of stress we are observing. Based on the results from the mass spectroscopy and the wafer curvature measurements, it was assumed that hydrogen desorption from pore walls is responsible for the change in PSi film stress from compressive to tensile. This effect takes place due to Si surface reconstruction in the absence of sufficient oxygen (Halimaoui et al. 1995b; Norenberg and Briggs 1999; Kim et al. 2002). Further annealing with $N_2$ switched off exposes the pore surface to the atmospheric ambient, resulting in the formation of an oxide layer, which is under compression (Ohring 1991). Kim et al. (2002) proposed to use these changes for controlling the stress in the PSi films. A controlled annealing cycle, such as the one shown in Figure 12.19b, renders the entire pore surface in a combination of three chemical states: H-terminated, reconstructed, and oxide covered. Kim et al. (2002) believe that adjusting the balance among the three

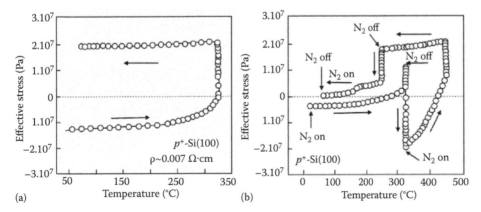

**FIGURE 12.19**    (a) Change of effective stress versus temperature for 50-μm-thick PSi samples with 51% porosity under $N_2$ ambient, and (b) thermal cycle behavior of 50-μm-thick PSi samples for achieving the stress control condition. Samples were fabricated with 25% HF:ethanol = 1:1 solution at $J$ = 50 mA/cm² for 20 min. (Reprinted with permission from Kim H.-S. et al., *Appl. Phys. Lett.* 80, 2287, 2002. Copyright 2002. American Institute of Physics.)

states allows controlling the macroscopic film stress. An intermediate oxidation step is used to keep the magnitude of the stress under a limit. This step is crucial for avoiding PSi film cracking and peeling when the stress exceeds the limit. A second oxidation step at a lower temperature is used to adjust the stress to zero in a controllable fashion. In this particular example, the room-temperature stress after the thermal cycle can be controlled to below 1 MPa.

Of course, there is the problem of preserving this state with low stress because the surface of PSi is very sensitive to heat treatment and any changes in the ambient atmosphere. This means that the porous region should then be encapsulated to ensure the long-term stability of the structure. Kim et al. (2002) believe that this problem is solved because the encapsulation can be done using many materials. For example, Kim et al. (2002) proposed using a 200-nm-thick plasma-enhanced chemical vapor-deposition nitride for this purpose.

## 12.10  RELEASE OF THE POROUS LAYER

At present, a combination of porosification and electropolishing modes is the most common approach to the formation of freestanding PSi-based membranes (Ohji et al. 1999; Lammel and Renaud 2000). A schematic diagram of current densities during the phases of porosification and electropolishing of Si is shown in Figure 12.20. As was shown by Lammel and Renaud (2000), depending on the mask technique used, the porous layer can be either held at the border or released. Metal masks result in an almost homogenous current flow during the electrochemical etch process. The current can leave through the mask and will flow mainly perpendicular to the wafer surface (see Figure 12.21a).

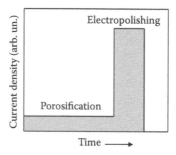

**FIGURE 12.20**  Schematic diagram of current densities during the phases of porosification and electropolishing of silicon. (Reprinted from *Sens. Actuators B* 85, Lammel G. and Renaud P., 356, Copyright 2000, with permission from Elsevier.)

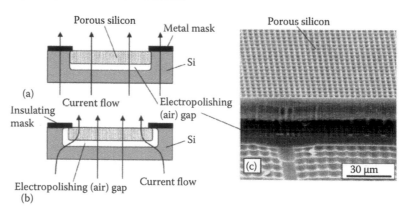

**FIGURE 12.21**  The forming of PSi with subsequent electropolishing gap formation: (a) structured using a metal mask; (b) structured using an insulating mask. Lines of electrical current flow are indicated. (c) Freestanding PSi block. The straight pores in *n*-type Si(100) with $10^{14}$ cm$^{-3}$ were made using 5% HF solution and the current density 32 mA/cm$^2$ for 10 min. In the following step, the current density was raised to 60 mA/cm$^2$ for 10 min to connect the pores under the block. ([a,b] Reprinted from *Sens. Actuators B* 85, Lammel G. and Renaud P., 356, Copyright 2000, with permission from Elsevier. [c] Reprinted from *Sens. Actuators A* 73, Ohji H. et al., 95, Copyright 1999, with permission from Elsevier.)

The layer thickness of the porous silicon in this case shows good uniformity. The lateral under-etch is less than the layer thickness. Electropolishing will release the layer from the substrate but not from the borders. Figure 12.21c shows the example of freestanding PSi blocks. A gap between the perforated blocks and the substrate and the depth of the block can be controlled by the etch time (Ohji et al. 1999). If the generation of gas bubbles during this subsurface electropolishing is too violent, the porous layer will break. A porous layer suspended at the borders can serve as a membrane. The thickness of this membrane can range from less than 1 µm to more than 50 µm. The integration of oxygen atoms can plug the pores if the porosity is not too high. Closing the pores by oxidation in an oxygen-atmosphere oven results in vacuum cavities under the membrane that could be used for ultra-miniaturized, surface-micromachined absolute pressure sensors.

In the case of insulating masks, the current coming from the backside plane is forced to squeeze through the mask openings, leading to a lateral current flow (see Figure 12.21). The current density is increased at the sides of the mask openings and the lateral underetch is nearly isotropic. Because of the lateral current, the subsequent electropolishing will also release the sidewalls of the porous layer. The porous part is then only held to the mask in the underetching zone, and has no contact to the substrate anymore. It was found that the porous layer is thicker at the borders of the mask openings and that the electropolishing is deeper there, resulting in a W-profile. This effect is less pronounced for highly doped, low resistive silicon because the current density can equalize better over the wafer.

Solanki et al. (2000) also proposed two-step electrochemical etching for separation of a thin porous silicon layer. However, to switch to electropolishing mode, they offered to change the concentration of HF. In this two-step approach, electrochemical etching is stopped just before lift-off of the PSi layer and the solution is changed from high HF concentration to low HF concentration (for instance to 10% HF) to switch the reaction from the porous Si formation regime to the electropolishing one. A proposed formation mechanism is shown in Figure 12.22. Solanki et al. (2000) believe that using this two-step approach it is possible to obtain PSi layers of any desirable area. In particular, in the indicated paper, they reported very homogeneous freestanding PSi films of 4 × 4 cm$^2$ area. In addition, Solanki et al. (2000) believe that this two-step approach gives possibility to use silicon substrate several times. Indeed no additional wet chemical cleaning step is needed for the Si substrate because it has been polished during the PSi layer formation step.

The oxygen plasma will also partly oxidize the porous silicon. The implanted oxygen atoms create a compressive strain in the porous plate and make it lift up. This effect could be explained

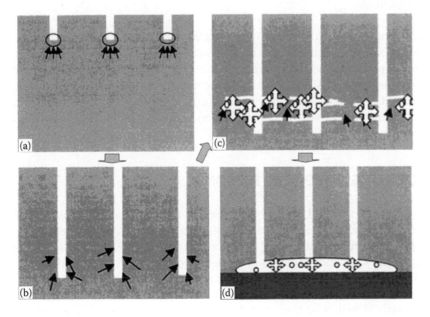

**FIGURE 12.22** Formation mechanism of two-step electrochemical etching for separation of a thin PSi layer: (a) PSi formation; (b) the shift in the reaction from porosification to electrochemical etching; (c) the branching of pores; and (d) the layer separation from substrate. (Solanki et al.: *Phys. Stat. Sol. (a)* 2000. 182. 97. Copyright Wiley-VCH Verlag GmbH & Co. KGaA. Reproduced with permission.)

by the fact that the current density at the bottom side of the perforated region is lower than at the top; therefore, the density of the porous silicon at the bottom is higher, leading to a stronger blow-up during oxidation. The strain-induced lift-up can be amplified by a metal layer on the silicon in the perforated region, which can also be used as an adhesion promoter for the photoresist mask. A layer of Cr and Au deposited on silicon by evaporation will contract at room temperature, which will pull up the plate when it is released.

In order to obtain large area flat membranes, a reducing of stress within the membranes is a mandatory requirement. X-ray diffraction experiments show that the lattice constant of PSi is larger than that of c-Si, and increases with the increasing of porosity (Papadimitriou et al. 1999). Because of the lattice mismatch between the c-Si substrate and PSi film, the unit cell of PSi is tetragonally distorted and the film is strained. This can affect the integration of PSi devices in microelectronic circuits because it can lead to the curvature of a wafer. If one cannot diminish the stress, the porous membranes are bent and they tend to be detached from the substrate. This effect is more severe when PSi layers are released from the c-Si substrate. In this case, even mechanical failure can occur due to the fast relaxation of the high stresses. It is known from the literature that annealing reduces the stress in the films of polycrystalline silicon. Conducted research has shown that the same approach could be used for porous layers (Papadimitriou et al. 2004). Stress-analysis based on the results of micro-Raman spectroscopy reveals that membranes, treated in an inert ambient at moderate temperatures, are less strained and do not break when they are released from the substrate.

## 12.11  POROUS SILICON REMOVAL

PSi can be removed using different methods depending on the feature size and the porosity of the layer (Prochaska et al. 2002). For example, nanoporous silicon can be removed by one of the following methods:

1. Immersion in $H_2O_2$:HF solution.
2. Transformation into porous silicon oxide and oxide removal.
3. Immersion in diluted (0.1–1.0%) hydroxide solutions (KOH, NaOH, $NH_4OH$, etc). Because of its high surface area, PSi dissolves very quickly even at room temperature.
4. Plasma etching.

Etching of nanoporous silicon in $H_2O_2$:HF solution exhibits very high selectivity with respect to bulk silicon (>20,000), which makes it the most suitable approach for PSi removal. Even for long etching time (several hours), virtually no bulk etching occurs, also very little surface roughness is observed after the removal. Macroporous silicon can be removed in alkaline solution. However, the etch selectivity with respect to bulk silicon is poor (~20–30). In case of mesoporous silicon, the best removal method is based on transformation into PSi oxide and subsequent oxide etching in HF solution.

## 12.12  THE FILLING OF POROUS SILICON

The open porous structure and the very large specific surface area of PSi have motivated scientists to introduce different materials into the pores, forming composite structures devoted to different applications relying on the luminescence properties (Bsiesy et al. 1995; Herino 1997), electrophysical and magnetic properties of formed heterostructures (Aravamudhan et al. 2007), or the sensing capability of the resulting nanostructures (Herino 2000). As it is known, nanocomposites offer a broad avenue of new and interesting properties depending on the kind of involved materials as well as on their morphology (Lai and Riley 2008). There are some reports on creation PSi-based nanocomposites including polymers, metals, semiconductors, organic dye molecules, and metal oxides. Porous silicon's filling also allows resolving a problem of increasing in the hardness and in the thermal conductivity of PSi, with no apparent changes in the PL characteristics (Dattagupta et al. 1997a). Several studies dealing with filling of the porous matrix have been carried out to shift the emission wavelength (Lopez et al. 1999). Other studies reported the interest of filling PSi with conjugated polymers to get a new all-optically active material (Errien et al. 2003). Recently, the filling of PSi structures with metals and other materials was investigated with respect to applications for electronic devices (Kleimann et al. 2001; Zacharatos and Nassiopoulou 2008). In particular, Kleimann et al. (2001) proposed to use such structures for fabrication of an X-ray imaging pixel detector. Devices related to the magnetic behavior of the composite system ferromagnetic materials are also under investigation (Granitzer et al. 2008a; Prischepa et al. 2014). Another interesting potential application of PSi with intermediate pore sizes is its use as a template or host membrane for the nanofabrication of different materials. For example, nanoporous silicon layers were also used as templates to prepare well-defined arrays of polypyrrole nanowires (Kleimann et al. 2001).

It was established that for PSi filling, various methods such as dry processes, impregnation by contact with a liquid, chemical bath deposition, or electrochemical deposition could be used. The most detailed analysis of those methods can be found in review papers (Herino 2000; Lai and Riley 2008; Granitzer and Rumpf 2010).

## 12.12.1  DRY PROCESSES

Dry processes such as evaporation, chemical vapor deposition (CVD), and derived techniques are very much welcome, as they are well controlled in the environment of the silicon technology. However, it appears quite difficult to obtain material incorporation inside the pores by simple evaporation. They suffer from the difficulty of finding the proper conditions to get efficient pore penetration because such a process generally leads to the pore mouth blockage (Herino 2000). Therefore, only a few reports related to those methods are found in the literature. In particular, small Ge clusters imbedded in PSi have been obtained by evaporation and inert gas condensation (Liu et al. 1996; Liu and Wang 1998). Another successful attempt of Ge incorporation has been reported by Halimaoui et al. (1995a), which used chemical vapor deposition under ultra high vacuum conditions. It has been shown that, by working at low growth rates, it was possible to completely fill the pores following an epitaxial regime leading to the formation of monocrystalline germanium. Unfortunately, the process leads to the total quenching of the PSi luminescence, which likely results from the thermal treatment under high vacuum, and no further application of this nice nanocomposite has been proposed since. There are also attempts to use such techniques as atomic layer epitaxy (ALE) (Aylett et al. 1996; Ducso et al. 1996; Utrainen et al. 1997) for PSi's filling by Sn and Ga oxides. However, there is no information about the optical and electrical properties of the so-formed nanostructure. Besides, we need to note that the above-mentioned techniques are expensive.

## 12.12.2  PORE IMPREGNATION

The deposition methods that involve the contact of a porous material with a liquid are expected to be quite effective because there is, in principle, no great problem of mass transport inside the pores, which may limit the process. The simplest process is to dip the PSi sample in a solution of the species to be deposited, and after a given time of contact, to evaporate the solution. One can find many reports in the literature on using this technique, mostly for the incorporation of metals or polymers, with the aim to realize intimate electrical contacts with the volume of the sample (Herino 1997, 2000).

Research has shown that the introduction of any materials into the pores of PSi from solution depends on the surface properties and chemistry. For example, Vellutini et al. (2007) concluded that attempts to fill the 10,12-docosadiynedioic acid inside the PSi-H pores probably failed because of the weak affinity of the hydrophilic monomer for the hydrophobic substrate. As it is known, freshly etched $SiH_x$-terminated porous silicon (PSi-H) is hydrophobic. Consequently, it is necessary to modify the chemical nature of the pore surface to optimize the filling of the monomer into the porous structure. For resolving this problem, the hydrogen-terminated porous silicon (PSi-H) was oxidized with a piranha solution to generate the silanol groups.

It must be noticed that, in the case of metals, deposition often results from a chemical reaction in which metallic ions are reduced through an electroless mechanism. The process is characterized by coupled redox reactions, one corresponding to the metal formation, the other to the silicon atom oxidation. Consequently, the reaction is inhibited after deposition of one or two monolayers, when the silicon surface atoms are no longer in contact with the solution, and there is simultaneous oxidation of the porous layer. In most cases, introduction of metals is accompanied by a strong quenching of the PL, likely because the resulting surface modification introduces nonradiative defects (Herino 2000) and only a few reports concern affecting the luminescence (Huang 1996).

However, without going into details, one can say that the incorporation of metals is generally successful, with more or less important in-depth concentration gradients, but complete pore filling is rarely reported. It seems that in many cases the deposit is noncontinuous, but rather corresponds to the formation of microcrystalline clusters. Using this method, metals such as Ni, Au, Cu, Ag, Al, Fe, and Pt were incorporated in the pores of PSi (see Table 12.4) (Herino 2000).

Various conducting polymers have been added to porous silicon (Table 12.5) using different solutions, but in many reports, there is no indication of the pore penetration (Herino 2000). Polymers have been placed in a PSi template by in situ polymerization (Yoon et al. 2003), injection molding (Li et al. 2003), or solution casting (Li et al. 2005). However, a commonly used way is to deposit a

**TABLE 12.4 List of the Main Studies Reporting Metal Incorporation by Contact with a Liquid**

| Element | Deposition Method | Deposit Characteristics | Luminescence |
|---|---|---|---|
| Ni | Electroless | Pore filling | PL quenched |
| Ni, Au, Cu | Electroless | Pore filling | PL quenched |
| Ag, Au, Cu | Immersion | Metal incorporation to depth of 0.3–0.4 μm | PL quenched |
| Cu | Electroless | 1.7–6.1 at.% in the whole depth | Red PL band of PSi near IR band assigned to deep levels |
| Cu | Immersion in $CuCl_2$ | In-depth gradients | PL quenched |
| Au, Pt | Electroless | Strong gradients | – |
| Al | Immersion in toluene solution | – | Improved PL |
| Fe | PSi formation in presence of $Fe^{3+}$ | – | Improved PL |

*Source:* Reprinted from *Mater. Sci. Eng. B*, 69–70, Herino R., 70, Copyright 2000, with permission from Elsevier.

**TABLE 12.5 List of the Main Studies Reporting Polymer Incorporation by Contact with a Liquid**

| Polymer | Deposition Method | Main Characteristics |
|---|---|---|
| Polyaniline | Spin-on coating | Improved rectifying properties |
| Polypyrrole | Spin-on coating | Improved rectifying properties |
| MDDO-PPV | Spin-on coating | Improved electroluminescence (EL) |
| PCDM | Spin-on coating | Improved EL |
| Polyaniline | Immersion under ultrasonication | Rectifying contact |
| Polyamide, polystyrene, polymethylmethacrylate, polyvinyl chloride | Immersion for a period of 2–3 days | Improved hardness and thermal conductivity, PL unaffected |

*Source:* Reprinted from *Mater. Sci. Eng. B*, 69–70, Herino R., 70, Copyright 2000, with permission from Elsevier.

solution of the polymer over the porous surface by a spin-coating method. It is clear that the viscosity and the rotation speed are determinant parameters for the pore penetration, but this question has never been detailed. Concerning the immersion of the porous sample into the polymer solution, this method can be successful only if the large polymer molecules can penetrate the pore network. One method has been to add ultrasonic agitation in order to introduce polyaniline by this way (Halliday et al. 1995), but the authors did not perform analysis to quantify the resulting efficiency on pore penetration. Other authors have chosen to favor the diffusion of the polymer molecules into the pores by substantially increasing the contact time with the solution to several days.

As a rule, the goal of most studies related to PSi/polymer composites was to improve an electrical contact with the porous material, and the authors are more concerned by the resulting electrical characteristics (Tada et al. 1997; Fan et al. 1998; Shen and Wan 1998). For example, Nguyen et al. (2000) have found that devices using composites as emitting layers showed improved efficiency, due to a better conductivity as well as to the modified morphology of the films, enhancing the contact area between the electrode and the emitter. However, the potential of such hybrid materials in sensing and in controlled release drug delivery (with the freestanding films after removal from the crystalline Si substrate) has also been demonstrated (Li et al. 2003). It was also demonstrated that the PL of PSi can be manipulated and enhanced by infiltrating the pores with polymers of varying dielectric constant (Lopez et al. 1999). In particular, blue shift of PL spectra was observed in PSi-polymer composites. At that, it was found that the degree of modification in PL depends on the dielectric constant of the polymers. As it is seen in Figure 12.23a, larger PL blue shifts were observed when higher dielectric constant polymers were introduced into the PSi pores (PAN and PVF). In addition, it was shown that infiltration is more efficient in PSi-polymer nanocomposites formed by *in situ* polymerization as opposed to polymer diffusion, resulting in larger PL blue shifts (up to ~220 meV) (Figure 12.23b). Lopez et al. (1999) believed that the larger shift was observed for the *in situ* polymerized composites because of the higher infiltration of the polymerized polymer that results from the more effective diffusion of the small molecules (acrylonitrile) as opposed to the larger polymer (PAN). Segal et al. (2007) have shown that porous Si can also be used as a template for the formation of various nanometer-scale hydrogels with distinct properties. As the porosity and average pore diameter of the PSi template can be easily tuned by adjusting the electrochemical preparation conditions, this system provides a convenient platform for studying confined materials.

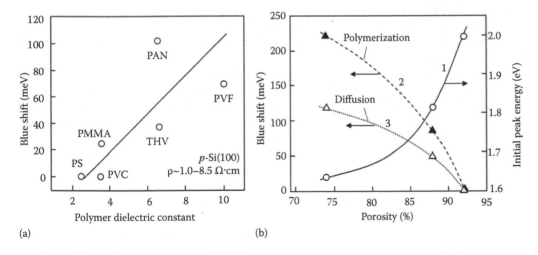

(a) (b)

**FIGURE 12.23** (a) PL blue shift versus polymer dielectric constant at a frequency of 60 Hz. PSi-polymer nanocomposites were produced (74% porosity) by diffusion of the polymers for a period of 2–3 days. Initial peak PL energy before infiltration was 1.63 eV. PAN = polyacrylonitrile, PMMA = polymethyl methacrylate, PS = polystyrene, PVC = polyvinyl chloride, PVF = polyvinylidene fluoride, and THV = fluorinated terpolymer. (b) PL blue shift of PSi-PAN composites prepared by diffusing and polymerizing the polymer in the pores as a function of PSi porosity (size). The PL peak energy of the initial PSi is also shown. (Reprinted from *J. Lumin.* 80, Lopez H.A. et al., 115, Copyright 1999, with permission from Elsevier.)

### 12.12.3 CHEMICAL BATH DEPOSITION

In this case, the deposition results from a chemical reaction in solution, which may either involve the surface silicon atoms or involve different reactants in the solution leading to the formation of a new compound which then is physisorbed on the pore walls. In the first case, the reaction is most often limited to the formation of one monolayer, and the resulting structure may not be considered a nanocomposite. A large variety of chemical groups has been grafted to the silicon surface, and the interested reader can refer to recent reviews focusing on the surface chemistry of PSi (Chazalviel and Ozanam 1997; Ozanam 1999). A good example of chemical bath deposition into the pores is given by the formation of CdS–PSi composite structures using successive ionic layer deposition method (Gros-Jean et al. 1998). However, this analysis and infrared spectroscopy show that the process is accompanied by significant oxidation of the silicon nanocrystallites and by the loss of the passivating Si-H surface covering. The PL after deposition appears to be partially quenched, indicating that the surface modification and the interactions between CdS and the host material introduce nonradiative recombination centers.

Another liquid phase reactive process that has been used is polymerization. For example, in the case of PMMA and polystyrene, the porous layer is immersed in a solution of the monomers (MMA or styrene) in the presence of the polymerization initiator ($FeCl_3$:HCl), and the polymerization is performed in situ by subsequent mild heat treatment (Dattagupta et al. 1997b). It is found that this method leads to a greater polymer density inside the material than simple immersion in the polymer solution. Another approach used for polyaniline has been the chemical polymerization of the monomer layer by layer, which allows a good inner surface coverage (Bsiesy et al. 1995). It is shown that five successive layers of polyaniline lead to the complete filling of 80% porosity mesoporous $p$-type samples. However, partial oxidation of the layer was also found.

### 12.12.4 ELECTROCHEMICAL PROCESSES

One great difficulty in the use of the chemical bath deposition techniques is that the deposition reaction occurs inside the pores. Such a problem is not expected from processes involving an electrochemical reaction, as it can be thought that charge carriers are provided by the bulk silicon, and that the electrochemical reaction will take place preferentially at the pore bottom, as it is the case during the formation of the porous layer. Therefore, at present one can find a large number of publications reporting on the incorporation of different materials like metals, polymers, or semiconductors into the pores using electrodeposition (Herino 2000; Granitzer and Rumpf 2010; Prischepa et al. 2014).

The first published investigations have been devoted to metal electroplating, still with the aim to realize deep electrical contact with the pore walls. The filling of macroporous silicon with Cu by electrodeposition has been investigated by the group of Föll (Fang et al. 2007a), Fukami et al. (2008b), and other groups (Sasano et al. 2000; Bandarenka et al. 2008; Zacharatos and Nassiopoulou 2008). Föll and co-workers (Fang et al. 2007a) reported the successful complete filling of macropores exhibiting an aspect ratio of 75 (pore-diameter 2 μm, pore-length 150 μm). The PSi templates have been post-treated by sonication in $CuSO_4$-solution and subsequently dipped in piranha solution leading to a quite thick oxide layer of about 10 nm. Electrodeposition of Au, Pt, and Pd into macropores has been shown by Fukami et al. (2008b), whereas Pd and Pt starts growing from the pore bottom but for the deposition of Au no condition could be found to achieve this behavior. The deposition is performed under cathodic conditions and choosing an appropriate supporting electrolyte to shift the open-circuit potential to negative values, which inhibits oxidation of the silicon. Since the equilibrium potential of Au is more positive compared with Pt and Pd, the rate of hole injection is faster, which is a result of the greater difference between the equilibrium potential and the silicon valence band. Thus, Au cannot be deposited from the bottom. In the case of Pt, the process begins at the pore bottom due to the high electric field strength at the tips of the pores if the proper electrolyte composition is employed (addition of NaCl). In the case of using a $Na_2SO_4$ solution, the nucleation begins at the pore opening.

Very different situations are found concerning the in-depth profile concentration of the plated metals. Homogeneous filling is found in the case of macropores (Ronkel et al. 1996), strong gradients are often observed with mesoporous and nanoporous materials, spotlike deposition throughout the volume is also reported (Ronkel et al. 1996; Ito et al. 1997). These differences can be assigned to changes in the pore shape, which may lead to pore blockage in certain circumstances. As a result, the pore is often closed by electrodeposits themselves. Experiment has shown that maximum inhomogeneity of the metal deposit is observed for pore shape like dendritic branches. Electrodeposition reactions, taking place not only at the pore bottom but also on the wall, lead to the complex distribution of electric field and concentration profiles in the pore. If the exchange of electrolyte is insufficient along the entire pore length, pores are also blocked and homogeneous metal-filling along the whole length is inhibited. Inhomogeneities can also arise from hydrogen evolution, which results from the proton reduction, which is a competing reaction that can hardly be eliminated. Thus, homogeneous filling of the pores is extremely difficult to obtain. It was found that besides pore shapes, the degree of pore filling is dependent on the current density as well as on the pulse duration. This means that this mode of application needs strong optimization of fabrication process (Granitzer et al. 2008b).

Taking into account the above-mentioned, one can come to the conclusion that there are possibilities to form various morphologies of deposits by controlling the electrodeposition reaction on the pore wall. Ogata and co-workers (Harraz et al. 2005; Kobayashi et al. 2006; Fukami et al. 2008b) reported that shapes of deposits could be controlled as rods or tubes in ordered macroporous *p*-type silicon (see Figure 12.24). They have shown that the mechanism of the macropore filling depends on various factors, for instance, the equilibrium potential of metal, the depth of the macropore, the supporting electrolyte for electrodeposition, and the displacement deposition. For example, Ogata et al. (2006) have found that metal deposition on silicon in general and in particular the deposition of copper forms microrods within macroporous silicon, while the deposition of nickel results in microtubes, covering the pore walls of the PSi structure. Metal deposition forms a Schottky barrier between silicon and electrolyte when the work function of the metal is higher than the one of silicon, in the case of *n*-type and smaller in the case of *p*-type silicon. In

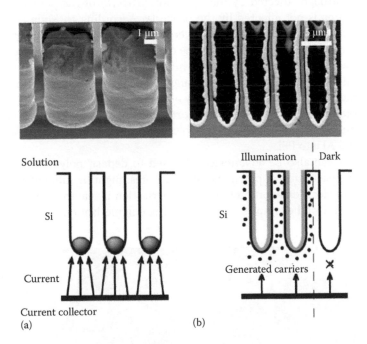

**FIGURE 12.24**  Scanning electron micrographs showing (a) the Cu-deposition within *p*-type macroporous silicon forming rods and (b) the deposition of Ni, which covers the pore-walls: (a) Metal grows from the pore bottom in the dark when the valence band process is possible (noble metals). Preferential growth at the bottom results mainly from the current distribution. Microrods are obtained after dissolving the silicon substrate. (b) Electrodeposition does not take place for less-noble metals in the dark. Illumination generates electrons over the wall area and enables the reduction reaction on the wall leading to the formation of a microtube structure. (Reprinted from *Curr. Opin. Solid State Mater. Sci.* 10, Ogata Y.H. et al., 163, Copyright 2006, with permission from Elsevier.)

the case of depositing noble metals when the electron transfer via valence band takes place, metal nucleation starts at the pore bottom. For less noble metals, the occurrence of the precipitation has to be supported by illumination, which permits the reduction at the pore walls leading to the formation of metal microtubes. Other groups also reported the filling of macroporous silicon by applying a damped or pulsed current (Fang et al. 2007a,b; Granitzer et al. 2007).

As to the electrodeposition in mesoporous or microporous silicon, a limited number of studies were reported. Hamm et al. (2005) concluded that PSi with smaller pore sizes (<50 nm) cannot be used as templates, especially for metal nanofabrication, because of mass-transfer limitations within these mesopores. However, Fukami et al. (2009a), studying copper electrodeposition from an aqueous solution, have shown that mesoporous Si with pore diameter ~40 nm can be filled. The results reported in this paper indicated that the mass transfer is still important, and much more critical than in the case of macropore filling. But Fukami et al. (2009a) have found that when the copper electrodeposition was carried out at a very small constant current density (−6.4 μA/cm²), the mesopores with 4 μm depth were filled with copper continuously from the bottom to the opening. When the electrodeposition current was set at an absolute value twice as large as in the above condition, the isolated particles were electrodeposited in the mesopores. The depth also affected the filling behavior. The pores 8 μm in depth were not continuously filled with copper even in the condition at which the pores 4 μm in length were completely filled. The numerical simulation suggested that the diffusion-limited electrodeposition could be achieved in mesopores at a very small current, at which the diffusion-limited condition had never been realized on a planar electrode. Herino et al. (1985) also reported micropore filling with nickel in an ethylene glycol based solution, and analyzed by X-ray microanalysis and SIMS. Jeske et al. (1995) reported about the micropore filling with various metals and analyzed by XPS. However, both studies did not show the morphologies of the filled pores.

Koda et al. (2012) studied electrodeposition of platinum and silver into chemically modified microporous silicon and established that when the pore diameter of PSi is extremely small, platinum deposition is strongly affected by the hydration property of the pore wall. On the other hand, silver deposition is not affected by the property. According to Koda et al. (2012), these results could be explained by the difference in overpotential between the deposition on the metal surface and that on the silicon surface, and by the difference in intrinsic displacement deposition rate of metal. It was also established that when metal electrodeposition is carried out in microporous silicon, the effect of the overpotential and the displacement deposition rate should be considered together with the hydration property of the metal ions.

We need to note that in most cases, it is reported that metal incorporation totally quenches the PSi luminescence. An increase of the electroluminescence has been observed only after the deposition of In and Al into PSi.

Electrochemistry can also be advantageously used to deposit polymers (see Figure 12.25). Because the direct penetration of polymers into the pores might be difficult due to size effects, a good solution is to electrically promote the polymerization inside the pores wetted by a solution of the monomer. Good examples are given by the polymerizations of aniline (Matveeva et

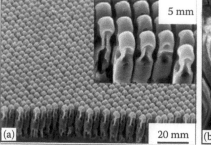
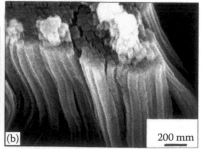

**FIGURE 12.25** SEM images of PPy nano- and microstructures after removal of PSi templates. Templates used were (a) ordered macropores and (b) medium-sized pores. All the electropolymerizations were performed at 127 μA/cm². (Reprinted from *Electrochem. Commun.* 10, Fukami K. et al., 56, Copyright 2008, with permission from Elsevier.)

al. 1993) and bithiophene (Jung et al. 1995), which have been reported to start at the bottom of the pores and to propagate toward the surface, but it was also found that anodic oxidation of the porous layer simultaneously occurs. Polypyrrole has been deposited by applying galvanostatic anodic permanent (Koshida et al. 1993) or pulsed (Moreno et al. 1997) current to a PSi electrode immersed in a solution of the monomer in acetonitrile. However, it is found that the degree of pore filling is highly dependent on the current density of the pulse, and that the process requires optimization.

In several papers, authors indicated that uniform filling of PSi with electrochemical reactions is sometimes difficult because of plugging at pore openings (Fang et al. 2007a,b). However, Harraz and co-workers (Harraz 2006; Fukami et al. 2008a) have succeeded in electropolymerization of polypyrrole (PPy) into PSi with mesopores (20 nm). Electropolymerization of PPy was performed immediately after the preparation of the PSi templates to avoid surface oxidation of PSi. An acetonitrile solution with 0.1 M PPy + 0.1 M tetraethylammonium perchlorate (TEAP) was used for PPy electropolymerization. The electropolymerization was carried out using a conventional 3-electrode system with an RE (Ag/Ag$^+$ in acetonitrile and TEAP). The filling of PPy was always continuous from the pore bottom to the top surface of PSi. Other groups also reported successful electropolymerization of conductive polymers into PSi without plugging problems (Onganer et al. 1996; Lara and Kathirgamanathan 2000; Sharma et al. 2007). Recently, Harraz et al. (2008) informed that they have also succeeded in filling micropores (less than 2 nm) using lightly doped $p$-type Si. Fukami et al. (2009b) have shown that electropolymerization can also be used for preparing through-tubes of PPY (see Figure 12.26). The tubular structure of PPy means that the PSi wall is covered at the initial stage of the electropolymerization, and then the electropolymerization proceeds on PPy leading to the thickening of the polymer film. In this formation process, the most important factor is that the electropolymerization of PPy in the pore prefers to occur not on PPy itself but on silicon.

Generally, different authors do not report in greater detail on the incorporated polymer amount, neither on the in-depth concentration variations, except for the polymerization of dithiophene, after which almost homogeneous sulfur concentration of 5 at.% is found (Jung et al. 1995; Schultze and Kung 1995). In many cases, it is observed that the in situ polymerization of a conductive material leads to improved electrical characteristics of the resulting diodes, attesting for the efficiency of the process. Several papers have shown improvement in the electroluminescence output and stability, although there is some light absorption by the polymer itself (Herino 2000).

Electrochemistry can also be used to introduce II–VI semiconductor compounds inside the pores. Once again, the main aim is to get an electrical contact with the volume of the PSi sample. The interest here is that it is possible to choose compounds with an energy gap large enough to be transparent toward the emission spectra of PSi. In the literature, one can find information about incorporation of CdTe and ZnSe (Herino et al. 1997; Montes et al. 1997). It is shown that the deposition can be localized at the bottom of the porous layer by choosing that the rate-determining step of the electrochemical reactions should be the charge carrier supply from the substrate. The

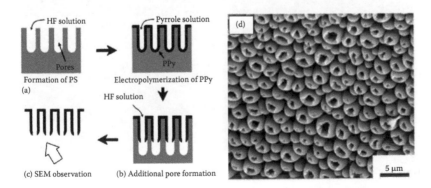

**FIGURE 12.26**  (a–c) A schematic illustration explaining how to make through-tubes of PPy, and (d) SEM images of PPY microtubes after PSi-template removal. (Fukami K. et al.: *Phys. Stat. Sol.* (a) 2009. 206(6). 1259. Copyright Wiley-VCH Verlag GmbH & Co. KGaA. Reproduced with permission.)

deposit is obtained by cathodic codeposition of both elements, which is made possible by the gain in the free energy of formation of the compound, which allows the underpotential deposition of zinc. While working with an *n*-type porous layer, a strong thickness gradient is found with preferential deposition in the upper part of the layer. In this case, the deposition reaction is limited by the diffusion of the active species toward the electrode.

### 12.12.5 ELECTROLESS PROCESS

Metal-particle-enhanced hydrofluoric acid (HF) etching is an electroless method that can produce porous Si by immersing metal-particle-modified Si in an HF solution without a bias. Experiment has shown that this method can be used for filling of silicon pores as well. For example, using this technology Yae et al. (2009) produced metal-filled Si nanopores. An electroless process used for PSi filling consisted of three steps: (1) displacement deposition of metal nanoparticles (Ag or Au); (2) Si nanopore formation by metal-particle enhanced HF etching; and (3) metal filling in nanopores by autocatalytic deposition. The important feature of this process is that the metal nanoparticles, that is, the initiation points of the autocatalytic metal deposition, are present on the bottoms of the Si nanopores. Metal nanoparticles that remain on the bottoms of the Si nanopores exhibit catalytic activity for the initiation of autocatalytic metal deposition. Thus, Si nanopores can be completely filled with the metal from their bottoms. Using this approach, Yae et al. (2009) fabricated PSi layers filled with a Co or a Co–Ni alloy.

## 12.13 POROUS SILICON SINTERING

Sintering is a heat treatment applied to a powder or porous system, which is accompanied by structural changes such as densification and surface area loss, taking place due to mass transfer. In the case of crystalline solid such as silicon, the crystal structure is preserved during sintering and the atoms solely exchange their position within the crystal lattice (Banerjee et al. 2008). Morphological changes in PSi layers during sintering are illustrating in Figure 12.27.

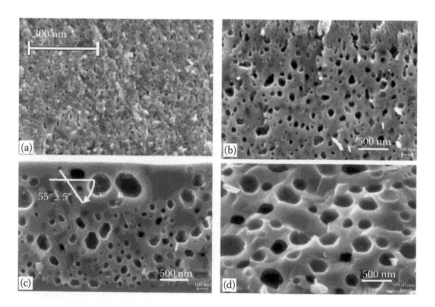

**FIGURE 12.27**    Cross-sectional SEM images of (a) an as-prepared 20% mesoporous silicon sample and after annealing at (b) 1000°C, (c) 1200°C, and (d) 1300°C for 60 min in a 1 bar $H_2$ atmosphere. The etching direction of $p^+$-Si:B(100) wafers is from the top to the bottom. (Muller G. and Brendel R.: *Phys. Stat. Sol.* (a) 2000. 182. 313. Copyright Wiley-VCH Verlag GmbH & Co. KGaA. Reproduced with permission.)

It was established that in the absence of a liquid phase, five different transport mechanisms are possible during sintering (Muller and Brendel 2000; Muller 2002; Muller et al. 2003; Djohari et al. 2009):

- Volume diffusion (migration of vacancies)
- Grain-boundary diffusion
- Surface diffusion
- Viscous or plastic flow (caused by surface tension or internal stresses)
- Evaporation/condensation of atoms on surfaces

However, one should note that the determination of a predominant mechanism, especially in the early stage of sintering, is a hard task because this mechanism depends on the composition and initial structure of sintered material, temperature, and environment. Experiment has shown that the higher the sintering temperature, the shorter is the sintering time required to achieve a desired degree of bonding between the crystallites in the porous matrix. Therefore, the temperature used for sintering is high but is below the melting point of the major constituent of material. For example, for PSi sintering the temperatures in the range of 1000–1100°C are usually used (Labunov et al. 1986; Muller 2002; Ott et al. 2003; Wolf 2007; Banerjee et al. 2008). Although these temperatures are well below the melting point of 1412°C, the increased thermal activation is enough for silicon atoms to reconfigure. One should note that structural changes in PSi were observed at low temperatures (Herino et al. 1984; Ogata et al. 2001; Joshi et al. 2010). For example, Herino et al. (1984) indicated that even at a temperature as low as 450°C, the structure of a PSi layer coarsens.

The driving force for these sintering phenomena is the minimization of the free surface energy, which is linked to the internal surface area (Ogata et al. 2001; Muller et al. 2003; Ott et al. 2004). The thermal activation permits the atoms to diffuse to an energetically more favorable position, thereby reducing the surface-to-volume ratio. This means that the described reorganization leads to a reduction of the internal surface area and thus fewer but larger voids (see Figure 12.28). For example, Labunov et al. (1986, 1987) observed that after heat treatment, PSi layers had comparatively large spherical cavities separated from each other and from the sample surface. In addition, the cavity sizes and their density were almost independent of the anodization regimes and were determined by the heat treatment conditions (Figure 12.28). The cavity diameter was 0.05–0.15 μm for annealing at 1000°C and increases up to 0.1–0.5 μm for annealing at 1200°C.

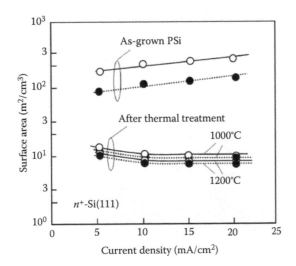

**FIGURE 12.28** The variation in the specific surface area $S$ with current density $J$ for as-grown and heat-treated PSi formed in 12% HF (O) and 48% HF (●). (Reprinted from *Thin Solid Films* 137, Labunov et al., 123, Copyright 1986, with permission from Elsevier.)

For comparison, the pores in as-prepared PSi samples had diameters of 3–16 nm. According to Muller et al. (2003), dependence of average pore size on the parameters of thermal treatments in $H_2$ atmosphere in the range of temperatures from 900 to 1300°C can be described by Equations 12.3 and 12.4.

(12.3)
$$d_{pore} \propto \exp\left(-\frac{E_a}{kT_{an}}\right)$$

(12.4)
$$d_{pore} \propto t_{an}^{1/4}$$

where $T$ and $t$ are the temperature and time of annealing, correspondingly, and $E_a$ is the activation energy equaled $E_a \approx (0.16 \pm 0.04)$ eV.

Besides a drastic increase of the pore size, the minimization of the surface energy also induces faceting of the pores. According to Muller et al. (2003) and Ott et al. (2004), this effect emerges at temperatures above 900°C and increases with the annealing time. Another observed phenomenon is the closure of the macroscopic surface, as first reported by Labunov et al. (1986, 1987). This closed surface layer forms during sintering because the macroscopic surface acts as a sink for vacations. Pores initially located close to the surface dissolve into the gas phase, leaving behind several 10-nm-thick monocrystalline layers of different porosity (see Figure 12.29). In this case, vacancies diffuse from the lowly into the highly porous region resulting in an increased porosity of the highly porous region and the formation of a thin pore-free layer at the interface. Both effects are important for the PSi layer-transfer technology, which is of great interest for a number of applications such as silicon on insulator technology, active matrix displays, and many other devices (Bergmann et al. 2001; Brendel 2001; Etter et al. 2012). In particular, exactly sintered PSi (SPSi) constitutes a key tool for the layer transfer process used for the separation of the solar cells from the growth substrate (Bergmann et al. 2001). SPSi plays several important roles in this process. First, the SPSi layer serves as a seed layer for the epitaxial growth of the monocrystalline silicon film (Sato et al. 1995). It was found that the top layer of SPSi is a quasi-monocrystalline silicon layer with the surface more or less pore-free. In addition, the surface of these annealed samples is uniform with significantly reduced roughness as compared to the as-anodized samples (Banerjee et al. 2008). Second, its mechanical properties control the subsequent transfer of the film to a device carrier.

It should be noted that in the bulk of a homogeneous sample, no vacancy sink is available and the number of vacancies per volume is preserved. Therefore, the porosity of the sintered samples is equivalent to the initial porosity of the as-etched state. Taking into account this effect, a porous double-layer structure is usually used for realization of the layer-transfer technology (Brendel 2001; Muller 2002; Wolf 2007). This structure consists of the low-porosity starting layer (10–20% porous) on top and the highly porous separation layer (50–70% porous) underneath the starting layer (see Figure 12.30a). The step in the porosity is realized by adjusting the etching current density during anodization of Si wafers. For example, for preparing such a double layer, Wolf (2007) used the following anodization process in HF(50%):ethanol = 2:1 electrolyte: for the formation

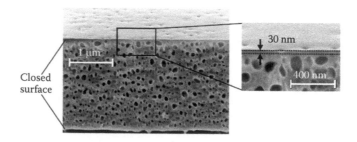

**FIGURE 12.29**  SEM image of a 2-μm-thick sintered PSi film (initial $P = 30\%$ and $d_{pore} = 48$ nm). During annealing, two 10-nm-thick pore-depleted surface layer forms on both sides of the sample. The micrograph is a composite of two images from the same sample. Two different contrast settings were necessary for the cross-section and the layer surface. (Reprinted with permission from Wolf A. et al., *J. Appl. Phys.* 104, 033106, 2008. Copyright 2008. American Institute of Physics.)

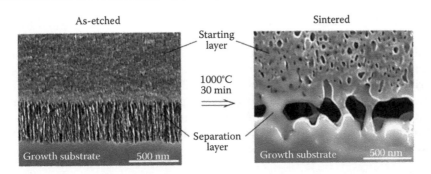

**FIGURE 12.30**   Cross-sectional SEM images of a porous double layer in the as-etched state and after annealing at 1000°C for 30 min in a 1 bar $H_2$ atmosphere. (Muller G. and Brendel R.: *Phys. Stat. Sol.* (a) 2000. 182. 313. Copyright Wiley-VCH Verlag GmbH & Co. KGaA. Reproduced with permission. Ott N. et al.: *Phys. Stat. Sol.* (a) 2003. 197. 93. Copyright Wiley-VCH Verlag GmbH & Co. KGaA. Reproduced with permission.)

of the starting layer in $p^+$-Si(100) substrate with a thickness of 1 μm and porosity $P = 21\%$, a low current density of 5 mA/cm² was applied during 100 s. For the formation of the highly porous separation layer underneath the starting layer, the current density was abruptly increased up to values between 150 and 250 mA/cm². This layer had a thickness of several 100 nm and a porosity that exceeded 40%. Wolf (2007) has also shown that during annealing at 1100°C in 1 bar $H_2$ atmosphere, the porous layers reorganized. As a result, the top surface of the low-porosity PSi layer became smooth due to the closing of pores at the surface and provided a perfect seed layer for the epitaxial deposition. Moreover, during sintering, the porosity of the separation layer increased due to the above-described transport of material from the highly porous layer to the adjacent low porosity regions, causing formation of big voids. Due to this process, the low-porosity layer and the Si substrate remain connected only through thin and weak Si pillars (see Figure 12.30b). Such Si pillars can easily be broken by applying a small mechanical force, providing detaching of the starting layer and the epitaxial film from the growth substrate.

One should note that freestanding PSi layers can also be prepared using indicated layer-transfer technology (Wolf and Brendel 2006) (see Figure 12.31).

It is important to note that for the reorganization of PSi, the internal surface must be free of native oxide because the presence of an oxide layer drastically reduces the mobility of the silicon atoms (Herino et al. 1984). However, it is known that if the PSi sample is exposed to air, the

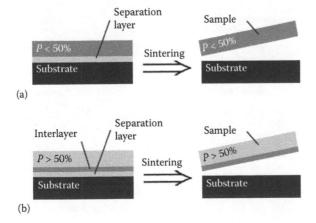

**FIGURE 12.31**   Preparation of freestanding PSi samples using the layer transfer technology. A highly porous separation layer that dissolves during sintering permits the removal of the sample from the substrate. (a) and (b) show the fabrication of samples with $P < 50\%$ and $P > 50\%$, respectively, where $P$ denotes the porosity. For the high porosity samples, the introduction of a low porosity interlayer is necessary to permit separation, as shown in (b). (Reprinted from *Thin Solid Films* 513, Wolf A. and Brendel R. 385, Copyright 2006, with permission from Elsevier.)

internal surface is rapidly oxidized. One way to remove this native oxide is to perform the sintering process in the reducing ambience of a hydrogen atmosphere. Such an atmosphere is also used for preconditioning in silicon epitaxy (Sato et al. 1995) because it prevents the formation of an oxide during the process. Therefore, as a rule after loading the samples into the tube, the standard sintering process starts with preheating to 100°C (Wolf 2007). Meanwhile, the chamber is evacuated and subsequently filled with nitrogen. This process is repeated a second time to minimize the oxygen content in the chamber. After switching from nitrogen to hydrogen supply, the temperature is raised to the desired value. During cooling at a temperature of 200°C, the control unit initiates the nitrogen purge of the chamber and, subsequently, the samples can be unloaded.

## 12.14 PSi NANOPARTICLES PREPARING

Experiment has shown that for preparing PSi nanoparticles aimed for various applications such as drug delivery, fluorescent labeling in biological imaging, and as therapeutic and diagnostic agents in biomedical in vivo applications, different top-down and bottom-up processes can be used (Wilson et al. 1993; Bley and Kauzlarich 1996; Lie et al. 2002, 2004; Wang et al. 2004; Tilley et al. 2005; Liu et al. 2014). One should note that methods to prepare micron-scale and smaller particles of PSi were developed soon after the discovery of photoluminescence from PSi in the early 1990s (Heinrich et al. 1992; Wilson et al. 1993). The ultrasonication route was the first method used to prepare small particles of PSi (Heinrich et al. 1992; Bley et al. 1996). This method is considerably simpler, less expensive, and less time-consuming than lithographic methods (Godin et al. 2012). Schematic process flow and SEM images of PSi nanoparticles prepared using photolithograpy are shown in Figure 12.32. The general approach to ultrasonication involves three steps (Anglin et al. 2008; Perrone Donnorso et al. 2012; Qin et al. 2014): (1) First, etching a PSi layer in a silicon substrate. Single crystalline Si wafer typically is etched with hydrofluoric acid electrolyte solution. Alternatively, chemical oxidant such as nitric acid can also be used to etch the Si surface to produce PSi. However, electrochemical etching provides better control over pore size and the structure of porous layer; (2) removing this layer from the silicon wafer as a freestanding porous film (referred to as "lift-off"); and then (3) placing the film in a liquid (usually ethanol or water) and fracturing it through the action of an ultrasonic cleaner. When a freestanding porous layer is subjected to ultrasonication, fracture occurs preferentially along the pore axis in the film. According to Qin et al. (2014), PSi manoparticles with size in the range 160–350 nm

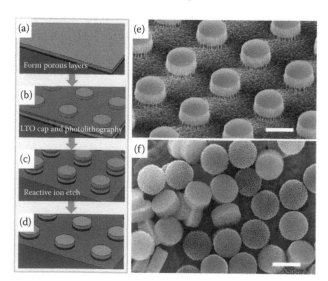

**FIGURE 12.32** Fabrication of discoidal porous silicon particles. (a–d) Schematic process flow. (e) SEM image of 1000 nm × 400 nm discoidal PSi particle array retained on a wafer. (f) Released monodispersed 1000 nm × 400 nm discoidal porous silicon particle. Scale bars are 1 μm. (Godin B. et al.: *Adv. Mater.* 2012. 22. 4225. Copyright Wiley-VCH Verlag GmbH & Co. KGaA. Reproduced with permission.)

can be prepared by pulsed electrochemical etching of single crystal silicon wafers, followed by ultrasonic fracture of the freestanding porous layer.

It was shown that a mechanochemical top-down technique like the ball-milling process could also be used as an alternative production method (Nychyporuk 2010; Russo et al. 2011; Zhang et al. 2014). For example, Russo et al. (2011) reported that when using this method, PSi powders with nanometric (<50 nm) crystallites and high surface area (from 29 up to 100 m²/g) have been produced. The milling process, which operates at room temperature and atmospheric pressure, could be appealing from a large-scale fabrication point of view, when industrial milling systems are adopted, and competitive with respect to other synthesis methods. Powders from PSi membranes are reduced in nanometric agglomerates in a very short time without destroying the starting silicon skeleton. It was found that the size and the size distribution of PSi nanoparticles depend on the porosity of the PSi layer. Contrary to Si nanoparticles extracted from nanoporous layers, Si nanoparticles obtained from mesoporous PSi layers present a well-pronounced crystalline order (Nychyporuk 2010). Russo et al. (2011) established that prepared nanoparticles showed a strong attitude to aggregate even after low power sonication. More samples that are homogeneous can be obtained by high power sonication. In this case, the nanoparticles observed by TEM are well under 100 nm and have spheroidal shapes. High power sonication also reduced the aggregation behavior.

At the final steps, prepared PSi nanoparticles can be suspended in a solvent (like ethanol, hexane), filtered or centrifugated in order to obtain the desired size distribution, and then eventually can be deposited on a substrate, mixed with other substances, or incorporated into different dielectric or polymer matrices (Švrček et al. 2002, 2004; Klangsin et al. 2008; Zhang et al. 2014). The process of PSi nanoparticle incorporation in polymer matrix, designed by Zhang et al. (2014), is shown in Figure 12.33.

Before using PSi nanoparticles, their properties usually are stabilized (Salonen et al. 2008). For these purposes, three main stabilization treatments of PSi can be used: oxidation (see Section 12.5), hydrosilylation, and thermal carbonization (see Chapter 13).

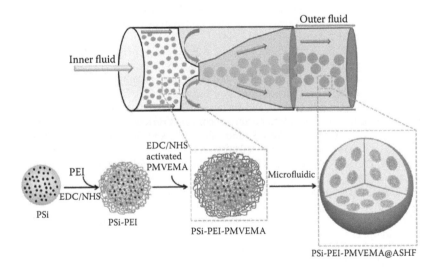

**FIGURE 12.33** Schematic illustration of the fabrication process of the PSi-PEI-PMVEMA@ASHF composites. The PSi-PEI-PMVEMA NPs were prepared by conjugation with the polyethyleneimine (PEI) and poly(methyl vinyl ether-co-maleic acid) (PMVEMA) polymers via 1-ethyl-3-(3-dimethylaminopropyl) carbodiimide/N-hydroxysuccinimide (EDC/NHS) chemical reaction. Then, the PSi-PEI-PMVEMA NPs were encapsulated in the pH-responsive hydroxypropylmethylcellulose acetate succinate based polymer (ASHF) with microfluidics. The 5-FU@PSi-PEI-PMVEMA NPs were dispersed in the inner fluid together with ASHF and drug celecoxib (CEL) in ethyl acetate (EA). The outer continuous fluid contained 2% of Poloxamer 407 (P-407). The PSi NPs (average particle size of 151.1 ± 4.0 nm and zeta-potential of −36.0 ± 0.6 mV) were fabricated using electrochemical anodization, followed by thermal hydrocarbonization, surface modification with undecylenic acid and milling. (Zhang H. et al.: *Adv. Mater.* 2014. 26. 4497. Copyright Wiley-VCH Verlag GmbH & Co. KGaA. Reproduced with permission.)

## ACKNOWLEDGMENTS

This work was supported by the Ministry of Science, ICT and Future Planning (MSIP) of the Republic of Korea, and partly by the National Research Foundation (NRF) grants funded by the Korean government (No. 2011-0028736 and No. 2013-K000315).

## REFERENCES

Aggarwal, G., Mishra, P., Joshi, B., Harsh, and Islam, S.S. (2014). Porous silicon surface stability: A comparative study of thermal oxidation techniques. *J. Porous Mater.* 21, 23–29.

Ahmad, M. and Naddaf, M. (2011). Investigation of MeV-Cu implantation and channeling effects into porous silicon formation. *Nucl. Instr. Meth. Phys. Res. B* 269, 2474–2478.

Allan, G., Delerue, C., Lannoo, M., and Martin, E. (1995). Hydrogenic impurity levels, dielectric constant, and Coulomb charging effects in silicon crystallites. *Phys. Rev. B* 52(16), 11982–11988.

Amato, G. and Brunetto, N. (1996). Porous silicon via freeze drying. *Mater. Lett.* 26, 295–298.

Amato, G., Bullara, V., Brunetto, N., and Boarino, L. (1996). Drying of porous silicon: A Raman, electron microscopy, and photoluminescence study. *Thin Solid Films* 276, 204–207.

Amato, G., Delerue, C., and von Bardeleben, H.J. (eds.) (1997). *Structural and Optical Properties of Porous Silicon Nanostuctures.* Gordon & Breach Sci., Amsterdam.

Ambrazevicius, G., Zicevas, G., Jasutis, V. et al. (1994). Layered structure of luminescent porous silicon. *J. Appl. Phys.* 76, 5442.

Anderson, R.C., Muller, R.S., and Tobias, C. (1991). Investigations of the electrical properties of porous silicon. *J. Electrochem. Soc.* 138, 3406–3411.

Anglin, E.J., Cheng, L., Freeman, W.R., and Sailor, M.J. (2008). Porous silicon in drug delivery devices and materials. *Adv. Drug Deliv. Rev.* 60, 1266–1277.

Aravamudhan, S., Luongo, K., Poddar, P., Srikanth, H., and Bhansali, S. (2007). Porous silicon templates for electrodeposition of nanostructures. *Appl. Phys. A* 87, 773–780.

Arens-Fischer, R., Kruger, M., Thonissen, M. et al. (2000). Formation of porous silicon filter structures with different properties on small areas. *J. Porous Mater.* 7, 223–225.

Astrova, E.V., Lebedev, A.A., Remenyuk, A.D., and Rud, Y.V. (1995). Optical and electrical properties of porous silicon and stain-etched films. *Thin Solid Films* 255, 196–199.

Astrova, E.V. and Tolmachev, V.A. (2000a). Effective refractive index and composition of oxidized porous silicon films. *Mater. Sci. Eng. B* 69–70, 142–148.

Astrova, E.V., Voronkov, V.B., Grekov, I.V., Nashchekin, A.V., and Tkachencko, A.G. (2000b). Deep diffusion doping of macroporous silicon. *Phys. Stat. Sol. (a)* 182, 145–150.

Aylett, B.J., Harding, I.S., Earwaker, L.G., Forcey, K., and Giaddui, T. (1996). Metallisation of porous silicon by chemical vapour infiltration and deposition. *Thin Solid Films* 276, 253–256.

Bandarenka, H., Balucani, M., Crescenzi, R., and Ferrari, A. (2008). Formation of composite nanostructures by corrosive deposition of copper into porous silicon. *Superlattices Microstruct.* 44, 583–587.

Banerjee, M., Bontempi, E., Tyagi, A.K., Basu, S., and Saha, H. (2008). Surface analysis of thermally annealed porous silicon. *Appl. Surf. Sci.* 254, 1837–1841.

Barbour, J.C., Dimos, D., Guilinger, T.R., and Kelly, M.J. (1992). Control of photoluminescence from porous silicon. *Nanotechnology* 3, 202–204.

Beckmann, K.H. (1965). Investigation of the chemical properties of stain films on silicon by means of infrared spectroscopy. *Surf. Sci.* 3, 314–332.

Bellet, D. (1997). Drying of porous silicon. In: Canham L. (ed.) *Properties of Porous Silicon.* INSPEC, London, pp. 38–43 and pp. 127–131.

Bellet, D. and Canham, L. (1998). Controlled drying: The key to better quality porous semiconductors. *Adv. Mater.* 10(6), 487–590.

Belmont, O., Bellet, D., and Brochet, Y. (1996). Study of the cracking of highly porous *p*+ type silicon during drying. *J. Appl. Phys.* 79, 7586.

Beloto, A.F., Ueda, M., Abramof, E. et al. (2001). Porous silicon implanted with nitrogen by plasma immersion ion implantation. *Nucl. Instr. Meth. Phys. Res. B* 175–177, 224–228.

Bergmann, R.B., Berge, C., Rinke, T.J., and Werner, J.H. (2001). Monocrystalline Si films from transfer process for thin film devices. *Mater. Res. Soc. Symp. Proc.* 685E, D2.1.1–D2.1.6.

Bisi, O., Ossicini, S., and Pavesi, L. (2000). Porous silicon: A quantum sponge structure for silicon based optoelectronics. *Surf. Sci. Rep.* 38, 1–126.

Bley, R.A. and Kauzlarich, S.M. (1996). A low-temperature solution phase route for the synthesis of silicon nanoclusters. *J. Am. Chem. Soc.* 118, 12461–12462.

Bley, R.A., Kauzlarich, S.M., Davis, J.E., and Lee, H.W.H. (1996). Characterization of silicon nanoparticles prepared from porous silicon. *Chem. Mater.* 8, 1881–1888.

Bomchil, G., Halimaoui, A., and Herino, R. (1989). Porous silicon: The material and its applications in silicon-on-insulator technologies. *Appl. Surf. Sci.* 41/42, 604–613.

Bondarenko, V., Dolgyi, L., Dorofeev, A. et al. (1997). Porous silicon as low-dimensional host material for erbium-doped structures. *Thin Solid Films* 297, 48–52.

Bondarenko, V., Kazuchits, N., Volchek, S. et al. (2003). Super fine structure of photoluminescence spectra from erbium co-incorporated with iron in porous silicon. *Phys. Stat. Sol. (a).* 197(2), 441–445.

Bouchaour, M., Ould-Abbas, A., Diaf, N., and Chabane Sari, N. (2004). Effect of drying of porous silicon. *J. Therm. Anal. Calorim.* 76, 677–684.

Breese, M.B.H., Champeaux, F.J.T., Teo, E.J., Bettiol, A.A., and Blackwood, D.J. (2006). Hole transport through proton-irradiated *p*-type silicon wafers during electrochemical anodization. *Phys. Rev. B* 73, 035428.

Brendel, R. (2001). Review of layer transfer processes for crystalline thin-film silicon solar cells. *Jpn. J. Appl. Phys.* 40, 4431–4439.

Bsiesy, A., Gaspard, A., Herino, R., Lingeon, M., Muller, F., and Oberlin, J.C. (1991). Anodic oxidation of porous silicon layers formed on lightly p-doped substrates. *J. Electrochem. Soc.* 138, 3450–3456.

Bsiesy, A., Nicolau, Y.F., Ermolieff, A., Muller, F., and Gaspard, F. (1995). Electroluminescence from *n*⁺-type porous silicon contacted with layer-by-layer deposited polyaniline. *Thin Solid Films* 255, 43–48.

Canham, L.T. (1990). Silicon quantum wire array fabrication by electrochemical and chemical dissolution of wafers. *Appl. Phys. Lett.* 57, 1046.

Canham, L. (ed.) (1997a). *Properties of Porous Silicon.* INSPEC, London.

Canham, L.T. (1997b). Storage of porous silicon. In: Canham L. (ed.) *Properties of Porous Silicon.* INSPEC, London, pp. 44–50.

Canham, L.T., Houlton, M.R., Leong, W.Y., Pickering, C., and Keen, J.M. (1991). Atmospheric impregnation of porous silicon at room temperature. *J. Appl. Phys.* 70(1), 422–431.

Canham, L.T., Cullis, A.G., Pickering, C., Dosser, O.D., Cox, T.I., and Lynch, T.P. (1994). Luminescent anodized silicon aerocrystal networks prepared by supercritical drying. *Nature* 368, 133–135.

Canham, L.T., Loni, A., Calcott, P.D.J. et al. (1996). On the origin of blue luminescence arising from atmospheric impregnation of oxidized porous silicon. *Thin Solid Films* 276, 112–115.

Cantin, J.L., Schoisswohl, M., Grosman, A., Lebib, S., Ortega, C., and von Bardeleben, H.J. (1996). Anodic oxidation of *p*- and *p*⁺-type porous silicon: Surface structural transformations and oxide formation. *Thin Solid Films* 276, 76–79.

Charrier, J., Lupi, C., Haji, L., and Boisrobert, C. (2000). Optical study of porous silicon buried waveguides from *p*-type silicon. *Mat. Sci. Semicond. Proc.* 3, 351–355.

Chazalviel, J.N. and Ozanam, F. (1997). Surface modification of porous silicon. In: Canham L.T. (ed.) *Properties of Porous Silicon*, INSPEC. The Institution of Electrical Engineers, London, pp. 59–65.

Chelyadinsky, A.R., Dorofeev, A.M., Kazuchits, N.M. et al. (1997). Deformation of porous silicon lattice caused by absorption/desorption processes. *J. Electrochem. Soc.* 144(4), 1463–1468.

Ciurea, M.L., Baltog, I., Lazar, M., Iancu, V., Lazanu, S., and Pentia, E. (1998). Electrical behaviour of fresh and stored porous silicon films. *Thin Solid Films* 325, 271–277.

Clerc, P.-A., Dellmann, L., Gretillat, F. et al. (1998). Advanced deep reactive ion etching: A versatile tool for microelectromechanical systems. *J. Micromech. Microeng.* 8, 272–278.

Cullis, A.G., Canham, L.T., and Calcott, P.D.J. (1997). The structural and luminescence properties of porous silicon. *J. Appl. Phys.* 82, 909–965.

Dattagupta, S.P., Chen, X.L., Jenekhe, S.A., and Fauchet, P.M. (1997a). Microhardness of porous silicon films and composites. *Solid State Commun.* 101, 33–37.

Dattagupta, S.P., Fauchet, P.M., Chen, X.L., and Jenekhe, S.A. (1997b). Fabrication and characterization of light-emitting porous silicon and polymer nanocomposites. *Mater. Res. Soc. Symp. Proc.* 452, 473–478.

Debarge, L., Stoquert, J.P., Slaoui, A., Stalmans, L., and Poortmans, J. (1998). Rapid thermal oxidation of porous silicon for surface passivation. *Mat. Sci. Semicon. Proc.* 1, 281–285.

Demidov, E.S., Rassolova, I.S., Gorshkov, O.N. et al. (2008). Photoluminescence and EPR of porous silicon formed on *n*⁺ and *p*⁺ single crystals implanted with boron and phosphorus ions. *Phys. Solid State* 50(8), 1565–1569.

Deresmes, D., Marissael, V., Stievenard, D., and Ortega, C. (1995). Electrical behaviour of aluminium-porous silicon junctions. *Thin Solid Films* 255, 258–261.

Diaz-Herrera, B., Gonzalez-Diaz, B., Guerrero-Lemus, R., Hernandez-Rodriguez, C., Mendez-Ramos, J., and Rodriguez V.D. (2009). Photoluminescence of porous silicon stain etched and doped with erbium and ytterbium. *Physica E* 41, 525–528.

Di Francia, G., De Filippo, F., La Ferrara, V. et al. (1998). Porous silicon layers for the detection at RT of low concentrations of vapoura from organic compounds. In: *Proceeding of European Conference Eurosensors XII*, September 13–16, 1998, Southampton, UK. IOP, Bristol, vol. 1, pp. 544–547.

Djohari, H., Martinez-Herrera, J.I., and Derby, J.J. (2009). Transport mechanisms and densification during sintering: I. Viscous flow versus vacancy diffusion. *Chem. Eng. Sci.* 64(17), 3799–3809.

Dolino, G. and Bellet, D. (1997). Strain in porous silicon. In: Canham L.T. (ed.) *Properties of Porous Silicon.* IEE INSPEC, London, 1997, pp. 118–124.

Dorofeev, A.M., Gaponenko, N.V., Bondarenko, V.P. et al. (1995). Erbium luminescence in porous silicon doped from spin-on films. *J. Appl. Phys.* 77, 2679.

Ducso, C., Khanh, N.Q., Horvath, Z. et al. (1996). Deposition of tin oxide into porous silicon by atomic layer epitaxy. *J. Electrochem. Soc.* 143, 683–687.

Earwaker, L.G., Briggs, M.C., Nasir, M.I., Fair, J.P.G., and Keen, J.M. (1991). Analysis of porous silicon silicon-on-insulator materials. *Nucl. Instrum. Methods B* 56/57, 855–859.

El-Bahar, A., Stolyarova, S., and Nemirovsky, Y. (2000). N-type porous silicon doping using phosphorous oxychloride (POCl₃). *IEEE Electron Dev. Lett.* 21(9), 436–438.

Errien, N., Joubert, P., Chaillou, A. et al. (2003). Electrochemical growth of poly(3-dodecylthiophene) into porous silicon: A nanocomposite with tubes or wires? *Mater. Sci. Eng. B* 100, 259–262.

Errien, N., Vellutini, L., Louarn, G., and Froyer, G. (2007). Surface characterization of porous silicon after pore opening processes inducing chemical modifications. *Appl. Surf. Sci.* 253, 7265–7271.

Etter, D.B., Zimmermann, M., Ferwana, S., Hutter, F.X., and Burghartz, J.N. (2012). Low-cost CMOS compatible sintered porous silicon technique for microbolometer manufacturing. In: *Proceedings of the 25th International Conference on Micro Electro Mechanical Systems, IEEE MEMS 2012*, January 29–February 2, Paris, France, pp. 273–276.

Fan, J., Wan, M., and Zhu, D. (1998). Studies on the rectifying effect of the heterojunction between porous silicon and water-soluble copolymer of polyaniline. *Synth. Metals* 95, 119–124.

Fan, H.J., Kuok, M.H., Ng, S.C. et al. (2003). Effects of natural and electrochemical oxidation processes on acoustic waves in porous silicon films. *J. Appl. Phys.* 94, 1243–1247.

Fang, C., Foca, E., Xu, S., Carstensen, J., and Föll, H. (2007a). Deep silicon macropores filled with copper by electrodeposition. *J. Electrochem. Soc.* 154, D45–D49.

Fang, C., Foca, E., Sirbu, L., Carstensen, J., Foell, H., and Tiginyanu, I.M. (2007b). Formation of metal wire arrays via electrodeposition in pores of Si, Ge and III–V semiconductors. *Phys. Stat. Sol. (a)* 204, 1388–1393.

Fauchet, P.M., Chan, S., Lopez, H.A., and Hirschman, K.D. (2000). Silicon light emitters: preparation, properties, limitations, and integration with microelectronic circuitry. In: Pavesi L. and Buzaneva E. (eds.) *Frontiers of Nano-Optoelectronic Systems*. Kluwer Academic Publishers, Dordrecht, the Netherlands, pp. 99–119.

Fricke, J. (ed.) (1986). *Aerogels.* Springer Verlag, Berlin.

Fried, M., Lohner, T., Polgar, O. et al. (1996). Characterization of different porous silicon structures by spectroscopic ellipsometry. *Thin Solid Films* 276, 223–227.

Friedersdorf, L.E., Searson, P.C., Prokes, S.M., Glembocki, O.J., and Macauley, J.M. (1992). Influence of stress on the photoluminescence of porous silicon structures. *Appl. Phys. Lett.* 60, 2285.

Frohnhoff, St., Marso, M., Berger, M.G., Thonossen, M., Luth, H., and Munder, H. (1995a). An extended quantum model for porous silicon formation. *J. Electrochem. Soc.* 142, 615–620.

Frohnhoff, St., Arens-Fischer, R., Heinrich, T., Fricke, J., Arntzen, M., and Theiss, W. (1995b). Characterization of supercritically dried porous silicon. *Thin Solid Films* 255, 115–118.

Frotscher, U., Rossow, U., Ebert Mpietryga, C. et al. (1996). Investigation of different oxidation processes for porous silicon studied by spectroscopic ellipsometry. *Thin Solid Film* 276, 36–39.

Fukami, K., Harraz, F.A., Yamauchi, T., Sakka, T., and Ogata, Y.H. (2008a). Fine-tuning in size and surface morphology of rod-shaped polypyrrole using porous silicon as template. *Electrochem. Commun.* 10, 56–60.

Fukami, K., Kobayashi, K., Matsumoto, T., Kawamura, Y.L., Sakka, T., and Ogata, Y.H. (2008b). Electrodeposition of noble metals into ordered macropores in *p*-type silicon. *J. Electrochem. Soc.* 155, D443–D448.

Fukami, K., Tanaka, Y., Chourou, M.L., Sakka, T., and Ogata, Y.H. (2009a). Filling of mesoporous silicon with copper by electrodeposition from an aqueous solution. *Electrochim. Acta* 54, 2197–2202.

Fukami, K., Sakka, T., Ogata, Y.H., Yamauchi, T., and Tsubokawa, N. (2009b). Multistep filling of porous silicon with conductive polymer by electropolymerization. *Phys. Stat. Sol. A* 206(6), 1259–1263.

Fürjes, P., Kovács, A., Dücso, Cs., Ádám, M., Müller, B., and Mescheder, U. (2003). Porous silicon-based humidity sensor with interdigital electrodes and internal heaters. *Sens. Actuators B* 95, 140–144.

Gelloz, B., Kojima, A., and Koshida, N. (2005). Highly efficient and stable luminescence of nanocrystalline porous silicon treated by high-pressure water vapor annealing. *Appl. Phys. Lett.* 87, 031107 (1–3).

Godin, B., Chiappini, C., Srinivasan, S. et al. (2012). Discoidal porous silicon particles: Fabrication and biodistribution in breast cancer bearing mice. *Adv. Funct. Mater.* 22, 4225–4235.

Gokarna, A., Bhave, T.M., Bhoraskar, S.V., and Kanjilal, D. (1999). Effect of swift high energy phosphorous ions on the optical and electrical properties of porous silicon. *Nucl. Instr. Meth. Phys. Res. B* 156(1999), 100–104.

Gole, J.L., Seals, L.T., and Lillehei, P.T. (2000). Patterned metallization of porous silicon from electroless solution for direct electrical contact. *J. Electrochem. Soc.* 147(10), 3785–3789.

Gondek, C., Lippold, M., Röver, I., Bachmann, T., and Kroke, E. (2012). HF-H₂O₂-H₂O-mixtures for cleaning sg-silicon: Removal of metal contaminations and formation of porous silicon. In: *Proceedings of 27th European Photovoltaic Solar Energy Conference and Exhibition*, September 22–23, 2012, Frankfurt, Germany, pp. 1109–1112.

Granitzer, P. and Rumpf, K. (2010). Porous silicon—A versatile host material. *Materials* 3, 943–998.

Granitzer, P., Rumpf, K., Pölt, P., Reichmann, A., and Krenn, H. (2007). Microrod and microtube formation by electrodeposition of metal into ordered macropores prepared in *p*-type silicon. *Physica E* 38, 205–210.

Granitzer, P., Rumpf, L., Pölt, P., Simic, S., and Krenn, H. (2008a). Three-dimensional quasi-regular arrays of Ni nanostructures grown within the pores of a porous silicon layer—Magnetic characteristics. *Phys. Stat. Sol. (c)* 5, 3580–3583.

Granitzer, P., Rumpf, K., Pölt, P., Simic, S., and Krenn, H. (2008b). Formation of self-assembled metal/silicon nanocomposites. *Phys. Stat. Sol. (a)* 205, 1443–1446

Grivickas, V. and Basmaji, P. (1993). Optical absorption in porous silicon of high porosity. *Thin Solid Films* 235, 234–238.

Gros-Jean, M., Herino, R., and Lincot, D. (1998). Incorporation of cadmium sulfide into nanoporous silicon by sequential chemical deposition from solution. *J. Electrochem. Soc.* 145, 2448–2452.

Grosman, A. and Ortega, C. (1997). Dopants in porous silicon. In: Canham L. (ed.) *Properties of Porous Silicon.* INSPEC, London, 328–335.

Guerrero-Lemus, R., Ben-Hander, F.A., Moreno, J.D. et al. (1999). Anodic oxidation of porous silicon bilayers. *J. Lumin.* 80, 173–178.

Halimaoui, A., Campidelli, Y., Badoz, P.A., and Bensahel, D. (1995a). Covering and filling of porous silicon pores with Ge and Si using chemical vapor deposition. *J. Appl. Phys.* 78, 3248.

Halimaoui, A., Campidelli, Y., Larre, A., and Bensahel, D. (1995b). Thermally induced modifications in the porous silicon properties. *Phys. Stat. Sol. (b)* 190, 35–40.

Halliday, D.P., Holland, E.R., Eggleston, J.M., Adams, P.N., Cox, S.E., and Monkman, A.P. (1995). Electroluminescence from porous silicon using a conducting polyaniline contact. *Thin Solid Films* 276, 299–302.

Hamm, D., Sakka, T., and Ogata, Y.H. (2005). Specific deposition of copper onto porous silicon. *Phys. Stat. Sol. (c)* 2, 3334–3338.

Harraz, F.A. (2006). Electrochemical polymerization of pyrrole into nanostructured p-type porous silicon. *J. Electrochem. Soc.* 153, C349–C356.

Harraz, F.A., Kamada, K., Sasano, J., Izuo, S., Sakka, T., and Ogata, Y.H. (2005). Pore filling of macropores prepared in p-type silicon by copper deposition. *Phys. Stat. Sol. (a)* 202, 1683–1687.

Harraz, F.A., Salem, M.S., Sakka, T., and Ogata, Y.H. (2008). Hybrid nanostructure of polypyrrole and porous silicon prepared by galvanostatic technique. *Electrochim. Acta* 53(10), 3734–3740.

Heinrich, J.L., Curtis, C.L., Credo, G.M., Kavanagh, K.L., and Sailor, M.J. (1992). Luminescent colloidal silicon suspensions from porous silicon. *Science* 255, 66–68.

Herino, R. (1997). Impregnation of porous silicon. In: Canham L. (ed.) *Properties of Porous Silicon.* INSPEC, London, UK, pp. 66–76.

Herino, R. (2000). Nanocomposite materials from porous silicon. *Mater. Sci. Eng. B,* 69–70, 70–76.

Herino, R., Perio, A., Barla, K., and Bomchil, G. (1984). Microstructure of porous silicon and its evolution with temperature. *Mater. Lett.* 2, 519–523.

Herino, R., Jan, P., and Bomchil, G. (1985). Nickel plating on porous silicon. *J. Electrochem. Soc.* 132, 2513–2514.

Herino, R., Gros-Jean, M., Montes, L., and Lincot, D. (1997). Electrochemical and chemical deposition of II-VI semiconductors in porous silicon. *Mater. Res. Soc. Symp. Proc.* 452, 467.

Huang, Y.M. (1996). Photoluminescence of copper-doped porous silicon. *Appl. Phys. Lett.* 69, 2855–2857.

Ito, T., Ooiwa, T., Nagao, T., and Hatta, A. (1997). Local structure of indium-plated porous silicon. *Mater. Res. Soc. Symp. Proc.* 452, 485.

Jacobsohn, L.G., Bennett, B.L., Cooke, D.W., Muenchausen, R.E., and Nastasi, M. (2006). Ion irradiation of porous silicon: The role of surface states. *Nucl. Instr. Meth. Phys. Res. B* 242, 164–166.

James, T.D., Steer, A., Musca, C.A., Parish, G., Keating, A., and Faraone, L. (2006). Investigation of aging effects in porous silicon. In: *Proceedings of IEEE Conference on Optoelectronic and Microelectronic Materials and Devices, COMMAD 2006,* December 6–8. Perth, Australia, pp. 290–293.

Jeske, M., Schultze, J.W., Thonissen, M., and Munder, H. (1995). Electrodeposition of metals into porous silicon. *Thin Solid Films* 255, 63–66.

Joshi, M.B., Hu, S.J., and Goorsky, M.S. (2010). Porous silicon films for thin film layer transfer applications and wafer bonding. *ECS Trans.* 33(4), 195–206.

Jumayev, B.R., Tam, H.L., Cheah, K.W., and Korsunska, N.E. (2002). Influence of different atmospheres on the life time of porous silicon light-emitting devices. In: *MRS Proc.* 737, doi:http://dx.doi.org/10.1557/PROC-737-F8.14.

Jung, K.G., Schultze, J.W., Thonissen, M., and Munder, H. (1995). Deposition of electrically conducting polybithiophene into porous silicon. *Thin Solid Films* 255, 317–324.

Kao, D.-B., McVittie, J.P., Nix, W.D., and Saraswat, K.S. (1987). Two-dimensional thermal oxidation of silicon-I. Experiments. *IEEE T. Electron. Dev.* 34(5), 1008–1017.

Kazuchits, N., Dolgiy, L., Bondarenko, V. Leshok, A., Dorofeev, A., and Borisenko, V. (1995). Strong 1.54 μm luminescence of porous silicon electrochemically doped with erbium. In: *Nanomeeting 95: Physics, Chemistry and Applications of Nanostructures,* Borisenko V., Filonov A., Gaponenko S., and Gurin V. (eds.). World Scientific, Singapore, pp. 83–86.

Kim, K.H., Bai, G., and Nicolet, M.-A. (1991). Strain in porous Si with and without capping layers. *J. Appl. Phys.* 69, 2201–2205.

Kim, Y.M., Noh, K.Y., Park, J.Y. et al. (2001). Fabrication of oxidized porous silicon (OPS) air-bridge for RF application using micromachining technology. *J. Korean Phys. Soc.* 39, S268–S270.

Kim, H.-S., Zouzounis, E.C., and Xie, Y.-H. (2002). Effective method for stress reduction in thick porous silicon films. *Appl. Phys. Lett.* 80, 2287–2289.

Kimura, T., Yokoi, A., Horiguchi, H., Saito, R., Ikoma, T., and Sato, A. (1994). Electrochemical Er doping of porous silicon and its room-temperature luminescence at ~1.54 μm. *Appl. Phys. Lett.* 65, 983.

Kistler, S.S. (1931). Coherent expanded aerogels and jellies. *Nature* 127, 741.

Klangsin, J., Marty, O., Munguía, J. et al. (2008). Structural and luminescent properties of silicon nanoparticles incorporated into zirconia matrix. *Phys. Lett. A* 372, 1508–1511.

Kleimann, P., Linnros, J., Frojdh, C., and Petersson, C.S. (2001). An X-ray imaging pixel detector based on scintillator filled pores in a silicon matrix. *Nucl. Instr. Meth. Phys. Res. A* 460, 15–19.

Kleps, I., Angelescu, A., Miu, M., Ghita, M., and Bercu, M. (1997). Porous silicon surface stabilization by anodic oxidation for optoelectronic applications. In: *Proceedings of International Semiconductor Conference, CAS'97*, Sinae, Romania. IEEE, Vol. 1. pp. 519–522.

Kobayashi, K., Harraz, F.A., Izuo, S., Sakka, T., and Ogata, Y.H. (2006). Microrod and microtube formation by electrodeposition of metal into ordered macropores prepared in *p*-type silicon. *J. Electrochem. Soc.* 153, C218–C222.

Kochergin, V.R. and Foell, H. (2006). Novel optical elements made from porous Si. *Mater. Sci. Eng. R* 52, 93–140.

Koda, R., Fukami, K., Sakka, T., and Ogata, Y.H. (2012). Electrodeposition of platinum and silver into chemically modified microporous silicon electrodes. *Nanoscale Res. Lett.* 7, 330.

Koshida, N., Koyoma, H., Yamamoto, Y., and Collins, G.J. (1993). Visible electroluminescence from porous silicon diodes with an electropolymerized contact. *Appl. Phys. Lett.* 63, 2655–2657.

Kovalevskii, A.A., Glukhmanchuk, V.V., Tarasikov, M.V., and Sorokin, V.M. (2004). Plasma-assisted diffusion doping of porous-silicon films. *Russ. Microelectron.* 33(1), 13–17.

Labunov, V., Bondarenko, V., Glinenko, L., Dorofeev, A., and Tabulina, L. (1986). Heat treatment effect on porous silicon. *Thin Solid Films* 137, 123–134.

Labunov, V., Bondarenko, V., Borisenko, E., and Dorofeev, A. (1987). High-temperature treatment of porous silicon. *Phys. Stat. Sol. (a)* 102, 193–198.

Lai, M. and Riley, D.J. (2008). Templated electrosynthesis of nanomaterials and porous structures. *J. Colloid Interf. Sci.* 323, 203–212.

Lammel, G. and Renaud, P. (2000). Free-standing, mobile 3D porous silicon microstructures. *Sens. Actuators B* 85, 356–360.

Langner, A., Muller, F., and Gosele, U. (2011). Macroporous silicon. In: Hayden O. and Nielsch K. (eds.) *Molecular- and Nano-Tubes*. Springer, New York, pp. 431–460.

Lara, J.A. and Kathirgamanathan, P. (2000). White light electroluminescence from PSi devices capped with poly(thiophene)(s) as top contact. *Synthetic Met.* 110, 233–240.

Lee, M.-K. and Tu, H.-F. (2007). Stabilizing light emission of porous silicon by *in-situ* treatment. *Jpn. J. Appl. Phys.* 46, 2901–2903.

Lehmann, V. (2007). Alkaline etching of macroporous silicon. *Phys. Stat. Sol. (a)* 204(5), 1318–1320.

Lehmann, V., Mitani, K., Feijoo, D., and Goesele, U. (1991). Implanted carbon: An effective etch-stop in silicon. *J. Electrochem. Soc.* 138, L3–L4.

Lehmann, V., Jobst, B., Muschik, T., Kux, A., and Petrova-Koch, V. (1993). Correlation between optical properties and crystallite size in porous silicon. *Jpn. J. Appl. Phys.* 32(Part 1)(5A), 2095–2099.

Leisenberger, F., Duschek, R., Czaputa, R., Netzer, R., Beamson, F.P., and Matthew, J.A.D. (1997). A high resolution XPS study of a complex insulator: The case of porous silicon. *Appl. Surf. Sci.* 108, 273–281.

Len'shin, A.S., Kashkarov, V.M., Turishchev, S.Y., Smirnov, M.S., and Domashevskaya, E.P. (2011). Effect of natural aging on photoluminescence of porous silicon. *Tech. Phys. Lett.* 37(9), 789–792.

Lerondel, G., Amato, G., Parisini, A., and Boarino, L. (2000). Porous silicon nanocracking. *Mater. Sci. Eng. B* 69–70, 161–166.

Li, Y.Y., Cunin, F., Link, J.R. et al. (2003). Polymer replicas of photonic porous silicon for sensing and drug delivery applications. *Science* 299, 2045–2047.

Li, X., He, Y., and Swihart, M.T. (2004). Surface functionalization of silicon nanoparticles produced by laser-driven pyrolysis of silane followed by HF–HNO$_3$ etching. *Langmuir* 20, 4720–4727.

Li, Y.Y., Kollengode, V.S., and Sailor, M.J. (2005). Porous-silicon/polymer nanocomposite photonic crystals formed by microdroplet patterning. *Adv. Mater.* 17, 1249–1251.

Li, S., Ma, W., Zhou, Y. et al. (2014). Influence of fabrication parameter on the nanostructure and photoluminescence of highly doped *p*-porous silicon. *J. Lumin.* 146, 76–82.

Liao, L.-S., Bao, X.-M.B., Yang, Z.F., and Min, N.-B. (1995). Intense blue emission from porous β-SiC formed on C+-implanted silicon. *Appl. Phys. Lett.* 66(18), 2382.

Lie, L.H., Duerdin, M., Tuite, E.M., Houlton, A., and Horrocks, B.R. (2002). Preparation and characterisation of luminescent alkylated-silicon quantum dots. *J Electrochem Chem.* 538, 183–190.

Linsmeier, J., Wüst, K., Schenk, H. et al. (1997). Chemical surface modification of porous silicon using tetraethoxysilane. *Thin Solid Films* 297, 26–30.

Liu, F., Liao, L., Wang, G., Cheng, G. and Bao, X. (1996). Experimental observation of surface modes of quasifree clusters. *Phys. Rev. Lett.* 76, 604–607.

Liu, F.-Q. and Wang, Z.-G. (1998). Photoluminescence from Ge clusters embedded in porous silicon. *J. Appl. Phys.* 83, 3435–3437.

Liu, C., Sui, X., Yang, F. et al. (2014). Fluorescence of silicon nanoparticles prepared by nanosecond pulsed laser. *AIP Adv.* 4, 031332.

Loni, A. (1997). Capping of porous silicon. In: Canham L. (ed.) *Properties of Porous Silicon.* INSPEC, London, pp. 51–58.

Loni, A., Simons, A.J., Canham, L.T., Phillips, H.J., and Earwaker, L.G. (1994). Compositional variations of porous silicon layers prior to and during ion-beam analysis. *J. Appl. Phys.* 76(5), 2825–2832.

Loni, A., Canham, L.T., Berger, M.G. et al. (1996). Porous silicon multilayer optical waveguides. *Thin Solid Films* 276, 143–146.

Loni, A., Simons, A.J., Calcott, P.D.J., Newey, J.P., Cox, T.I., and Canham, L.T. (1997). Relationship between storage media and blue photoluminescence for oxidized porous silicon. *Appl. Phys. Lett.* 71(1), 107–109.

Lopez, H.A. (2001). Porous Silicon Nanocomposites for Optoelectrnic and Telecommunication Applications. PhD Thesis, University of Rochester, Rochester, NY.

Lopez, H.A. and Fauchet, P.M. (2000). 1.54 μm film EL from Erbium-doped porous Si composites. *Phys. Stat. Sol. (a)* 182, 413–418.

Lopez, H.A., Chen, X.L., Jenekhe, S.A., and Fauchet, P.M. (1999). Tunability of the photoluminescence in porous silicon due to different polymer dielectric environments. *J. Lumin.* 80, 115–118.

Maccagnani, P., Angelucci, R., Pozzi, P. et al. (1998). Thick oxidized porous silicon layer as a thermo-insulating membrane for high-temperature operating thin- and thick-film gas sensors. *Sens. Actuators B* 49, 22–29.

Manuaba, A., Pinter, I., Szilagyi, E. et al. (1997). Plasma immersion ion implantation of nitrogen into porous silicon layers. *Mater. Sci. Forum* 248–249, 233–236.

Manuaba, A., Paszti, F., Battistig, G., Ortega, C., and Grossman, A. (1998). Grazing irradiation of porous silicon by 500 keV He ions. *Vacuum* 50, 349–351.

Manuaba, A., Paszti, F., Ortega, C. et al. (2001). Effect of MeV energy He and N pre-implantation on the formation of porous silicon. *Nucl. Instr. Meth. Phys. Res. B* 179, 63–70.

Mares, L.J., Kristofik, J., Pangrac, J., and Hospodkova, A. (1993). On the transport mechanism in porous silicon. *Appl. Phys. Lett.* 63, 180.

Mason, M.D., Sirbuly, D.J., and Buratto, S.K. (2002). Correlation between bulk morphology and luminescence in porous silicon investigated by pore collapse resulting from drying. *Thin Solid Films* 406, 151–158.

Mattei, G., Alieva, E.V., Petrov, J.E., and Yakovlev, V.A. (2000). Quick oxidation of porous silicon in presence of pyridine vapor. *Phys. Stat. Sol. (a)* 182, 139–143.

Matveeva, E., Parkhutic, V.P., Diaz Calleja, R., and Martinez-Duart, J.M. (1993). Growth of polyaniline films on porous silicon layers. *J. Lumin.* 57, 175–180.

Mawhinney, D.B., Glass, J.A., and Yates, J.T. (1997). FTIR study of the oxidation of porous silicon. *J. Phys. Chem. B* 101, 1202–1206.

Michel, J., Benton, J.L., Ferrante, R.F. et al. (1991). Impurity enhancement of the 1.54-μm $Er^{3+}$ luminescence in silicon. *J. Appl. Phys.* 70, 2672.

Mizuno, H., Koyama, H., and Koshida, N. (1996). Oxide-free blue photoluminescence from photochemically etched porous silicon. *Appl. Phys. Lett.* 69(25), 3779–3781.

Montes, L., Muller, F., and Herino, R. (1997). Investigation on the electrochemical deposition of cadmium telluride in porous silicon. *Thin Solid Films* 297, 35–38.

Moreno, J.D., Agullo-Rueda, F., Guerrero-Lemus, R. et al. (1997). Deposition of polypyrrole into porous silicon. *Mater. Res. Soc. Symp. Proc.* 452, 479.

Mula, G., Setzu, S., Manunza, G., Ruffilli, R., and Falqui, A. (2012). Optical, electrochemical, and structural properties of Er-doped porous silicon. *J. Phys. Chem. C* 116, 11256–11260.

Muller, G. (2002). Restrukturierung von Porosem Silizium durch Temperatur behandlung, PhD Thesis, Institute of Applied Physics, University of Erlangen-Nurnberg, Germany.

Muller, G. and Brendel, R. (2000). Simulated annealing of porous silicon. *Phys. Stat. Sol. (a)* 182, 313–318.

Muller, G., Nerding, M., Ott, N., Strunk, H.P., and Brendel, R. (2003). Sintering of porous silicon, *Phys. Stat. Sol. (a)* 197, 83–87.

Mulloni, V. and Pavesi, L. (2000). Electrochemically oxidised porous silicon microcavities. *Mater. Sci. Eng. B* 69–70, 59–65.

Munder, H., Berger, M.G., Frohnhoff, S., Luth, H., and Thonissen, N. (1993). A non-destructive study of the microscopic structure of porous Si. *J. Lumin.* 57, 5–8.

Najar, A., Lorrain, N., Ajlani, H., Charrier, J., Oueslati, M., and Haji, L. (2009). Er$^{3+}$ doping conditions of planar porous silicon waveguides. *Appl. Surf. Sci.* 256, 581–586.

Namavar, F., Lu, F., Perry, C.H., Cremins, A. et al. (1995). Er-implanted porous silicon: A novel material for Si-based infrared LEDs. *Mat. Res. Soc. Symp. Proc.* 358, 375.

Naveas, N., Hernandez-Montelongo, J., Pulido, R. et al. (2013). Fabrication and characterization of a chemically oxidized-nanostructured porous silicon based biosensor implementing orienting protein A. *Colloids Surf. B* 115C, 310–316.

Nguyen, T.P., Le Rendu, P., Lakehal, M., Joubert, P., and Destruel, P. (2000). Poly(*p* phenylene vinylene):porous silicon composites. *Mater. Sci. Eng. B* 69–70, 177–181.

Nishimura, K., Nagao, Y., and Ikeda, N. (1996). Thermal diffusion of antimony into nanostructured porous silicon. *Jpn. J. Appl. Phys.* 35, L1145–1147.

Norenberg, H. and Briggs, G.A.D. (1999). The Si(001) c(4×4) surface reconstruction: A comprehensive experimental study. *Surf. Sci.* 430, 154–164.

Nychyporuk, T. (2010). Nouvelles morphologies du Silicium nanostructuré issues de l'anodisation électrochimique: élaboration, propriétés physico-chimiques et applications. PhD thesis. Edition Universitaires Européennes (In French).

Ogata, Y., Niki, H., Sakka, T., and Iwasaki, M. (1995). Oxidation of porous silicon under water vapor environment. *J. Electrochem. Soc.* 142, 1595–1601.

Ogata, Y.H., Kato, F., Tsuboi, T., and Sakka, T. (1998). Changes in the environment of hydrogen in porous silicon with thermal annealing. *J. Electrochem. Soc.* 145, 2439–2444.

Ogata, Y.H., Tsuboi, T., Sakka, T., and Naito, S. (2000). Oxidation of porous silicon in dry and wet environments under mild temperature conditions. *J. Porous. Mater.* 7, 63–66.

Ogata, Y.H., Yoshimi, N., Yasuda, R., Tsuboi, T., Sakka, T., and Otsuki, A. (2001). Structural change in p-type porous silicon by the thermal annealing. *J. Appl. Phys.* 90, 6487–6492.

Ogata, Y.H., Kobayashi, K., and Motoyama, M. (2006). Electrochemical metal deposition on silicon. *Curr. Opin. Solid State Mater. Sci.* 10, 163–172.

Ohji, H., Trimp, P.J., and French, P.J. (1999). Fabrication of free standing structure using single step electrochemical etching in hydrofluoric acid. *Sens. Actuators A* 73, 95–100.

Ohring, M. (1991). *The Materials Science of Thin Films.* Academic Press, San Diego, CA.

Onganer, Y., Saglam, M., Turut, A., Efeoglu, H., and Tuzemen S. (1996). High barrier metallic polymer/*p*-type silicon Schottky diodes. *Solid State Electron.* 39, 677–737.

Ott, N., Nerding, M., Müller, G., Brendel, R., and Strunk, H.P. (2003). Structural changes in porous silicon during annealing. *Phys. Stat. Sol. (a)* 197(1), 93–97.

Ott, N., Nerding, M., Muller, G., Brendel, R., and Strunk, H.P. (2004). Evolutiom of the microstructure during annealing of porous silicon multilayers. *J. Appl. Phys.* 95, 497–503.

Ow, Y.S., Liang, H.D., Azimi, S., and Breese, M.B.H. (2011). Modification of porous silicon formation by varying the end of range of ion irradiation. *Electrochem. Sol.-St. Lett.* 14(5) D45–D47.

Ozanam, J.N. (1999). Surface chemistry of porous silicon. *Mater. Res. Soc. Symp. Proc.* 536, 155–166.

Ozanam, F. and Chazalviel, J.N. (1993). In-situ infrared characterization of the electrochemical dissolution of silicon in a fluoride electrolyte. *J. Electron. Spectrosc. Rel. Phenom.* 64–65(1), 395–402.

Ozdemir, S. and Gole, J.L. (2007). The potential of porous silicon gas sensors. *Curr. Op. Solid State Mater. Sci.* 11, 92–100.

Pap, A.E., Kordas, K., George, T.F., and Leppavuori, S. (2004). Thermal oxidation of porous silicon: Study on reaction kinetics. *J. Phys. Chem. B* 108, 12744–12747.

Pap, A.E., Kordás, K., Tóth, G. et al. (2005). Thermal oxidation of porous silicon: Study on structure. *Appl. Phys. Lett.* 86(4), 041501/1–3.

Papadimitriou, D., Bitsakis, J., López-Villegas, J.M., Samitier, J., and Morante, J.R. (1999). Depth dependence of stress and porosity in porous silicon: A micro-Raman study. *Thin Solid Films* 349, 293–297.

Papadimitriou, D., Tsamis, C., and Nassiopoulou, A.G. (2004). The influence of thermal treatment on the stress characteristics of suspended porous silicon membranes on silicon. *Sens. Actuators B* 103, 356–361.

Park, J.Y. and Lee, J.H. (2003). Characterization of 10 μm thick porous silicon dioxide obtained by complex oxidation process for RF application. *Mater. Chem. Phys.* 82, 134–139.

Parkhutik, V.P. (2000). AC impedance study of porous silicon aging in HF solution. *J. Porous Mater.* 7, 97–101.

Paszti, F. and Battistig, G. (2000). Ion beam characterisation and modification of porous silicon. *Phys. Stat. Sol. (a)* 182, 271–278.

Paszti, F., Manuaba, A., Szilagyi, E., Vazsonyi, E., and Vertesy, Z. (1996). Densification and noble gas retention of ion-implanted porous silicon. *Nucl. Instr. Meth. Phys. Res. B* 117, 253–259.

Paszti, F., Szilagyi, E., Manuaba, A., and Battistig, G. (2000). Application of resonant backscattering spectrometry for determination of pore structure changes. *Nucl. Instr. Meth. Phys. Res. B* 161–163, 963–968.

Pavesi, L. and Mulloni, V. (1999). All porous silicon microcavities: Growth and physics. *J. Lumin.* 80, 43–52.

Pavesi, L., Giebel, G., Ziglio, F. et al. (1994). Nanocrystal size modifications in porous silicon by preanodization ion implantation. *Appl. Phys. Lett.* 65(17), 2182.

Peng, C., Fauchet, P.M., Rehm, J.M., McLendon, G.L., Seiferth, F., and Kurinec, S.K. (1994). Ion implantation of porous silicon. *Appl. Phys. Lett.* 64(10), 1259–1261.

Perrone, Donnorso, M., Miele, E., De Angelis, F. et al. (2012). Nanoporous silicon nanoparticles for drug delivery applications. *Microelectron. Eng.* 98, 626–629.

Petrovich, V., Volchek, S., Dolgyi, L. et al. (2000a). Deposition of erbium containing film in porous silicon from ethanol solution of erbium salt. *J. Porous Mater.* 7(1–3), 37–40.

Petrovich, V., Volchek, S., Dolgyi, L. et al. (2000b). Formation features of deposits during a cathode treatment of porous silicon in aqueous solutions of erbium salts. *J. Electrochem. Soc.* 147(2), 655–658.

Pirasteh, P., Charrier, J., Soltani, A. et al. (2006). The effect of oxidation on physical properties of porous silicon layers for optical applications. *Appl. Sur. Sci.* 253, 1999–2002.

Piryatinskii, Y.P., Klyui, N.I., and Rozhin, A.G. (2000). Photoluminescence of porous silicon layers formed in ion-implanted silicon wafers. *Techn. Phys. Lett.* 26(11), 944–946.

Polman, A. (1997). Erbium implanted thin film photonic materials. *J. Appl. Phys.* 82, 1–39.

Prabakaran, R., Kesavamoorthy, R., Amirthapandian, S., and Xavier, F.P. (2003). Raman scattering and photoluminescence studies on $O^+$ implanted silicon. *Physica B* 337, 36–41.

Prabakaran, R., Kesavamoorthy, R., Amirthapandian, S., and Ramanand, A. (2004). Raman scattering and photoluminescence studies on $O^+$ implanted porous silicon. *Mater. Lett.* 58, 3745–3750.

Prischepa, S.L., Dolgiy, A.L., Bandarenka, A.V. et al. (2014). Synthesis and properties of Ni nanowires in porous silicon tempaltes. In: Wilson L.J. (ed.) *Nanowires: Synthesis, Electrical Properties and Uses in Biological Systems.* Nova Science Publishers, New York, pp. 89–129.

Prochaska, A., Mitchell, S.J.N., and Gamble, H.S. (2002). Porous silicon as a sacrificial layer in production of siliocn diaphragms by precision grinding. In: Tay, F.E.H. (ed.) *Materials and Process Integration for MEMS.* Kluwer Academic Publisher, the Netherlands, pp. 27–50.

Qin, Z., Joo, J., Gu, L., and Sailor, M.J. (2014). Size control of porous silicon nanoparticles by electrochemical perforation etching. *Part. Part. Syst. Charact.* 31, 252–256.

Ramos, A.R., Conde, O., Paszti, F. et al. (2000). Ion beam synthesis of chromium silicide on porous silicon. *Nucl. Instr. Meth. Phys. Res. B* 161–163, 926–930.

Riikonen, J., Salomaki, M., van Wonderen, J. et al. (2012). Surface chemistry, reactivity, and pore structure of porous silicon oxidized by various methods. *Langmuir* 28, 10573–10583.

Ronkel, F., Schultze, J.W., and Arens-Fisher, R. (1996). Electrical contact to porous silicon by electrodeposition of iron. *Thin Solid Films* 276, 40–43.

Roussel, P., Lysenko, V., Remaki, B., Delhomme, G., Dittmar, A., and Barbier, D. (1999). Thick oxidised porous silicon layers for the design of a biomedical thermal conductivity microsensor. *Sens. Actuators A* 74, 100–103.

Roy A., Jayaram K., and Sood A.K. (1994). Origin of visible photoluminesecece from porous silicon as studies by Raman spectroscopy *Bull. Mater. Sci.* 17, 513–522.

Russo, L., Colangelo, F., Cioffi, R., Rea, I., and De Stefano, L. (2011). A mechanochemical approach to porous silicon nanoparticles fabrication. *Materials.* 4, 1023–1033.

Sakly, H., Mlika, R., Chaabane, H., Beji, L., and Ben Ouada, H. (2006). Anodically oxidized porous silicon as a substrate for EIS sensors. *Mater. Sci. Eng. C* 26, 232–235.

Salonen, J., Lehto, V.P., and Laine, E. (1997a). The room temperature oxidation of porous silicon. *Appl. Surf. Sci.* 120, 191–198.

Salonen, J., Lehto, V.P., and Laine, E. (1997b). Thermal oxidation of free-standing porous silicon films. *Appl. Phys. Lett.* 70, 637–639.

Salonen, J., Lehto, V.-P., Björkqvist, M., Laine, E., and Niinistö, L. (2000). Studies of thermally-carbonized porous silicon surfaces. *Phys. Stat. Sol. (a)* 182, 123–126.

Salonen, J., Kaukonen, A.M., Hirvonen, J., and Lehto, V.P. (2008). Mesoporous silicon in drug delivery applications. *J. Pharm. Sci.* 97, 632–653.

Sasano, J., Jorne, J., Yoshimi, N., Tsuboi, T., Sakka, T., and Ogata, Y.H. (2000). Effects of chloride ion on copper deposition into porous silicon. In: Matlosz, M., Landolt, D., Agoaki, R., Sato, Y., and Talbot, J.B. (eds.) *Fundamental Aspects of Electrochemical Deposition and Dissolution.* The Electrochemical Society, Pennington, NJ, pp. 84–90.

Sato, N., Sakaguch, K., Yamagata, K., Fujiyama, Y., and Yonehara, T. (1995). Epitaxial growth on porous Si for a new bond and etchback silicon-on-insulator. *J. Electrochem. Soc.* 142, 3116–3122.

Sato, K., Shikida, M., Yamashiro, T., Asaumi, K., Iriye, Y., and Yamamoto, M. (1999). Anisotropic etching rates of single-crystal silicon for TMAH water solution as a function of crystallographic orientation. *Sens. Actuators B* 73, 131–137.

Scherer, G.W. (1992). Recent progress in drying of gels. *J. Non-Cryst. Solids* 147/148, 363–374.

Schultze, J.W. and Kung, K.G. (1995). Regular nanostructured systems formed electrochemically: Deposition of electroactive polybithiophene into porous silicon. *Electrochim. Acta* 40, 1369–1383.

Schwartz, C. and Schaible, P.M. (1979). Reactive ion etching of silicon. *J. Vac. Sci. Technol.* 16, 410–413.

Segal, E., Perelman, L.A., Cunin, F. et al. (2007). Confinement of thermoresponsive hydrogels in nanostructured porous silicon dioxide templates. *Adv. Funct. Mater.* 17, 1153–1162.

Setzu, S., Letant, S., Solsona, P., Romenstain, R., and Vial, J.C. (1998). Improvement of the luminescence in p-type as-prepared or dye impregnated porous silicon microcavities. *J. Lumin.* 80, 129–132.

Sham, T.K., Feng, X.H., Jiang, D.T. et al. (1992). Si K-edge X-ray absorption fine structure studies of porous silicon. *Can. J. Phys.* 70, 813–818.

Sharma, R.K., Rastogi, A.C., and Desu, S.B. (2007). Nano crystalline porous silicon as large-area electrode for electrochemical synthesis of polypyrrole. *Physica B* 388, 344–349.

Shen, Y. and Wan, M. (1998). Heterojunction diodes of soluble conducting polypyrrole with porous silicon. *Synth. Metals* 98, 147–152.

Shen, Q. and Toyoda, T. (2002). Characterization of thermal properties of porous silicon film/silicon using photoacoustic technique. *J. Therm. Anal. Cal.* 69, 1037–1044.

Shih, S., Tsai, C., Jung, K.-H., Campbell, J.C., and Kwong, D.L. (1992). Control of porous silicon photoluminescence through dry oxidation. *Appl. Phys. Lett.* 60(5), 633–635.

Shriver, D.F., Atkins, P.W., and Langford, C.H. (1990). *Inorganic Chemistry*. Oxford University Press, New York.

Simon, A., Paszti, F., Manuaba, A., and Kiss, A.Z. (1999). Three-dimensional scanning of ion-implanted porous silicon. *Nucl. Instr. Meth. Phys. Res. B* 158, 658–664.

Solanki, C.S., Bilyalov, R.R., Bender, H., and Poortmans, J. (2000). New approach for the formation and separation of a thin porous silicon layer. *Phys. Stat. Sol. (a)* 182, 97–101.

Sorokin, L.M., Ratnikov, V.V., Kalmykov, A.E., and Sokolov, V.I. (2010). Evolution of porous silicon crystal structure during storage in ambient air. *J. Phys.: Conference Ser.* 209, 012059.

Stepanov, A.L., Osin, T.N., Trifonov, A.A., Valeev, F., and Nuzhdin, V.I. (2014). New approach to the synthesis of porous silicon with silver nanoparticles using ion implantation technique. *Nanotechnol. Rus.* 9(3–4), 163–167.

Suda, Y., Ban, T., Koizumi, T. et al. (1994). Surface structures and photoluminescence mechanisms of porous Si. *Jpn. J. Appl. Phys.* 33, 581–585.

Sugiyama, H. and Nittono, O. (1990). Microstructure and lattice distortion of anodized porous silicon layers. *J. Cryst. Growth* 103, 156–163.

Sundaram, K.B., Ali, S.A., Peale, R.E., and McClintic, Jr. W.A. (1997). Photoluminescence studies of thermal impurity diffused porous silicon layers. *J. Mater. Sci.: Mater. Electron.* 8, 163–169.

Švrček, V., Pelant, I., Rehspringer, J.-L. et al. (2002). Photoluminescence properties of sol–gel derived SiO$_2$ layers doped with porous silicon. *Mater. Sci. Eng. C* 19, 233–236.

Švrček, V., Slaoui, A., and Muller, J.-C. (2004). *Ex situ* prepared Si nanocrystals embedded in silica glass: Formation and characterization. *J. Appl. Phys.* 95, 3158–3163.

Tada, K., Hamaguchi, M., Hosono, A., Yura, S., Harada, H., and Yoshino, K. (1997). Light emitting diode with porous silicon/conducting polymer heterojunction. *Jpn. J. Appl. Phys.* 36, L418–L420.

Teschke, O., Galembeck, F., Goncalves, M.C., and Davanzo, C.U. (1994). Photoluminescence spectrum redshifting of porous silicon by a polymeric carbon layer. *Appl. Phys. Lett.* 64(26), 3590–3592.

Tilley, R.D., Warner, J.H., Yamamoto, K., Matsui, I., and Fujimori, H. (2005). Micro-emulsion synthesis of monodisperse surface stabilized silicon nanocrystals. *Chem. Commun.* 2005, 1833–1835.

Tischler, M.A., Collins, R.Y., Stathis, J.H., and Tsang, J.C. (1992). Luminescence degradation in porous silicon. *Appl. Phys. Lett.* 60, 639–641.

Trifonov, T., Garin, M., Rodriguez, A., Marsal, L.F., Pallares, J., and Alcubilla, R. (2007a). Towards more complex shapes of macroporous silicon. In: *Proceeding of Spanish Conference on Electron Devices*, January 31–February 2, 2007, Madrid, Spain. IEEE, pp. 9–12.

Trifonov, T., Garín, M., Rodríguez, A., Marsal, L.F., and Alcubilla, R. (2007b). Tuning the shape of macroporous silicon. *Phys. Stat. Sol. (a)* 204(10), 3237–3242.

Tserepi, A., Tsamis, C., Gogolides, E., and Nassiopoulou, A.G. (2003). Dry etching of porous silicon in high density plasmas. *Phys. Stat. Sol. (a)* 197(1), 163–167.

Unagami, T. (1980). Oxidation of porous silicon and properties of its oxide film. *Jpn. J. Appl. Phys.* 19, 231–241.

Unagami, T. (1997). Intrinsic stress in porous silicon layers formed by anodization in HF solution. *J. Electrochem. Soc.* 144, 1835–1838.

Utrainen, M., Lehto, S., Niinisto, L. et al. (1997). Porous silicon host matrix for deposition by atomic layer epitaxy. *Thin Solid Films* 297, 39–42.

Vellutini, L., Errien, N., Froyer, G. et al. (2007). Polymerization of supramolecular diacetylenic monomer embedded in porous silicon matrix. *Chem. Mater.* 19, 497–502.

Von Behren, J., Fauchet, P.M., Chimowitz, E.H., and Lira, C.T. (1997). Optical properties of free-standing ultrahigh porosity silicon films prepared by supercritical drying. *Mater. Res. Soc. Symp. Proc.* 452, 565–570.

Wang, L., Reipa, V., and Blasic, J. (2004). Silicon nanoparticles as a luminescent Label to DNA. *Bioconj. Chem.* 15, 409–412.

Wang, Q., Fu, S.Y., Qu, S.L., and Liu, W.J. (2007). Enhanced photoluminescence from Si$^+$ and C$^+$ ions co-implanted porous silicon formed by electrochemical anodization. *Solid State Commun.* 144, 277–281.

Wilson, W.L., Szajowski, P.F., and Brus, L.E. (1993). Quantum confinement in size-selected, surface-oxidized silicon nanocrystals. *Science* 262, 1242–1244.

Wolf, A. (2007). *Sintering of Porous Silicon*, PhD Thesis, University of Hannover, Germany.

Wolf, A. and Brendel, R. (2006). Thermal conductivity of sintered porous silicon films. *Thin Solid Films* 513, 385–390.

Wolf, A., Terheiden, B., and Brendel, R. (2008). Light scattering and diffuse light propagation in sintered porous silicon. *J. Appl. Phys.* 104, 033106.

Wolkin, M.V., Jorne, J., and Fauchet, P.M. (1999). Electronic states and luminescence in porous silicon quantum dots: The role of oxygen. *Phys. Rev. Lett.* 82(1), 197–200.

Wu, X.L., Yan, F., Bao, X.M. et al. (1996). Raman scattering of porous structure formed on $C^+$-implanted silicon. *Appl. Phys. Lett.* 68(15), 2091.

Wu, X., White, R., Hijmmerich, U., Namavar, F., and Cremins-Costa, A.M. (1997). Time-resolved photoluminescence spectroscopy of Er-implanted porous silicon. *J. Lumin.* 71, 13–20.

Xie, Y.H., Wilson, W.L., Ross, F.M. et al. (1992). Luminescence and structural study of porous silicon films. *J. Appl. Phys.* 71, 2403–2407.

Yae, S., Hirano, T., Matsuda, T., Fukumuro, N., and Matsuda, H. (2009). Metal nanorod production in silicon matrix by electroless process. *Appl. Surf. Sci.* 255, 4670–4672.

Yakovtseva, V., Dolgyi, L., Vorozov, N. et al. (2000). Oxidized porous silicon: From dielectric isolation to integrated optical waveguides. *J. Porous Mater.* 7, 215–222.

Yamana, M., Kashiwazaki, N., Kinoshita, A., Nakano, T., Yamamoto, M., and Walton, C.W. (1990). Porous silicon oxide layer formation by the electrochemical treatment of a porous silicon layer. *J. Electrochem. Soc.* 137, 2925–2927.

Yon, J.J., Barla, K., Herino, R., and Bomchil, G. (1987). The kinetics and mechanism of oxide layer formation from porous silicon formed on *p*-Si substrates. *J. Appl. Phys.* 62, 1042–1048.

Yoon, M.S., Ahn, K.H., Cheung, R.W. et al. (2003). Covalent crosslinking of 1-D photonic crystals of microporous Si by hydrosilylation and ring-opening metathesis polymerization. *Chem. Commun.* 2003, 680.

Youssef, G.M. (2001). Effect of etching sequence and aging on photoluminescence of porous silicon. *Egypt. J. Sol.* 24(2), 227–234.

Zacharatos, F. and Nassiopoulou, A.G. (2008). Copper-filled macroporous Si and cavity underneath for microchannel heat sink technology. *Phys. Stat. Sol. (a)* 205, 2513–2517.

Zhang, H., Liu, D., Shahbazi, M.-A., Mäkilä, E. et al. (2014). Fabrication of a multifunctional nano-in-micro drug delivery platform by microfluidic templated encapsulation of porous silicon in polymer matrix. *Adv. Mater.* 26, 4497–4503.

Zhu, W.X., Gao, Y.X., Zhang, L.Z. et al. (1992). Time evolution of the localized vibrational mode infrared absorption of porous silicon in air. *Superlattices Microstruct.* 12(3), 409–412.

Zuk, J., Ochalski, T.J., Kulik, M., Liskiewicz, J., and Kobzev, A.P. (1999). Effect of oxygen implantation on ionoluminescence of porous silicon. *J. Lumin.* 80, 187–192.

Zur Muhlen, E., Chang, D., Rogaschewski, S., and Niehus, H. (1996). Morphology and luminescence of *p*-doped porous silicon. *Phys. Stat. Sol. (b)* 198, 673–686.

# Surface Chemistry of Porous Silicon

**Yannick Coffinier and Rabah Boukherroub**

13

## CONTENTS

## 13.1 INTRODUCTION

Porous silicon (PSi) films are typically prepared by electrochemical etching of crystalline Si (c-Si) in HF-based solutions. This process is compatible with the fabrication of standard silicon-based microelectronic devices. The pore morphology, diameter, and pore direction can all be controlled by the c-Si wafer surface orientation, doping level and type, the etching solution, and the current density. Under these preparation conditions, the freshly prepared PSi surface is covered with silicon-hydrogen (Si–$H_x$) bonds. The hydrogen-terminated porous silicon (PSi-H) film is of good electronic quality, but the Si–$H_x$ bonds formed on its surface do not protect against photoluminescence (PL) quenching, leading to slow degradation of PL upon exposure to air and concomitant degradation of the electronic properties of the material. This limitation restricts the use of PSi in the fabrication of commercial devices. Moreover, the poor aqueous stability of PSi remains a major challenge in its medical and biological applications. Indeed, the PSi-H surface reacts in ambient air to form an oxide sub-monolayer. Although, the Si–O bond is relatively stable in air, noncontrolled oxidation of the PSi surface introduces surface defects, responsible for the PL quenching. Furthermore, the Si–O bond slowly hydrolyzes and dissolves in water; this phenomenon is accelerated in alkaline media. These chemical transformations, occurring upon exposure of PSi surface to ambient air and/or aqueous-based solutions, affect the optical/electronic but also the mechanical integrity of the porous matrix.

In the past decades, many efforts have been made to stabilize the PSi-H surface. Initially, deliberate oxidation under controlled thermal or electrochemical conditions has been the most investigated method to stabilize the porous matrix. Given the parallel that exists between the chemistry of Si–H bonds in solution phase and on solid surfaces, there has been much effort devoted toward PSi passivation using chemical derivatization of the freshly prepared surfaces by replacing the Si–$H_x$ bonds with more stable Si–C or Si–O–C bonds, under various conditions.

Chemical functionalization of PSi surfaces not only allowed surface passivation and stabilization, but also offered interesting possibilities for the development of new strategies for immobilization of either chemical or biological species on the PSi surface and detection of diverse interactions characterizing biological and chemical systems. To introduce specific functional groups on the PSi surface, chemical modification can be performed either on oxide-surrounding PSi or on hydrogen-terminated PSi (H-PSi). The choice is often related to the targeted application.

In this chapter, we will focus on the various strategies reported in the literature for the chemical functionalization of H-PSi surfaces, including controlled oxidation, hydrosilylation, electrochemical alkylation, and thermal hydrocarbonization.

## 13.2 REACTIVE SPECIES ON POROUS SILICON

### 13.2.1 HYDROGEN-TERMINATED POROUS SILICON (H-PSi)

Since this chapter is dedicated to the surface chemistry of PSi, the details for the preparation of hydrogenated PSi surface are not developed herein. PSi layers are commonly prepared by chemical or electrochemical etching of crystalline silicon in HF-based solutions. The freshly prepared PSi surfaces are hydrogenated, that is, covered with a monolayer of Si–H$_x$ ($x$ = 1, 2, 3) bonds.

FTIR spectroscopy characterization shows that the surface comprises absorptions due to Si–H$_x$ stretching modes (2088 cm$^{-1}$ for Si–H$_1$, 2117 cm$^{-1}$ for Si–H$_2$, and 2138 cm$^{-1}$ for Si–H$_3$), Si–H$_2$ scissor mode $\delta_{Si-H_2}$ at 912 cm$^{-1}$ and $\delta_{Si-H_x}$ at 669 and 629 cm$^{-1}$ (Figure 13.1). A small peak at ~1037 cm$^{-1}$ (Si–O–Si stretching mode), which is present in all PSi samples, results most likely from a small oxidation of the reactive surface or is due to interstitial oxygen in the original silicon substrate lattice (Matsumoto et al. 1997).

The hydrogen-passivated PSi film is of good electronic quality, but the Si–H termination does not fully protect the surface from oxidation and corrosion. The hydrogen-terminated PSi surface reacts in ambient air to form an oxide submonolayer, which introduces surface defects responsible for the degradation of the optical and electronic properties of the material. Hydrogen-termination can be easily regenerated through HF treatment.

### 13.2.2 DEUTERIUM-TERMINATED POROUS SILICON (D-PSi)

Deuterium-terminated porous silicon surfaces (D-PSi) can be easily prepared under electrochemical anodization by replacing the reagents and solvents with deuterated ones: DF-ethanol-D6 or 48% DF/D$_2$O/EtOD (Matsumoto et al. 1997; Bateman et al. 2000; Belogorokhov et al. 2005; de Smet et al. 2005). The FTIR spectrum of a D-PSi sample prepared by electrochemical anodization of $p$-type Si (5–10 Ω cm) at a current density of 100 mA/cm$^2$ in DF:C$_2$D$_5$OD:D$_2$O = 1:1:2 is depicted in Figure 13.1. It shows stretching modes at 1518, 1529, and 1554 cm$^{-1}$ due to Si–D, Si–D$_2$, and SiD$_3$, respectively. This is in a good agreement with the theoretically predicted values in which the ratio of the Si–H and Si–D stretching frequencies must be equal to 1.376 (Bateman et al. 2000).

Similar to hydrogen termination, the D-PSi surface slowly oxidizes upon exposure to ambient air. In fact, during aging of the D-PSi in air, the FTIR spectra changes related to formation of silicon

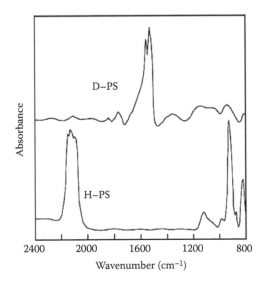

**FIGURE 13.1** FTIR absorption spectra of hydrogen- and deuterium-terminated PSi surfaces. (Reprinted from *Thin Solid Films* 297, Matsumoto T. et al., 31, Copyright 1997, with permission from Elsevier.)

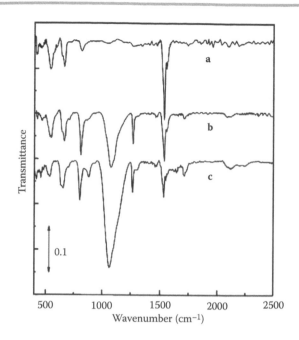

**FIGURE 13.2**    FTIR spectra of the D-PS samples before and after exposure to ambient air: (a) as-prepared, (b) 1 week in air, (c) 3 months in air. (Belogorokhov A.I. et al.: *Phys. Stat. Sol. (a)* 2005.202.1581. Copyright Wiley-VCH Verlag GmbH & Co. kGaA. Reproduced with permission.)

oxide were observed (Figure 13.2). The oxidation results in an appearance of $O_3$–Si–D stretching (1630 cm$^{-1}$), Si–O–Si asymmetric stretching (1063 cm$^{-1}$), or $SiO_2$ lattice optical transverse phonon (TO) and Si–O–Si symmetric stretching (800 cm$^{-1}$) modes. According to the mechanism of room temperature oxidation of the Si surface, the oxygen is originated from chemisorbed $H_2O$. $H_2O$ adsorbs dissociatively producing Si–H or Si–OH bonds (Belogorokhov et al. 2005).

### 13.2.3 HALOGEN-TERMINATED POROUS SILICON (X-PSi, X = Cl, Br, I)

Halogenation of H-PSi has been described in several reports. Because of the high polarity of the Si–X bond, halogenated PSi surfaces are very reactive in ambient air and should be manipulated under inert atmosphere. They are hydrolyzed very quickly in contact with water to form an oxide submonolayer. Halogen termination is commonly produced via chemical treatment of H-PSi with halide gas ($X_2$, X = Cl, Br, I) or alkyl halides (R-X) under inert atmosphere or vacuum (Lauerhaas and Sailor 1993; Lavine et al. 1993; Seo et al. 1993; Hory et al. 1995; Joy and Mandler 2002; Gun'ko et al. 2003). Lauerhaas and Sailor (1993) showed that the direct exposure of H-PSi surface to molecular iodine (<1 torr of $I_2$ (g) for 1 min) results in Si–I bond formation (Figure 13.3). XPS analysis confirmed the presence of Si–I bonds on the PSi surface (peak of I $3d_{5/2}$ at ~619 eV). The reaction occurs by initial attack of $I_2$ at the weak Si-Si bonds rather than at Si–H bonds, as evidenced by the unchanged Si–H and Si–$H_2$ infrared stretching modes on $I_2$ exposure. Subsequent hydrolysis of the I-PSi surface results in a hydrophilic material. Its PL is quenched largely by water as compared to that of H-PSi surface.

Joy and Mandler (2002) described an original technique for *in situ* iodination of H-PSi surfaces. They employed visible light (15 W ordinary white lamp for approximately 6 h) or mild heat

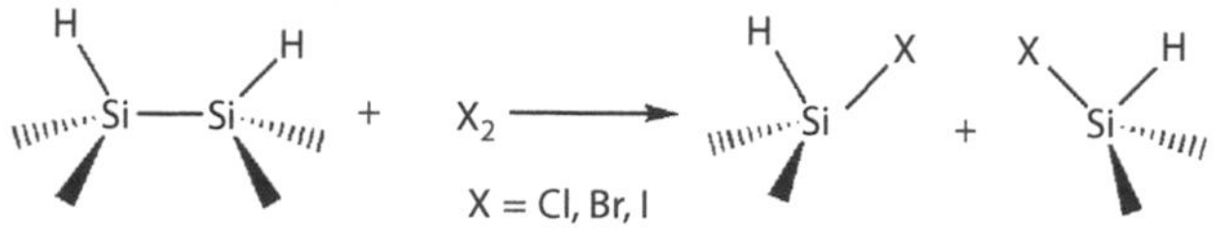

**FIGURE 13.3**    Preparation of X-PSi (X = Cl, Br, I).

(45°C for approximately 8 h) to induce the decomposition of iodoform (CHI$_3$) in dry toluene, as a means of generating iodine radicals. The generation of iodine radicals takes place according to

(13.1)
$$CHI_3 \xrightarrow{hv} CHI_2^* + I*$$

(13.2)
$$Si–H + I* \rightarrow Si* + HI$$

(13.3)
$$Si* + CHI_3 \rightarrow Si–I + CHI_2^*$$

(13.4)
$$Si* + I* \rightarrow Si–I$$

(13.5)
$$I* + I* \rightarrow I_2$$

(13.6)
$$2CHI_2^* \rightarrow C_2H_2I_4$$

In contrast to H-PSi iodination with I$_2$(g) (Lauerhaas and Sailor 1993), the reaction of H-PSi with CHI$_3$ takes place with Si–H consumption, in accordance with a free-radical mechanism.

Chlorine-terminated porous silicon (Cl-PSi) surfaces were prepared by ultraviolet (UV) irradiation of H-PSi in the presence of Cl$_2$ gas (Gun'ko et al. 2003). The Cl-PSi surface was exploited for the covalent grafting of ferrocene groups through the reaction with lithiated ferrocene in THF at the ambient temperature. The formation of Cl-PSi was also described in several reports, although the presence of Si–Cl bonds was not evidenced, but only hypothesized through Si–H consumption and rapid oxidation of the modified surface as compared to the parent H-PSi surface. Different conditions such as exposure of H-PSi to CCl$_4$ vapor for 6 h at 150°C (Lavine et al. 1993), boiling in CCl$_4$ for 30 min (Hory et al. 1995), and chemical treatment with trichloroethylene (0–10 min) were found to promote hydrogen removal (Seo et al. 1993). This process led to a significant decrease of the PL intensity as compared to that of the initial H-PSi via introduction of nonradiative surface defects.

Partially halogenated PSi surfaces were produced through the direct reaction of H-PSi with alkyl halides under microwave irradiation (Guo et al. 2005; Petit et al. 2008). The presence of bromine (Br 3p at 75.6 eV and Br 3d at 188.7 eV) in the XPS spectra of H-PSi functionalized with alkyl bromides under microwave irradiation was attributed to Si–Br formation (Guo et al. 2005). Petit et al. (2008) found that the direct reaction of H-PSi with bromodecane at 120°C for 30 min under microwave irradiation produced a surface terminated with decyl groups and silicon oxide (Figure 13.4a). Indeed, FTIR analysis showed vibrations due to C–H stretching of the alkyl chain

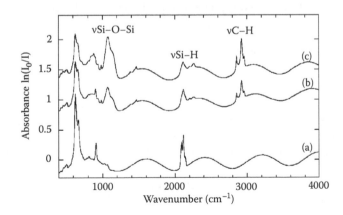

**FIGURE 13.4**   Transmission-mode FTIR spectra of freshly prepared PSi surfaces before (a) and after chemical derivatization with bromodecane under microwave irradiation for 30 min at 120°C (b) and 30 min at 150°C (c). (Reprinted with permission from Petit A. et al., *J. Phys. Chem. C* **112**, 16622, 2008. Copyright 2008. American Chemical Society.)

around 2900 cm$^{-1}$ and peaks due to Si–O–Si stretching at 1100 cm$^{-1}$ (Figure 13.4b). Peaks associated with chemical oxidation of the PSi surface were also observed at 2200 and 2250 cm$^{-1}$ and unambiguously assigned to stretching of Si–H bonds bearing oxygen in their back-bonds with different oxidation states (Petit et al. 2008). Increasing the temperature to 150°C and keeping the time constant yielded a comparable surface with a higher concentration of grafted alkyl chains and a significant surface oxidation. One can notice the complete disappearance of the SiH$_2$ scissor mode at 910 cm$^{-1}$ (Figure 13.4c). A similar behavior has been observed during the thermal treatment of $p$-type hydrogen-terminated crystalline silicon Si(111) with tetradecyl bromide (Fellah et al. 2004). Alkyl iodides gave comparable results.

The reaction of freshly prepared PSi-H with alkyl halides under microwave irradiation takes place with Si–H consumption and surface oxidation. Different reaction pathways are expected to lead to such a result in a parallel process.

The first process would lead to grafting of the alkyl part of the reactant according to

$$(13.7) \qquad \equiv\text{Si–H} + \text{RX} \rightarrow \equiv\text{Si–R} + \text{HX}$$

The second one would lead to the formation of Si–X bonds, which are prone to hydrolysis and responsible for the surface oxidation

$$(13.8) \qquad \equiv\text{Si–H} + \text{RX} \rightarrow \equiv\text{Si–X} + \text{RH} \xrightarrow{\text{H}_2\text{O}} \text{SiO}_2$$

However, these two processes may take place simultaneously because of a homolytic decomposition of the alkyl halide to yield both alkyl and halide radicals. These radicals are able to react with the Si–H bonds terminating the PSi surface by hydrogen abstraction followed by chemical bond formation of Si–R or Si–X (according to the previous equations). Because of the high polarity of the Si–X bond, simple exposure to ambient air or solvent rinsing will lead to surface oxidation. It is well known that silicon surfaces terminated with halogens react almost spontaneously with moisture present in air to yield an oxidized surface.

The reaction of free silyl radicals with organic halides in solution is well documented in the literature (Chatgilialoglu 1995). Silanes are used as effective reagents to reduce alkyl halides to the corresponding alkanes in a free radical process. In other words, reduction of alkyl halides with silanes leads to the formation of silyl halides in a free-radical process or under metal catalysis (Boukherroub et al. 1996). If a parallel exists between molecular solution and the surface-related chemistry, one would expect to obtain a fully halogenated PSi surface when the PSi-H surface is reacted with alkyl halides. However, the presence of both alkyl chains and oxidized silicon-silicon bonds (resulting most likely from hydrolysis of Si–X bonds) on the PSi surface excludes the reaction pathway proposed for free-radical reduction of alkyl halides with silanes. This is in agreement with the conclusions of our previous report (Fellah et al. 2004).

### 13.2.4 OXIDATION OF H-PSi SURFACES

A common surface stabilization method for PSi is oxidation, which is usually performed thermally in gas phase (Pap et al. 2004; Jarvis et al. 2010). It enables an efficient coverage of most of the surface area, provides a relatively stable chemical termination, and confers hydrophilic properties to the PSi surface (Björkqvist et al. 2003). Once oxidized, the SiO$_2$ layer readily dissolves in aqueous media (Dancil et al. 1999), and surfactants or nucleophiles accelerate the process (Bjorklund et al. 1997; Canaria et al. 2002a).

Si–O bonds are easy to prepare on PSi by oxidation, and a variety of chemical or electrochemical oxidants can be used. The most used is the thermal oxidation in air that tends to produce a relatively stable oxide layer (Petrova-Koch et al. 1992), in particular if the reaction is performed at temperatures >600°C (Pacholski et al. 2005). Room temperature oxidation such as UV/ozone treatment forms a more hydrated oxide that dissolves quickly in aqueous media. Nitric acid is also used as a strong oxidant (Xu et al. 2004; Kolasinski 2005). Milder chemical oxidants such as dimethyl sulfoxide (DMSO) (Song and Sailor 1998a), benzoquinone (Harper and Sailor 1997), or pyridine (Mattei et al. 2000) can also be used for this chemical transformation. Mild oxidants are sometimes preferred because they can improve the mechanical stability of highly PSi films, which are typically quite fragile. Electrochemical oxidation, in which a PSi sample is anodized in the presence of sulfuric acid, leads to a stable oxide layer (Létant et al. 2000).

## 13.3  CHEMICAL ACTIVATION/PASSIVATION OF PSi SURFACES

The surface chemistries of silicon and silicon dioxide surfaces are well known and fully described in the literature (Buriak 2002; Haensch et al. 2010). The main class of chemical reactions presented in this chapter are adaptable to any silicon surface morphology (nonexhaustive list), that is, flat Si, silicon nanowires, nano- or micropillars, PSi, silicon particles, and PSi particles. Silanization and reaction with organophosphonate-based compounds can also be used for 2D metal oxide surfaces or structured ones (0D or 1D objects).

The chemical modification of PSi surfaces can be achieved depending on their surface termination. All these reactions permit the activation/passivation of PSi by introducing a desired chemical function or tail group on the PSi surface and thus conferring it new surface wetting properties, surface resistance against corrosion, oxidation, degradation, and stability of probe immobilization under experimental conditions over time. The tail group is introduced via chemical compounds composed by an anchoring moiety (silane [silanization] or alkene/alkyne [hydrosilylation]), a spacer (alkyl, aryl, ethylene glycol chains, etc.), and the tail group. Furthermore, by controlling PSi surface functionalization and its wetting properties, it is also possible to inhibit nonspecific interactions of proteins and other species that can lead to opsonization or encapsulation, biocompatibility, blood circulation, and increase lifetime imaging (Anglin et al. 2008; Gu et al. 2013). The reaction of polyethylene glycol (PEG) linkers on a PSi surface has been employed to this end. The distal end of the PEG linker can be modified to allow coupling of other species, such as drugs, cleavable linkers, or targeting moieties to the material (Kilian et al. 2007).

## 13.4  MODIFICATION OF NATIVE OXIDE SiO$_x$/PSi (Si−O−R)

### 13.4.1  SILANIZATION REACTION

The silanization reaction involves linking molecules through the intermediate oxide sheath that typically surrounds air-exposed PSi surfaces, with the molecule being anchored via siloxane bonds (Si−O−Si) (Figure 13.5). Silanes are more commonly used on oxide surfaces, where they can covalently bind to the surface by the transfer of a proton from the surface hydroxyl group to a silane leaving group, eliminating an alcohol (in the case of methoxy or ethoxysilanes) or HCl (in case of chlorosilanes).

However, care must be taken to limit formation of 3D silane networks by siloxane cross-linking that can predominate over surface attachment (Silverman et al. 2005). Indeed, the degree of siloxane cross-condensation depends critically on the water content of the deposition solvent. One method to overcome organosilane condensation on the silicon surface is to perform the reaction in the vapor phase (Densmore et al. 2009; García-Rupérez et al. 2010; Shi et al. 2012). Indeed, a good comparison between vapor and liquid phase silanization was made by Hunt et al. (2010) for the functionalization of flat SiO$_x$/Si with APTES and found that chemical vapor deposition provides more ordered monolayers. However, such a protocol cannot be applied to all silane compounds because it is dependent on their vapor pressure.

In addition, if the number of available surface OH groups is limited, low yields of direct surface attachment can result (Gawalt et al. 2003). For this reason and immediately before silanization, PSi should be cleaned with oxidant media to remove organic pollutants and to increase the hydroxyl density on the surface ($\approx 10^{15}$ cm$^{-2}$) (Aswal et al. 2006). A cleaning process to generate reactive hydroxyl groups is critical for the effective immobilization of silanes. There are several types of Si−OH groups that can be formed on SiO$_x$ surfaces. Some (germinal and isolated silanols) are reactive, whereas others (vicinal silanols and siloxane groups) are not. The most widely used oxidants are oxygen plasma (De Vos et al. 2009) and piranha solution (Hunt et al.

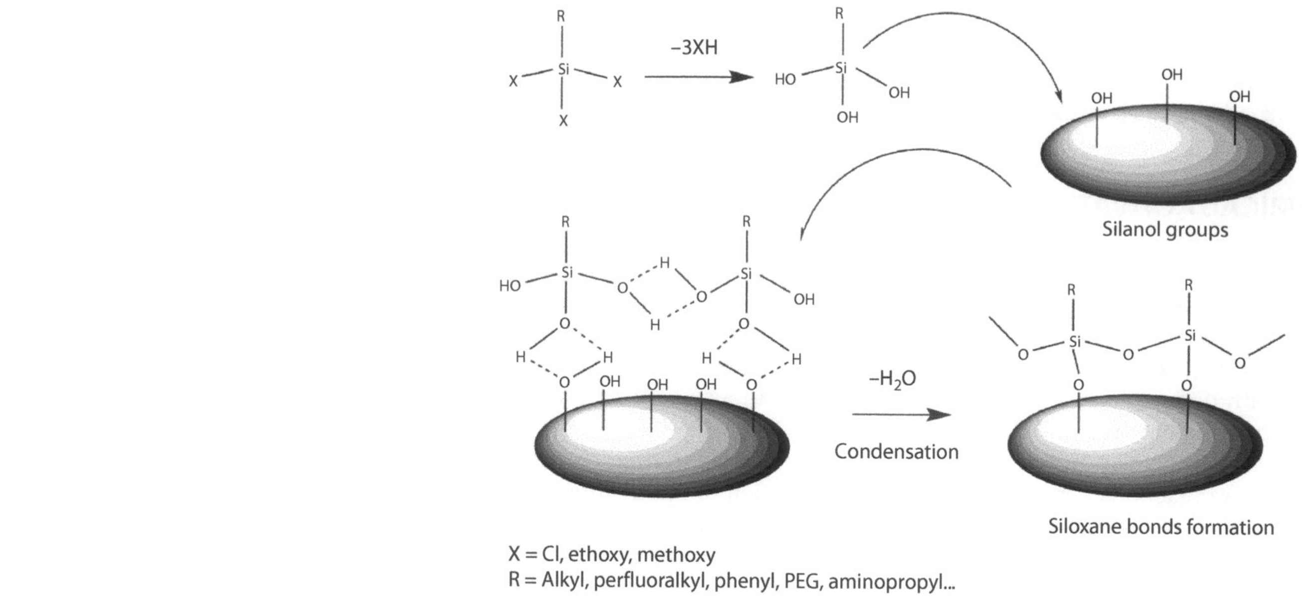

X = Cl, ethoxy, methoxy
R = Alkyl, perfluoralkyl, phenyl, PEG, aminopropyl...

**FIGURE 13.5**   Silanization reaction pathways on oxidized PSi.

2010) (consisting of a concentrated sulfuric acid/hydrogen peroxide mixture at different ratios). This treatment is well performed at room temperature or by heating, but usually for only a few minutes. The literature also describes other oxidants and cleaning agents comprising UV-ozone (Piret et al. 2010), sodium hydroxide (Ramachandran et al. 2008), ammonia/hydrogen peroxide mixture (Sharma et al. 2004; Lee et al. 2010), nitric acid, hydrochloric acid (Zlatanovic et al. 2009), sulfuric acid (Schneider et al. 2000), chromic acid (Xu et al. 2007), or mineral acids with hydrogen peroxide (Sharma et al. 2004). Sometimes more than one of these treatments is combined and sequentially applied to the surface (Zlatanovic et al. 2009).

In addition, some silane films have been shown to be hydrolytically unstable in aqueous base and in biological media (Xiao et al. 1998). This can be a challenge for applications that involve ambient conditions or a biological environment. The choice of the tail group of the organic layer depends on the targeted applications.

By tuning the anodization conditions and surface modification, the PSi surface can be made to display superhydrophobic behavior because of the obtained macro-/nanostructured porous matrix. Interestingly, altered anodization conditions can provide a surface with different wetting properties ranging from highly wetting to water-repellent surfaces. The reason for the changed wetting behavior is governed by the obtained morphology (pore size and distance between adjacent pores, pore shape, orientation, and level of branching) (Ressine et al. 2007). Surface silanization with *1H,1H,2H,2H*-perfluorooctyltrichlorosilane improves the water repellency of the PSi. After modification, water static contact angle increases with values ranging from 115°C to 175°C depending on surface morphology. The obtained structures showed good mechanical and chemical stability and exhibited no decay in contact angle value over 7 months.

Such superhydrophobic PSi surfaces can be used for sample droplet confinement, leading to high concentration of analytes on restricted areas on PSi. Then, further analysis of compounds by mass spectrometry via a new method of ionization developed by the group of Siuzdak and known as DIOS (desorption/ionization on silicon) can be performed (Wei et al. 1999). Another example of DIOS-MS application was published by Lowe et al. (2010). They have immobilized anti-cocaine antibody (Ab) on PSi for cocaine enrichment from a mixture containing different illicit drugs prior to its detection by mass spectrometry. To do so, they have silanized the PSi surface with isocyanate trimethoxysilane to introduce an isocyanate function for further reaction with amino groups on Ab (Lowe et al. 2010).

Di Francia et al. (2005) have derivatized oxidized PSi surface using trimethoxy-3-bromoacetamidopropylsilane in order to link single strand DNA (ss-DNA). Here, ss-DNA was immobilized by means of the bromoacetamido moiety.

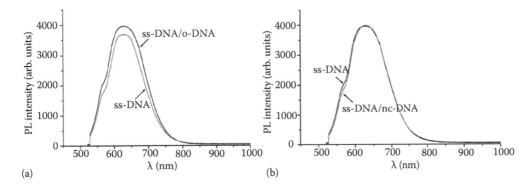

**FIGURE 13.6**  Evolution of PL following hybridation of single strained DNA (ss-DNA) with complementary DNA (c-DNA) (a) and noncomplementary DNA (nc-DNA) (b).

They have found that the derivatized samples exhibit a photoluminescence that is stable in time and is not modified after exposure to noncomplementary DNA strand. On the other hand, a sensible enhancement of the light emission has been observed when the derivatized samples react with the complementary strand, showing that the specific ss-DNA/complementary DNA (c-DNA) interaction can be optically detected without using further labeling steps (Figure 13.6) (Di Francia et al. 2005).

Recently, De Stefano et al. (2013) have performed direct solid phase synthesis of oligonucleotides on aminated PSi. They have investigated the passivation ability of oxidized PSi multilayered structures by two aminosilane compounds, 3-aminopropyltriethoxysilane (APTES) and 3-aminopropyldimethylethoxysilane (APDMES), for optical label-free oligonucleotide biosensor fabrication. Despite this not negligible distinction, they found that both silane-modified PSi devices showed good chemical resistance to reagents used for *in situ* synthesis, phosphates deprotection, and hybridization of a polythymine oligonucleotide. In particular, APDMES passivation, owing to less steric hindrance of the pores, resulted in a better functionalization quality with respect to APTES. Finally, hybridization with a target sequence of oligonucleotide synthesized on APDMES modified porous silica optical structure has been demonstrated by complementary optical techniques, such as spectroscopic reflectivity and fluorescence microscopy (De Stefano et al. 2013).

## 13.4.2  REACTION WITH ORGANOPHOSPHONATES

Although the dominant strategy for the functionalization of oxide-based surfaces is based on silanization, metal-oxide ($Al_2O_3$, $TiO_2$) (Silverman et al. 2005; Hauffman et al. 2008) or silicon oxide ($Si/SiO_2$) surfaces can also be chemically modified with organophosphonates (Midwood et al. 2004). The chemical modification with organophosphates provides stable and elegant systems that can be used to bond biological systems to native oxide surfaces and thus obviates many disadvantages of silanization such as limited hydrolytic stability, low surface coverage, critical dependence on available hydroxyl binding sites on the $SiO_2$, and the intrinsic risk of multilayer formation (see Section 13.4.1) (Hanson et al. 2003; Silverman et al. 2005). Phosphonate organic layer formation involves two steps: first, the phosphonic acid is adsorbed on the oxide surface and then converted to a phosphonate organic layer by heating at 120–140°C (Figure 13.7). This method is called "T-BAG" and was developed by Hanson et al. (2003).

In contrast to silanization, where only surface OH groups react, both surface OH and bridging surface oxide groups can react during this process. Phosphonate layers adhere strongly to the substrate surface and are homogeneous and versatile for further chemical modification. These organophosphonates have superior physicochemical properties. Relative to silanes, phosphonate layers can form densely packed monolayers with higher surface coverage (Cattani-Scholz et al. 2009) and are much more stable in both acidic and alkaline solutions (Silverman et al. 2005; Hoque et al. 2006; Dubey et al. 2010) and in electronically active environments (McDermott et al. 2007). Previous studies have demonstrated the efficacy of phosphonate chemistry in the

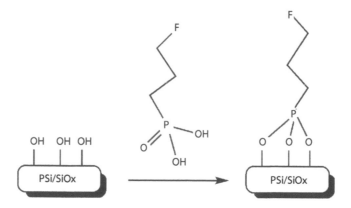

**FIGURE 13.7** Functionalization of silicon oxide surface using phosphonic acid derivatives (F = functional group).

fabrication of complementary circuits and transistors (Klauk et al. 2007; Zschieschang et al. 2008), modification of DNA biosensors (Cattani-Scholz et al. 2009), and preparation of cell adhesion substrates (Midwood et al. 2004; Adden et al. 2006; Luo et al. 2008).

However, to the best of our knowledge there is no example in the literature of such PSi functionalization, despite the huge potential of such functionalization. Cattani-Scholz et al. (2008) have used such a chemical strategy to introduce hydroxyl groups onto silicon nanowire surface using 11-hydroxyundecylphosphonate (HUP). Then, thiolated PNA molecules were immobilized by using a maleimide heterobifunctional cross-linker for label-free detection of DNA via electrical measurements. They have shown that such modification can be an interesting alternative to native oxide modification through silanization (Cattani-Scholz et al. 2008). In addition, Shang et al. (2012) have demonstrated a conjugation strategy based on an organophosphonate surface coating to functionalize silicon resonators for sensing carbohydrate-protein interactions.

## 13.5 FUNCTIONALIZATION OF HYDROGEN-TERMINATED PSi SURFACES THROUGH Si–C BOND FORMATION

Generally, freshly prepared PSi surfaces are terminated with silicon-hydride (Si–$H_x$) bonds. Even though the hydrogen-termination confers good electronic and optical properties to the material, Si–$H_x$ are prone to oxidation under ambient conditions and are reactive toward various chemicals, leading to the deterioration of the PSi properties. Thus, it is mandatory to passivate the PSi surface through functionalization and stabilization treatments (Canham et al. 1991; Canham 1997; Lees et al. 2003). Various methods, including hydrosilylation, halogenation/alkylation (Grignard reaction), arylation via aryl diazonium salt, electrografting, and so on, have been developed for the formation of organic monolayers on PSi–H through Si–C bonds. All these reactions, as for oxide-based reactions, can be used for different types of silicon such as flat, nanowire, or porous. In contrast to silane-based monolayers, those prepared on PSi–H are highly stable thanks to the nonpolar Si-C bond. In the following section, we will describe the most adopted techniques for the formation of organic layers on hydrogenated PSi surfaces.

### 13.5.1 HYDROSILYLATION

Hydrosilylation reaction corresponds to a chemical process in which a silicon-hydrogen bond adds across a double or a triple bond. H–PSi surface chemically modified by hydrosilylation is chemically robust in harsh environments, such as boiling KOH solution (pH = 12) (Buriak and Allen 1998). This high level of stability against external media makes this type of chemical modification well adapted for biomedical applications and for designing stable sensors. The hydrosilylation reaction consists of the use of organic molecular layers bearing unsaturated alkene or alkyne bonds that can be linked directly to H–PSi as first demonstrated by Buriak (1999, 2002)

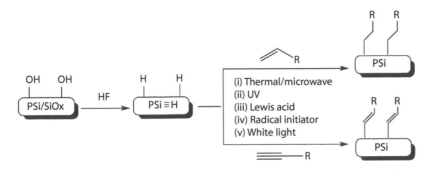

**FIGURE 13.8**  Functionalization of hydrogen-terminated PSi surface through hydrosilylation with alkenes or alkynes.

and extensively studied by Canham, Boukherroub, Zhuilof, and others (Sieval et al. 1998, 2000; Canham et al. 2000; Boukherroub et al. 2000, 2001, 2002, 2003; Mattei and Valentini 2003). The reaction can proceed under UV (Effenberger et al. 1998; Boukherroub and Wayner 1999; Cicero et al. 2000; Coffinier et al. 2005; Langner et al. 2005; Wang et al. 2010; Kolasinski 2013) or visible (Sun et al. 2005; Scheres et al. 2009) light irradiation, thermal (150–200°C) (Linford and Chidsey 1995; Sung et al. 1997; Sieval et al. 1998, 1999), by peroxide activation (radical initiator) (Linford et al. 1993; Linford et al. 1995), or by Lewis acid catalysts (Boukherroub et al. 1999; Buriak and Allen 1999; Buriak et al. 1999). All these processes lead to the formation of strong Si–C bonds without an intervening oxide (Figure 13.8). The method yields surfaces with improved stability and higher reproducibility of modification (Sung et al. 1997; Stewart and Buriak 1998; Boukherroub and Wayner 1999; Cicero et al. 2000; Strother et al. 2000a,b; Buriak 2002; Lin et al. 2002; Yang et al. 2002).

However, the two main and most used activation means for hydrosilylation of H–PSi surfaces are photo- and thermal activations.

### 13.5.1.1 PHOTO-ACTIVATION

Buriak (2014) published an interesting paper based on the variety of mechanisms involved during light-promoted hydrosilylation of flat and nanostructured (particles and porous) silicon surfaces. Even though the results are the same, for example, formation of Si–C bonds, light-promoted hydrosilylation (white and UV) performed under gas or liquid phases are achieved via different mechanisms. In addition, light-promoted hydrosilylation can involve several mechanisms in the same chemical reaction (Buriak 2014). Among the mechanisms, we can find Si–H cleavage on UV irradiation (185 and 254 nm). In that case, photon energy is sufficient to produce a dangling bond (Si°, radical) which can then react with α-carbon of an alkene molecule. Then, the surface coverage is performed via a radical mechanism. However, Effenberger et al. (1998) have shown that the lowest wavelength (higher photon energy) did not lead to the highest coverage and that the best one was obtained at 385 nm and at 60°C. The same mechanism is involved in light-promoted hydrosilylation of H–PSi surfaces (Hua et al. 2005).

White light was also used to promote hydrosilylation of PSi. That was shown in 1998 on photo-luminescent H–PSi yieding alkyl and alkenyl functionalized surfaces (Stewart and Buriak 1998). As white light is not sufficiently energetic to cleave Si–H bond, the mechanism proposed is that an exciton is formed, that is, a hole/electron pair, thanks to the absorption of photons by the nanocrystallites inside PSi network (Figure 13.9). Then, the attack of photogenerated holes on the surface (silylium ion Si⁺) in a nucleophilic fashion by primary aliphatic alkenes or alkynes was proposed as a possible mechanistic route to Si–C bond formation. It is only effective when PSi exhibits strong photoluminescence (Stewart and Buriak 2001).

Further evidence for an exciton-based mechanism was provided through the addition of low concentrations of energy- or charge-quenching agents that would rapidly annihilate the exciton before it could become involved in hydrosilylation (Stewart and Buriak 2001). In addition, if oxidizing agents, that is, compounds extracting electrons from electron/hole pair, are added to the reaction media during white light promoted hydrosilyation, an increase of the rate (speed) and

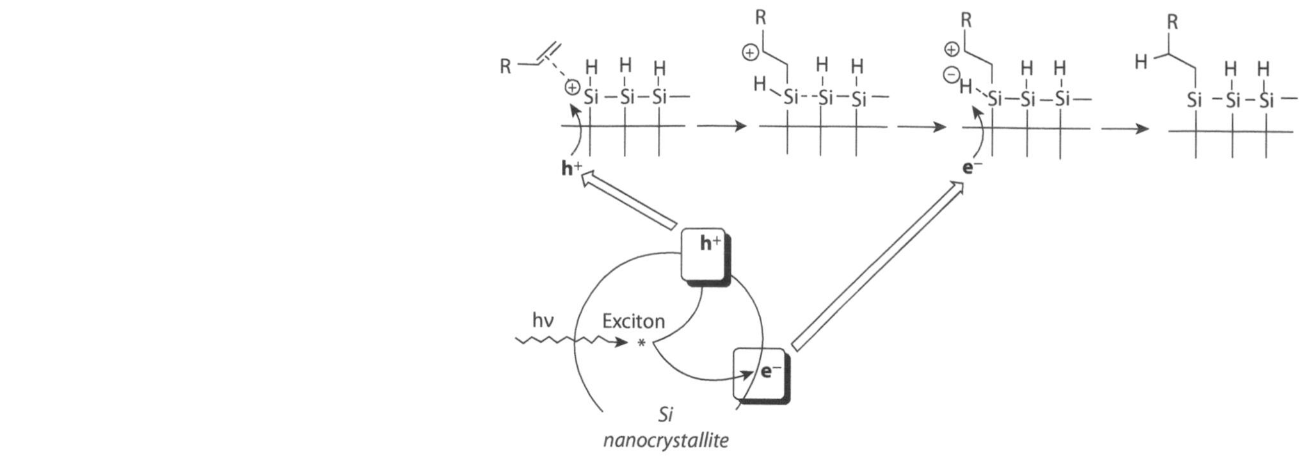

**FIGURE 13.9**    Mechanism proposed for light-promoted hydrosilylation of an alkene on H–PSi.

a yield (surface coverage) of reaction are observed. Such reaction was also achieved on flat and Si nanocrystals with higher reaction time (>10 h) needed for flat Si (Sun et al. 2004; Kelly and Veinot 2010). However, a similar mechanism for PSi surface was then proposed and confirmed by the group of Chabal by a nice study (Langner et al. 2005). They used two H-terminated flat silicon surfaces: one was a hydrogenated silicon surface and the other was an Si(100/SiO$_2$) surface terminated by Si–H through silanization. In this latter case, no hydrosilylation is expected if the grafting occurs via an exciton-based mechanism. Indeed, since no electron/hole pair can pass through the dielectric layer (SiO$_2$), the exciton-based reaction is prevented. However, on flat Si, the reaction mechanism was refined to involve radical and hole instead of electron and hole for PSi (Sun et al. 2005).

Guan et al. (2012) presented a generic strategy toward achieving depth resolved functionalization of the external and internal porous surfaces by a simple change in the wavelength of the light being used to promote surface chemical reactions. UV-assisted hydrosilylation, limited by the penetration depth of UV light, is used to decorate the outside of the mesoporous structure with carboxylic acid molecules, and white light illumination triggers the attachment of dialkyne molecules to the inner porous matrix (Guan et al. 2012).

Another mechanism, proposed by the group of Hamers, consists of photoemission-driven hydrosilylation. In that case, UV photon energy ($\lambda$ = 254 nm) is sufficient to extract an electron directly from the conduction band followed by its capture by an electron acceptor molecule in the solution phase. As a result, a positive charge remains on the substrate that can be attacked by the alkene, leading to a silicon-carbon bond formation. Electron acceptors, present as additive in solution (trifluoroacetamide, phenylchloride, paradichlorobenzene, etc.) or as part of alkene to graft (trifluroacetamide-alkene) presenting lower energy levels, led to better rate and coverage reactions than in the absence of those low energy level electron acceptor groups (Wang et al. 2010; Huck and Buriak 2012).

Advantages of light-promoted hydrosilylation are that it can be performed at room temperature, and is ideal for temperature-sensitive devices where silicon has to be passivated or functionalized. In addition, chemical photopatterning can be achieved through a classical optical mask used in the photolithography process to control the location of specific functions over the surface (Stewart and Buriak 1998).

### 13.5.1.2 THERMAL ACTIVATION

Thermally initiated hydrosilylation is widely used for organic functionalization of H–PSi. It is not influenced by particle size and shape; it affords comparatively high yields and there is no need to remove trace catalyst impurities (Yang et al. 2012). A commonly accepted thermal hydrosilylation mechanism requires sufficient heat ($T \geq 150^\circ$C) to homolytically cleave Si–H bonds to create silyl radicals that subsequently react with neat alkenes or alkynes (Figure 13.10). Once initiated, it is assumed that the reaction propagates via a surface chain reaction similar to that proposed for bulk systems (Linford et

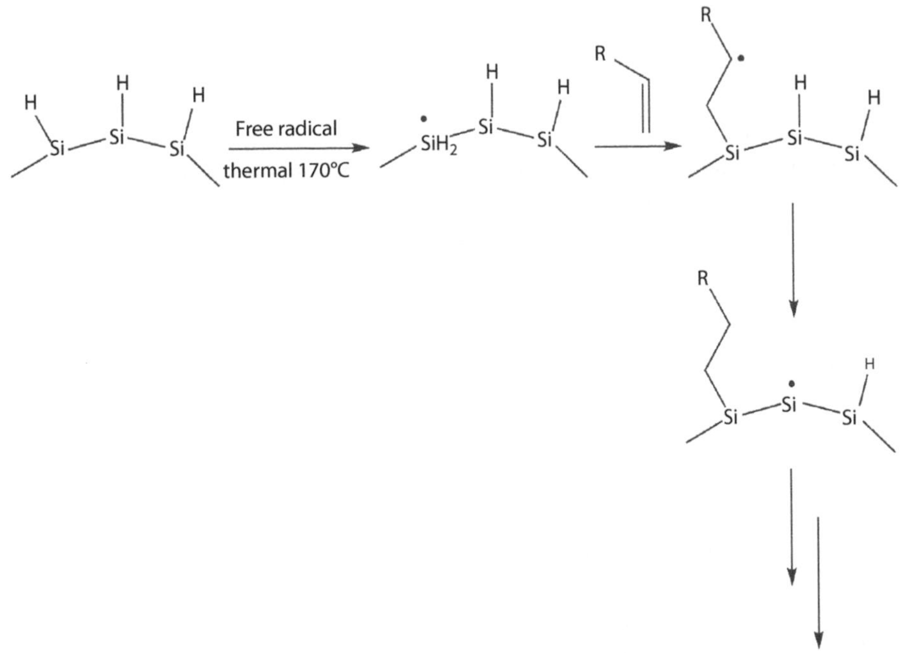

**FIGURE 13.10**    Mechanism proposed for thermal-activated hydrosilylation of alkene on H–PSi.

al. 1995). However, thermal hydrosilylation can also occur at much lower temperatures (e.g., 110°C), suggesting Si–H thermal cleavage is not the only mode of activation. Woods et al. (2005) have invoked a kinetic model for hydrosilylation in which reaction initiation occurs via hydrogen abstraction by trace oxygen to explain low-temperature initiation by creating silyl radicals.

Some examples of PSi surfaces produced by the thermal method are depicted in Figure 13.11. Coffinier et al. (2013) reported on the wetting properties of silicon-based materials as a function of their roughness and chemical composition. The investigated surfaces consist of hydrogen-terminated and chemically modified atomically flat crystalline silicon, porous silicon, and silicon nanowires. The hydrogenated surfaces have been functionalized with 1-octadecene or undecyle-nic acid under thermal conditions. Whereas the roughness of all considered surfaces was measured by atomic force microscopy (AFM), the changes occurring upon surface functionalization

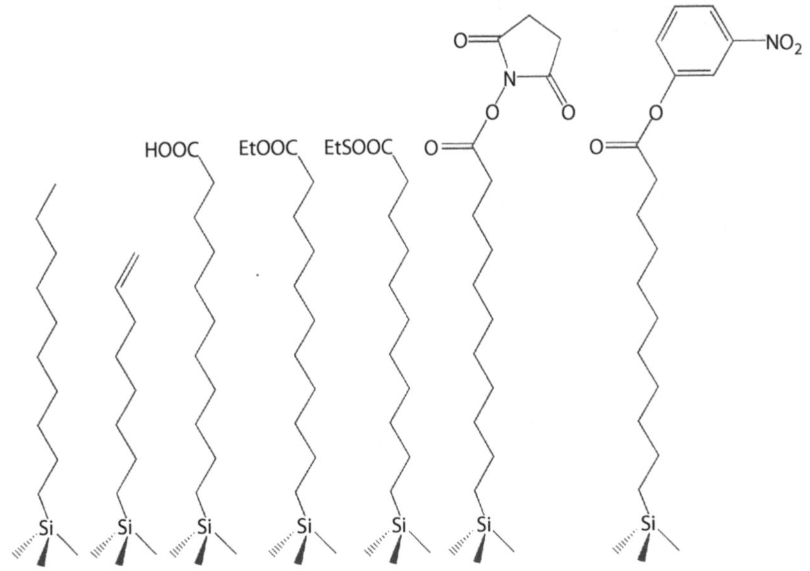

**FIGURE 13.11**    Examples of PSi surfaces produced through thermal reaction of H–PSi with simple and functional alkenes and dienes.

were characterized using Fourier transform infrared (FTIR) spectroscopy, X-ray photoelectron spectroscopy (XPS), and static water contact angle (SWCA) measurements. By increasing the surface roughness, the SWCA increases. The combination of high surface roughness with chemical functionalization with water-repellent coating (1-octadecene) enables reaching superhydrophobicity with SWCA greater than 150° for silicon nanowires and 105° and 122° for flat Si and PSi, respectively. Undecylenic acid reaction with flat silicon, PSi, and silicon nanowires led to SWCA values of 58°, 60°, and 124°, respectively (Coffinier et al. 2013).

To reduce nonspecific adsorption of biomolecules, oligo(ethylene glycol) (EG) moieties can be incorporated into the monolayer by different chemical strategies (single or multistep chemical modification) using thermal activation. It was shown that direct grafting of OEG compounds by hydrosilylation on PSi could be achieved to limit protein biofouling on the optical PSi-based sensors (Kilian et al. 2009).

### 13.5.1.2.1 Microwave-Assisted Thermal Hydrosilylation

An alternative route to classical thermal hydrosilylation is the use of microwave activation. Microwave chemical hydrosilylation is based on the irradiation of H–PSi surface in neat alkene with a microwave source. The reaction is faster than the one promoted by thermal hydrosilylation minimizing the risk of the reoxidation of PSi. Boukherroub et al. (2003) have intensively studied the microwave chemical modification of PSi and have successfully achieved H–PSi functionalization with alkene, aldehyde, and alkyl halide molecules. Although this microwave activated reaction resulted in higher reaction efficiency than the classical thermal hydrosilylation, a specific equipment is needed (microwave furnace) and safety rules should be respected and performed with appropriate protection equipment.

### 13.5.1.2.2 Sonochemical Hydrosilylation

The acoustic cavitation phenomenon at the solid–liquid interface induces the bubble collapsing causing intense localized heating and producing hot spots where the temperature could increase up to 5000°C.

Application of ultrasound to activate or modify the rate of chemical reactions in liquid media was used and it has proven effective to speed up chemical reactions (Suslick 1990). Moreover, hydrosilylation of simple and bifunctional 1-alkenes was demonstrated by Zhong et al. (2011). The principle is the use of an ultrasonic bath in which the reaction tube is immerged. They used this mild reaction to graft undecenol and undecylenic acid. More importantly, a thermally labile and UV-sensitive alkene, bearing an activated leaving group (*N*-succinimidyl undecylenate) could also be tethered on Si without any degradation of compounds. Recently, Pace et al. (2013) used ultrasounds to fabricate and functionalize PSi microparticles with olefins (undecenoic acid and 1-dodecene) (2 h of reaction). In that case, silicon microparticles were functionalized at room temperature with no need for heating or UV treatment (Pace et al. 2013).

Results of H–PSi hydrosilylation using different activation means are summarized in Table 13.1.

### 13.5.1.3 LEWIS ACID AND METAL-ASSISTED FUNCTIONALIZATION

The group of Buriak was the first to report on Lewis acid mediated functionalization of H–PSi with substituted alkenes and alkynes (Figure 13.12) (Buriak and Allen 1998; Buriak et al. 1999). In this work, the choice of $EtAlCl_2$ as Lewis acid over $AlCl_3$ was made based on its higher solubility in nonpolar solvents.

Hydrosilylation of alkynes on H-PSi produces surface-bound vinyl groups as revealed by the observation of a strong vibration in the transmission FTIR spectrum at around 1595 cm$^{-1}$, indicative of a monosilyl-substituted carbon-carbon double bond $\nu(Si_{C=C})$. To clearly demonstrate that the absorption band at 1595 cm$^{-1}$ is due to an olefinic stretch, the alkenyl-terminated PSi surface was subjected to hydroboration (Figure 13.13). FTIR spectroscopy was used to follow the changes that occurred on the PSi surface after reaction with $BH_3THF$ under $N_2$ atmosphere. The quantitative disappearance of the stretch at 1595 cm$^{-1}$ and appearance of a new stretching mode at 1334 cm$^{-1}$ due to B–O stretching frequency is clear evidence of the presence of the olefinic band.

The technique is tolerant of a wide variety of functional groups as has been demonstrated through the hydrosilylation of functional alkynes bearing nitrile, hydroxyl, and ester groups (Figure 13.14).

**TABLE 13.1    Hydrosilylation of H–PSi Using Different Activation Means**

| Hydrosilylation | Photoactivation | | Thermal | | | Lewis Acid | Heavy Metal |
| | UV | Visible | Reactor | Sono-chemistry | Microwave | | |
| --- | --- | --- | --- | --- | --- | --- | --- |
| Equipment | Lamp | Lamp | GlassW + heater | US bath | MW | GlassW | GlassW |
| Easy to handle | Yes | Yes | Yes | Yes | Yes | Yes | Yes |
| Hazardous | Yes | No | No | Yes | Yes | Yes | Yes |
| Contamination of substrate | No | No | No | No | No | Yes | Yes |
| Patterning | Yes | Yes | No | No | No | No | No |
| Reaction time | 2 h | Minutes to hours (flat Si) | 2–3 h | 2 h | Minutes | Minutes to hours (alkynes or alkenes) | Minutes to hours |
| Improved by | Electron acceptor | Electron acceptor 5-fold increase | – | – | – | – | – |
| Mechanism of activation | Photoemission, radicals | Excitons | Radical | Radical | Radical | EtAlCl$_2$ | Karstedt's catalyst, Rhodium complex |

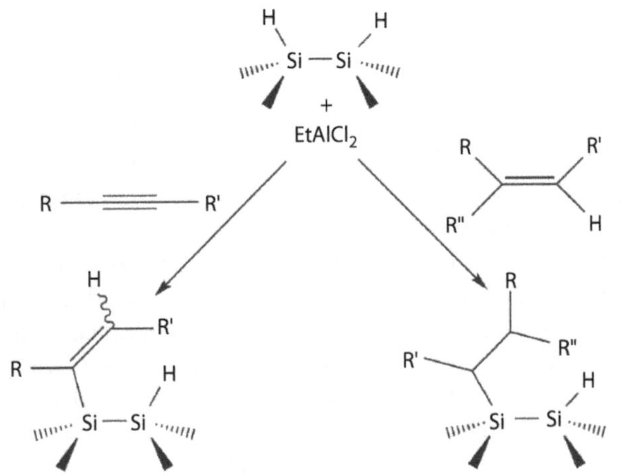

**FIGURE 13.12**    EtAlC$_2$ mediated hydrosilylation of H–PSi with substituted alkynes and alkenes.

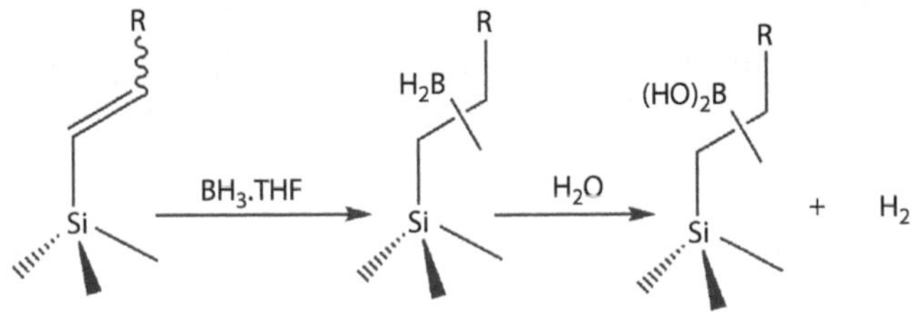

**FIGURE 13.13**    Hydroboration of alkenyl-terminated PSi surface.

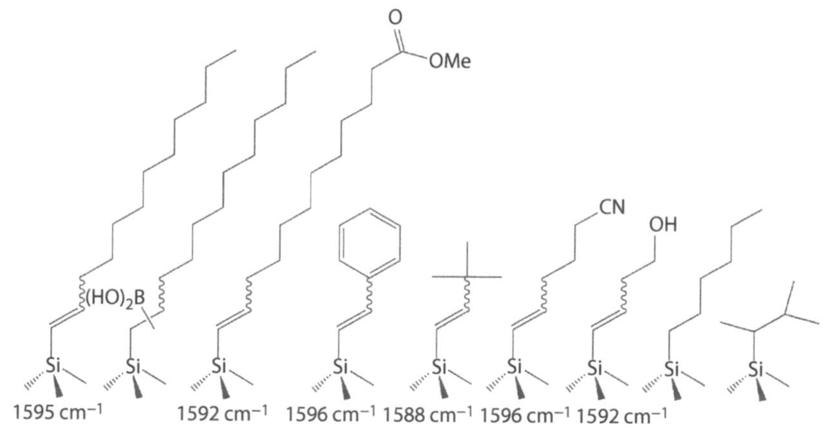

**FIGURE 13.14** Different organic layers produced through EtAlCl$_2$ mediated hydrosilylation of H–PSi with various substituted alkynes and alkenes.

The PSi surfaces functionalized with alkyl and alkenyl hydrophobic groups using this technique exhibit a remarkable stability in harsh environments such as HF/EtOH rinsing, boiling in aerated water, and boiling aqueous KOH (pH 10). Moreover, stability tests performed using chemography experiments based on silver halide reduction by emitted silane gas due to hydrolysis of an Si-SiH$_3$ bond on the PSi surface, releasing SiH$_4$, were performed on the functionalized surfaces (Buriak et al. 1999). Reduction of silver halide on photographic plates generates images in which the optical density will depend on the rate of silane release. In this study, hydrogen- and dodecenyl-terminated PSi were exposed to 100% humidity. After 1 h exposure, the hydride-terminated PSi surface heavily reduced the photographic plate, leading to a dark spot, while the dodecenyl-modified PSi surface produced a very faint trace. The results clearly indicate that the rate of silane release from the dodecenyl-terminated PSi surface is very slow, suggesting that the chemical modification consumed a large fraction of SiH$_3$ and thus reduced the rate of surface hydrolysis and oxidation.

The effect of the different surface functionalities, introduced through Lewis acid mediated hydrosilylation, on the photoluminescence (PL) properties of PSi was investigated in detail (Buriak and Allen 1999). It was found that the PSi surfaces functionalized with aromatic alkynes resulted in quantitative PL quenching (>95% quenching). Similarly, PSi surface modification with alkenes or alkynes induced about 80% PL quenching.

Transition metal-mediated processes can be also used for PSi surface modification as demonstrated by Saghatelian et al. (2001). They achieved chemical modification of PSi by using two different processes. One was based on rhodium-catalyzed metallocarbene insertion reaction of diazo compounds into exposed Si-H on a PSi surface (Figure 13.15a). This reaction was achieved in 10 min. The second method employed a platinum-based hydrosilylation reaction with alkenes (Karstedt's catalyst) and was achieved in 20 h (Figure 13.15b). Both methods created stable Si–C bonds and were highly chemoselective under mild conditions (Saghatelian et al. 2001).

The presence of alkyl species on the PSi surface was confirmed by FTIR spectroscopy. However, a substantial amount of oxidation accompanied the hydrosilylation process as evidenced by the increase in intensity of the broad absorption peak at 1070 cm$^{-1}$. Interestingly, oxygen was found to play a key role in the process. In fact, when the hydrosilylation reaction was carried out under N$_2$ atmosphere, the PSi surface remained unchanged as neither oxidation nor alkyl species incorporation was observed. It was also found that upon oxidation of the surface Si–H to OSi–H bonds (5 days at 100°C in air), the hydrosilylation occurred and under these experimental conditions the presence of oxygen is not necessary. The role of oxygen in these reactions is stabilizing Pt colloids.

### 13.5.2 HALOGENATION/ALKYLATION FOLLOWED BY GRIGNARD REACTION

An alternative approach to the formation of alkylated PSi surfaces is via alkyl Grignard (R-MgX) and alkyl or aryllithium reagents (Linford and Chidsey 1993). First, a halogenation step is performed

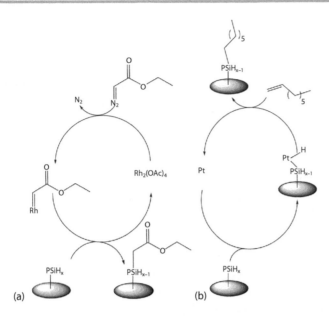

**FIGURE 13.15** Metal-assisted chemical modification of H–PSi with (a) rhodium catalyzed metallocarbene insertion of diazo compounds and (b) hydrosilylation platinum-based reaction.

to obtain (Cl, Br, I) terminated PSi surfaces that can be prepared via a two-step process: (1) initial removal of the surface oxide using aqueous HF, followed by (2) treatment with an appropriate halogenation reagent (Figure 13.16) (Bansal et al. 1996). However, it was also shown that direct reaction of alkyl Grignard and aryllithium compounds with the hydrogen-terminated silicon surface could be achieved without the need of the halogenation step (Kim and Laibinis 1998; Song and Sailor 1998b).

The reaction mechanism of the direct functionalization of H–PSi surfaces through Si–C bond formation using organolithium (R-Li) and Grignard (R-MgX) reagents involves nucleophilic attack of the Si–Si bond by the carbanion R⁻, as depicted in Figure 13.17.

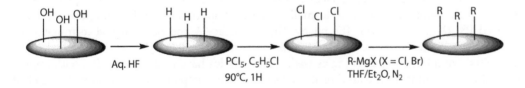

**FIGURE 13.16** Halogenation/alkylation.

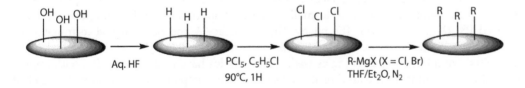

**FIGURE 13.17** Reaction mechanism for the formation of organic layers on PSi surfaces through the direct reaction of H–PSi with alkyllithium reagents at room temperature.

### 13.5.3 ELECTROGRAFTING

#### 13.5.3.1 ANODIC AND CATHODIC GRAFTING OF 1-ALKYNES

Electrografting provides another method for the direct (Si–C) covalent functionalization of silicon surfaces. The formation of alkyl and alkenyl layers via electrochemical grafting has been demonstrated on H–PSi (Robins et al. 1999). Robins et al. (1999) reported a system for electrochemically grafting terminal alkynes to H–PSi, which gives two distinct surface derivatizations, depending on the polarity of the surface bias. Cathodic electrografting (CEG) allows direct grafting of alkynes to the surface, whereas anodic electrografting (AEG) yields an alkylated surface (Figure 13.18) (Robins et al. 1999). The different mechanisms proposed are represented in Figure 13.19.

The technique is quite simple and versatile. FTIR analysis of the PSi surface functionalized with phenylacetylene through CEG reveals Si–$H_x$ stretching modes, which are broadened and decreased in integrated intensity compared to unmodified PSi. The absence of a $\nu_{(\equiv CH)}$ mode around 3300 cm$^{-1}$ and the presence of a sharp silylated alkyne $\nu_{(C\equiv C)}$ at 2159 cm$^{-1}$ is consistent with the formation of an Si-alkynyl surface and not simple physisorption. For instance, the $\nu_{(C\equiv C)}$ of 1-phenyl-2-(trimethylsilyl)acetylene appears at 2160 cm$^{-1}$ while that of phenyl-acetylene is observed at 2110 cm$^{-1}$. Figure 13.20a displays the FTIR spectrum of the pentynyl terminated PSi surface (surface 3, Figure 13.21) with the $\nu_{(C\equiv C)}$ at 2179 cm$^{-1}$. To confirm the presence of the silylated C≡C triple bond, the PSi surface was hydroborated with a 0.5 M THF solution of disiamylborane. The appearance of a broad band at 1580 cm$^{-1}$ and the concomitant consumption of the $\nu_{(C\equiv C)}$ is indicative of a silylated, borylated double bond, which was verified by hydroboration of 1-trimethylsilyldodec-1-yne and subsequent FTIR analysis [$\nu_{(C=C)}$ at 1584 cm$^{-1}$].

In contrast to CEG, the FTIR spectrum of a PSi surface modified with dodec-1-yne through AEG displays features related only to aliphatic C–H bonds (Figure 13.20b). The functionalized PSi surfaces showed good stability on boiling in CHCl$_3$ or in aqueous NaOH solution (pH 10), suggesting that the organic layers are grafted through Si–C covalent bonding.

The technique was successfully applied for H–PSi surface functionalization with various alkynes through CEG, as depicted in Figure 13.21.

Interestingly, the photoluminescence (PL) properties of the modified PSi surfaces are dependent on the chemical composition. The PL spectra of CEG samples exhibited varying intensities depending on the surface type. Surfaces 2–4 (alkynyl groups) retained between approximately 5 and 15% of the light emission, with a small red shift of approximately 10 nm relative to freshly prepared PSi ($\lambda_{max}$ = 663 nm). The phenylethynyl surface 1 and other arynyl terminated surfaces (surfaces 5 and 6) showed no light emission. Similarly, diphenylphosphinoethynyl surface 7 exhibited complete PL quenching as well. AEG samples have more intense PL, with approximately 20% remaining for the alkyl protected surface 9 as compared to the freshly etched H–PSi surface.

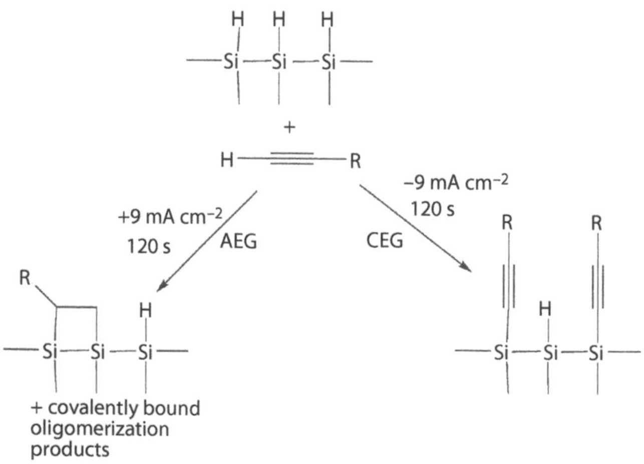

**FIGURE 13.18** Cathodic and anodic electrografting on H–PSi.

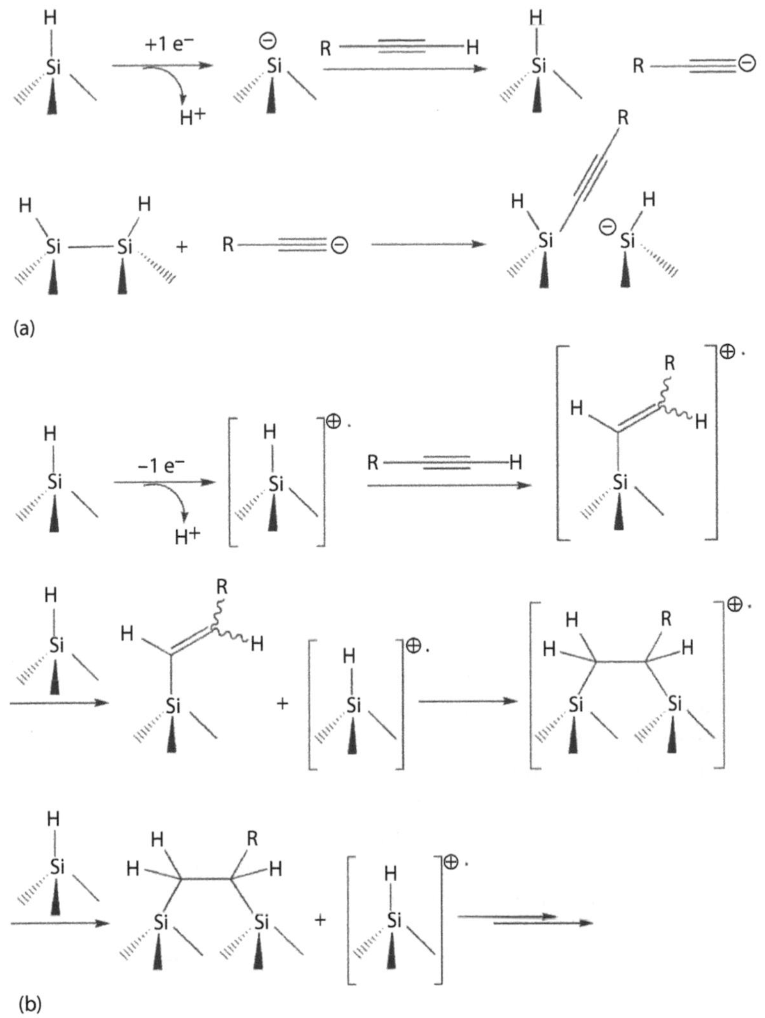

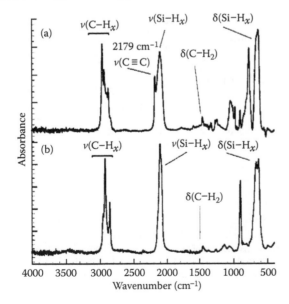

**FIGURE 13.19**   Proposed CEG (a) and AEG (b) mechanisms.

**FIGURE 13.20**   FTIR spectra of derivatized PSi with (a) CEG of pent-1-yne and (b) AEG of dodec-1-yne. (From Robins E.G. et al., *Chem. Commun.* 24, 2479, 1999. Reproduced by permission of the Royal Society of Chemistry.)

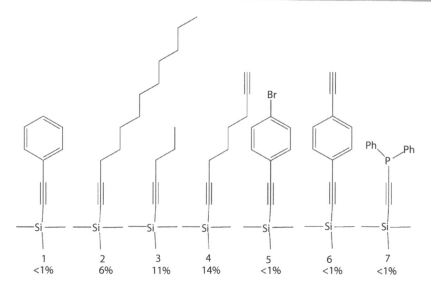

**FIGURE 13.21**    Different alkynes on PSi through CEG modifications (surfaces 1 to 7). Percentages represent the photoluminescence remaining after CEG functionalization.

Mattei and Valentini (2003) reported an *in situ* functionalization scheme of PSi during its electrochemical formation process using organic molecules that are hydrofluoric acid resistant and containing a triple terminal bond. In contrast to the technique described above, anodic polarization of PSi (100) wafers in an electrochemical cell containing an HF (50%)/EtOH solution (1:2 by volume) and a concentration of 1-heptyne of 0.72 M gives an organic layer through Si–H addition onto the C≡C triple bond (hydrosilylation) (Figure 13.22).

The absence of $\nu_{C\equiv C}$ triple bond stretching (around 2200 cm$^{-1}$) and the presence of $\nu_{C=C}$ double bond stretching (around 1600 cm$^{-1}$) accompanied by a strong decrease of the signal due to the

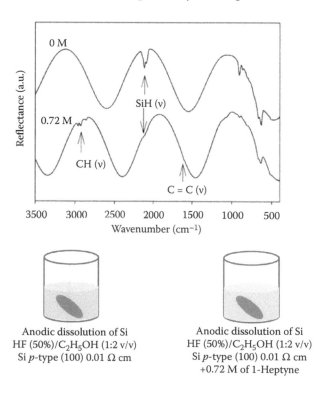

**FIGURE 13.22**    *In situ* functionalization of PSi during its electrochemical formation. (Reprinted with permission from Mattei G. and Valentini V., *J. Am. Chem. Soc.,* **125,** 9608, 2003. Copyright 2003. American Chemical Society.)

Si-H species in the FTIR spectrum of the derivatized PSi surface is good evidence of the *in situ* hydrosilylation reaction. Furthermore, the direct reaction of (0.7 M) 1-heptyne HF ethanoic solution with freshly prepared H-PSi surface for a time as long as 1000 s did not show any surface functionalization. This demonstrates that the grafting process is electrochemical in nature. A surface coverage of 50–60% was reported using this method.

The same process, that is, *in situ* functionalization of H-PSi during its electrochemical formation, was extended to 1-alkynes bearing a reactive functional group such as carboxylic acid (6-heptynoic acid) (Blackwood and Akber 2006). FTIR analysis of the modified PSi surfaces indicated that the process occurred through Si-H addition to the C≡C triple bond without affecting the terminal –COOH groups, as previously reported for the thermal reaction of undecylenic acid with H-PSi (Boukherroub et al. 2002, 2003). The functionalized PSi samples were subjected to additional chemical transformations to confirm that the organic molecules were covalently bonded with the Si atoms and not merely physisorbed on the PSi surface. For example, the Si–C=C double bond of the PSi surface that has been modified with 1-heptyne was successfully hydrated in acidic medium. Similarly, the COOH terminal groups of the PSi layer functionalized with 6-heptynoic acid were converted to COO–Me by reaction with methanol at a temperature >100°C.

However, some disadvantages of *in situ* functionalization were also noted, such as a reduction of the thickness and porosity of the PSi layer, compared to etched PSi in the same conditions. This resulted in a reduction of the PL intensity but did not result in a wavelength shift. Another disadvantage of the technique is related to the incomplete coverage of the PSi; Si–H stretches still being observed on the FTIR spectra. This is most likely due to the competition between the electrochemical functionalization and etching processes, with it being likely that the formed Si–C bonds are sufficiently stable and do not weaken the silicon back bonds such that the silicon atom on which they form can no longer take part in the etching process.

## 13.5.3.2 ELECTROCHEMICAL REDUCTION OF ALKYL HALIDES

Electrochemical reduction of organohalides has also been demonstrated as an effective grafting technique (Gurtner et al. 1999). Electrochemical reduction of alkyl iodides, alkyl bromides, and benzyl bromides on *p*- or *n*-type H–PSi surfaces gave organic layers covalently bonded through Si-C bonds with high surface coverage (Figure 13.23).

The method led to 20–40% Si–H consumption during the electrochemical process. The formation of organic layers through Si–C bonds was confirmed by the presence of Si–C stretching mode at 766 cm$^{-1}$ in the FTIR spectrum of the methyl-terminated PSi surface obtained by reduction of methyl iodide. Isotopic labeling ($^{13}$C and $^{2}$H) studies were further used to identify unambiguously species characteristics of the Si–C stretching vibrational modes (Canaria et al. 2002b).

The proposed mechanism relies on the reduction of alkyl or benzyl halides to produce alkyl or benzyl radical species. These radicals react with the surface Si–H bonds to generate silicon radicals. The formation of an organic layer may take place by the simple reaction of the silicon radical with an alkyl or phenyl radical as outlined in Equations 13.9–13.11. The surface silyl radical may also be reduced to silyl anion (Equation 13.12), which in turn can react with R-X (nucleophilic attack) to generate the Si-R termination (Equation 13.13). Another plausible mechanism involves

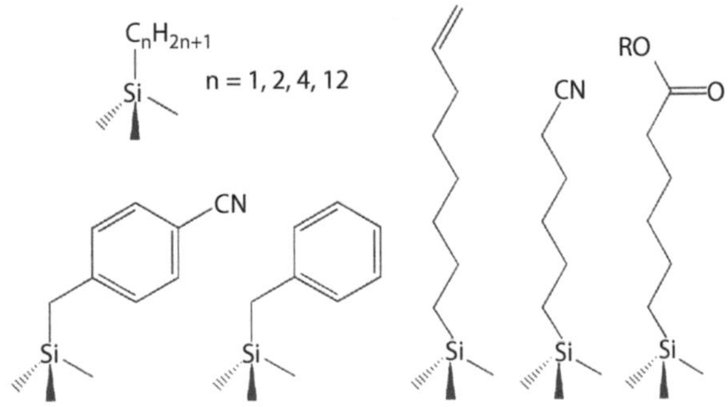

**FIGURE 13.23** Functional groups grafted onto H–PSi by electrochemical reduction of organo halides.

the *in situ* reduction of the alkyl or benzyl radicals to carbanions (R⁻), which can attack the weak Si–Si bonds to generate Si–R species and silicon anions, in a similar reaction pathway like for Grignard and alkyllithium reagents (Figure 13.17) (Equations 13.14 and 13.15).

(13.9)   $R–X \; + \; e^- \; \rightleftharpoons \; (R–X)^{\cdot-} \; \longrightarrow \; R^\cdot \; + \; X^-$

(13.10)

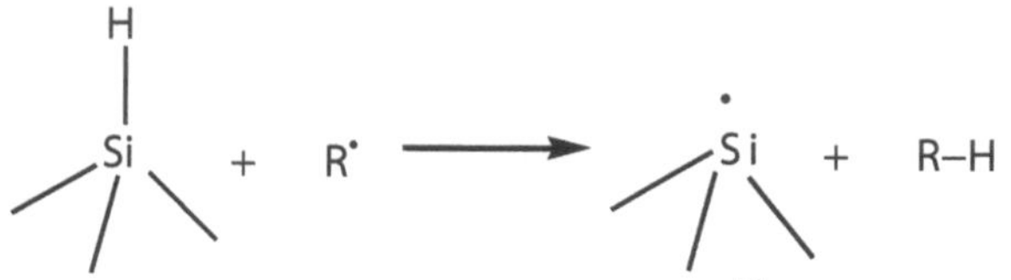

(13.11)

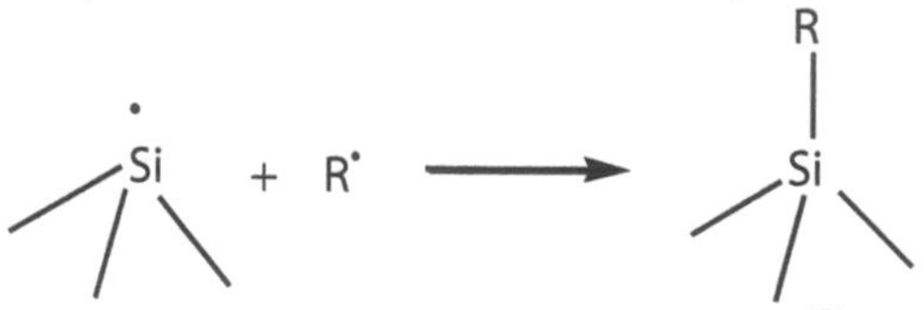

(13.12)

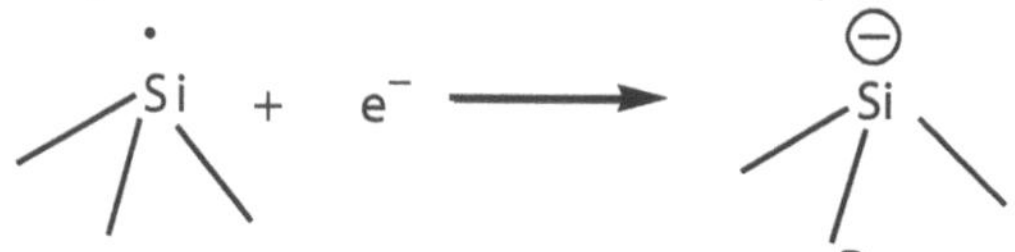

(13.13)

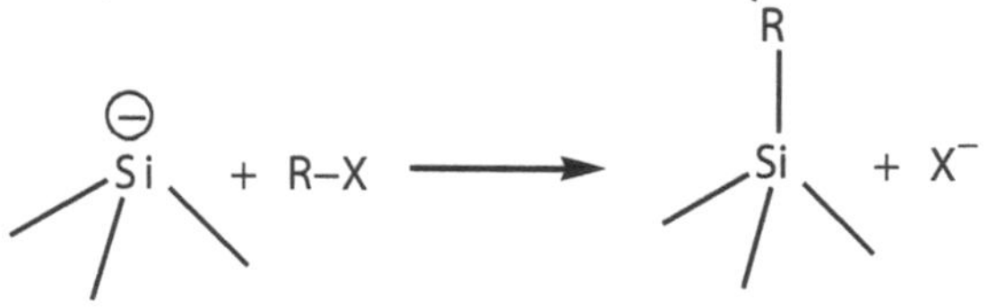

(13.14)   $R^\cdot \; + \; e^- \; \rightleftharpoons \; R–$

(13.15)

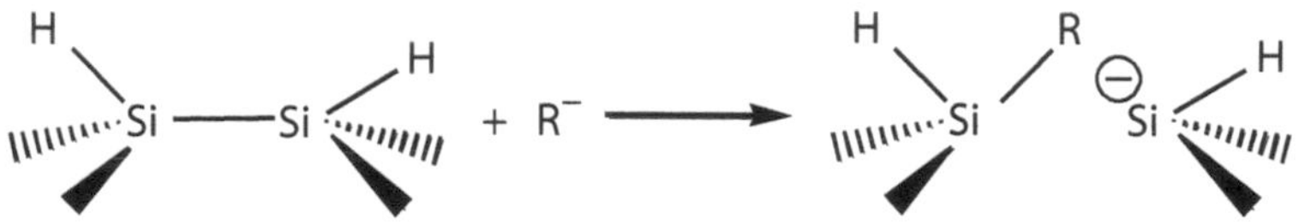

Electrochemical oxidation of methyl Grignard reagent and electrochemical reduction of phenyldiazonium salts on H–PSi have been shown to yield dense monolayers of methyl and phenyl groups, respectively (Dubois et al. 1997).

### 13.5.3.3 ELECTRON-BEAM ASSISTED Si–C BOND FORMATION

Electron beam lithography (EBL) can also be used to form Si–C bonds on PSi surface. It was shown that patterns consisting of lines of 700 nm in width of alkene molecules on a PSi surface could be achieved. In that case, an Si–H dissociation is obtained by using high electron energy (25 KeV, 5 nA, and an optimal dose of 120,000 μC/cm²). Then, subsequently, the surface was submitted to alkenes (gas or liquid phase) during a short period (~3–30 min.), allowing the grafting (Rocchia et al. 2003).

### 13.5.4 ARYLATION VIA DIAZONIUM SALT

Spontaneous reduction of diazonium salt on H–PSi can also be performed. In that case, the silicon surface is a source of electrons (from the conduction band) to reduce the diazonium salts into aryl radicals leaving behind a hole. Then, surface silicon radicals are formed on H–PSi (Figure 13.24). The key of reaction is the 1 e⁻ reduction of the diazonium cation, leading to N₂ loss, formation of the aryl radical, and a hole, or positive charge, on the silicon. This mechanism involving

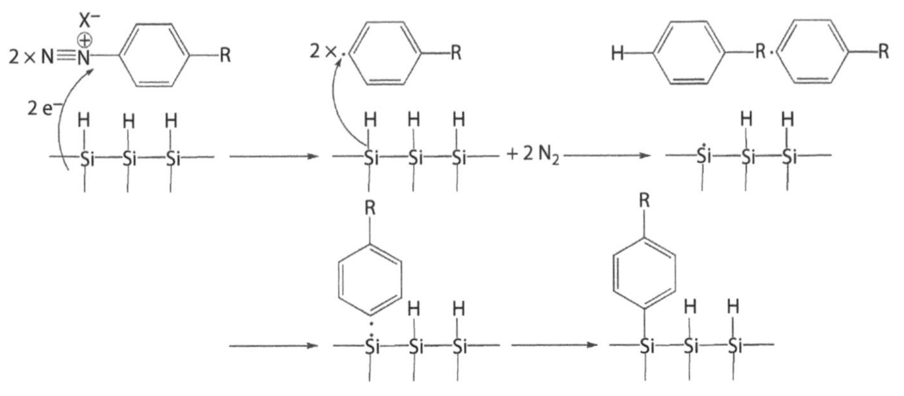

**FIGURE 13.24**    Arylation of H-PSi *via* diazonium salt.

radicals to form an Si–C bond using diazonium reagents was confirmed by Wang et al. by using radical traps (Wang and Buriak 2006).

## 13.6  THERMAL FUNCTIONALIZATION OF HYDROGEN-TERMINATED PSi SURFACE VIA Si–O–C BOND FORMATION: REACTION WITH ALCOHOLS, ALDEHYDES, AND KETONES

### 13.6.1  REACTION WITH ALCOHOLS

Glass et al. (1995) first studied the thermal reaction of a hydride-terminated PSi surface with methanol. They found that the initial reaction of methanol with the porous layer occurs only at temperatures $\geq 600$ K. At temperatures above 600 K, hydrogen desorption takes place to generate open sites for methanol reaction. In excess of $CH_3OH$ (g), incorporation of oxygen (Si–O–Si) and carbon (Si–C) were observed in addition to $Si–OCH_3$, $Si–CH_3$, and Si–H surface species. The observed Si–O–Si and $Si–CH_3$ originate most likely from the decomposition of surface $Si–O–CH_3$ species at high temperatures. Increasing the annealing temperature for the reaction of 9 Torr of $CH_3OH$ with H–PSi surfaces generated more intense Si–O–Si, Si–C, and $Si–CH_3$ features. However, when H–PSi surfaces were immersed in boiling methanol for 15 min, a slight surface methoxylation was observed whereas a significant oxidation occurred (Hory et al. 1995). The relatively high level of oxidation was assigned to the reaction of the surface with residual water present in the solvent.

Other compounds bearing alcohol function have been investigated by Kim and Laibinis (1997). Indeed, they have studied the thermal reaction of aromatic and long alkyl chain alcohols with hydride-terminated PSi surfaces. Direct reaction of the H–PSi surface in neat alcohols or a solution in anhydrous dioxane (0.1 M) at temperatures between 20°C and 90°C were performed under $N_2$ atmosphere for 0.5–24 h. The method was successfully used to graft a range of alcohols onto PSi, including phenol, 3-phenylpropanol, 10-undecenol, 11-bromoundecanol, ethyl glycolate, and ethyl 6-hydroxyhexanoate. These compounds were covalently attached to the PSi surface through Si–O–C bonds (Kim et al. 1997). FTIR analysis was consistent with the formation of $O–SiH_x$ species and a relative decrease of the $Si–H_3$ species at 2139 cm$^{-1}$ relative to the changes for Si–H (2089 cm$^{-1}$) and $Si–H_2$ species (2115 cm$^{-1}$) on reaction with alcohol. This observation suggests that the $SiH_3$ groups are either more reactive or removed from the PSi surface during reaction.

The same group has further studied mechanistic aspects of the reaction between PSi-H and alcohols by using deuterated ethanol ($CH_3CH_2OD$). Diffuse reflectance infrared Fourier transform (DRIFT) analysis of PSi surface, exposed to the deuterated alcohol for 1 h at 45°C, displayed peaks at 1517–1636 cm$^{-1}$ corresponding to Si–D bonds. The presence of Si–D vibration modes implies a reaction between the PSi–H surface and alcohols occurring with cleavage of Si–Si bonds (Kim and Laibinis 1997).

The effect of the derivatization on the photoluminescent properties of the PSi was also examined. In general, the thermal treatment with alcohols produced no change in the PL intensity or frequency, regardless of the alcohol (Kim and Laibinis 1997). For the PSi samples modified in boiling methanol, there was almost no change in the PL intensity, although a small blue shift and a noticeable increase in the decay time were observed. However, it is hard to assign precisely the efficiency of the process because the treatment was accompanied by a significant surface oxidation (Hory et al. 1995).

The thermal reaction of halogenated PSi with alcohols can also take place. In that case, the reaction was achieved in two steps: (1) exposition of PSi surface to $Br_2$ (bromination) during 5 min., followed by (2) reaction with alcohols (~10 min), leading to the attachment of the organic layers via covalent Si–O–C bonds. The mechanism also involved a breaking of Si-Si bond to form Si-X bonds instead. Then, Si-X bonds, extremely reactive, are able to achieve nucleophilic reaction with alcohols leading to organic layer grafting via Si–O–C bond formation (Lee et al. 1995).

### 13.6.2 REACTION WITH ALDEHYDES AND KETONES

The thermal hydrosilylation of an aldehyde (decanal) produced organic layers covalently bonded to PSi through Si–O–C bonds (Boukherroub et al. 2000, 2001). However, contrary to the reaction of alcohols with H-PSi surfaces, no Si–Si bond breaking occurred. The efficiency of the reaction, as determined by the $Si–H_x$ stretch that was consumed during the chemical process, was 30–50%; no apparent surface oxidation was observed during hydrosilylation reaction, as evidenced by the weak Si–O–Si stretching peak intensity.

Ketones react as well with H-PSi surfaces under microwave irradiation (Boukherroub et al. 2004). Indeed, the reaction of PSi–H with 2-decanone at 180°C for 15 min gave an organic layer covalently attached to the PSi surface through Si–O–C bonds. IR analysis showed a similar spectrum to that obtained for decanal, but with a higher degree of surface oxidation. Raman spectroscopy showed that the PSi structure was not affected by the thermal process (Boukherroub et al. 2001).

The PSi surfaces derivatized with aldehydes are chemically robust and can stand the following sequential treatments: sonication in $CH_2Cl_2$, boiling in $CHCl_3$, boiling in water, immersion in 1.2 N HCl at 75°C, and immersion in 48% HF for 65 h at room temperature without apparent chemical degradation of the monolayer as evidenced by FTIR spectroscopy. The hydrophobic character of the alkyl chain and the high surface coverage prevents the permeation and diffusion of molecules inside the porous layer (Boukherroub et al. 2001). The chemical treatment did not have a significant effect on the PL intensity and peak position. The PSi sample modified with octyl aldehyde exhibited an orange-red PL comparable to that of a hydride-terminated sample. Samples modified with decanal showed a reduction in the PL intensity of ~60%. The PL was stable during aging the derivatized PSi samples in air for several months. On the other hand, aging for 3 days in 100% humidity air at 70°C induced an increase of the PL intensity by about 40%. However, aging the freshly etched sample under the same conditions resulted in PL intensity increase by a factor of 50. This behavior was assigned to a gradual oxidation of the PSi skeleton, as evidenced by XPS and FTIR spectroscopies (Maruyama and Ohtani 1994). Treatment of the modified PSi surfaces with HF had no effect on the PL intensity, which is in agreement with an effective passivation of the surface by the organic coating layer.

## 13.7 DEPROTECTION

When bifunctional molecules such as ω-amino alkenes react with hydrogenated silicon surfaces, both functional groups—alkene and amino—will competitively react with the Si–H bonds, resulting in disordered monolayers (Sieval et al. 2001). Thus, the introduction of the required functionality can be achieved by using protecting groups. Figure 13.25 illustrates the commonly employed protecting groups. Amino or semicarbazide groups can be introduced by using a *t*-butyloxy-carbonyl (*t*-Boc) protected amine (Coffinier et al. 2005; Stern et al. 2007; Haensch et al. 2009). Alternatively, phthalimide or acetamide moieties also serve as effective protecting groups for the introduction of well-ordered amino functionalities (Haensch et al. 2009). Ester-terminated monolayers can also be easily modified to obtain a variety of functional groups: reduction with $NaBH_4$ or $LiAlH_4$ results in alcohol termination, acid hydrolysis leads to carboxylic acid formation, and

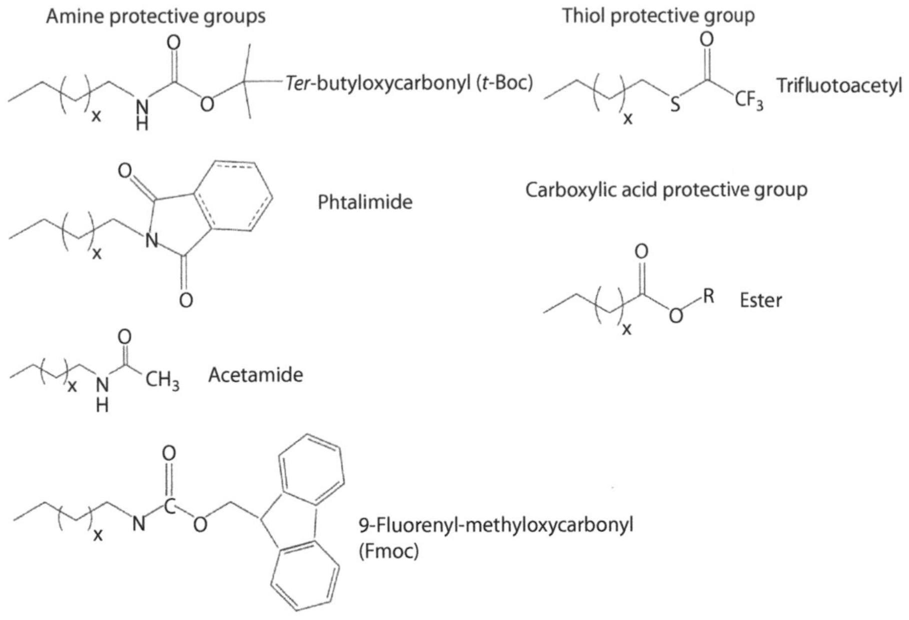

**FIGURE 13.25** Protective groups.

reaction with alkyl Grignard reagents produces a tertiary alcohol (Sieval et al. 1998; Boukherroub and Wayner 1999; Shao et al. 2009). Trifluoroacetyl groups are also effective protecting agents for thiol groups (Böcking et al. 2007). Most protective groups described here have been used on crystalline silicon, but might be used on PSi and nanowires as well.

Indeed, a recent example from McInnes and Voelcker (2012) described the preparation of several types of PSi microparticles as supports for the solid-phase synthesis of oligonucleotides. To do so, fresh PSi microparticles were hydrosilylated with fmoc-11-aminoundecene in mesitylene. Then the fmoc-protecting group was removed with 25% piperidine in DMF prior to the *in situ* synthesis of oligonucleotides.

## 13.8  POSTCHEMICAL MODIFICATION

After the functionalization of PSi (oxidized or not), a second chemical modification is often needed to proceed postmodification promoting biomolecule/probe attachment. The same strategies or cross-linkers can be used on either oxidized PSi or H-PSi because the same chemical functionalities will be present. Coffinier et al. (2012) have used NHS ester-activated PSi to immobilize amino-NTA-Ni$^{2+}$ complex ligands and performed His*6-Tag-peptide enrichment. In this process, H–PSi (1) surfaces were reacted with undecylenic acid (UA) via hydrosilylation reaction initiated under thermal conditions (~150°C) to yield carboxylic acid terminated PSi. Then, the terminal carboxylic group was converted into an amino-reactive linker, NHS-ester (2), allowing the immobilization of N-(5-amino-1-carboxylpentyl)iminodiacetic acid (NH$_2$–NTA) via amide bond formation (3). Finally, the NTA–Ni$^{2+}$ complex was formed by nickel loading (4) allowing capture of the His-tag-peptide (5) (Figure 13.26).

Sam et al. (2010) reported on a multistep route fully compatible with the mild conditions required for anchoring amino acids onto the PSi surface. A dense monolayer of carboxylic acid chains is first grafted via thermal hydrosilylation of undecylenic acid at the surface of H–PSi and subsequently transformed into succinimidyl ester termination, *N*-hydroxysuccinimide (NHS). Then the "activated ester" terminated PSi was made to react with the amino end of the amino acid, allowing for its covalent attachment through amide bond formation (Sam et al. 2010). The same chemical strategy was used for the immobilization of antibodies for the detection of B-type natriuretic peptide (BNP), an important biomarker in early diagnosis of congestive heart failure. Antibodies on PSi allowed the enrichment of BNP from human plasma and its subsequent detection by MALDI-ToF with a detection limit as low as 10 pg/mL (Chen et al. 2008).

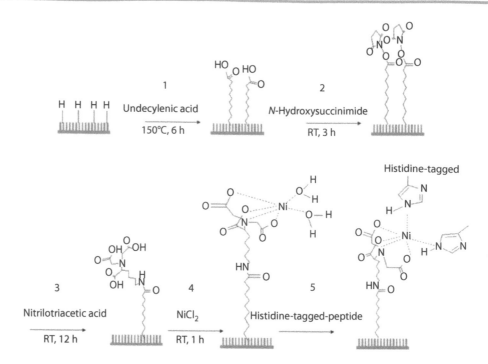

**FIGURE 13.26** Chemical pathways for the introduction of nitrilotriacetic acid compounds for metallic ions complexation and interaction with tagged peptide. (From Coffinier Y. et al., *Analyst*, **137**, 5527, 2012. Reproduced by permission of the The Royal Society of Chemistry.)

Sometimes, the active sites of a substantial population of immobilized molecules are not accessible to targets in the solution phase (Cha et al. 2005). To ensure a specific orientation on the surface, various covalent chemoselective and site-specific immobilization strategies have been developed on flat surfaces. Indeed, it has been demonstrated that the direct interface between solid surfaces and probes can affect the quality of the molecular interactions (e.g., the catalytic efficiency of an enzyme toward its substrate, or antibody–antigen recognition) (Zhu et al. 2001; Wacker et al. 2004) and that site-specific and -oriented immobilization can improve the detection (Stewart and Buriak 2001; Bonroy et al. 2006; Alonso et al. 2008). Among them, we can mention Schiff-base site-specific ligation methods such as oxime (Scheibler et al. 1999; Dendane et al. 2007) or α-oxo-semicarbazone ligations (Olivier et al. 2006), the Staudinger ligation (Kohn et al. 2003; Soellner et al. 2003), the Diels–Alder reaction (Houseman et al. 2002), the "click" chemistry (Devaraj et al. 2006), and native chemical ligation (NCL) (Wojtyk et al. 2002; Dendane et al. 2012).

However, only a few examples in the literature were reported on the site-specific immobilization of biological probes via such chemical strategies despite their advantages (Soeriyadi et al. 2014; Zhu et al. 2014). Among them, Zhu et al. (2014) have reported on PSi rugate filters modified with alkyne-terminated monolayers. By using a combination of photolithography and "click" chemistry, two chemical functionalities were obtained by conjugating ethylene glycol moieties containing two different terminal groups. From this, specific patterns containing enzyme substrate (gelatin) were achieved. Then, protease activity (Subtilisin) was monitored by optical reflectivity measurements (Zhu et al. 2014).

## 13.9 HYBRIDE PSi

As previously mentioned, one of the common methods for PSi stabilization is oxidation (Pap et al. 2004; Jarvis et al. 2010), which is usually done thermally in a gaseous phase. It enables an efficient coverage of most of the surface area and provides a relatively stable and hydrophilic surface (Björkqvist et al. 2003).

However, other methods such as thermal carbonization of PSi with acetylene were developed notably by the group of Salonen (Salonen et al. 2000, 2004). This method is based on the adsorption of acetylene onto the surface of PSi and its subsequent absorption into the silicon structure under thermal treatment, providing a nonstoichiometric silicon carbide (SiC) layer (Salonen et al. 2005). Utilizing the dissociation temperature of adsorbed acetylene and hydrogen desorption (Dufour et al. 1997), the thermal carbonization process is divided into two distinct treatments, yielding at lower temperatures thermally hydrocarbonized PSi (THCPSi) and at higher temperatures thermally carbonized PSi (TCPSi) (Salonen et al. 2004, 2008). Thermal carbonization provides almost complete surface coverage. As a result, this treatment gave rise to a relatively nontoxic material (Santos et al. 2010) suitable for gas sensing, drug delivery, and radiolabeling purposes (Björkqvist et al. 2006; Kilpeläinen et al. 2011; Sarparanta et al. 2011) with high chemical stability even in basic solutions (Björkqvist et al. 2003). The method may also be used in postfabrication for pore size modification (Fang et al. 2010).

Recently, Li and Sailor (2014) showed that the deposition of $TiO_2$ by sol-gel method on thermally oxidized PSi could improve its stability. The modified surface displayed greater aqueous stability in the pH range 2–12 relative to a $PSi:SiO_2$ surface. A label-free optical interference immunosensor (reflective interferometric Fourier transform [RIFT]) based on the $TiO_2$-coated PSi film is demonstrated by real-time monitoring of the physical adsorption of protein A, followed by the specific binding of rabbit anti-sheep immunoglobulin (IgG) and then specific capture of sheep IgG (Li and Sailor 2014).

Another PSi post-modification by materials is nitridation. Silicon nitride is chemically stable even at high temperatures and is a better diffusion barrier than $SiO_x$. In addition, PSi is stabilized against further oxidation increasing its optical properties. To perform PSi nitridation, two major reactions can be processed that are based on plasma or thermal treatment (~1000°C) under ammonia or molecular nitrogen atmosphere (Morazzani et al. 1996). The main problem is that elevated temperature can lead to the coarsening of pores (sometimes used to make bigger pores) (Salonen et al. 2009). To solve this problem, James et al. (2009) described a low temperature process. They have shown that even at low temperature, Si–N bonds can be formed. The nitridation process can stabilize PSi surface despite the silicon nitride will oxidize to ambient air. Then, oxide-based functionalization (silanization) or even hydrosilylation (after hydrogenation of $Si_xN_y$) can be performed (Arafat et al. 2004; Coffinier et al. 2007). However, despite all the advantages, nitridation was not intensively studied and used.

## 13.10 CONCLUSION

PSi is a fascinating material with a huge variety of applications in different fields. Among PSi surfaces, we find porous silicon surface, but also porous silicon nanoparticles, porous nanofilaments, nanorods, and so on. Shape and dimension are chosen depending on the targeted application. In this chapter, we focused on the description of the main chemical methods to modify porous silicon that is a crucial point for using this material. Indeed, surface chemistry is mainly used to either introduce new chemical functionalities for further grafting of biomolecules for sensing or to protect PSi from its external environment (from oxidation) keeping luminescence/electroluminescence properties or both of them.

Although the surface chemistry of silicon surfaces is very well described in the literature, this field is still buoyant especially for porous silicon. Indeed, there are still reports on the mechanisms of functionalization of porous silicon surfaces as recently shown by Kosinlinksi in 2013 and Buriak in 2014.

## REFERENCES

Addenm, N., Gamble, L.J., Castner, D.G., Hoffmann, A., Gross, G., and Menzel, H. (2006). Phosphonic acid monolayers for binding of bioactive molecules to titanium surfaces. *Langmuir* **22**, 8197–8204.

Alonso, J.M., Reichel, A., Piehler, J., and del Campo, A. (2008). Photopatterned surfaces for site-specific and functional immobilization of proteins. *Langmuir* **24**, 448–457.

Anglin, E.J., Cheng, L., Freeman, W.R., and Sailor, M.J. (2008). Porous silicon in drug delivery devices and materials. *Adv. Drug Deliv. Rev.* **60**, 1266–1277.

Arafat, A., Schroën, K., de Smet, L.C.P.M., Sudhölter, E.J.R., and Zuilhof, H. (2004). Tailor-made functionalization of silicon nitride surfaces. *J. Am. Chem. Soc.* **126**, 8600–8601.

Aswal, D.K., Lenfant, S., Guerin, D., Yakhami, J.V., and Vuillaume, D. (2006). Self assembled monolayers on silicon for molecular electronics. *Anal. Chim. Acta* **568**, 84–108.

Bansal, A., Li, X., Lauermann, I., Lewis, N.S., Li, X., Yi, S.I., and Weinberg, W.H. (1996). Alkylation of Si surfaces using a two-step halogenation/Grignard route. *J. Am. Chem. Soc.* **118**, 7225–7226.

Bateman, J.E., Eagling, R.D., Horrocks, B.R., and Houlton, A. (2000). A deuterium labeling, FTIR, and Ab initio investigation of the solution-phase thermal reactions of alcohols and alkenes with hydrogen-terminated silicon surfaces. *J. Phys. Chem. B* **104**, 5557–5565.

Belogorokhov, A.I., Gavrilov, S.A., Kashkarov, P.K., and Belogorokhov, I.A. (2005). FTIR investigation of porous silicon formed in deutrofluoric acid based solutions. *Phys. Stat. Sol. (a)* **202**, 1581–1585.

Bjorklund, R.B., Zangooie, S., and Arwin, H. (1997). Adsorption of surfactants in porous silicon films. *Langmuir* **13**, 1440–1445.

Björkqvist, M., Salonen, J., Laine, E., and Niinistö, L. (2003). Comparison of stabilizing treatments on porous silicon for sensor applications. *Phys. Stat. Sol. (a)* **197**, 374–377.

Björkqvist, M., Paski, J., Salonen, J., and Lehto, V.P. (2006). Studies on hysteresis reduction in thermally carbonized porous silicon humidity sensor. *IEEE Sens. J.* **6**, 542–547.

Blackwood, D.J. and Akber, M.F.B.M. (2006). In situ electrochemical functionalization of porous silicon. *J. Electrochem. Soc.* **153**, G976–G980.

Böcking, T., Salomon, A., Cahen, D., and Gooding, J.J. (2007). Thiol-terminated monolayers on oxide-free Si: Assembly of semiconductor–alkyl–S–metal junctions. *Langmuir* **23**, 3236–3241.

Bonroy, K., Frederix, F., Reekmans, G. et al. (2006). Comparison of random and oriented immobilization of antibody fragments on mixed self-assembled monolayers. *J. Immunol. Methods* **312**, 167–181.

Boukherroub, R. and Wayner, D.D.M. (1999). Controlled functionalization and multistep chemical manipulation of covalently modified Si(111) surfaces. *J. Am. Chem. Soc.* **121**, 11513–11515.

Boukherroub, R., Chatgilialoglu, C., and Manuel, G. (1996). Palladium dichloride catalyzed reduction of organic halides by triethylsilane. *Organometallics* **15**, 1508.

Boukherroub, R., Morin, S., Bensebaa, F., and Wayner, D.D.M. (1999). New synthetic routes to alkyl monolayers on the Si(111) surface. *Langmuir* **15**, 3831–3835.

Boukherroub, R., Morin, S., Wayner, D.D.M., and Lockwood, D.J. (2000). Thermal route for chemical modification and photoluminescence stabilization of porous silicon. *Phys. Stat. Sol. (a)* **182**, 117–121.

Boukherroub, R., Morin, S., Wayner, D.D.M. et al. (2001). Ideal passivation of luminescent porous silicon by thermal, non catalytic reaction with alkenes and aldehydes. *Chem. Mater.* **13**, 2002–2011.

Boukherroub, R., Wojtyk, J.T.C., Wayner, D.D.M., and Lockwood, D.J. (2002). Thermal hydrosilylation of undecylenic acid with porous silicon. *J. Electrochem. Soc.* **149**, 59–63.

Boukherroub, R., Petit, A., Loupy, A., Chazalviel, J.N., and Ozanam, F. (2003). Microwave-assisted chemical functionalization of hydrogen-terminated porous silicon surfaces. *J. Phys. Chem. B* **107**, 13459–13462.

Boukherroub, R., Petit, A., Loupy, A., Chazalviel, J.N., and Ozanam, F. (2004). Organic functionalization of porous silicon nanostructures. *ECS Conf. Proc.* **2004–19**, 13–22.

Buriak, J.M. (1999). Organometallic chemistry on silicon surfaces: Formation of functional monolayers bound through Si–C bonds. *Chem. Commun.* **12**, 1051–1060.

Buriak, J.M. (2002). Organometallic chemistry on silicon and germanium surfaces. *Chem. Rev.* **102**, 1271–1308.

Buriak, J.M. (2014). Illuminating silicon surface hydrosilylation: An unexpected plurality of mechanisms. *Chem. Mater.* **26**, 763–772.

Buriak, J.M. and Allen, M.J. (1998). Lewis acid mediated functionalization of porous silicon with substituted alkenes and alkynes. *J. Am. Chem. Soc.* **120**, 1339–1340.

Buriak, J.M. and Allen, M.J. (1999). Photoluminescence of porous silicon surfaces stabilized through Lewis acid mediated hydrosilylation. *J. Lumin.* **80**, 29–35.

Buriak, J.M., Stewart, M.P., Geders, T.W. et al. (1999). Lewis acid mediated hydrosilylation on porous silicon surfaces. *J. Am. Chem. Soc.* **121**, 11491–11502.

Canaria, C.A., Huang, M., Cho, Y. et al. (2002a). The effect of surfactants on the reactivity and photophysics of luminescent nanocrystalline porous silicon. *Adv. Funct. Mater.* **12**, 495–500.

Canaria, C.A., Lees, I.N., Wun, A.W., Miskelly, J.M., and Sailor, M.J. (2002b). Characterization of the carbon-silicon stretch in methylated porous silicon—Observation the anomalous isotope shift in FTIR spectrum. *Inorg. Chem. Commun.* **5**, 560–564.

Canham, L.T. (1997). Storage of porous silicon. In: Canham, L., (ed.) *Properties of Porous Silicon*. INSPEC, London, pp. 44–50.

Canham, L., Houlton, M., Leong, W., Pickering, C., and Keen, J. (1991). Atmospheric impregnation of porous silicon at room temperature. *J. Appl. Phys.* **70**, 422–431.

Canham, L.T., Stewart, M.P., Buriak, J.M. et al. (2000). Derivatized porous silicon mirrors: Implantable optical components with slow resorbability. *Phys. Stat. Sol. (a)* **182**, 521–525.

Cattani-Scholz, A., Pedone, D., Dubey, M. et al. (2008). Organophosphonate-based PNA-functionalization of silicon nanowires for label-free DNA detection. *ACS Nano* **2**, 1653–1660.

Cattani-Scholz, A., Pedone, D., Blobner, F. et al. (2009). PNA-PEG modified silicon platforms as functional bio-interfaces for applications in DNA microarrays and biosensors. *Biomacromolecules* **10**, 489–496.

Cha, T.W., Guo, A., and Zhu, X.Y. (2005). Enzymatic activity on a chip: The critical role of protein orientation. *Proteomics* **5**, 416–419.

Chatgilialoglu, C. (1995). Structural and chemical properties of silyl radicals. *Chem. Rev.* **95**, 1229.

Chen, Y.Q., Bi, F., Wang, S.Q., Xiao, S.J., and Liu, J.N. (2008). Porous silicon affinity chips for biomarker detection by MALDI-TOF–MS. *J. Chromatogr. B* **875**, 502–508.

Cicero, R.L., Linford, M.R., and Chidsey, C.E.D. (2000). Photoreactivity of unsaturated compounds with hydrogen-terminated silicon(111). *Langmuir* **16**, 5688–5695.

Coffinier, Y., Olivier, C., Perzyna, A. et al. (2005). Semicarbazide-functionalized Si(111) surfaces for the site-specific immobilization of peptides. *Langmuir* **21**, 1489–1496.

Coffinier, Y., Boukherroub, R., Wallart, X. et al. (2007). Covalent functionalization of silicon nitride surfaces by semicarbazide group. *Surf. Sci.* **601**, 5492–5498.

Coffinier, Y., Nguyen, N., Drobecq, H., Melnyk, O., Thomy, V., and Boukherroub, R. (2012). Affinity surface-assisted laser desorption/ionization mass spectrometry for peptide enrichment. *Analyst* **137**, 5527–5532.

Coffinier, Y., Piret, G., Das, M.R., and Boukherroub, R. (2013). Effect of surface roughness and chemical composition on the wetting properties of silicon-based substrates. *C.R. Chim.* **16**, 65–72.

Dancil, K.P.S., Greiner, D.P., and Sailor, M.J. (1999). A porous silicon optical biosensor: Detection of reversible binding of IgG to a protein A-modified surface. *J. Am. Chem. Soc.* **121**, 7925–7930.

Dendane, N., Hoang, A., Guillard, L., Defrancq, E., Vinet, F., and Dumy, P. (2007). Efficient surface patterning of oligonucleotides inside a glass capillary through oxime bond formation. *Bioconjugate Chem.* **18**, 671–676.

Dendane, N., Melnyk, O., Xu, T. et al. (2012). Direct characterization of native chemical ligation of peptides on silicon nanowires. *Langmuir* **28**, 13336–13344.

Densmore, A., Vachon, M., Xu, D.X. et al. (2009). Silicon photonic wire biosensor array for multiplexed real-time and label-free molecular detection. *Opt. Lett.* **34**, 3598–3600.

de Smet, L.C.P.M., Zuilhof, H., Sudholter, E.J.R., Lie, L.H., Houlton, A., and Horrocks, B.R. (2005). Mechanism of the hydrosilylation reaction of alkenes at porous silicon: Experimental and computational deuterium labeling studies. *J. Phys. Chem. B* **109**, 12020–12031.

De Stefano, L., Oliviero, G., Amato, J. et al. (2013). Aminosilane functionalization of mesoporous oxidized silicon for oligonucleotide synthesis and detection. *J. R. Soc. Interface* **10**, 1–7.

Devaraj, N.K., Dinolfo, P.H., Chidsey, C.E.D., and Collman, J.P. (2006). Selective functionalization of independently addressable microelectrodes by electrochemical activation and deactivation of a coupling catalyst. *J. Am. Chem. Soc.* **128**, 1794–1795.

De Vos, K., Girones, J., Popelka, S., Schacht, E., Baets, R., and Bienstman, P. (2009). SOI optical microring resonator with poly(ethylene glycol) polymer brush for label-free biosensor applications. *Biosens. Bioelectron.* **24**, 2528–2533.

Di Francia, G., La Ferrara, V., Manzo, S., and Chiavarini, S. (2005). Towards a label-free optical porous silicon DNA sensor. *Biosens. Bioelectron.* **21**, 661–665.

Dubey, M., Weidner, T., Gamble, L.J. and Castner, D.G. (2010). Structure and order of phosphonic acid-based self-assembled monolayers on Si(100). *Langmuir* **26**, 14747–14754.

Dubois, T., Ozanam, F., and Chazalviel, J.N. (1997). Stabilization of the porous silicon surface by grafting of organic groups: Direct electrochemical methylation. *Electrochem. Soc. Proc.* **97–7**, 296–310.

Dufour, G., Rochet, F., Stedile, F. et al. (1997). SiC formation by reaction of Si(001) with acetylene: Electronic structure and growth mode. *Phys. Rev. B* **56**, 4266–4282.

Effenberger, F., Gotz, G., Bidlingmaier, B., and Wezstein, M. (1998). Photoactivated preparation and patterning of self-assembled monolayers with 1-alkenes and aldehydes on silicon hydride surfaces. *Angew. Chem. Int. Ed.* **37**, 2462–2464.

Fang, D.Z., Striemer, C.C., Gaborski, T.R., McGrath, J.L., and Fauchet, P.M. (2010). Pore size control of ultrathin silicon membranes by rapid thermal carbonization. *Nano Lett.* **10**, 3904–3908.

Fellah, S., Boukherroub, R., Chazalviel, J.N., and Ozanam, F. (2004). Hidden electrochemistry in the thermal grafting of Si surfaces from Grignard reagents. *Langmuir* **20**, 6359–6364.

García-Rupérez, J., Toccafondo, V., Bañuls, M.J. et al. (2010). Label-free antibody detection using band edge fringes in SOI planar photonic crystal waveguides in the slow-light regime. *Opt. Express* **18**, 24276–24286.

Gawalt, E.S., Avaltroni, M.J., Danahy, M.P. et al. (2003). Bonding organics to Ti alloys: Facilitating human osteoblast attachment and spreading on surgical implant materials. *Langmuir* **19**, 200–204.

Glass, J.A., Wovchko, E.A., and Yates, J.T. (1995). Reaction of methanol with porous silicon. *Surf. Sci.* **338**, 125–137.

Gu, L., Hall, D.J., Qin, Z. et al. (2013). In vivo time-gated fluorescence imaging with biodegradable luminescent porous silicon nanoparticles. *Nature Commun.* **4**, 2326–2333.

Guan, B., Ciampi, S., Luais, E., James, M., Reece, P.J., and Gooding, J.J. (2012). Depth-resolved chemical modification of porous silicon by wavelength-tuned irradiation. *Langmuir* **28**, 15444–15449.

Gun'ko, Y.K., Perova, T.S., Balakrishnan, S., Potapova, D.A., Moore, R.A., and Astrova, E.V. (2003). Chemical modification of silicon surfaces with ferrocene functionalities. *Phys. Stat. Sol. (a)* **197**, 492–496.

Guo, D.J., Xiao, S.J., Xia, B. et al. (2005). Reaction of porous silicon with both end-functionalized organic compounds bearing α-bromo and ω-carboxy groups for immobilization of biomolecules. *J. Phys. Chem. B* **109**, 20620–20628.

Gurtner, C., Wun, A.W., and Sailor, M.J. (1999). Surface modification of porous silicon by electrochemical reduction of organo halides. *Angew. Chem. Int. Ed.* **38**, 1966–1968.

Haensch, C., Erdmenger, T., Fijten, M.W.M., Hoeppener S., and Schubert, U.S. (2009). Fast surface modification by microwave assisted click reactions on silicon substrates. *Langmuir* **25**, 8019–8024.

Haensch, C., Hoeppener, S., and Schubert, U.S. (2010). Chemical modification of self-assembled silane based monolayers by surface reactions. *Chem. Soc. Rev.* **39**, 2323–2334.

Hanson, E.L., Schwartz, J., Nickel, B., Koch, N., and Danisman, M.F. (2003). Bonding self-assembled, compact organophosphonate monolayers to the native oxide surface of silicon. *J. Am. Chem. Soc.* **125**, 16074–16080.

Harper, T.F. and Sailor, M.J. (1997). Using porous silicon as a hydrogenating agent: Derivatization of the surface of luminescent nanocrystalline silicon with benzoquinone. *J. Am. Chem. Soc.* **119**, 6943–6944.

Hauffman, T., Blajiev, O., Snauwaert, J., van Haesendonck, C., Hubin, A., and Terryn, H. (2008). Study of the self-assembling of n-octylphosphonic acid layers on aluminum oxide. *Langmuir* **24**, 13450–13456,

Hoque, E., DeRose, J.A., Kulik, G., Hoffmann, P., Mathieu, H.J., and Bhushan, B.J. (2006). Alkylphosphonate modified aluminum oxide surfaces. *J. Phys. Chem. B* **110**, 10855–10861.

Hory, M.A., Hérino, R., Ligeon, M. et al. (1995). Fourier transform IR monitoring of porous silicon passivation during post-treatments such as anodic oxidation and contact with organic solvents. *Thin Solid Films* **255**, 200–203.

Houseman, B.T., Huh, J.H., Kron, S.J., and Mrksich, M. (2002). Peptide chips for the evaluation of protein kinase activity. *Nature Biotechnol.* **20**, 270–274.

Hua, F., Swihart, M.T., and Ruckenstein, E. (2005). Efficient surface grafting of luminescent silicon quantum dots by photoinitiated hydrosilylation. *Langmuir* **21**, 6054–6062.

Huck, L.A. and Buriak, J.M. (2012). UV-initiated hydrosilylation on hydrogen-terminated silicon (111): Rate coefficient increase of two orders of magnitude in the presence of aromatic electron acceptors. *Langmuir* **28**, 16285.

Hunt, H.K., Soteropolos, C., and Armani, A.M. (2010). Bioconjugation strategies for microtoroidal optical resonators. *Sensors* **10**, 9317–9336.

James, T.D., Keating, A., Parish, G., and Musca, C.A. (2009). Low temperature N2-based passivation technique for porous silicon thin films. *Solid State Commun.* **149**, 1322–1324.

Jarvis, K.L., Barnes, T.J., and Prestidge, C.A. (2010). Thermal oxidation for controlling protein interactions with porous silicon. *Langmuir* **26**, 14316–14322.

Joy, V.T. and Mandler, D. (2002). Surface functionalization of H-terminated silicon surfaces with alcohols using iodoform as an in situ iodinating agent. *Chem. Phys. Chem.* **3**, 973–975.

Kelly, J.A. and Veinot, J.G.C. (2010). An investigation into near UV hydrosilylation of freestanding silicon nanocrystals. *ACS Nano* **4**, 4645–4656.

Kilian, K.A., Böcking, T., Gaus, K., Gal, M., and Gooding, J.J. (2007). Peptide-modified optical filters for detecting protease activity. *ACS Nano* **1**, 355–361.

Kilian, K.A., Böcking, T., and Gooding, J.J. (2009). The importance of surface chemistry in mesoporous materials: Lessons from porous silicon biosensors. *Chem. Commun.* **14**(6), 630–640.

Kilpeläinen, M., Mönkäre, J., Vlasova, M.A. et al. (2011). Nanostructured porous silicon microparticles enable sustained peptide (Melanotan II) delivery. *Eur. J. Pharm. Biopharm.* **77**, 20–25.

Kim, N.Y. and Laibinis, P.E. (1997). Thermal derivatization of porous silicon with alcohols. *J. Am. Chem. Soc.* **119**, 2297–2298.

Kim, N.Y. and Laibinis, P.E. (1998). Derivatization of porous silicon by Grignard reagents at room temperature. *J. Am. Chem. Soc.* **120**, 4516–4517.

Klauk, H., Zschieschang, U., Pflaum, J., and Halik, M. (2007). Ultralow-power organic complementary circuits. *Nature* **445**, 745–748.

Kohn, M., Wacker, R., Peters, C. et al. (2003). Staudinger ligation: A new immobilization strategy for the preparation of small-molecule arrays. *Angew. Chem. Int. Ed.* **42**, 5830–5834.

Kolasinski, K.W. (2005). Silicon nanostructures from electroless electrochemical etching. *Curr. Opin. Solid State Mater. Sci.* **9**, 73–83.

Kolasinski, K.W. (2013). The mechanism of photohydrosilylation on silicon and porous silicon surfaces. *J. Am. Chem. Soc.* **135**, 11408–11412.

Langner, A., Panarello, A., Rivillon, S., Vassylyev, O., Khinast, J.G., and Chabal, Y.J. (2005). Controlled silicon surface functionalization by alkene hydrosilylation. *J. Am. Chem. Soc.* **127**, 12798–12799.

Lauerhaas, J.M. and Sailor, M.J. (1993). Chemical modification of the photoluminescence quenching of porous silicon. *Science* **261**, 1567–1568.

Lavine, J.M., Sawan, S.P., Shieh, Y.T., and Bellezza, A.J. (1993). Role of Si-H and Si-H$_2$ in the photoluminescence of porous Si. *Appl. Phys. Lett.* **62**, 1099–1101.

Lee, E.J., Ha, J.S., and Sailor, M.J. (1995). Chemical modification of the porous silicon surface. *Mater. Res. Soc. Symp. Proc.* **358**, 387–392.

Lee, S., Eom, S.C., Chang, J.S., Huh, C., Sung, G.Y., and Shin, J.H. (2010). Label-free optical biosensing using a horizontal air-slot SiN$_x$ microdisk resonator. *Opt. Express* 18, 20638–20644.

Lees, I.N., Lin, H., Canaria, C.A., Gurtner, C., Sailor, M.J., and Miskelly, G.M. (2003). Chemical stability of porous silicon surfaces electrochemically modified with functional alkyl species. *Langmuir* **19**, 9812–9817.

Létant, S.E., Content, S., Tan, T.T., Zenhausern, F., and Sailor, M.J. (2000). Integration of porous silicon chips in an electronic artificial nose. *Sens. Actuators B* **69**, 193–198.

Li, J. and Sailor, M.J. (2014). Synthesis and characterization of a stable, label-free optical biosensor from TiO$_2$-coated porous silicon. *Biosens. Bioelectron.* **55**, 372–378.

Lin, Z., Strother, T., Cai, W., Cao, X., Smith, L.M., and Hamers, R.J. (2002). DNA attachment and hybridization at the silicon (100) surface. *Langmuir* **18**, 788–796.

Linford, M.R. and Chidsey, C.E.D. (1993). Alkyl monolayers covalently bonded to silicon surfaces. *J. Am. Chem. Soc.* **115**, 12631–12632.

Linford, M.R., Fenter, P., Eisenberger, P.M. and Chidsey, C.E.D. (1995). Alkyl monolayer on silicon prepared from 1-alkene and hydrogen terminated silicon. *J. Am. Chem. Soc.* **117**, 3145–3155.

Lowe, R.D., Szili, E.J., Kirkbride, P., Thissen, H., Siuzdak, G., and Voelker, N.H. (2010). Combined immunocapture and laser desorption/ionization mass spectrometry on porous silicon. *Anal. Chem.* **82**, 4201–4208.

Luo, W., Westcott, N.P., Pulsipher, A., and Yousaf, M.N. (2008). Renewable and optically transparent electroactive indium tin oxide (ITO) surfaces for chemoselective ligand immobilization and biospecific cell adhesion. *Langmuir* **24**, 13096–13101.

Maruyama, T. and Ohtani, S. (1994). Photoluminescence of porous silicon exposed to ambient air. *Appl. Phys. Lett.* **65**, 1346–1348.

Matsumoto, T., Masumoto, Y., Nakashima, S., and Koshida, N. (1997). Luminescence from deuterium-terminated porous silicon. *Thin Solid Films* **297**, 31–34.

Mattei, G. and Valentini, V. (2003). *In situ* functionalization of porous silicon during the electrochemical formation process in ethanoic hydrofluoric acid solution. *J. Am. Chem. Soc.* **125**, 9608–9609.

Mattei, G., Alieva, E.V., Petrov, J.E., and Yakovlev, V.A. (2000). Quick oxidation of porous silicon in presence of pyridine vapor. *Phys. Stat. Sol. (a)* **182**, 139–143.

McDermott, J.F., McDowell, M., Hill, I.G. et al. (2007). Organophosphonate self-assembled monolayers for gate dielectric surface modification of pentacene-based organic thin-film transistors: A comparative study. *J. Phys. Chem. A* **111**, 12333–12338.

McInnes, S.J.P. and Voelcker, N.H. (2012). Porous silicon-based nanostructured microparticles as degradable supports for solid phase synthesis and release of oligonucleotides. *Nanoscale Res. Lett.* **7**, 385.

Midwood, K.S., Carolus, M.D., Danahy, M.P., Schwarzbauer, J.E., and Schwartz, J. (2004). Easy and efficient bonding of biomolecules to an oxide surface of silicon. *Langmuir* **20**, 5501–5505.

Morazzani, V., Cantin, J.L., Ortega, C. et al. (1996). Thermal nitridation of *p*-type porous silicon in ammonia. *Thin Solid Films* **276**, 32–35.

Olivier, C., Perzyna, A., Coffinier, Y. et al. (2006). Detecting the chemoselective ligation of peptides to silicon with the use of cobalt-carbonyl labels. *Langmuir* **22**, 7059–7065.

Pace, S., Sciacca, B., and Geobaldo, F. (2013). Surface modification of porous silicon microparticles by sonochemistry. *RSC Adv.* **3**, 18799–18802.

Pacholski, C., Sartor, M., Sailor, M.J., Cunin, F., and Miskelly, G.M. (2005). Biosensing using porous silicon double layer interferometers: Reflective interferometric Fourier transform spectroscopy. *J. Am. Chem. Soc.* **127**, 11636–11645.

Pap, A.E., Kordás, K., George, T.F., and Leppävuori, S. (2004). Thermal oxidation of porous silicon: Study on reaction kinetics. *J. Phys. Chem. B* **108**, 12744–12747.

Petit, A., Delmotte, M., Loupy, A., Chazalviel, J.N., Ozanam, F., and Boukherroub, R. (2008). Microwave effects on chemical functionalization of hydrogen-terminated porous silicon nanostructures. *J. Phys. Chem. C* **112**, 16622–16628.

Petrova-Koch, V., Muschik, T., Kux, A., Meyer, B.K., Koch, F., and Lehmann, V. (1992). Rapid thermal oxidized porous Si—The superior photoluminescent Si. *Appl. Phys. Lett.* **61**, 943–945.

Piret, G., Drobecq, H., Coffinier, Y., Melnyk, O., and Boukherroub, R. (2010). Desorption/ionization mass spectrometry on silicon nanowire arrays prepared by chemical etching of crystalline silicon. *Langmuir* **26**, 1354–1361.

Ramachandran, A., Wang, S., Clarke, J. et al. (2008). A universal biosensing platform based on optical micro-ring resonators. *Biosens. Bioelectron.* **23**, 939–944.

Ressine, A., Marko-Varga, G., and Laurell, T. (2007). Porous silicon protein microarray technology and ultra-/superhydrophobic states for improved bioanalytical readout. *Biotechnol. Ann. Rev.* **13**, 1–52.

Robins, E.G., Stewart, M.P., and Buriak, J.M. (1999). Anodic and cathodic electrografting of alkynes on porous silicon. *Chem. Commun.* **24**, 2479–2480.

Rocchia, M., Borini, S., Rossi, A.M., Boarino, L., and Amato, G. (2003). Submicrometer functionalization of porous silicon by electron beam lithography. *Adv. Mater.* **15**, 1465–1469.

Saghatelian, A., Buriak, J.M., Lin, V.S.Y., and Ghadiri, M.R. (2001). Transition metal mediated surface modification of porous silicon. *Tetrahedron* **57**, 5131–5136.

Salonen, J., Lehto, V.P., Björkqvist, M., Laine, E., and Niinistö, L. (2000). Studies of thermally-carbonized porous silicon surfaces. *Phys. Stat. Sol. (a)* **182**, 123–126.

Salonen, J., Björkqvist, M., Laine, E., and Niinistö, L. (2004). Stabilization of porous silicon surface by thermal decomposition of acetylene. *Appl. Surf. Sci.* **225**, 389–394.

Salonen, J., Laitinen, L., Kaukonen, A.M. et al. (2005). Mesoporous silicon microparticles for oral drug delivery: Loading and release of five model drugs. *J. Control. Release* **108**, 362–374.

Salonen, J., Kaukonen, A.M., Hirvonen, J., and Lehto, V.P. (2008). Mesoporous silicon in drug delivery applications. *J. Pharm. Sci.* **97**, 632–653.

Salonen, J., Mäkilä, J.E., Riikonen, J., Heikkilä, H., and Lehto, V.P. (2009). Controlled enlargement of pores by annealing of porous silicon. *Phys. Stat. Sol. (a)* **206**, 1313–1317.

Sam, S., Chazalviel, J.N., Gouget-Laemmel, A.C. et al. (2010). Covalent immobilization of amino acids on the porous silicon surface. *Surf. Interface Anal.* **42**, 515–518.

Santos, H.A., Riikonen, J., Salonen, J. et al. (2010). *In vitro* cytotoxicity of porous silicon microparticles: Effect of the particle concentration, surface chemistry and size. *Acta Biomater.* **6**, 2721–2731.

Sarparanta, M., Mäkilä, E., Heikkilä, T. et al. (2011). (18)F labeled modified porous silicon particles for investigation of drug delivery carrier distribution in vivo with positron emission tomography. *Mol. Pharm.* **8**, 1799–1806.

Scheibler, L., Dumy, P., Boncheva, M. et al. (1999). Functional molecular thin films: Topological templates for the chemoselective ligation of antigenic peptides to self-assembled monolayers. *Angew. Chem. Int. Ed.* **38**, 696–699.

Scheres, L., Achten, R., Giesbers, M. et al. (2009). Covalent attachment of Bent-core mesogens to silicon surfaces. *Langmuir* **25**, 1529–1533.

Schneider, B.H., Dickinson, E.L., Vach, M.D., Hoijer, J.V., and Howard, L.V. (2000). Highly sensitive optical chip immunoassays in human serum. *Biosens. Bioelectron.* **15**, 13–22.

Seo, Y.H., Lee, H.J., Jeon, H.I. et al. (1993). Photoluminescence, Raman scattering, and infrared absorption studies of porous silicon. *Appl. Phys. Lett.* **62**, 1812–1814.

Shang, J., Cheng, F., Dubey, M. et al. (2012). An organophosphonate strategy for functionalizing silicon photonic biosensors. *Langmuir* **28**, 3338–3344.

Shao, M.W., Wang, H., Fu, Y., Hua, J., and Ma, D.D.D. (2009). Surface functionalization of HF-treated silicon nanowires. *J. Chem. Sci.* **121**, 323–327.

Sharma, S., Johnson, R.W., and Desai, T.A. (2004). XPS and AFM analysis of antifouling PEG interfaces for microfabricated silicon biosensors. *Biosens. Bioelectron.* **20**, 227–239.

Shi, C., Mehrabani, S., and Armani, A.M. (2012). Leveraging bimodal kinetics to improve detection specificity. *Opt. Lett.* **37**, 1643–1645.

Sieval, A.B., Demirel, A.L., Nissink, J.W.M. et al. (1998). Highly stable Si-C linked functionalized monolayers on the silicon (100) surface. *Langmuir* **14**, 1759–1768.

Sieval, A.B., Vleeming, V., Zuilhof, H., and Sudholter, E.J.R. (1999). An improved method for the preparation of organic monolayers of 1-alkenes on hydrogen-terminated silicon surfaces. *Langmuir* **15**, 8288–8291.

Sieval, A.B., Linke, R., Zuilhof, H., and Sudholter, E.J.R. (2000). High-quality alkyl monolayers on silicon surfaces. *Adv. Mater.* **12**, 1457–1460.

Sieval, A.B., Linke, R., Heij, G., Meijer, G., Zuilhof, H., and Sudholter, E.J.R. (2001). Amino-terminated organic monolayers on hydrogen-terminated silicon surfaces. *Langmuir* **17**, 7554–7559.

Silverman, B.M., Wieghaus, K.A., and Schwartz, J. (2005). Comparative properties of siloxane *vs* phosphonate monolayers on a key titanium alloy. *Langmuir* **21**, 225–228.

Soellner, M.B., Dickson, K.A., Nilsson, B.L., and Raines, R.T. (2003). Site specific protein immobilization by Staudinger ligation. *J. Am. Chem. Soc.* **125**, 11790–11791.

Soeriyadi, A.H., Gupta, B., Reece, P.J., and Gooding, J.J. (2014). Optimising the enzyme response of a porous silicon photonic crystal via the modular design of enzyme sensitive polymers. *Polym. Chem.* **5**, 2333.

Song, J.H. and Sailor, M.J. (1998a). Dimethyl sulfoxide as a mild oxidizing agent for porous silicon and its effect on photoluminescence. *Inorg. Chem.* **37**, 3355–3360.

Song, J.H. and Sailor, M.J. (1998b). Functionalization of nanocrystalline porous silicon surfaces with aryllithium reagents: Formation of silicon–carbon bonds by cleavage of silicon–silicon bonds. *J. Am. Chem. Soc.* **120**, 2376–2381.

Stern, E., Klemic, J.F., Routenberg, D.A. et al. (2007). Label-free immunodetection with CMOS-compatible semiconducting nanowires. *Nature* **445**, 519–522.

Stewart, M.P. and Buriak, J.M. (1998). Photopatterned hydrosilylation on porous silicon. *Angew. Chem. Int. Ed.* **37**, 3257–3260.

Stewart, M.P. and Buriak, J.M. (2001). Exciton mediated hydrosilylation on nanocrystalline silicon surfaces. *J. Am. Chem. Soc.* **123**, 7821–7830.

Strother, T., Cai, W., Zhao, X., Hamers, R.J., and Smith, L.M. (2000a). Synthesis and characterization of DNA-modified silicon (111) surfaces. *J. Am. Chem. Soc.* **122**, 1205–1209.

Strother, T., Hamers, R.J., and Smith, L.M. (2000b). Covalent attachment of oligodeoxyribonucleotides to amine-modified Si (001) surfaces. *Nucleic Acids Res.* **28**, 3535–3541.

Sun, Q.Y., de Smet, L., van Lagen, B., Wright, A., Zuilhof, H., and Sudhölter, E.J.R. (2004). Covalently attached monolayers on crystalline hydrogen-terminated silicon: Extremely mild attachment by visible light. *Angew. Chem. Int. Ed.* **43**, 1352.

Sun, Q.Y., de Smet, L., van Lagen, B. et al. (2005). Covalently attached monolayers on crystalline hydrogen-terminated silicon: Extremely mild attachment by visible light. *J. Am. Chem. Soc.* **127**, 2514–2523.

Sung, M.M., Kluth, G.J., Yauw, O.W., and Maboudian, R. (1997). Thermal behavior of alkyl monolayers on the Si(100) surface. *Langmuir* **13**, 6164–6168.

Suslick, K.S. (1990). Sonochemistry. *Science* **247**, 1439–1445.

Wacker, R., Schroder, H., and Niemeyer, C.M. (2004). Performance of antibody microarrays fabricated by either DNA-directed immobilization, direct spotting, or streptavidin–biotin attachment: A comparative study. *Anal. Biochem.* **330**, 281–287.

Wang, D. and Buriak, J.M. (2006). Trapping silicon surface-based radicals. *Langmuir* **22**, 6214–6221.

Wang, X.Y., Ruther, R.E., Streifer, J.A., and Hamers, R.J. (2010). UV-induced grafting of alkenes to silicon surfaces: Photoemission versus excitons. *J. Am. Chem. Soc.* **132**, 4048–4049.

Wei, J., Buriak, J.M., and Siuzdak, G. (1999). Desorption-ionization mass spectrometry on porous silicon. *Nature* **20**, 243–246.

Wojtyk, J.T.C., Morin, K.A., Boukherroub, R., and Wayner, D.D.M. (2002). Modification of porous silicon surfaces with activated ester monolayers. *Langmuir* **18**, 6081–6087.

Woods, M., Carlsson, S., Hong, Q. et al. (2005). A kinetic model of the formation of organic monolayers on hydrogen-terminated silicon by hydrosilylation of alkenes. *J. Phys. Chem. B* **109**, 24035–24045.

Xiao, S.J., Textor, M., Spencer, N.D., and Sigrist, H. (1998). Covalent attachment of cell-adhesive, (Arg-Gly-Asp)-containing peptide to titanium surfaces. *Langmuir* **14**, 5507–5516.

Xu, S., Pan, C., Hu, L. et al. (2004). Enzymatic reaction of the immobilized enzyme on porous silicon studied by matrix-assisted-laser desorption/ionization-time of flight-mass spectrometry. *Electrophoresis* **25**, 3669–3676.

Xu, J., Suarez, D., and Gottfried, D.S. (2007). Detection of avian influenza virus using an interferometric biosensor. *Anal. Bioanal. Chem.* **389**, 1193–1199.

Yang, W., Auciello, O., Butler, J.E. et al. (2002). DNA-modified nanocrystalline diamond thin-films as stable, biologically active substrates. *Nature Mater.* **1**, 253–257.

Yang, Z., Dobbie, A.R., Cui, K., and Veinot, J.G.C. (2012). A convenient method for preparing alkyl-functionalized silicon nanocubes. *J. Am. Chem. Soc.* **134**, 13958–13961.

Zhong, X., Qu, Y., Lin, Y.C., Liao, L., and Duan, X. (2011). Unveiling the formation pathways of single crystalline porous silicon nanowires. *ACS Appl. Mater. Interfaces* **3**, 261–270.

Zhu, H., Bilgin, M., Bangham, R. et al. (2001). Global analysis of protein activities using proteome chips. *Science* **293**, 2101–2105.

Zhu, Y., Soeriyadi, A.H., Parker, S.G., Reece, P.J., and Gooding, J.J. (2014). Chemical patterning on preformed porous silicon photonic crystals: Towards multiplex detection of protease activity at precise positions. *J. Mater. Chem. B* **2**, 3582.

Zlatanovic, S., Mirkarimi, L.W., Sigalas, M.M. et al. (2009). Photonic crystal microcavity sensor for ultracompact monitoring of reaction kinetics and protein concentration. *Sens. Actuators B* **141**, 13–19.

Zschieschang, U., Halik, M., and Klauk, H. (2008). Microcontact-printed self-assembled monolayers as ultrathin gate dielectrics in organic thin-film transistors and complementary circuits. *Langmuir* **24**, 1665–1669.

# Contacts to Porous Silicon and PSi-Based *p-n* Homo- and Heterojunctions

**14**

Jayita Kanungo and Sukumar Basu

CONTENTS

## 14.1 INTRODUCTION

For its multifarious applications, porous silicon (PSi) is getting popular attention from the scientific and technological arena (Canham 1997; Pavesi and Dubos 1997; Tsamis et al. 2002). As it was testified before, the main attractions behind the research and development of porous silicon are very large surface to volume ratio of PSi, tailormade technology to control the surface morphology by varying the formation parameters, and relatively easy formation. However, most the important factor for its large-scale applications is the compatibility to silicon IC technology leading to the development of effective devices for electrical and microelectronics applications, photonics, sensors, optoelectronic system, and so on. Another technological advantage of PSi is the tailormade morphology for the desired applications. Recently, the eminent researchers working with PSi proposed the field to be an educational vehicle for introducing nanotechnology and interdisciplinary material science. However, for developing PSi-based devices and their integration to the electronic circuits, the most important requirement is the presence of low resistance and stable electrical contacts. As it is known, one of the main outstanding problems of PSi is the instability of its native interface due to the metastable Si–H$_x$ termination (Tsai et al. 1991) that causes spontaneous oxidation in the ambient atmosphere and results in the degradation of surface structures. As a result, a stable ohmic contact formation (Deresmes et al. 1995; Stievenard and Deresmes 1995) becomes difficult and makes the commercial applications problematic. Therefore, the stabilization of the PSi surfaces and fabrication of reliable electrical contacts to PSi are the primary needs of the PSi-based devices.

## 14.2 METAL-POROUS SILICON JUNCTIONS

### 14.2.1 OHMIC AND RECTIFYING BEHAVIOR OF M-PSi CONTACTS

Similarly to bulk silicon, the conventional methods like evaporation, sputtering, electroless deposition, electroplating, and screen-printing are usually used to fabricate the metal contacts to porous silicon. Table 14.1 gives the summary of contact to PSi with different metals and metal alloys.

Thus far, the metals like Al, Ag, Au, Ni, Cu, Pd, and metal alloys are reported to have been tried as the contact materials to porous silicon. Some of them show ohmic contacts and some of them behave rectifying as indicated by the I-V measurements. However, the ohmic and rectifying nature depend on some properties of PSi like type of conductivity, porosity, and resistivity. For example, it was found that while PSi with low porosity and low resistivity forms ohmic contact with Al, both $n$- and $p$-type PSi with high resistivity forms Al-PSi contact of rectifying behavior (Simons et al. 1995; Zimin et al. 1995, 1998; Kanungo et al. 2009a). There are reports on the ohmic behavior of Al to PSi after the surface modification, presumably due to the reduction of the porosity through surface passivation (Kanungo et al. 2009a; Maji et al. 2010). Ag has also been shown to behave as the ohmic junction to PSi (Vinod 2005, 2009, 2013). An interesting observation has been reported on Ni contact to PSi (Dhar and Chakrabarti 1996). Electroless deposited and annealed Ni thin film shows an excellent ohmic behavior to PSi on $n$-type Si substrate, but rectifying behavior was observed for PSi formed on $p$-type silicon. In addition, the electroless Ni on PSi showed ohmic nature at low bias voltage (Kanungo et al. 2006). Further research on the ohmic behavior of the electroless Ni on PSi deposited on $p$-type Si revealed the formation of nickel silicide by heat treatment (Andersson et al. 2008). It was concluded that both rectifying and ohmic contacts could be formed between electroless deposited Ni and PSi depending on the heat treatment conditions.

Most of the reports available on gold, copper, palladium, indium, and titanium as contact metals to PSi are found rectifying (Han et al. 1994; Jeske et al. 1995; Simons et al. 1995; Diligenti et al. 1996; Ichinohe et al. 1996; Angelescu and Kleps 1998; Matsumoto et al. 1998; Skryshevsky et al. 1998; Slobodchikov et al. 1998, 1999; Lue et al. 1999; Bhattacharya et al. 2000; Vikulov et al. 2000; Ghosh et al. 2002a; Rabinal and Mulimani 2007; Gallach et al. 2012). The metal alloys like Au-In, In-Sn, and so on showed rectifying behavior (Angelescu and Kleps 1998). The alloyed PtSi/porous Si contact shows a good ohmic characteristic with a low contact resistance (Ichinohe et al. 1996) but the PtSi/PSi/PtSi structure produces a nonlinear relation for carrier transport as confirmed

**TABLE 14.1    Ohmic and Rectifying Contacts to Porous Silicon Formed by Different Metals and Metal Alloys**

| Metal | Si Type | Metallization Technique | References |
|---|---|---|---|
| | | | **Ohmic Contacts** |
| Al | *p* | Evaporation | Dhar and Chakrabarti 1996; Zimin and Komarov 1998; Kanungo et al. 2009a; Maji et al. 2010 |
| | *n* | Evaporation | Simons et al. 1995; Zimin et al. 1995; Zimin and Komarov 1998 |
| Ag | *p* | Screen printing | Vinod 2005, 2009, 2013 |
| Ni | *p* | Electroless | Kanungo et al. 2006; Andersson et al. 2008 |
| | *n* | Electroless | Dhar and Chakrabarti 1996 |
| PtSi | *p* | Sputtering | Ichinohe et al. 1996 |
| | | | **Rectifying Contacts** |
| Al | *p* | Evaporation | Deresmes et al. 1995; Stievenard and Deresmes 1995; Zimin et al. 1995; Angelescu and Kleps 1998; Skryshevsky et al. 1998; Martin-Palma et al. 1999; Rabinal and Mulimani 2007; Cherif et al. 2013 |
| | | Sputtering | Anderson et al. 1991; Zimin and Bragin 1999 |
| | *n* | Evaporation | Zimin et al. 1995; Diligenti et al. 1996 |
| | | Sputtering | Anderson et al. 1991; Zimin and Bragin 1999 |
| Au | *p* | Evaporation | Angelescu and Kleps 1998; Matsumoto et al. 1998; Lue et al. 1999; Bhattacharya et al. 2000; Rabinal and Mulimani 2007 |
| | | Sputtering | Han et al. 1994; Ichinohe et al. 1996; Gallach et al. 2012 |
| | | Electroless | Jeske et al. 1995 |
| | *n* | Evaporation | Simons et al. 1995; Diligenti et al. 1996 |
| Ag | *p* | Co$_2$ SCF | Lin et al. 2006 |
| Ni | *p* | Electroless | Jeske et al. 1995; Dhar and Chakrabarti 1996; Andersson et al. 2008 |
| Cu | *p* | Electroless | Jeske et al. 1995 |
| | | Sputtering | Ghosh et al. 2002 |
| Pd | *p* | Evaporation | Slobodchikov et al. 1998; Vikulov et al. 2000 |
| | *n* | Electroless | Slobodchikov et al. 1999 |
| In | *p* | Evaporation | Angelescu and Kleps 1998 |
| | *n* | Evaporation | Diligenti et al. 1996 |
| Ti | *p* | Evaporation | Skryshevsky et al. 1998 |
| PtSi | *p* | Evaporation | Banihashemian et al. 2010 |
| Au-In, In-Sn | *p* | Evaporation | Angelescu and Kleps 1998 |

by I-V curve measurement (Banihashemian et al. 2010). Au/porous Si contact shows the Schottky behavior with a high resistance, even under the forward bias. The rectifying nature of *p*-PSi/Si interface has been studied and found to be independent of the metal/PSi contact and dependent on the thickness of the PSi layer (Giebel and Pavesi 1995).

### 14.2.2 I-V CHARACTERISTICS AND BAND-DIAGRAM OF M-PSi JUNCTIONS: GENERAL CONSIDERATION

For understanding the nature of the effects observed in the M-PSi contacts, let us consider their I-V characteristics. As a rule, the expression for the current-voltage characteristics of metal-semiconductor contacts (Schottky barriers) is described by the thermionic emission model

$$(14.1) \qquad I = I_s \exp\left[ \frac{(qV - IR_s)}{kT} - 1 \right]$$

where $R_s$ is the series resistance, and $I_s$ is the saturation current, which can be expressed as

$$(14.2) \qquad I_s = A^* T^2 \exp\left[\frac{-\Phi_b}{kT}\right]$$

In this equation, $A^*$ is the Richardson constant, and $\Phi_b$ is the Schottky barrier height.

According to the Schottky-Mott rule, the barrier between a metal and semiconductor should be proportional to the difference of the metal-vacuum work function and the semiconductor-vacuum electron affinity.

$$(14.3) \qquad \Phi_b = \Phi_M - \chi_S$$

where $\Phi_M$ is the work function of the metal, and $\chi_S$ is the electron affinity of the semiconductor. So, the barrier height can be controlled by selecting the appropriate pair of metal and semiconductor. Experiment has shown that for bulk Si, this model works and usually for Si clear $\Phi_b$ dependence on metal work function is observed. This means that in the frame of this model rectifying properties of Si-based Schottky diodes will be determined by the properties of the metal used, namely its work function. Based on work function consideration, Al and Ag should give ohmic contact to $p$-Si and $n$-Si, respectively, and Pd, Pt, Cu, In, and so on normally give rectifying contacts to silicon.

However, for porous silicon, such a situation is not realized. It was established that for porous silicon, $\Phi_b$ is only weakly dependant on $\Phi_M$, and, as shown previously, the material parameters of porous silicon are more important factors affecting the properties of the M-PSi contacts (Scheen et al. 2012; Skryshevsky 2014). Therefore, I-V characteristics differ significantly from the ideal behavior and the more general I-V relation with the ideality factor ($n$) is applicable. Therefore, the I-V characteristics of the porous silicon hetero junction, described by the thermo-ionic emission model, may describe the current as a function of the applied voltage to the junction and it can be expressed as

$$(14.4) \qquad I = I_s \exp\left[\frac{(qV - IR_s)}{nkT} - 1\right]$$

where $I_s$, $R_s$, and $n$ are the saturation current, the series resistance, and the ideality factor of the junction, respectively. $I_s$ can be expressed as

$$(14.5) \qquad I_s = A^* T^2 \exp\left[\frac{-\Phi_b}{nkT}\right]$$

where $A^*$ is the Richardson constant.

The deviation from the ideality primarily occurs because of the presence of the dielectric layer between the metal and the semiconductor and the high concentration of defects at the interface. The large amount of defect states at the interface may be responsible for Fermi level pinning, which generally increases or decreases the Schottky barrier height compared to the ideal model. This variation is dependent on the energetic position of the defects.

The interface charge density can be changed by controlling the metal work function and so the Fermi level can be unpinned by reducing the density of surface states. Then the barrier height is simply controlled by the difference between the work function of the metal and the flat-band energy of the semiconductor. When the metal-semiconductor junction has high series resistance and large density of interface states, the ideality factor ($n$) becomes higher than unity and then the recombination and tunneling components of the current increase.

The high ideality factor can be attributed to the sum of the ideality factors of the PSi/Si heterojunction and the metal/PSi Schottky barrier. In addition, the porous silicon has a very large effective surface area and, consequently, a large concentration of dangling bonds that may act as the high concentration of carrier traps in PS (Remaki et al. 2003; Skryshevsky 2014). Therefore, the porous layers with high resistance impart a major contribution to the large value of dynamic series resistance of the metal/PSi/c-Si/metal structure.

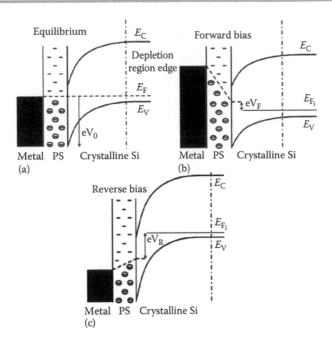

**FIGURE 14.1** A schematic band diagram of a metal-PSi-Si device following Ben-Chorin et al. in (a) equilibrium, (b) forward, and (c) reverse bias. The dashed lines represent the Fermi level. The PSi surface states are shown by filled circles and single dashes corresponding to occupied and free states, respectively. (Reprinted with permission from Ben-Chorin M. et al., *J. Appl. Phys.* 77, 4482, 1995. Copyright 1995. American Institute of Physics.)

Moreover, the large density of states associated with the mid-gap defects present in PSi restricts the band bending at the metal-PSi interface. The significant photo response signals are reported to observe only when the silicon-PSi junction is illuminated (Pulsford et al. 1993). As the PSi Fermi level is pinned at the mid-gap, the band bending and the depletion essentially occur inside silicon and the associated energy barrier obtained from the photocurrent experiments is about 0.6 eV (Ben-Chorin et al. 1995a,b). Considering all these factors, the schematic band diagram of a diode under neutral, forward, and reverse conditions is shown in Figure 14.1 (Ben-Chorin et al. 1995a).

If the electroluminescence is considered for this kind of diode, Figure 14.1 demonstrates that the hole injection is efficient while the electron injection becomes complicated because the excess energy of electrons at the Fermi level is appreciably low. In such cases, holes need to travel all the way up to the metal contact to recombine with an electron.

### 14.2.3 SPECIFIC CONTACT RESISTANCE OF M-PSi CONTACTS

Generally, the specific contact resistance, $\rho_c$ $(\Omega \cdot cm^2)$, is measured to find out the nature of ohmic contact to semiconductors. Reports are available on the measurement of $\rho_c$ of the PSi surfaces prepared from both *p*-type and *n*-type silicon wafers and with Al contacts (Zimin et al. 1995, 1998). The specific contact resistance of Al and Ni contact to *p*-type PSi (55% porosity) (Kanungo et al. 2006, 2009a) and Al contact to macroporous silicon (Maji et al. 2010) are reported. Similarly, the quantitative measurement of the specific contact resistance has been studied (Vinod 2005, 2009, 2013) for silver ohmic contact to *p*-type porous silicon. Table 14.2 gives the summary of the reported work on specific contact resistance or transition specific resistivity measurements.

The studies do not give a specific experimental evidence of a unique method to achieve a low specific contact resistance of a metal-PSi junction. However, Ag contact to PSi by screen-printing method followed by annealing may be an effective method for low resistance ohmic contact to *p*-Si. However, Ag is a relatively expensive metal for commercial applications. Therefore, Al is preferred to make a compromise between the low specific contact resistance and the low cost. Further studies may be necessary to find out a new metal or metal alloy that can satisfy much lower specific resistance, low cost, and improved performance as ohmic contact to the metal-PSi junction devices.

**TABLE 14.2    Specific Contact Resistance or Transition Specific Resistivity of the Metal Contacts to Porous Silicon Made by Different Methods**

| Metal | Si Type | Method of Contact Formations | Specific Contact Resistance/Transition Specific Resistivity ($\Omega \cdot cm^2$) | References |
|---|---|---|---|---|
| Al | $n$ | Evaporation | $8 \times 10^{-3} - 1.2 \times 10^{-2}$ | Zimin et al. 1995 |
|  |  | Evaporation | 0.6–18 | Zimin et al. 1998 |
|  | $p$ | Evaporation | 0.6–1.3 k | Zimin et al. 1998 |
|  |  | Evaporation | $1.51 \times 10^{-1}$ | Kanungo et al. 2009a |
|  |  | Evaporation | 0.57 (unannealed sample) 0.1 (annealed sample) | Maji et al. 2010 |
| Ag | $p$ | Screen printing | 0.0736 | Vinod 2005 |
|  |  | Screen printing → firing at 800°C in air → annealing at 450°C in $N_2$ → Ag electroplating | $1.01 \times 10^{-6}$ | Vinod 2009 |
|  |  | Screen printing → baking at 240°C in inert ambient → firing at 725°C in air → annealing at 450°C in $N_2$ → Ag electroless | $1.025 \times 10^{-4}$ (before electroless Ag deposition) $3.25 \times 10^{-5}$ (after electroless Ag deposition) | Vinod 2013 |
| Ni | $p$ | Electroless deposition | $1.76 \times 10^{-2}$ | Kanungo et al. 2006 |

### 14.2.4  AGING EFFECT OF M-PSi CONTACTS

Several reports indicate that the performance of metal-PSi junction degrades with time due to the slow oxidation processes because of the available surface and interface states of PSi (Tsai et al. 1991). Metal/PSi structure, as reported in the literature, may keep the nature of the contact the same with the atmosphere for a long time but the value of the current decays with time. It was observed by the series resistance as well as the ideality factor increases with the time of storage in ambient air and this fact can be explained as being due to the gradual oxidation of the porous silicon skeleton. This results in a more insulating PSi layer and in an increase of the additional voltage drop across this layer yielding higher ideality factor. It was also observed that for the PSi surface through a stabilization treatment, decrease in current is slower compared to the untreated samples (Martin-Palma et al. 1999). Perhaps the time dependent variation of the series resistance ($R_S$) and the ideality factor ($n$) has the same origin, namely the oxidation of the Si skeleton, the passivation of the surface states during the oxidation, and the absorption of other species from the atmosphere. These phenomena can decrease the conductivity of the PSi layer and increase the additional voltage drop in the interfacial region yielding higher ideality factors. In both the cases, after a long time of exposure to the ambient air, there is a tendency of $R_S$ and $n$ to reach a saturation value as shown in Figure 14.2, implying a saturation of the amount of impurities absorbed by the porous structure.

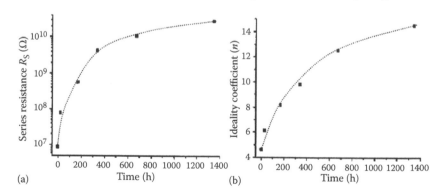

**FIGURE 14.2**   Values of (a) the series resistance ($R_s$) and (b) the ideality coefficient ($n$) after different times of exposition to the atmosphere. (Reprinted with permission from Martin-Palma R.J. et al., *J. Appl. Phys.* 85(1), 583, 1999. Copyright 1999. American Institute of Physics.)

## 14.2.5 SURFACE STABILIZATION AND STABLE METAL CONTACTS TO PSi

The huge numbers of surface states, interface states, and volume traps, of the order of $10^{15}$ cm$^{-3}$ eV$^{-1}$ are present in the quasi-Fermi level of PSi layer (Matsumoto et al. 1998), mainly originated from its nanostructure. These defect states are responsible for the creation of the recombination centers, the slow uncontrolled surface oxidation (Beckmann 1965; Petrova et al. 2000; Astrova et al. 2002; Karacali et al. 2003), Fermi-level pinning (Deresmes et al. 1995; Kanungo et al. 2009a), and hinder the electrical response of the PSi–Si interface. Therefore, the surface passivation of PSi is an important step to make a stable device for practical application.

In 1965, it was observed by Beckmann (1965) the PSi film is oxidized when kept in ambient air for a long time. This chemical oxidation is a slow process like aging of Si wafers (Harrison and Dimitriev 1991) and the formation of native oxide layer occurs on the surface of the pores. As a result, PSi showed a continuous change in the structural (Astrova et al. 2002), electrical, and optical properties (Karacali et al. 2003) with the time of storage until the completion of the growth of the native oxide after about one year (Petrova et al. 2000). To avoid slow and long aging period, chemical, anodic, dry, and wet thermal oxidation processes were investigated by Rossi et al. (2001). Rapid thermal oxidation was found to be more effective than anodic oxidation (Bsiesy et al. 1991) and provided the PSi better electronic surface passivation and improved stability for optical measurements (Petrova-Koch et al. 1992). The alternate possibility for the stabilization of the porous silicon surface is derivatization by organic groups and polymer (Lees et al. 2003; Mandal et al. 2006). There are reports on nitradation (Anderson et al. 1993) and halogenation (Lauerhaas and Sailor 1993) to stabilize the PSi surface. To stabilize its photoluminescence properties, metals like Cu, Ag, In, and so on were used to modify the porous silicon surface (Andsager et al. 1994; Steiner et al. 1994).

However, most of the available reports are on the passivation of PSi surface for the optical measurements (Bsiesy et al. 1991; Petrova-Koch et al. 1992; Anderson et al. 1993; Lauerhaas and Sailor 1993; Andsager et al. 1994; Steiner et al. 1994; Rossi et al. 2001; Lees et al. 2003; Mandal et al. 2006). Therefore, it is a challenge for the last two decades to attain a low resistance ohmic contact to porous silicon for its applications for electronic, optoelectronic, and sensor devices (Burrows et al. 1996).

Very few reports are available on surface passivation for stable electrical contact formation. While Jeske et al. (1995) showed a method of passivating PSi surface by oxidation during metallization, Bhattacharya and co-workers (2000) applied hydrogen plasma before depositing Au metal to improve the junction properties of PSi. Rabinal and Mulimani (2007) reported on the covalent attachment of 1-dodecyne molecules to PSi to passivate the surface for improved electronic charge transport of Al and Au contacts to surface passivated PSi. They explained that the density of surface states on PSi could be reduced through surface passivation making the interface between PSi and metal more sensitive to the work function of the contacting metal. It has already been mentioned earlier in this chapter that Martin-Palma et al. (1999) found a rectifying behavior of Al/PSi/Si structures even after prolonged exposure (~1400 h) to the ambient air with the decaying current and the surface of PSi layer, chemically etched with HNO$_3$, registers more stable current reading compared to the untreated samples. The controlled preannealing of the PSi surface before metallization was reported by Maji et al. (2010) to facilitate the growth and coalescence of the shallow nanocrystallites that help in removing the nanopores on macroporous silicon surfaces and yield much better ohmic contact. Mahmoudi et al. (2007) reported the sensitive, direct detection of CO$_2$ and propane by means of photoluminescence-quenching of CH$_x$ modified porous Si. CH*x* films were coated over the PSi samples by RF plasma decomposition of a methane–argon mixture followed by a thermal treatment. They also observed intense blue emission from the carbon derivatized PSi surface bounds species. Table 14.3 presents the reported results of the surface passivation of porous silicon to improve the electrical contact.

The modification of porous silicon surface by electroless dispersion of the noble metals like palladium (Pd), ruthenium (Ru) and platinum (Pt) on PSi surface using the aqueous acidic solutions of the corresponding metal chlorides at room temperature was reported by Kanungo et al. (2009b). Metal ions in the aqueous solution are reduced to metal islands by a chemical reduction process and $h^+$ released thereby oxidizes PSi surface to an ultra thin layer of SiO$_2$.

**TABLE 14.3    Different Ways of Porous Silicon Surface Modifications before Metal Contact Formation**

| Modification Procedure | Metal on PSi | Type of Contact | References |
|---|---|---|---|
| Simultaneous oxidation during metallization | Au, Cu, Ni | Rectifying | Jeske et al. 1995 |
| Hydrogen plasma modification | Au | Rectifying | Bhattacharya et al. 2000 |
| Surface modification using 1-Dodecyne organic molecule | Au, Al | Rectifying | Rabinal and Mulimani 2007 |
| HNO$_3$ treatment | Al | Rectifying | Martin-Palma et al. 1999 |
| Electroless dispersion of Pd | Al | Ohmic | Kanungo et al. 2009a |
| Heat treatment | Al | Ohmic | Maji et al. 2010 |
| CHx films were coated over the PSi samples by RF plasma decomposition of a methane–argon mixture followed by a thermal treatment | Ag | Rectifying | Mahmoudi et al. 2007 |

I-V measurements were repeated at different intervals of time for a period of 28 days to verify the quality and the stability of Al contact to modified porous silicon. Figure 14.3 shows that there is practically an overlap of I-V curves signifying a stable Al contact. Specific contact resistance of Al to Pd modified PSi was determined by TLM method and found to be $1.51 \times 10^{-1}\,\Omega \cdot cm^2$.

Luminescence property of PSi is also very sensitive to its surface properties and it might be improved by surface modification. The deposition of some alkali metals on PSi surface by immersion plating using XNO$_3$ solutions (X = Li, Na, K) are reported in the literature. Immersion plating of alkali metals oxidizes the PSi surface and PL intensity is enhanced for the critical metallization times. The energy states due to the newly formed oxygen or alkali metal oxide may give rise to a blue shift in the PL spectra. Thus, the surface modification by alkali metallization using immersion plating method leads to a stable and efficient photoluminescence from PSi (Esmer and Kayahan 2009; Kayahan et al. 2012).

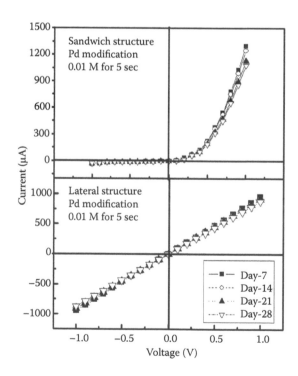

**FIGURE 14.3** I-V characteristics of Pd modified PSi surface with Al contacts for lateral and sandwich structures, measured for 28 days at different intervals of time.

## 14.3 SEMICONDUCTOR—POROUS SILICON JUNCTION

### 14.3.1 c-Si/PSi JUNCTION

c-Si/PSi heterojunction is produced by the formation of nanocrystalline porous Si layers on c-Si by electrochemical etching. The junction between porous silicon and crystalline silicon has been confirmed by the detailed study of metal/PSi/c-Si structure (Bhattacharya et al. 2000). The study on the carrier transport of metal/PSi/Si structure reveals that the Al/PSi junctions are nonrectifying and quasilinear where as Al/PSi/c-Si junctions are weakly rectifying. Therefore, the rectifying behavior is believed to be due to PSi/c-Si heterojunction.

It was observed by the researchers that the forward and reverse current of metal/PSi/c-Si structure at low bias is limited by the resistivity of the oxidized PSi layers. The distribution of localized states is determined by the trap-filled space charge limited current (SCLC) observed in these structures under high bias. There is a current spreading over the large areas of device structures due to oxidized PSi, similar to a device fabricated on the highly resistive substrate. Highly conductive inversion (*n*-type) layer that is formed at the *c*-Si substrate of the PSi/c-Si heterojunction may be the reason of the spreading current. The current-spreading effect creates high reverse current and high capacitance of a device structure. The aging of oxidized sandwich structures may remove the inversion layer and increase the resistivity of the PSi layer (Balagurov et al. 2001a,b). Further investigation also reports that the diode ideality factor (*n*) is almost 8 for the PSi/c-Si heterojunction at ≤0.5 V forward bias voltage (and about 50 for the bias ≤5 V) and nearly 1 for ≤0.5 V reverse bias. In contrast to the conventional expectation, the barrier height measured from I-V data at ≤0.5 V is higher for the forward bias than that for the reverse bias. The results of the I-V studies with PSi/c-Si junctions are explained as due to a multitunneling recombination model for the forward bias and a carrier generation and barrier lowering effects for reverse bias. Valuable information on the current transport across the c-Si/PSi junction is obtained from the reverse bias I-V characteristics. As already mentioned, the carrier generation-recombination mechanism in the depletion region formed on the PSi side of the PSi/c-Si heterojunction dominates the reverse bias current transport mechanism. However, the barrier lowering effect at the relatively higher reverse bias governs the reverse current transport. The behavior is more like a Schottky junction with the Fermi level pinned to the defect energy levels of the c-Si/PSi interface. The c-Si Fermi level is pinned at the defect levels of the interface. In this case, the reverse bias barrier height is found to be equal to the band offset at the conduction band edges. Based on the I-V data analysis, the energy band diagram of the c-Si/PSi heterojunction (Figure 14.4) is presented next (Islam et al. 2007).

### 14.3.2 POROUS SILICON *p-n* HOMOJUNCTION

Porous silicon *p-n* homojunction is an interesting structure that can produce different novel electronic devices. Since sufficient information on this junction is not available in the literature, there is a scope for more investigations. Porous silicon *p-n* junction is fabricated by diffusion or by ion-implantation technique followed by the formation of porous configuration over the junction.

Figure 14.5 shows the schematic of the junction with the different layers on the substrate. p-type porous silicon is first formed by electrochemical etching and then doping of the PSi layer by diffusion or by ion-implantation to make the surface *n*-type so that PSi *n-p* junction is fabricated. Au and Al metal contacts are made by thermal evaporation or sputtering.

Electroluminescence effect in this kind of homojunction can be explained from the concept of electron transport through the *p-n* junction following the idea of the modified quantum confinement effect (Bisi et al. 2000). In such cases, the *p*+ or *n*+-PSi top layer is necessarily highly doped so that the diffusion of the doping impurities can occur from all the sides. Under the influence of an electric field, both carriers start flowing possibly by injection. At low voltage and current, it is possible for the carriers to undergo the radiative recombination in the thinned silicon wires and to create the bright luminescence.

Figure 14.6a depicts the proposed band diagram at the thermal equilibrium of a light-emitting PSi based *p-n* junction device. According to this model, there exists an internal electric field within the PSi layer. Assuming that both *n*- and *p*-type PSi have approximately the same carrier concentration,

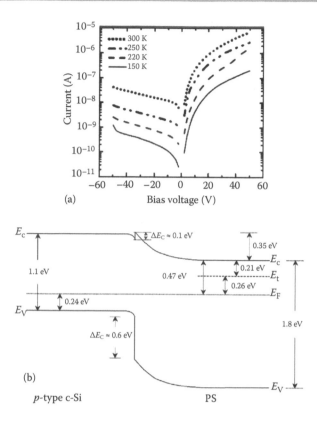

**FIGURE 14.4**   I-V characteristics and energy band diagram of the PSi/c-Si junction with Al as the contact metal. (From Islam M.N. et al., *J. Phys. D: Appl. Phys.* 40, 5840, 2007. Copyright 2007: The Institute of Physics. With permission.)

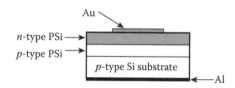

**FIGURE 14.5**   Schematic of the junction with the different layers on the substrate.

the Fermi levels are nearly 0.5 eV above and below the intrinsic level, respectively, and thus a band bending of 1 eV is expected within the PSi layer. Figure 14.6b exhibits the forward bias J–V characteristics as reported in the literature. The ideality factor at room temperature, calculated from the inset of Figure 14.6b, is 2.1 for the applied bias less than 0.5 V. This value of the ideality factor may be explained from the mechanism of carrier recombination in the depletion region, which is especially significant for the junction devices with the wide band gap semiconductors. However, for a large deviation from the ideal I-V behavior at higher voltages, only the series resistance ($R_S$) of the contact and the substrate region is not responsible. Therefore, a voltage dependent $R_S$ needs to be considered to devise an accurate model of the I-V characteristics over the entire forward bias voltage regime. The impact of carrier diffusion current, space charge current, and series resistance is to be taken into account. In fact, the voltage applied to the *p-n* junction device is distributed across the series combination of the above three components and the total voltage is the sum of the three like,

$$(14.6) \qquad\qquad V = V_1\,(I) + V_2\,(I) + V_3\,(I)$$

where $V_1$, $V_2$, and $V_3$ are the voltage drop across the *p-n* junction due to the carrier diffusion, the space charge region, and the series resistance, $R_S$, respectively. The above equation provides an excellent fit to the experimental I-V characteristics as shown in Figure 14.7, including the temperature dependence (140 K and 300 K). Thus, the existence of a *p-n* junction within the PSi layer is substantiated.

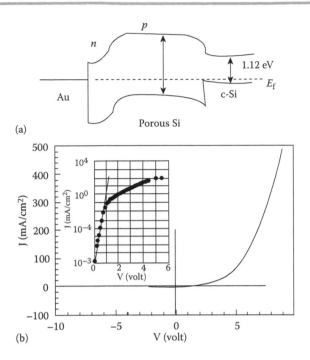

(a)

(b)

**FIGURE 14.6** (a) The band diagram of a PSi *p-n* device. (b) The J–V relation of a PSi *p-n* device. (Reprinted with permission from Peng C. et al., *J. Appl. Phys.* 80(1), 295, 1996. Copyright 1996. American Institute of Physics.)

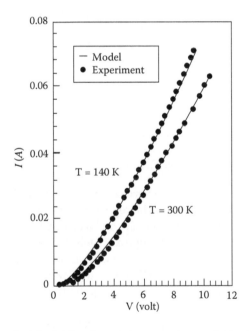

**FIGURE 14.7** The theoretical (solid lines) and the experimental (dots) I-V curves of a LEPSi *p-n* device at two different temperatures. (Reprinted with permission from Peng C. et al., *J. Appl. Phys.* 80(1), 295, 1996. Copyright 1996. American Institute of Physics.)

Different mechanisms have been proposed thus far to interpret the rectifying properties of PSi diodes and the three most common mechanisms are: (1) supply of the minority carriers from the Si substrate or from the barrier at the PSi/Si junction, (2) Schottky barrier at the contact/PSi junction, and (3) *p-n* junction within the PSi layer made from the c-Si (Shih et al. 1992; Muller et al. 1993). There is no single mechanism to explain the function of all PSi diodes and the rectification mechanism depends on the fabrication conditions of the PSi diode. Table 14.4 gives the electrical properties of porous silicon *p-n* homojunctions reported in the literature.

**TABLE 14.4** **Electrical Properties of Porous Silicon *p-n* Homojunctions Reported in the Literature**

| Si Type | Type of Junction | Application and Properties | References |
|---|---|---|---|
| Diffused *p⁺-n, n⁺-p, p⁺-p, n⁺-n* | *p⁺*-PSi/*n*-PSi, *n⁺*-PSi/*p*-PSi, *p⁺*-PSi/*p*-PSi, *n⁺*-PSi/*n*-Psi | *PL measurements* (Modified quantum confinement effect; top *p⁺* or *n⁺* PSi layer is highly doped so diffusion of impurities in silicon wires occurs from all sides of the wire; flow of carriers is possible by injection under an electric field; radiative recombination possible at low voltage and currents; produces bright luminescence.) | Gupta et al. 1995 |
| Diffused *n⁺-p* and *p⁺-p* | *n⁺*-PSi/*p*-Psi *p⁺*-PSi/*p*-Psi | *Visible light emitting diode* (Light emission is attributed to electron-hole recombination across the direct band gap of the monocrystalline quantum wire.) | Chen et al. 1993 |
| Diffused *p⁺-n, n⁺-p,* | Nanoporous *n*-layer between a mesoporous *p⁺*-capping layer and the macroporous *n* substrate | *Light emitting diode* (Bright red-orange light emission under forward bias; increased quantum efficiency.) | Steiner et al. 1993. |
| *p⁺* on *n* | *p*-PSi/*n*-PSi | *Gas sensor* (Relative current variation is a function of the diode bias; possible to tune the diode output by changing the polarization voltage.) | Barillaro et al. 2008 |
| Diffused *n* on *p* | Heavily doped *p-n* junction porous silicon | *Light emitting diode* (Electroluminescence observed under forward bias conditions; proposed light emission mechanism is due to electron-hole injection in the silicon nanostructures forming the porous silicon material.) | Das and McGinnis 1999 |

### 14.3.3 POROUS SILICON HETEROJUNCTION

#### 14.3.3.1 METAL OXIDE-PSi HETEROSTRUCTURES

The heterojunction formed at the interface between metal oxide and porous silicon is schematically represented in Figure 14.8. In most of the reported literature, porous silicon was fabricated from *p*-type c-Si material. In addition, porous silicon has a smaller band gap (1.2–1.8 eV, depending on the porosity) than the metal oxides (approximately ≥2.7 eV). Therefore, when these materials are brought into contact, depending on the position of the Fermi level, the electrons flow from the material with low work function to that with high work function until their Fermi levels equalize, which creates an electron depletion layer or potential barrier at the interface and bends the energy band.

Table 14.5 presents the electrical properties of porous silicon–metal oxide heterojunctions reported in the literature.

#### 14.3.3.2 POLYMER-PSi HETEROSTRUCTURES

Nanopolyaniline/*p*-type porous silicon (NPANI/PSi) heterojunction fabricated via in situ polymerization (El-Zohary et al. 2013) is reported in the literature. I-V measurements were done at different temperatures to calculate the ideality factor, barrier height, and series resistance of the junction. The nonideal diode behavior was confirmed from the calculated ideality factor. The series resistance was found to decrease with increasing temperature. As reported by the other group (Halliday et al. 1996), the work function of polyaniline is around 4.8 eV, a value between that of Au and Al. Therefore, it is expected that polyaniline deposited on a porous silicon substrate

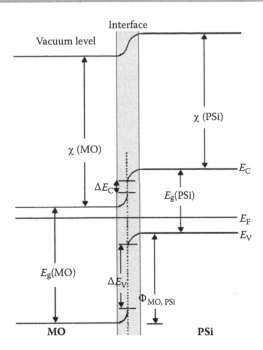

**FIGURE 14.8**    A schematic band diagram of metal oxide (MO)–porous silicon (PSi) junction.

would behave like a $p^+-n$ junction and form a rectifying contact (Li et al. 1994; Halliday et al. 1996; Fan et al. 1998). The rectification ratio measured at 5:5 V is 490. The I-V curves clearly demonstrate the formation of a rectifying contact. White light illumination of the diode during the I-V measurement showed an increase in the reverse bias current as due to the expected electron-hole pair generation in the depletion region. The forward bias characteristics show a hysteresis, probably due to the charge transfer and trapping in the polyaniline film.

Electro polymerization of pyrrole on PSi was utilized to fabricate a heterojunction of porous silicon with polypyrrole (PPy). I-V characteristics of different PPy/PSi structures were studied in dark and under illumination. PPy forms a rectifying contact with PSi layer on p-Si substrate (Shen and Wan 1998; Arenas et al. 2006). However, the photo voltage was obtained with the PPy/*n*-Si heterojunction. When porous silicon powder was added between *n*-Si and PPy layer, the photovoltaic performance of this novel junction was significantly improved. By fitting the I-V data with the modified diode equation, the series resistance of PPy/*n*-Si junction and PPy/PSi (powder)/*n*-Si junction are 10 KΩ and 1 KΩ, respectively (Arenas et al. 2006). Table 14.6 presents the electrical properties of PSi/polymer junction reported in the literature.

By considering the typical energy parameters, the energy band diagram of Al (Au)/Si–PSi–OMs–M junction was reported and is shown in Figure 14.9. PSi is a complex material made up of nanowires of silicon with large density of surface states. In the past, an average band gap of 2 eV with equilibrium Fermi-level close to the intrinsic value was accepted for many calculations. Experimentally, the observed band offsets between Si and PSi are $E_C$ = 0.29 eV and $E_V$ = 0.59 eV. The surface states of PSi are decreased drastically by chemical bonding with 1-dodecyne molecules and an ultra thin layer (a few angstroms) of molecular insulator with HUMO–LUMO gap of about 9 eV is formed. Considering this gap, the work function of Si and the work function of the contact metals, an energy band diagram for the nonintimate contacts is shown in Figure 14.9. Fermi-levels of both Al and Au are shown in the same figure. The charge transport through this junction occurs mainly by tunneling of the charge carriers in the presence of the thin molecular layer between the metal and PSi. With a small forward bias to the metal of Al-PSi junction, electrons directly face the conduction band of PSi and can easily tunnel through the barrier. This tunneling probability increases with the increasing applied voltage. On the other hand, the holes can tunnel from the metal to the valence band of PSi under the reverse bias, but this probability is very low because of the greater separation between the two levels. In case of Au-PSi junction, the situation is different and the metal Fermi-level is closer to the valence band of PSi. Therefore, under the forward bias, the electrons of the metal face a higher barrier to move into

**TABLE 14.5  Electrical Properties of Porous Silicon Heterojunction Reported in the Literature**

| Si Type | Type of Junction | Application and Properties | References |
|---|---|---|---|
| p- | Pd-SnO$_2$/PSi | *Luminescence H$_2$, CO$_2$, CO sensor* <br> (Blue emission originates from the electron transition from the shallow donor level of oxygen vacancies to the valance band; band gap can be tuned by varying the current density; brighter PL is at longer wavelengths.) | Ismail et al. 2013 |
| p- | SnO$_2$:F/SiO$_x$/PSi | *Electrical characterization* <br> (Structure consists of two opposing diodes; each diode is associated with an independent current source.) | Garces et al. 2012 |
| p- | SnO$_2$:F/PSi | *Light emitted diodes* <br> (Fluorinated species in SnO$_2$ help in the improvement of resistance to irradiation and degradation of carrier injection.) | Macedo et al. 2008 |
| p-, n- | p-Ag$_2$O/p-PSi/c-Si; <br> p-Ag$_2$O/n-PSi/c-Si | *Optoelectronic application* <br> (Films formed at thickness 100 nm showed [111] strong reflection along with weak reflections of [101] corresponding to the growth of single phase Ag$_2$O with cubic structure.) | Hassan et al. 2014 |
| p- | ITO/PSi | *Light emitted diodes* <br> (Electrical transport is limited by the carrier recombination in the depletion layer; recombination could be due to the high density of interface states related to grain boundaries and nanopores.) | Ghosh et al. 2002b |
| p- | ITO/PSi | *Light emitted diodes* <br> (Emits white light when biased in the opposite direction; it may be from the intra-oriented-band transitions of hot carriers generated in an avalanche process.) | Xie and Blackwood 2013 |
| p- | Va$_2$O$_5$/PSi/Si | *Ethanol sensor* <br> (C–V and G/ω–V characteristics confirm the presence of interface states.) | Chebout et al. 2013 |
| p- | ZnO/PSi | *Gas sensor* <br> (The PL test suggested containing some lattice defects in the electrodeposited ZnO on PSi.) | Yan et al. 2014 |
| p- | ZnO/PSi | *Gas sensor* <br> (Temperature stability of the sensor has been improved.) | Kanungo et al. 2010 |
| p- | WO$_3$/PSi | *Optoelectronic application* <br> (Evolution of PL peak, corresponding to band edge emission of WO$_3$, increases with annealing temperature under nitrogen atmosphere in contrast with air atmosphere.) | Mendoza-Agüero and Agarwal 2013 |
| p- | WO$_3$ NWs/PSi composite | *Gas sensor* <br> (Modulations of the potential barriers at the interface due to interaction with gas and sensor response take place at the relatively low operating temperature.) | Ma et al. 2014 |
| p- | CHx/PSi/Si | *Gas sensor* <br> (Junction shows a good rectifying behavior and reversible interaction with gases; CHx/PSi layers are responsible for strong band bending at the silicon surface.) | Belhousse et al. 2004, 2005; Gabouze et al. 2006; Mahmoudi et al. 2007 |

the conduction band of PSi. Therefore, the tunneling should occur at higher voltages compared to that in Al-junction. However, under the reverse bias, the situation is opposite to that of the Al-PSi junction and the tunneling barrier for the holes to move from metal to PSi is relatively low. Hence, higher reverse current can be expected to be observed. Therefore, for polymer/PSi junctions it can be generally inferred that Al acts as a better electron injector (minority carrier device) and Au as a better hole injector (majority carrier device) (Rabinal and Mulimani 2007).

**TABLE 14.6    Tabular Representation of the Electrical Properties of PSi/Polymer Junction Reported in the Literature**

| Si Type | Type of Junction | Application and Properties | References |
|---|---|---|---|
| *p-, n-* | Polypyrrole/PSi | *Opto-electrical effect of the junction*<br>(*p*-PSi/PPy forms a rectified and photoconductive junction; *n*-PSi/PPy shows photovoltaic phenomenon.) | Arenas et al. 2006 |
| *p-* | Polypyrrole/PSi | *Photoluminescence based gas sensor*<br>(Strong discrimination of PL response for polar alcohols.) | Vrkoslav et al. 2006 |
| *n-* | Polyaniline/PSi | *Light-emitting diodes*<br>(Rectifying junction; emits visible electroluminescence when forward biased.) | Halliday et al. 1996 |
| *n-* | Polyaniline/PSi | *Light-emitting diodes*<br>(Devices showed increased luminescence; polymer contact is relatively transparent at the emission wavelengths.) | Li et al. 1994 |
| *n-* | Polyaniline/PSi | *Light-emitting diodes*<br>(The contact has rectifying behavior; Schottky barrier model explained the I-V curve; the barrier height varies from 0.78 to 0.85 eV; the ideality factor ranges from 2.8–5.2; visible electroluminescence has been obtained under a forward bias.) | Halliday et al. 1997 |
| *p-* | Polyaniline/PSi | *Electrical characterization*<br>(Nonideal diode behavior; the series resistance was found to decrease with increasing temperature.) | El-Zohary et al. 2013 |
| *p-* | 1-Dodecyne organic molecules/PSi/Si | *Electrical characterization*<br>(PSi–M and Si–PSi interfaces are equally important in influencing the junction parameters when surface states are passivated.) | Burrows et al. 1996 |

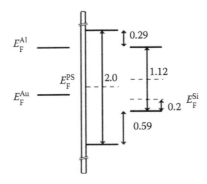

**FIGURE 14.9**    Schematic band diagram of a polymer-PSi heterostructures. (From Rabinal M.K. et al. *New J. Phys.* 9, 440, 2007. Copyright 2007: the Institute of Physics. With permission.)

## 14.4  SUMMARY

Like any other electronic devices, electrical contacts are important and critical with the PSi-based devices. Since there are pores between the tiny silicon particles in PSi, the contact formation becomes difficult, and inappropriate contacts affect the electrical connections in the circuit as well as the performance of the devices. Unfortunately, until now, the formation of high-quality contacts to PSi is a typical problem of materials science and engineering, and the solution of this problem requires additional research.

## ACKNOWLEDGMENTS

The authors are thankful to Prof. C.K. Sarkar, the coordinator of IC Design & Fabrication Center, Department of Electronics & Telecommunication Engineering, Jadavpur University, Kolkata,

India for providing the ambience and the facilities for writing the chapter. Dr. (Ms) J. Kanungo thankfully acknowledges CSIR, Government of India for the Senior Research Associateship.

## REFERENCES

Anderson, R.C., Muller, R.S., and Tobias, C.W. (1991). Investigation of electrical properties of porous silicon. *J. Electrochem. Soc.* 138, 3406–3411.

Anderson, R.C., Muller, R.S., and Tobias, C.W. (1993). Chemical surface modification of porous silicon. *J. Electrochem. Soc.* 140, 1393–1396.

Andersson, H.A., Thungstrom, G., and Nilsson, H. (2008). Electroless deposition and silicidation of Ni contacts into *p*-type porous silicon. *J. Porous Mater.* 15, 335–341.

Andsager, D., Hilliard, J., and Nayfeh, M.H. (1994). Behavior of porous silicon emission spectra during quenching by immersion in metal ion solutions. *Appl. Phys. Lett.* 64, 1141–1143.

Angelescu, A. and Kleps, I. (1998). Metallic contacts on porous silicon layers. In: *Proceedings of Semiconductor Conference, CAS '98*, Sinaia, Romania, 6–10 October, pp. 447–450.

Arenas, M.C., Hu H., Rio, J.A. et al. (2006). Electrical properties of porous silicon/polypyrrole heterojunctions. *Sol. Energ. Mater. Sol. Cells* 90, 2413–2420.

Astrova, E.V., Ratnikov, V.V., Remenyuk, A.D., and Shulpina, I.L. (2002). Starins and crystal lattoce defects arising in macroporous silicon under oxidation. *Semiconductors* 36, 1033–1042.

Balagurov, L.A., Bayliss, S.C., Kasatochkin, V.S., Petrova, E.A, Unal, B., and Yarkin, D.G. (2001a). Transport of carriers in metal/porous silicon/c-Si device structures based on oxidized porous silicon. *J. Appl. Phys.* 90, 4543–4548.

Balagurov, L.A., Bayliss, S.C., Andrushin, S.Y. et al. (2001b). Metal/PS/c-Si photodetectors based on unoxidized and oxidized porous silicon. *Solid-State Electron.* 45, 1607–1611.

Banihashemian, S.M., Hajghassem, H., Erfanian, A. et al. (2010). Observation and measurement of negative differential resistance on PtSi Schottky junctions on porous silicon. *Sensors* 10, 1012–1020. doi:10.3390/s100201012.

Barillaro, G., Diligentia, A., Strambinia, L.M., Cominib, E., and Faglia, G. (2008). NO$_2$ adsorption effects on $p^+$–$n$ silicon junctions surrounded by a porous layer. *Sens. Actuators B* 134, 922–927.

Beckmann, K.H. (1965). Investigation of the chemical properties of stain films on silicon by means of infrared spectroscopy. *Surf. Sci.* 3, 314–332.

Belhousse, S., Cheraga, H., Gabouze, N., and Outamzabet, R. (2004). Fabrication and characterization of a new sensing device based on hydrocarbon groups (CH$_x$) coated porous silicon. *Sens. Actuators B* 100, 250–255.

Belhousse, S., Gabouze, N., Cheraga, H., and Henda, K. (2005). CHx/PS/Si as structure for propane sensing. *Thin Solid Films* 482, 253–257.

Ben-Chorin, M., Moller, F., and Koch, F. (1995a). Band alignment and carrier injection at the porous silicon—Crystalline silicon interface. *J. Appl. Phys.* 77, 4482–4488.

Ben-Chorin, M., Moller, F., Koch, F., Schirmacher, W., and Eberhard, M. (1995b). Hopping transport on a fractal: Ac conductivity of porous silicon. *Phys. Rev. B* 51, 2199.

Bhattacharya, E., Ramesh, P., and Kumar, C.S. (2000). Studies on gold/porous silicon/crystalline silicon junctions. *J. Porous Mater.* 7, 299–301.

Bisi, O., Ossicini, S., and Pavesi, L. (2000). Porous silicon: A quantum sponge structure for silicon based optoelectronics. *Surf. Sci. Rep.* 38, 1–126.

Bsiesy, A., Vial, J.C., Gaspard, F. et al. (1991). Photoluminescence of high porosity and of electrochemically oxidized porous silicon layers. *Surf. Sci.* 254, 195–200.

Burrows, P.E., Shen, Z., Bulovic, V. et al. (1996). Relationship between electroluminescence and current transport in organic heterojunction light-emitting devices. *J. Appl. Phys.* 79, 7991–8006.

Canham, L. (ed.) (1997). *Properties of Porous Silicon*. INSPEC—The Institution of Electrical Engineers, London.

Chebout, K., Iratni A., Bouremana A., Sam S., Keffous A., and Gabouze N. (2013). Electrical characterization of ethanol sensing device based on vanadium oxide/porous Si/Si structure. *Solid State Ionics* 253, 164–168.

Chen, Z., Bosman, G., and Ochoa, R. (1993). Visible light emission from heavily doped porous silicon homojunction pn diodes, *Appl. Phys. Lett.* 62, 708–710.

Cherif, A., Jomni, S., Hannachi, R. et al. (2013). Electrical investigation of the Al/porousSi/$p^+$-Si heterojunction. *Phys B* 409, 10–15.

Das, B. and McGinnis, S.P. (1999). Porous silicon $p$-$n$ junction light emitting diodes. *Semicond. Sci. Technol.* 14, 988–993.

Deresmes, D., Marissael, V., Stievenard, D. et al. (1995). Electrical behaviour of aluminium-porous silicon junctions. *Thin Solid Films* 255, 258–261.

Dhar, S. and Chakrabarti, S. (1996). Electroless nickel plated contacts on porous silicon. *Appl. Phys. Lett.* 68(10), 1392–1393.

Diligenti, A., Nannini, A., Pennelli, G. et al. (1996). Electrical characterization of metal Schottky contacts on luminescent porous silicon. *Thin Solid Films* 276, 179–182.

El-Zohary, S.E., Shenashen, M.A., Allam, N., Okamoto, T., and Haraguchi, M. (2013). Electrical characterization of nanopolyaniline/porous dilicon heterojunction at high temperatures. *J. Nanomater.* 2013, 568175.

Esmer, K. and Kayahan, E. (2009). Influence of alkali metallization (Li, Na and K) on photoluminescence properties of porous silicon. *Appl. Surf. Sci.* 256, 1548–1552.

Fan, J., Wan, M., and Zhu, D. (1998). Studies on the rectifying effect of the heterojunction between porous silicon and water-soluble copolymer of polyaniline. *Synth. Metals* 95, 119–124.

Gabouze, N., Belhousse, S., Cheraga, H., Ghellai, N., Ouadah, Y., Belkacem, Y., and Keffous, A. (2006). $CO_2$ and $H_2$ detection with a $CH_x$/porous silicon-based sensor. *Vacuum* 80, 986–989.

Gallach, D., Torres-Costa, V., García-Pelayo, L. et al. (2012). Properties of bilayer contacts to porous silicon. *Appl. Phys. A* 107, 293–300.

Garces, F.A., Urteaga, R., Acquaroli, L.N., Koropecki, R.R., and Arce, R.D. (2012). Current-voltage characteristics in macroporous silicon/$SiO_x$/$SnO_2$:F heterojunctions. *Nanoscale Res. Lett.* 7, 419–424.

Ghosh, S., Hong, K., and Lee, C. (2002a). Structural and physical properties of thin copper films deposited on porous silicon. *Mater. Sci. Eng. B* 96, 53–59.

Ghosh, S., Kim, H., Hong, K., and Lee, C. (2002b). Microstructure of indium tin oxide films deposited on porous silicon by rf-sputtering *Mater. Sci. Eng. B* 95, 171–179.

Giebel, G. and Pavesi, L. (1995). About the I-V characteristics of metal porous silicon diode. *Phys. Stat. Sol. (a)* 151, 355–361.

Gupta, A., Jain, V.K., Jalwania, C.R. et al. (1995). Technologies for porous silicon devices. *Semicond. Sci. Technol.* 10, 698–702.

Halliday, D.P., Holland, E.R., Eggleston, J.M. et al. (1996). Electroluminescence from porous silicon using a conducting polyaniline contact. *Thin Solid Films* 276, 299–302.

Halliday, D.P., Eggleston, J.M., Adams, P.N., Pentland, I.A., and Monkman, A.P. (1997). Visible electroluminescence from a polyaniline–porous silicon junction. *Synthetic Metals* 85, 1245–1246.

Han, Z., Shi, J., Tao, H. et al. (1994). Photovoltaic effect of a metal/porous silicon/silicon structure. *Phys. Lett A* 186, 265–268.

Harrison, H.B. and Dimitriev, S. (1991). Ultra-thin dielectrics for semiconductor applications—Growth and characteristics. *Micrel. J.* 3–38.

Hassan, M.A.M., Agool, I.R., and Raoof, L.M. (2014). Silver oxide nanostructure prepared on porous silicon for optoelectronic application, *Appl. Nanosci.* 4, 429–447.

Ichinohe, T., Nozaki, S., and Morisaki, H. (1996). Visible light emission from the porous alloyed PtlSi contacts. *Thin Solid Films* 281–282, 610–612.

Islam, M.N., Ram, S.K., and Kumar, S. (2007). Band edge discontinuities and carrier transport in c-Si/porous silicon heterojunctions. *J. Phys. D: Appl. Phys.* 40, 5840–5846.

Ismail, R.A., Aseel, M., and Majeed, A. (2013). Investigation of Pd:$SnO_2$/ porous silicon structures for gas sensing. *Int. J. Appl. Innov. Eng. Manag. (IJAIEM)* 2, 339–343.

Jeske, M., Schultze, J.W., Thonissen, M. et al. (1995). Electrodeposition of metals into porous silicon. *Thin Solid Films* 255, 63–66.

Kanungo, J., Pramanik, C., Bandopadhyay, S. et al. (2006). Improved contacts on a porous silicon layer by electroless nickel plating and copper thickening. *Semicond. Sci. Technol.* 21, 964–970.

Kanungo, J., Maji, S., Saha, H., and Basu, S. (2009a). Stable aluminium Ohmic contact to surface modified porous silicon. *Solid-State Electron.* 53, 663–668.

Kanungo, J., Saha, H., and Basu, S. (2009b). Room temperature metal-insulator–semiconductor (MIS) hydrogen sensors based on chemically surface modified porous silicon. *Sens. Actuators B* 140, 65–72.

Kanungo, J., Saha, H., and Basu, S. (2010). Pd sensitized porous silicon hydrogen sensor—Influence of ZnO thin film. *Sens. Actuators B* 147, 128–136.

Karacali, T., Cakmak, B., and Efeoglu, H. (2003). Aging of porous silicon and the origin of blue shift. *Opt. Express* 11, 1237–1242.

Kayahan, E., Özerc, M., and Oral, A.Y. (2012). Effects of alkali metallization on the luminescence degradation of porous silicon. *Acta Phys. Pol. A* 121, 281–283.

Lauerhaas, J.M. and Sailor, M.J. (1993). The effects of halogen exposure on the photoluminescence of porous silicon. *Mater. Res. Soc. Symp. Proc.* 298, 259–263.

Lees, I.N., Lin, H., Canaria, C.A. et al. (2003). Chemical stability of porous silico surfaces electrochemically modified with functional alkyl species. *Langmuir* 19, 9812–9817.

Li, K., Diaz, D.C., He, Y. et al. (1994). Electroluminescence from porous silicon with conducting polymer film contacts. *Appl. Phys. Lett.* 64(18), 2394–2396.

Lin, J.C., Tsai, W.C., and Lee, W.S. (2006). The improved electrical contact between a metal and porous silicon by deposition using a supercritical fluid. *Nanotechnology* 17, 2968–2971.

Lue, J.T., Chang, C.S., Chen, C.Y. et al. (1999). The bistable switching property of a porous-silicon Schottky barrier diode during the charging period. *Thin Solid Films* 339, 294–298.

Ma, S., Hu, M., Zeng, P., Li, M., Yan, W., and Qin, Y. (2014). Synthesis and low-temperature gas sensing properties of tungstenoxide nanowires/porous silicon composite. *Sens. Actuators B* 192, 341–349.

Macedo, A.G., Vasconcelos, E.A., Valaski, R., Muchenski, F., Silva, E.F. Jr., Silva, A.F., and Roman, L.S. (2008). Enhanced lifetime in porous silicon light-emitting diodes with fluorine doped tin oxide electrodes. *Thin Solid Films* 517, 870–873.

Mahmoudi, B., Gabouze, N., Guerbous, L., Haddadi, M., Cheraga, H., and Beldjilali, K. (2007). Photoluminescence response of gas sensor based on $CH_x$/porous silicon—Effect of annealing treatment. *Mater. Sci. Eng. B* 138, 293–297.

Maji, S., Das, R.D., Jana, M. et al. (2010). Formation of ohmic contact by pre-annealing of shallow nanopores in macroporous silicon and its characterization. *Solid-State Electron.* 54, 568–574.

Mandal, N.P., Sharma, A., and Agarwal, S.C. (2006). Improved stability of nanocrystalline porous silicon after coating with a polymer. *J. Appl. Phys.* 100, 024308.

Martin-Palma, R.J., Perez-Rigueiro, J., Guerrero-Lemus, R. et al. (1999). Ageing of aluminum electrical contacts to porous silicon. *J. Appl. Phys.* 85(1), 583–586.

Matsumoto, T., Mimura, H., Koshida, N. et al. (1998). The density of states in silicon nanostructures determined by space-charge-limited current measurements. *J. Appl. Phys.* 84(11), 6157–6161.

Mendoza-Agüero, N. and Agarwal, N.V. (2013). Optical and structural characterization of tungsten oxide electrodeposited on nanostructured porous silicon: Effect of annealing atmosphere and temperature. *J. Alloys Compounds* 581, 596–601.

Muller, F., Herino, R., Ligeon, M. et al. (1993). Photoluminescence and electroluminescence from electrochemically oxidized porous silicon layers. *J. Lumin.* 57, 283–292.

Pavesi, L. and Dubos, P. (1997). Random porous silicon multilayers application to distributed Bragg reflectors and interferential Fabry Perot filters. *Semicond. Sci. Technol.* 12, 570–575.

Peng, C., Hirschman, K.D., and Fauchet, P.M. (1996). Carrier transport in porous silicon light-emitting devices. *J. Appl. Phys.* 80(1), 295–300.

Petrova, E.A., Bogoslovskaya, K.N., Balagurov, L.A., and Kochoradze, G.I. (2000). Room temperature oxidation of porous silicon in air. *Mater. Sci. Eng. B* 69–70, 152–156.

Petrova-Koch, V., Muschik, T., Kux, A., Meyer, B.K., Koch, F., and Lehmann, V. (1992). Rapid-thermal-oxidized porous Si—The superior photoluminescent Si. *Appl. Phys. Lett.* 61, 943–945

Pulsford, N.J., Rikken, G.L.J.A., Kessener, Y.A.R.R., Lous, E.J., and Venhuizenv, A.H.J. (1993). Carrier injection and transport in porous silicon Schottky diodes. *J. Lumin.* 57, 181–184.

Rabinal, M.K. and Mulimani, B.G. (2007). Transport properties of molecularly stabilized porous silicon schottky junctions. *New J. Phys.* 9, 440–448.

Remaki, B., Populaire, C., Lysenko, V., and Barbier, D. (2003). Electrical barrier properties of meso-porous silicon. *Mater. Sci. Eng. B* 101, 313–317.

Rossi, A.M., Amato, G., Camarchia, V., Boarnio, L., and Borini, S. (2001). High-quality porous-silicon buried waveguides. *Appl. Phys. Lett.* 78, 3003–3005.

Scheen, G., Bassu, M., and Francis, L.A. (2012). Metal electrode integration on macroporous silicon: Pore distribution and morphology. *Nanoscale Res. Lett.* 7, 395–399.

Shen, Y. and Wan, M. (1998). Heterojunction diodes of soluble conducting polypyrrole with porous silicon. *Synth. Metals* 98, 147–152.

Shih, S., Tsai, C., Li, K.H., Jung, K.H., Campbell, J.C., and Kwong, D.L. (1992). Control of porous Si photoluminescence through dry oxidation. *Appl. Phys. Lett.* 60, 633–635.

Simons, A.J., Cox, T.I., Uren, M.J. et al. (1995). The electrical properties of porous silicon produced from $n^+$ silicon substrates. *Thin Solid Films* 255, 12–15.

Skryshevsky, V.A. (2014). Basics of MIS-type gas sensors with thin nanoporous silicon. In: *Proceedings of IEEE XXXIV International Scientific Conference on Electronics and Nanotechnology (ELNANO)*, Kyiv, Ukraine, 15–18 April, pp. 78–82.

Skryshevsky, V.A., Strikha, V.I., Mamikin, A.V. et al. (1998). Availability of current-voltage characteristics for porous silicon gas sensors, discrete gas sensor. In: *Proceedings of Eorosensors XII*, 13–16 September, pp. 277–280.

Slobodchikov, S.V., Salikhov, Kh.M., and Russu, E.V. (1998). Current transport in porous *p*-Si and Pd-porous Si structures. *Semiconductors* 32(9), 960–962.

Slobodchikov, S.V., Goryachev, D.N., Salikhovand, Kh.M. et al. (1999). Electrical and photoelectric characteristics of n-Si/porous silicon/Pd diode structures and the effect of gaseous hydrogen on them. *Semiconductors* 33(3), 339–342.

Steiner, P., Kozlowski, F., and Lang, W. (1993). Light emitting porous silicon diode with an increased electroluminescence quantum efficiency. *Appl. Phys. Lett.* 62, 2700–2702.

Steiner, P., Kozlowski, F., Wielunski, M., and Lang, W. (1994). Enhanced blue-light emission from an indium-treated porous silicon device. *Jpn. J. Appl. Phys.* 33, 6075.

Stievenard, D. and Deresmes, D. (1995). Are electrical properties of an aluminium-porous silicon junction governed by dangling bonds? *Appl. Phys. Lett.* 67, 1570–1572.

Tsai, C., Li, K.H., Sarathi, J., Campbell, J.C., Hance, B.K., and White, J.M. (1991). Thermal treatment studies of the photoluminescence intensity of porous silicon. *Appl. Phys. Lett.* 59, 2814–2816.

Tsamis, C., Tsoura, L., Nassiopoulou, A.G. et al. (2002). Hydrogen catalytic oxidation reaction on Pd-doped porous silicon. *IEEE Sensors J.* 2, 89–95.

Vikulov, V.A., Strikha, V.I., and Skryshevsky, V.A. (2000). Electrical features of the metal–thin porous silicon–silicon structure. *J. Phys. D: Appl. Phys.* 33, 1957–1964.

Vinod, P.N. (2005). Specific contact resistance of the porous silicon and silver metal Ohmic contact structure. *Semicond. Sci. Technol.* 20, 966–971.

Vinod, P.N. (2009). Specific contact resistance and carrier tunneling properties of the silver metal/porous silicon/*p*-Si ohmic contact structure. *J. Alloys Compd.* 470, 393–396.

Vinod, P.N. (2013). The fire-through processed screen-printed Ag thick film metal contacts formed on an electrochemically etched porous silicon antireflection coating of silicon solar cells. *RSC Adv.* 3, 3618–3622.

Vrkoslav, V., Jelınek, I., Broncova, G., Kral, V., and Dian, J. (2006). Polypyrrole-functionalized porous silicon for gas sensing applications. *Mater. Sci. Eng. C* 26, 1072–1076.

Xie, Z. and Blackwood, D.J. (2013). White electroluminescence from ITO/porous silicon junctions. *J. Lumin.* 134, 67–70.

Yan, D., Hu, M., Li, S., Liang, J., Wu, Y., and Maa, S. (2014). Electrochemical deposition of ZnO nanostructures onto porous silicon and their enhanced gas sensing to NO$_2$ at room temperature. *Electrochim. Acta* 115, 297–305.

Zimin, S.P. and Komarov, E.P. (1998). Influence of short-term annealing on the conductivity of porous silicon and the transition resistivity of an aluminum-porous silicon contact. *Tech. Phys. Lett.* 24(3), 226–228.

Zimin, S.P. and Bragin, A.N. (1999). Conductivity relaxation in coated porous silicon after annealing. *Semiconductors* 33(4), 457–460.

Zimin, S.P., Kuznetsov, V.S., and Prokaznikov, A.V. (1995). Electrical characteristics of aluminum contacts to porous silicon. *Appl. Surf. Sci.* 91, 355–358.

# Index

Page numbers followed by f and t indicate figures and tables, respectively.

Milton Keynes UK
Ingram Content Group UK Ltd.
UKHW050131071024
449327UK00029B/2538